PROCESSING OF
REMOTE SENSING DATA

Processing of Remote Sensing Data

MICHEL-CLAUDE GIRARD
Professor of Soil Science at the Paris-Grignon INA

COLETTE GIRARD
Professor emeritus of Geobotany at the Paris-Grignon INA

In collaboration with
Dominique Courault
Jean-Marc Gilliot
Lionel Loubersac
Jean Meyer-Roux
Jean-Marie Monget
Bernard Seguin

Translated by
N. Venkat Rao
Professor of Geophysics
Osmania University
Hyderabad, India

A.A. BALKEMA PUBLISHERS LISSE / ABINGDON / EXTON (PA) / TOKYO

Published by arrangement with Dunod, Paris

Ouvrage publié avec le concours du Ministère français chargé de la Culture—Centre national du livre
Published with the support of the French Ministry of Culture

Translation of: *Traitement des données de télédétection*
ISBN 2 10 004185 1 © 1999 Dunod, Paris

Published by: A.A. Balkema, a member of Swets & Zeitlinger Publishers
www.balkema.nl and www.szp.swets.nl

ISBN 90 5809 232 1

Printed in India

Preface

'Who sees from a height, sees well; who sees from far, sees correctly'
Victor Hugo, *The art of being a grandfather, satisfied exile, 1877*

'I can make a circle around the Earth in twenty minutes'
William Shakespeare, *A midsummer night's dream, 1595*

After the 'Course on Photo-interpretation', published by ENSA of Grignon in 1970, and the books '*Applications of remote sensing in the biosphere (1975)* and '*Applied remote sensing—Temperate and intertropical zones (1989)'* published by Masson, this book is one of the most important to be released by Dunod this year. It covers a part of the preceding books, which are out of print; however, its novelty lies in the fact that it describes all the methods of digital data processing of remote sensing images.

In fact, today, image processing can be carried out on personal computers, which are accessible to all professionals, researchers and students, and the large number of existing software programs provides a wide choice. We have opted for one of the commercial softwares. Although it is not the most widely known, it is mathematically most explicit and gives many details about the quality of classifications. It also enables analyses of reflectance and digital computations, and is less expensive. Only a part of the possibilities of the *TeraVue* program is given in the CD supplied with this book.

Starting with a simplified version of this program, examples of SPOT and LANDSAT TM images are given, showing various phases of the methods, which constitute the justification for a new book. For all the methods described in this book, corresponding images are given in the CD, which can be analysed on personal computers.

For simplification, the given examples of data processing deal with one SPOT image of the Brienne region between the humid and chalky zones of Champagne. We have had the occasion to work with our friend F. Dumas, Director of the Chamber of Agriculture of Aube for studying the erosion-prone zones, among others.

A number of illustrations related to the various chapters, which can be analysed by the same program, are also given in the CD. Thus, the conventional colour illustrations of the remote sensing books are replaced by digital images in this book.

Obviously, it is assumed that the reader has a PC at his disposal, as it is a common practice to have one these days. If not, it is expected that all readers will have access to one while studying this book.

Remote sensing is a multispectral tool; for studying any problem, it is often very useful to have data in different bands of the electromagnetic spectrum, viz., visible, near and middle reflective infrared, thermal infrared and microwave.

This book is mainly concerned with the processing of data acquired in visible and near- and middle-infrared bands. In fact, these domains are the most often used, since the corresponding satellite images can be readily procured. Also, the methods of visual interpretation of images and aerial photographs have been updated, as they are still very useful. They are, perhaps, forgotten now, although they constituted the basis for any interpretation about twenty years ago. It was good to recollect them and adapt them to new remote sensing data.

As the data cannot be processed by various models underlying each method without validating the model, the theoretical and practical bases of estimating the quality of the results obtained have been reviewed.

In order to safeguard the main objective of image processing, viz., answering a relevant thematic question, a number of concrete and real examples of data processing of remote sensing are included at the end of the book.

As we did not wish to make this book a bibliographic collection on remote sensing and listing all the references would have required ten pages at the end of every chapter, these are restricted to the absolute minimum. The references cited constitute only guides to those that can be found in periodicals whose names are given in the appendix. The fact that a book is not cited does not indicate lack of interest. Such books include those authored by MM. Flouzat, Wilmet and a number of others.

This book is also addressed to teachers of secondary schools who have introduced in the first, second and final programs multidisciplinary topics on remote sensing. It ought to help them in presenting specific cases and in preparing the students for applied studies in mathematics, physics, geography or biology.

The images given in the CD are due to kind courtesy as acknowledged below:

—The two Thematic Mapper scenes of LANDSAT-5 acquired on 1 April 1990 (scene 197/26) and 15 May 1992 (scene 198/26), by ESA: @ **ESA (1990 and 1992), acquired by Fucino Station, Distributed by Eurimage-Geosys, by courtesy of Eurimage);**

—The image of Aube taken on 16 September 1996, by SPOT-Image: © **CNES 1996, Distribution SPOT-Image.**

Acknowledgements

We sincerely thank various colleagues with whom we worked and who have provided information on different aspects of remote sensing:

M.J.-M. Gilliot (INA PG) for studies on filters and geometric corrections (Chapters 12 and 13);

M.J. Meyer-Roux (CCR of ISPRA, European Union) for estimation of yields (Chapter 22);

Mrs. D. Courault, Ch. King, E. Vaudour, MM. R. Escadafal, B. Mougenot, C. Yongchalermchai, for the numerous studies on soils conducted in our laboratory of INA PG, most of which were borrowed for this book (Chapter 23);

M.J.-M. Monget (National Higher School of Mines, Paris) for applications in mineral prospecting Chapter 24);

M.L. Loubersac (IFREMER) for coastal-zone applications (Chapter 25);

Mrs. D Courault (INRA) and M. B. Seguin (INRA) for thermal-imagery applications (Chapter 26) and Mrs. Ch. King for her data on microwave remote sensing.

We thank P. Bertrand, F. Burlot, T. Francoual and S. Mollet who, during the preparation of their theses, have contributed to experimentations on various methods developed in our laboratory as part of their Master's program in 'Localised Information Systems for land-use management'.

Several parts of this book have been taken from our own investigations, published or not, as well as those of our research unit on 'Dynamics of Environment and Spatial Organisation' (Department of Agronomy-Environment) of the Paris-Grignon National Institute of Agronomy: Mmes. Ch. King, D. Courault and D. Orth, E. Vaudour and MM. J.-P. Rogala, D. King, R Escadafal, B. Mougenot, J.-M. Gilliot, C. Yongchalermchai, G. Belluzo who, although working in various organisations (INA PG, BRGM, ORSTOM, INRA, APIC-System), have constituted a scientific team for the last twenty years studying applications of remote sensing to soil and vegetation. We also thank P. Boissard, M. Bornand, J.-P. Lagouarde, B. Seguin (INRA) who since long have been associated with our studies on remote sensing.

We have taken several figures from MM. D'Allemand, Becker, Guyot, Lliboutry, Perrin de Brichambaut, Vossen and others, which we have modified to our requirement. Thanks are also due to P. Guillore (†) and J.-C. Carle for a significant number of drawings used in the book. We are thankful to D. Lepoutre and GeoSYS who have permitted us to use a digital elevation model for Brienne region.

We cannot forget that remote sensing studies were initiated by M.R. Chevallier, M. Guy and M. Jean Boulaine, who have provided the results of their research on the use of aerial photos in the Agricultural Academy of France. Our acknowledgements are also due to M. G. Brachet for his encouragement to continue our studies; to M.B. Cabrieres, the incharge of SPOT-4 system at CNES, for providing data on SPOT systems (a major part of which is given in the CD-ROM) and for carefully reading the material on this subject; to Mme. Lecochennec of SPOT-Image for reading the part concerning SPOT; M. Egels of IGN who has read Chapter 14 on fundamentals of photogrammetry; M. P. Fasquel for extending wise counsel on geometric quality of data (Chapter 17).

Practice of remote sensing techniques in administrative and consultancy agencies and various professional sectors has enabled us to apply our methods in real situations. We thank all those who have shared this challenge with us: M. V. Le Dolley who, as the district director of agriculture, encouraged processing of images of Yonne district; MM. F. Dumas and V. Ellisseeff of the Aube Chamber of Agriculture, M. F. Limaux of Lorraine Regional Chamber of Agriculture, M.J.-M. Vinatier and Mrs I. Boutefoy of the Soil Info Rhone Alps (SIRA), M. P. Juillet de Saint-Lager, Regional Director of Environment of Champagne-Ardenne, M. L. Lurton of the Inter-professional Committee on Vineyards of Cotes-du-Rhone and Rhone Valley.

In this book, we also address the secondary school teachers who have introduced multidisciplinary courses on remote sensing in their primary, secondary and final school programs. We had the occasion to work in this experiment due to some teachers, viz., Ch. Guisti, M. Dupuis and A. Herpe of La Queue les-yvelines high school and professors Monchamp and Vigneron of Plaisir high school. We have often thought of them and their students while preparing this book: it should provide concrete base for preparing to applied studies in mathematics, physics, geography or biology. We also wish that this book forms a useful tool for all other secondary teachers who want to engage themselves in this marvellous multidisciplinary adventure extending across processing and interpretation of images.

We are especially grateful to Lucien Bugeat who nominated one of us to the Remote Sensing Committee of the Agricultural Ministry and to Jean Dunglass and Maurice Dehegere who subsequently presided over it. We were thus informed of the several attempts in the agricultural domain and the numerous applications of remote sensing that were already operational.

Our thanks are also due to Mrs. S. Meriaux, secretary of section VII, our colleagues of section VII and the permanent secretary A. Cauderon of Agricultural Academy of France for organising meetings on remote sensing. Wish several other such meetings followed.

We are thankful to the national program of spatial remote sensing and CNES (stimulated action specific to use of SPOT images, headed by F. Blasco), Ministry of Agriculture and Fisheries (directorate of rural areas and forest, bureau of soils, crop inventory program, management and conservation of soils), IBM and INRA, who enabled progress of these studies through their scientific and financial support.

Lastly, we wish to acknowledge the person associated with the pedagogic aspect of this book, i.e., introduction of a CD-ROM as the heart of book. M. J.-M. Monget, our colleague at the National Higher School of Mines, developed the concepts and wrote the TeraVue programs. We can thank him only by assuring that our joint endeavour will continue through the proposed exciting program of establishing information exchange between the European Union and the United States.

Our thanks to J.-M. Monget cannot be complete without simultaneously mentioning Mrs. Druel, the incharge of La Boyere publications, who was responsible for the publication of the TeraVue image processing software and the production of CD-ROM.

Contents

A
Data Sources

D
Quality Assessment

E
Applications

A

DATA SOURCES

1

Physical Basis

The subject presented in this chapter provides the fundamentals necessary for understanding the techniques and applications of remote sensing. For more detailed information, the reader may refer to specialised books, in particular the one by Guyot (1997), from which some of the concepts presented here have been taken.

1.1 RADIATION

Remote sensing is based on the utilisation of emission and reflection properties of electromagnetic radiation, which carries energy and propagates without attenuation in vacuum but is more or less absorbed in different media. The solar radiation constitutes the external source of energy for the Earth. This energy is fixed by chlorophyll vegetation for producing living matter (primary producer). It will be seen later how this characteristic is used in remote sensing to study vegetation (Chap. 4).

Any object at a temperature higher than 0° K acts as a source of electromagnetic radiation by converting a part of the thermal energy into radiative energy. In addition, the object also receives energy from its environment, partly absorbing it and transforming it into thermal energy. The fraction of energy absorbed modifies internal energy of the object, which is represented by emission in a different wavelength. The blackbody is an ideal source. Its radiant exitance $M_e = d\Phi_e / dS$ (radiant energy flux Φ_e emitted by an extended source per unit area in a hemisphere) is independent of the angle of emission (*a*) and depends only on its temperature according to Planck's radiation law:

$$dM_{(\lambda,T)} / d\lambda = 2\pi hc^2\lambda^{-5} / [\exp(hc / \lambda kT) - 1] \text{ in W} \cdot \text{m}^{-2} \cdot \mu\text{m}^{-1} \quad (1)$$

where T is the absolute temperature in Kelvin, λ the wavelength in µm, h Planck's constant: 6.63×10^{-34} J·s, c the velocity of light: 3×10^8 m·s^{-1} and k the Boltzmann constant: 1.38×10^{-23} J·K^{-1}.

The exitance of a blackbody is related to the radiance L_e by the equation:

$$M_e = \pi \times L_e$$

Integration of equation (1) over the entire spectral region gives the total exitance emitted by a blackbody (Stefan-Boltzmann law):

$$M = \sigma T^4$$

where σ is the Stefan-Boltzmann constant: 5.67×10^{-8} W·m^{-2}·K^{-4}.

According to Planck's law, radiance decreases with temperature (T). The maximum wavelength of emission of a body is given by Wien's displacement law:

$$\lambda_{max} = 2897 / T$$

For the Sun, whose temperature is approximately 6000° K, λ_{max} = 0.48 µm and the solar energy flux occurs in the wavelength range of 0.15 to 4 µm (CD 1.1: see the CD enclosed with the book). For the Earth, whose temperature is about 300° K, λ_{max} = 9.6 µm and the energy flux lies in the wavelength region of 3 to 100 µm.

A comparison of the flux reflected by the Earth (assuming its reflectivity as 1) and the flux directly emitted by the Earth (assuming its emissivity as 1) is shown in Fig. 1.1. The atmospheric windows are shaded (Table 1.1 and section 1.4).

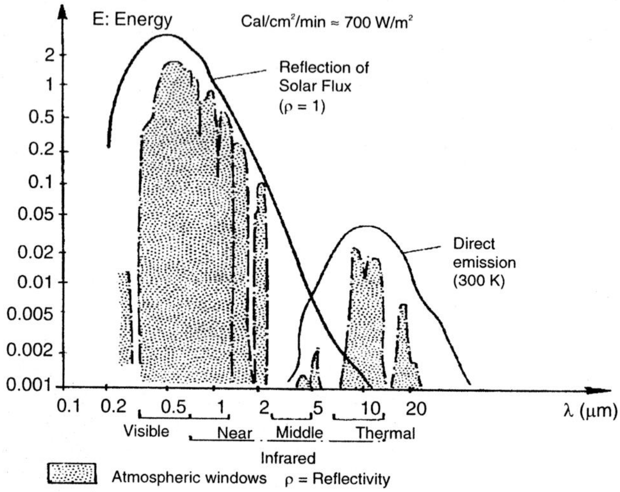

Fig. 1.1: Comparison between the fluxes reflected and emitted by the Earth (after Becker, 1978).

The Earth's surface behaves like a grey body. Spectral emissivity of a natural surface $\varepsilon_\lambda\ (\theta)$ in the direction θ is defined as the ratio of the radiance of the surface $L_\lambda(\theta)$ to that of the blackbody B_λ:

$$\varepsilon_\lambda(\theta) = L_\lambda(T,\theta) / B_\lambda(T)$$

For the entire spectrum, we get:

$$R = \varepsilon\sigma T_s^4 + (1-\varepsilon)R_{atm}$$

R_{atm} is the atmospheric radiation, which will be discussed in connection with thermal infrared remote sensing (Chap. 26 and Table 1.3).

Any radiation comprises a wide range of wavelengths:

$$\lambda = c/\nu$$

where c is the velocity of light: 3×10^8 m·s^{-1} and ν the frequency in s^{-1}.

Energy carried by a photon is: $e = h\nu$, where $\nu = c/\lambda$ and h Planck's constant. The total energy of a radiation is:

$$E = hc\,\Sigma(n_\lambda / \lambda)d\lambda$$

where n_λ is the photon density: $n_\lambda = N_{(\lambda,\Delta\lambda)} / \delta\lambda$, N the number of photons per unit time and $\Delta\lambda$ the spectral window or band.

The various domains of the electromagnetic spectrum and the main types of sensors used in remote sensing are shown in Table 1.1.

Table 1.1: Electromagnetic spectrum and remote sensing systems

SPECTRAL DOMAIN		λ	Remote sensing systems		
ULTRAVIOLET		290 nm 380 nm	UV scanner Photography		MULTISPECTRAL BANDS
VISIBLE		400 nm 700 nm	Colour panchromatic	500 mn Colour infrared 900 mn	
INFRARED	Near	800 nm 1500 nm	700 nm Black and white infrared 900 nm		
	Reflective middle	1.6 µm 2.2 µm			
	Middle	3 µm 5 µm			
	Thermal	8 µm 14 µm	Infrared radiometers		
MICROWAVE		1.6 µm 136 cm	Radar, backscatter sensors, radiometers		

1.2 SOURCE AND SENSOR PARAMETERS

The radiant intensity (I_e) is the energy flux emitted by a point source per unit solid angle in a given direction:

$$I_e(\theta, \varphi) = d\Phi_e(\theta, \varphi) / d\Omega \text{ in } W \cdot sr^{-1}$$

where θ is the angle of zenith, φ the azimuth and Ω the solid angle. If $d\Sigma$ is the surface situated at a distance r from the source and corresponding to the solid angle $d\Omega$, then $d\Omega = d\Sigma / r^2$ (Fig. 1.2).

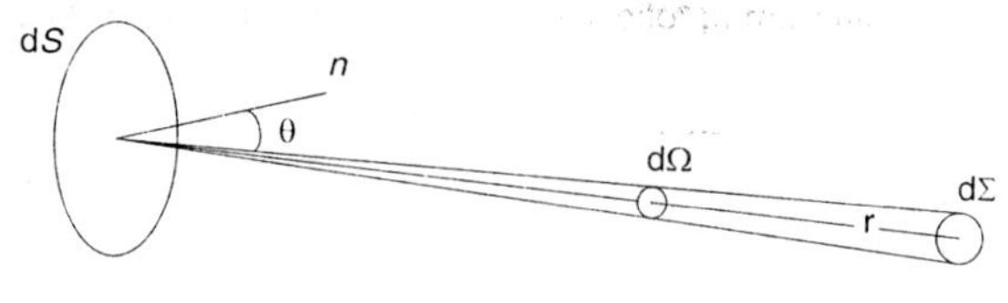

Fig. 1.2: Representation of the characteristics of a source.

If a point source is isotropic, it emits a radiant flux Φ_e whose intensity is independent of the orientation under consideration:

$$I_e = \Phi_e / 4\pi \text{ in } W \cdot sr^{-1}.$$

The concept of radiant intensity is valid only for a point source. In the case where the dimension of the source is small relative to the distance at which measurement is carried out (Sun for measurements on the Earth), the source can be considered a point.

Radiance (L_e) of a source corresponds to the total energy radiated by it in a given direction per solid angle per unit area of its apparent surface:

$$L_e(\theta, \varphi) = d^2\Phi_e(\theta, \varphi) / d\Omega \times dS \times \cos\theta \text{ in } W \cdot m^{-2} \cdot sr^{-1}.$$

Radiance can also be defined, with reference to a point on the surface in a given direction, as the ratio of the radiant intensity of an infinitely small surface element dS around the point and the area of orthogonal projection of this element on a plane perpendicular to this direction:

$$L_e(\theta, \varphi) = I_e(\theta, \varphi) / dS \times \cos\theta \text{ in } W \cdot m^{-2} \cdot sr^{-1}.$$

The radiant exitance (M_e) is the energy radiated by an extended source per unit area and in a hemisphere. At a given point, it is equal to the ratio of the flux emitted by an infinitely small element of the surface surrounding the point to the area dS of this element: $M_e = d\Phi_e / dS$ in $W \cdot m^{-2} \cdot sr^{-1}$.

For a radiation sensor, irradiance (E_e) is defined as the total energy received by a unit surface area of the sensor. At a point, this irradiance corresponds to the ratio of the radiant flux received $d\Phi_{er}$ by an infinitely small element centred on the point to the area dS of this element: $E_e = d\Phi_{er}$ dS in $W \cdot m^{-2} \cdot sr^{-1}$.

The *Bouguer law* represents the relationship between a source element and a receiver element. Let us assume two elementary surfaces, of areas dS and dS', separated by a distance r, whose normals make angles θ and θ' respectively, with the axis joining their centres. The radiant flux $d^2\Phi_e$ that reaches dS' from the source dS of radiance L_e, is equal to:

$$d^2\Phi_e = [(L_e \times dS \times \cos\theta) / r^2] \; (dS' \times \cos\theta).$$

If $d\Omega$ is the solid angle under which dS is seen from dS' and $d\Omega'$ the solid angle under which dS' is seen from dS,

$$d\Omega = (dS \times \cos\theta) / r^2$$

$$d\Omega' = (dS' \times \cos\theta') / r^2$$

It follows from the above that:

$$d^2\Phi_e = L_e \times dS \times \cos\theta \times d\Omega' = L_e \times dS' \times \cos\theta \times d\Omega.$$

In remote sensing the flux $d\Phi_{er}$, reaching a receiver from a source, is measured and the radiance is determined from it. This enables understanding the properties of surfaces.

Albedo corresponds to the fraction of solar radiation (direct and scattered radiations) reflected by a unit surface area into a hemisphere. It represents the mean hemispherical spectral reflectance in the range 0.3 to 3.0 µm.

The blackbody emits radiation in all directions and as seen earlier, radiance is independent of the angle of emission. Contrarily, other bodies radiate in preferred directions and their radiance depends on the angle of emission. According to *Lambert's law*:

$$I = I_e / dS = L_e \times \cos\theta$$

where I is the radiant intensity emitted, I_e the radiant intensity emitted in a given direction and L_e the radiance.

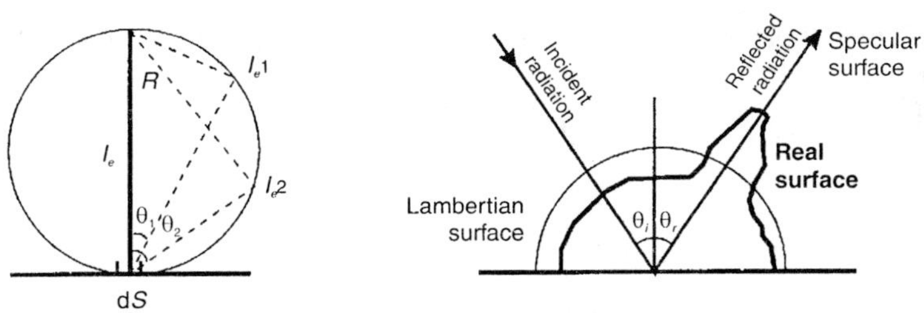

Radiant intensity of a Lambertian Surface

Radiance Characteristics of Lambertian, specular and real surfaces

Fig. 1.3: Characteristics of radiant intensity and radiance of different surfaces.

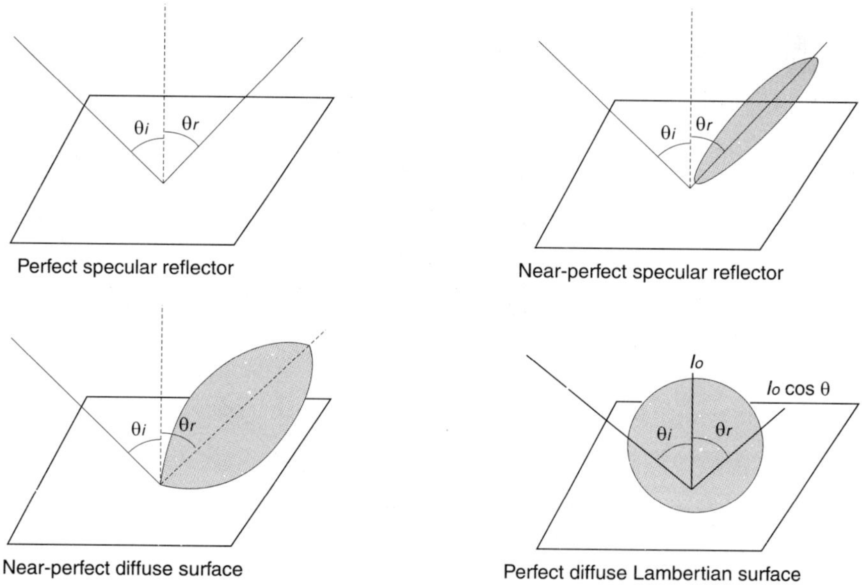

Perfect specular reflector

Near-perfect specular reflector

Near-perfect diffuse surface

Perfect diffuse Lambertian surface

Fig. 1.4: Examples of various radiant-intensity characteristics

The characteristic representation of angular variation of radiance (or radiant intensity) at a point on a surface is shown in Fig. 1.3. In polar co-ordinates, the geometric characteristic of radiant intensity of a Lambertian surface is a sphere tangential to the surface, of a diameter equal to I_e, while the characteristic of radiance is a hemisphere centred on the surface and of radius L_e. Natural surfaces generally do not follow Lambert's law and their radiance or radiant intensity varies with the angle of view. In the case of a perfect reflector, the angle of incidence (θ_i) and the angle of reflection (θ_r) are equal and situated in a plane perpendicular to the surface. Such a reflection is known as specular reflection (Fig. 1.4).

1.3 REFLECTANCE FACTOR

Spectral reflectance factor (R_λ) is the ratio of the radiant flux reflected by a surface element into a cone, with its apex on the surface element, to the radiant flux reflected into the same solid angle by a perfect diffuse reflector (white Lambertian surface) receiving the same irradiance (Fig. 1.5).

$$R_\lambda = \int_{\Omega_r} L_r(\theta_r, \varphi_r) \times \cos\theta_r \times d\Omega_r / L_w \int_{\Omega_r} \cos\theta_r \times d\Omega_r = \overline{L_r(\Omega_r)} / L_w$$

where Ω_r is the solid angle in which the radiant flux is measured and L_w the radiance of a white Lambertian reflector.

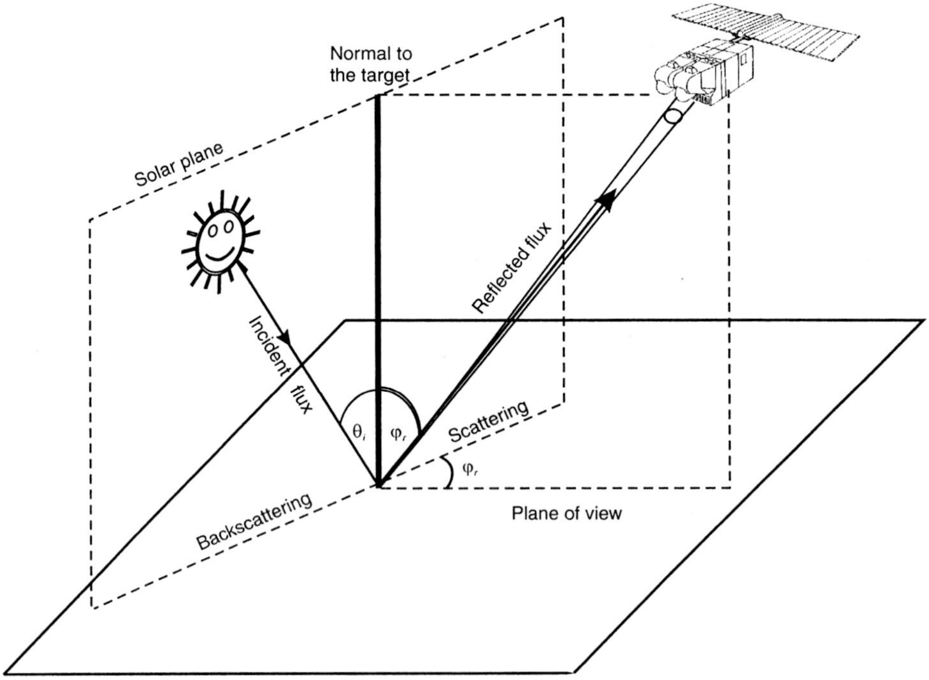

Fig. 1.5: Geometric configuration of reflectance.

In conclusion, the optical properties of natural surfaces are very different depending on the spectral domain under consideration. For example, fresh snow reflects 95% of solar radiation whereas it behaves like a blackbody in the thermal infrared band ($0.90 < \varepsilon < 0.99$). Similarly, albedo of bare soils may vary significantly with moisture content while their emissivity is less sensitive to water content.

1.4 SOLAR RADIATION AND ATMOSPHERIC PERTURBATIONS

This section presents the definitions and a review of concepts required for understanding certain data. More detailed information concerning the influence of atmosphere on radiation will be given in the section dealing with thermal infrared remote sensing.

True solar time is the dihedral angle made by the meridian in the Sun's direction with the observer's meridian. It is used to determine the position of the Sun and indicates solar dials. It is zero during the passage of the Sun in the meridian (solar noon).

Mean solar time is the hour angle observed for an apparent regular motion of the Sun.

Civilian time is the mean solar time enhanced by 12 hours.

Universal time (UT) is the civilian time of the Greenwich meridian chosen as origin.

Legal time or local standard time is constant on an hour's axis. It is obtained by adding or subtracting a certain number of hours to or from the universal time.

Irradiance of a plane surface, normal to the solar radiation and situated at the mean distance D_0 from the Earth to the Sun, is almost constant (with variations before solar activity). This is known as the solar constant E_s. The irradiance ($E_{e,h}$) of a horizontal surface situated at the boundary of the atmosphere depends on the distance from the Sun D_s and the zenith angle θ. The value of D_s varies between winter solstice and summer solstice, and the zenith angle varies with the latitude, hour and season. The greater the zenith angle θ, the smaller the irradiance (Fig. 1.6):

$$E_{e,h} = E_s [D_0 / D_s]^2 \times \cos \theta$$

1.4.1 Atmospheric absorption

The influence of atmosphere on solar radiation is shown in Fig. 1.7. The top curve represents the solar irradiance received at the boundary of the atmosphere, whereas the bottom curve the solar radiation

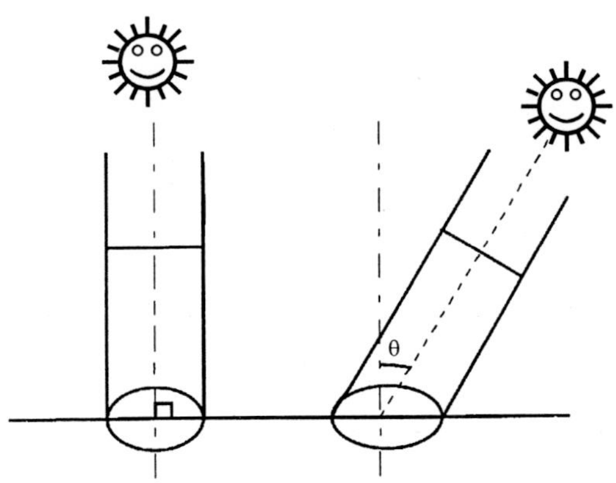

Fig. 1.6: Influence of zenith angle on the irradiance of a horizontal surface (after Guyot, 1997).

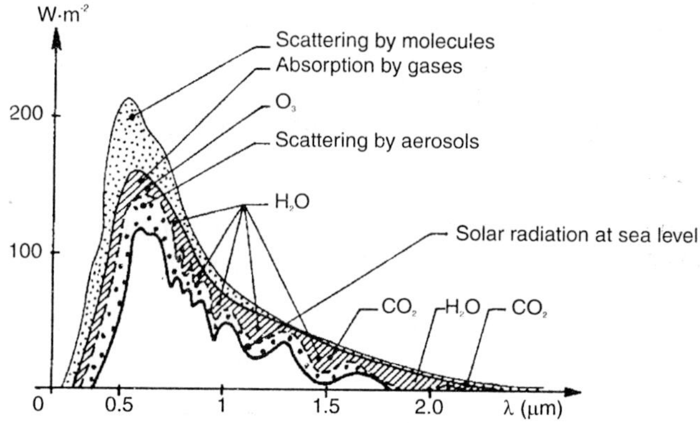

Fig. 1.7: Influence of atmosphere on direct solar radiation (after Perrin de Brichambaut, 1985).

received at sea level. The intermediate curves correspond to absorption by various gases. Ozone absorbs wavelengths below 290 nm and causes a little attenuation at 600 nm. Oxygen has an intense but very narrow absorption band at 760 nm. In the range from near-infrared to thermal infrared, water vapour, carbon dioxide and methane are mainly responsible for atmospheric absorption.

The larger the thickness of the atmosphere traversed by solar radiation, the weaker the direct irradiance. The spectral distribution is also modified (Fig. 1.8) with a greater decrease of visible radiation in the violet and blue spectral bands than in the orange and red bands.

The atmosphere transmits solar radiation only in a limited number of spectral bands known as atmospheric windows. The atmosphere is relatively transparent in the visible band and has a wide atmospheric window in the infrared (8 to 14 μm). It is opaque in the range 22 μm to 1 mm and hence this part of the remote sensing spectrum is not used. In the microwave region, the atmosphere is transparent beyond 3 cm but becomes opaque for wavelengths greater than 30 mm due to interaction with the ionosphere.

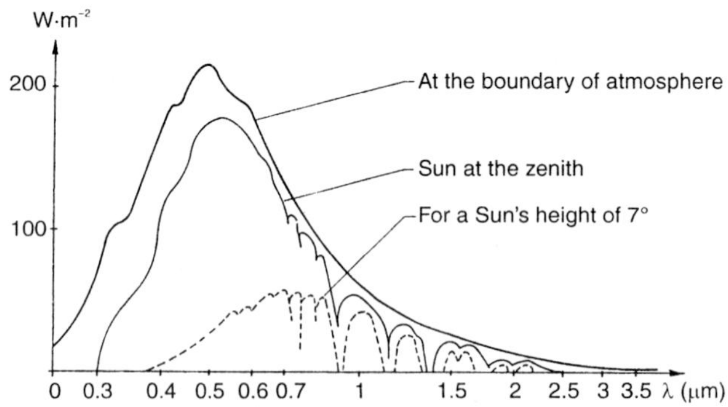

Fig. 1.8: Influence of Sun's height on the spectral distribution of direct solar irradiance on the ground (after Perrin de Brichambaut, 1985).

1.4.2 Atmospheric scattering

Scattering in the atmosphere corresponds to the action of molecules and particles (water droplets, dust, fumes etc.) on the wavelength (λ). The particle diameter *d* varies from:

1 to 500 nm	for charcoal dust
0.5 to 50 μm	for industrial fumes, fog, dust
10 to 100 μm	for pollens, ash
20 to 300 μm	for clouds, mist
0.5 to 5 mm	for raindrops.

The following cases may be distinguished:

- $\lambda > d$. In this case, the Rayleigh scattering is predominant due to interaction of photons with molecules. This scattering is proportional to λ^{-4}. It plays an important role for short wavelengths (λ < 550 nm) where it is associated with the action of oxygen and nitrogen molecules. This process of interaction explains the blue colour of the sky and the red colour of the rising or setting Sun. In fact, blue light (λ = 450 nm) is scattered six times more than red light (λ = 700 nm). When the Sun is low on the horizon (Fig. 1.8), the radiation traverses a thicker atmospheric layer than that when the Sun is at the zenith. Small wavelength radiation is hence eliminated and the Sun looks red.
- $\lambda \cdot 10^{-2} < d < \lambda \cdot 10^{2}$. In this case, the main type of scattering is the Mie scattering due to suspended aerosols in air (dust, microdroplets of water). This scattering is roughly proportional to λ^{-1} and manifests in the entire solar spectrum but has very small influence in the thermal infrared band. It is especially significant near industrial sites.

1.4.3 Atmospheric radiation

The atmosphere re-emits part of the radiation received, which contributes to increase the value of the flux measured by the sensor (Fig. 1.9). This atmospheric radiance adds to the flux reflected or emitted by the Earth's surface, which constitutes the object of study in remote sensing. It can be measured

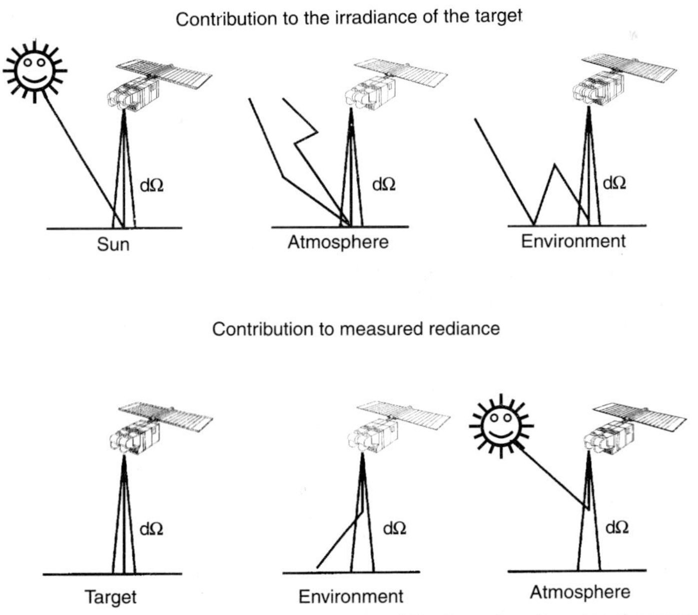

Fig. 1.9: Influence of the atmosphere and the environment of the target on the signal measured by a satellite.

from vertical profiles of temperature and humidity obtained by radio-soundings and TOVS sensors of NOAA satellites. It can also be modelled using different formulae (see infra) since the atmospheric effects are particularly significant on satellite-measured fluxes.

Atmospheric scattering enhances the reflectance observed at the top of the atmosphere (ρ') relative to that observed on the ground (ρ). An example of reflectance for a green wheat and a ripe wheat is shown in Fig. 1.10.

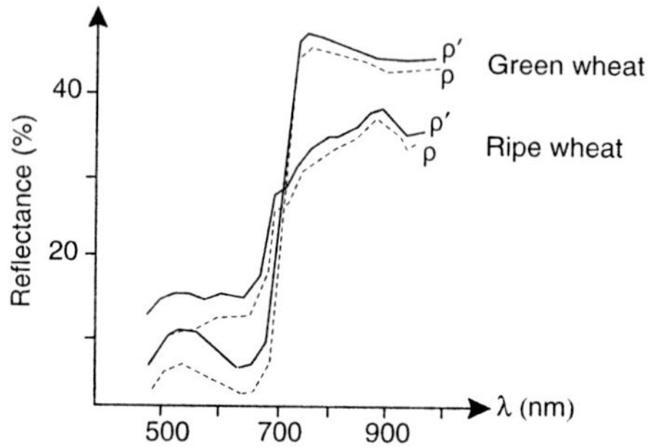

Fig. 1.10: Effect of atmospheric scattering on reflectance observed at the top of the atmosphere (ρ') and on the ground (ρ) for a green wheat and a ripe wheat (after Deschamps et al., 1984 in INRA Publications).

1.5 THERMAL INFRARED REMOTE SENSING

The thermal infrared band corresponds to a range of wavelengths varying from a few µm to several hundred µm, situated between middle infrared and microwave. Some instruments operate around 3 µm but the widest range used lies between 8 to 14 µm. Measurements in this band using airborne or satellite sensors enable determination of surface temperatures and furnish information on the aqueous and energy state of surfaces. Such measurements are hence significant in the agricultural sector for monitoring soils and crops.

1.5.1 Physical basis

It was indicated at the beginning of this chapter that any body with temperature different from absolute zero emits a radiation depending on its temperature. Emissivity is always less than 1 and generally greater than 0.9 for most natural objects. This parameter depends on the wavelength and the direction θ, as well as on the moisture content of surfaces. Its measurement still remains difficult and is not representative in natural conditions except very locally (Stoll, 1988). In practice, values found in publications are used. Its precise knowledge is however essential since a relative error of 1% in emissivity may result in an error of 0.8°C in surface temperature. Emissivity values for some surfaces are given in Table 1.2.

Brightness temperature (or apparent radiant temperature T_B) of a surface corresponds to the temperature of a blackbody which delivers the same radiance as that of the surface under investigation. It can be defined for a spectral band by the relation:

$$\int_{\lambda_1}^{\lambda_2} B_\lambda(T_B) = \int_{\lambda 1}^{\lambda_2} \varepsilon_\lambda B_\lambda(T) d\lambda \qquad (2)$$

Table 1.2: Emissivity values of some important types of surfaces (after Guyot, 1997)

Type of surface	Emissivity
Ice	0.92–0.97
Water	0.99
Bare soil (dry sand, wet sand)	0.94 (0.84–0.90; 0.91–0.94)
Grassland	0.98
Wheat	0.97
Maize	0.96
Grapevines, orchards	0.95
Conifers	0.98
Hardwood	0.97

or over the entire spectrum by:

$$\sigma T_B^4 = \varepsilon\sigma\, T^4 \tag{3}$$

Radiant (brightness) temperature is a directional parameter which varies with the direction θ. Hence, it varies according to the spectral bands of the satellites used. Conversion of this temperature to radiometric temperature requires a correction for atmospheric radiation, as given below.

In these wavelengths absorption of radiation by snow is very high and only the surface temperature affects the radiance. Emissivity of snow is close to 0.99 in this spectral domain.

The surface of the sea can be considered Lambertian in the thermal infrared band and the general equation for conversion of the radiance received by the sensor to sea surface temperature is based on Planck's law. The emissivity of water is very close to 1, and water surfaces are good accumulators of heat, exhibiting very small thermal amplitudes on the surface. This is due to convection motions existing between water masses of different temperatures situated at depth. The observed temperature is related only to the surficial layer of water over a thickness equivalent to the observation wavelength. For more details of software for computing temperature with AVHRR sensors of NOAA satellites, refer to Barton (1989).

1.5.2 Correction for atmospheric radiation

The radiance received at the point of the receiver L_λ depends on the emission by the surface as well as from the atmosphere (Fig. 1.11). The long wavelength atmospheric radiance (L_a) results from radiation emitted by particles in the atmosphere (see supra). Radiance contributions from water vapour, carbon dioxide and ozone are the largest. Suspended aerosols also participate in this thermal emission. The atmospheric radiance comprises different components:

—Its own radiation (thermal emission by particles present in it): On the ground, this radiation results from all the particles present on the entire profile and having varying temperatures according to their altitude. This flux comes down in the direction of the surface being irradiated by the atmosphere ($L_{a\lambda}\downarrow$);

—At the receiver, the resultant of contributions from particles, which is a flux directed towards the sky ($L_{a\lambda}\uparrow$).

Concentration of gases, aerosols and temperature can be determined by means of radio soundings. They can also be obtained from climatic atlases. The atmosphere attenuates the surface radiance by an absorption factor τ; as a result, the surface emission becomes $\varepsilon_\lambda B_\lambda(T_s)$ and the reflected

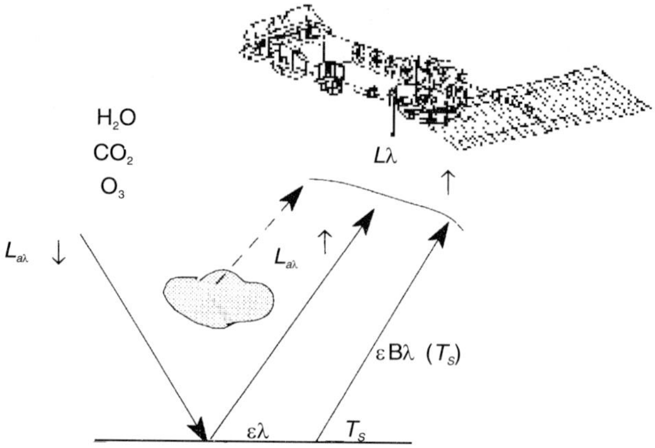

Fig. 1.11: Components of radiance received by the thermal infrared sensor.

atmospheric radiation $(1-\varepsilon_\lambda)L_{a\lambda}$. Thus the final equation for radiance at the receiver is obtained as:

$$L_\lambda = \tau[\varepsilon_\lambda B_\lambda(T_s) + (1-\varepsilon_\lambda)L_{a\lambda}\downarrow] + L_{a\lambda}\uparrow \tag{4}$$

where τ is the atmospheric absorption coefficient (Fig. 1.11).

Knowledge of emissivity of the surface and atmospheric radiance is hence necessary in order to determine the surface temperature. Several possibilities exist to determine the emissivity. It can be measured in the field although such measurements are highly local and cumbersome. Alternatively, emissivity data can be accessed from published literature. One can also use relations between emissivity and vegetation index (see Chap. 4) computed from satellite images in the visible and infrared bands (actes de télédétection IRT, 1993). To estimate atmospheric radiation and corrections for obtaining surface temperature, various methods are proposed depending on the type of measurement: single or multichannel thermal infrared sensors, radio soundings etc.

■ Atmospheric corrections for single-channel sensors

Satellites such as LANDSAT or METEOSAT have only one channel in the thermal infrared band. Hence atmospheric correction models such as LOWTRAN (Kneisys et al., 1983) or MODTRAN (1995) have to be used for them. These models are based on dividing the atmosphere into a multilayer system and solving the transport equation for each layer. The necessary data characterising the atmosphere (thermal profiles, humidity and pressure) are obtained either from radio soundings or from indirect measurements, from outputs of meteorological models or forecasts. Use of meteorological forecasts poses many problems, which have been discussed by Kerr (1991). Among these problems, we cite those concerned with the hypothesis of surface homogeneity, the cumbersome method of computations and errors introduced by these models.

Direct measurements by radio soundings or use of the TOVS sensor of NOAA, which provides vertical profiles of temperature and humidity, are recommended for single-channel sensors (Ottlé and Vidal-Madjar, 1992). Combining these data with a radiative transport model such as MODTRAN, it is possible to accurately correct for atmospheric effects on surface-temperature measurements. The results are generally satisfactory provided a radio sounding is available close to the region of study and to the time of acquisition of image. However, interpolation of radio soundings is often problematic. In France, only seven sites exist where radio soundings are carried out routinely every day at 12 and 24 hour (sites in important airports). Density of radio sounding sites is much smaller in other regions such as West Africa. This constitutes a major constraint in employing these correction methods. Data issued by TOVS soundings do not yet overcome the problems of representativeness since determinations of the characteristics of the lower atmospheric layers are still difficult and too few. In fact, as this probe does not provide access to direct measurements that are equivalent to radio soundings, it is necessary to reconstruct data acquired with low resolution (Ottlé and Vidal-Madjar, 1992).

■ Atmospheric corrections for two-channel sensors

Methods based on differential absorption can be used for atmospheric corrections in two neighbouring channels (such as Channels 4 and 5 of AVHRR-2). It is assumed that the spectral emissivity of the surface in the two spectral bands is identical. The apparent radiative temperature in such a case results from a linear summation of temperatures obtained in the two channels of the sensor:

$$T_s = a_0 + a_1\, T_{c\,4(10.8)} + a_2\, T_{c\,5(11.9)} \tag{5}$$

This method, known as Split-Window, was initially used to compute sea-surface temperatures for which it gives good results. However, it poses more problems for land surfaces because of differences in spectral emissivity between bare soils and vegetation covers (Kerr, 1991). The different values of coefficients are found to be more or less suitable or unsuitable for exposed ground. Kerr (1991) proposed a semiempirical method based on computing two temperatures from different values of coefficients: one corresponding to bare soils (T_{bs}) and the other for dense vegetation (T_{dv}). The two temperatures T_{bs} and T_{dv} are summed linearly as a function of a coefficient CV, representative of the percentage of soil cover, and divided by vegetation index (VI) to obtain the final surface temperature.

$$(T_{bs}) = 3.1 + 3.2\, T_{c\,4\,(10.8)} - 2.2\, T_{c\,5(11.2)}$$

$$(T_{dv}) = -\,2.4 + 3.6\, T_{c\,4(10.8)} - 2.6\, T_{c\,5(11.2)}$$

$$CV = \alpha\, / V$$

$$\alpha = (CV_1 - CV_0\,)\, /\, (IV_{max} - IV_{min})$$

$$T_s = CV\, T_{dv} + (1 - CV)\, T_{bs} \tag{6}$$

where α is a coefficient varying between 0 and 1, IV_{min} corresponds to bare soils or poorly covered surfaces, IV_{max} corresponds to dense vegetation cover or cultivated land. This method gives a precision of about 1.5 K.

Another method of atmospheric correction is based on using the data acquired under varying conditions of observation (e.g. ATSR sensor on ERS-1 which acquires two views) (Ottlé and François, 1993).

■ Sources of errors in T_s measurement by satellite

Let us now review various factors that disturb measurement of T_s. These factors may affect only certain sensors (e.g. saturation phenomenon) or certain satellites (drift acquisition in time).

□ Orbital drift

The local sun time on each pass of a satellite changes with time. Data issued from the NOAA-9 satellite launched in 1984 and coming to an end in 1988 exhibited a difference of 2 hours in acquisition time in 1988 relative to the initial time at beginning of launching (Privette et al., 1995). Hence, it is often necessary to apply corrections to bring temperatures to the same time in order to compare and to monitor temporal variations. The relationships between surface temperature acquired at different times can be used for this purpose. Based on field data obtained using a thermal radiometer, it is in fact shown that the relationships between these variables are linear. An example of such relationships is given by the following equation:

$$T_{s14h} = (1.03 \pm 0.01)\,(T_{s15} - T_{amax}) + (3.3 \pm 1.7) + T_{amax} \quad (7)$$

where T_{s14h} is the temperature of the surface under investigation, T_{s15} the temperature measured by NOAA-9 in 15 h and T_{amax} the maximum temperature of air measured over a meteorological site situated in the image.

This relationship was used to correct the NOAA-1988 surface temperatures in order to compare them (CD 26.1) with those of 1990 and 1991 (Courault et al., 1994).

□ Signal saturation

This factor also needs to be corrected due to the fact that the NOAA sensor gets saturated above 320 K (Kerr, 1991); however, it affects only some regions of the globe.

□ Angular effects

Several types of angular effects are identified:

—atmosphere-related effects (difference in optical trajectory),
—effects related to directional emissivity, not understood at present (Becker, 1987),
—effects associated with structure of plant canopies.

Models have been developed to ascertain the effects of vegetation structure, combining a description of thermal profiles inside vegetation with the structure of the canopy (angle of leaves, foliar index etc.; actes de télédétection IRT, 1993). Values of radiative temperatures acquired with thermal radiometers along different angles showed that differences in temperatures between a vertical view and an inclined view may reach up to 4 K over maize and 1.5 K over grasslands (actes de télédétection IRT, 1993). Temperature values of vertical measurement are generally higher due to the contribution of soil, which is inversely proportional to the density of vegetation.

1.5.3 Relationship between radiant temperature and aerodynamic temperature

A whole nomenclature exists in thermal infrared technique wherein the surface temperature (general term) may indicate various physical parameters. Norman and Becker (1995) discussed these differences in terminology of surface temperature and emissivity. In this section, only the relationships between the radiative temperature measured by a thermal infrared sensor and the aerodynamic temperature of plant canopies is reviewed.

■ Definitions

The *aerodynamic temperature* of a surface (T_{sa}) corresponds to the extrapolation of the temperature profile of air to a height of the apparent canopy given by the height of displacement (d) plus the length of roughness (z_0). It occurs in the estimation of sensitive heat flux (equation 3). It is combined with air temperature and aerodynamic resistance (ra) which is calculated according to the Monin-Obukhov theory. Variables related to the wind and temperature profiles such as thermal (z_{0t}) and mechanical (z_{0m}) roughness are used (Guyot, 1997).

The *radiative surface temperature* T_{sr} is defined from the radiation emitted by the surface L_λ (see above). It integrates the surface temperatures of all the elements of a canopy, such as leaves in shade and leaves in sunlight, viewed by the sensor. It depends on the structure of vegetation and wind profiles in the canopy. It is a directional parameter which corresponds to a given wavelength range depending on the receiver. If R_λ is the radiance measured by the sensor and ε_λ the emissivity of the surface, we get:

$$R_\lambda = L_\lambda + (1 - \varepsilon_\lambda) R_{atm} \downarrow$$

$$L_\lambda = \varepsilon_\lambda B_\lambda (T_{sr}) \quad (8)$$

If the surface temperature measured by satellite is to be used to compute the detectable heat flux for vegetation of strong vertical development, a correction has to be applied. It is generally considered that aerodynamic resistance $ra = r_1 + r_2$, where:

$$r_1 = (1/ku^*) \, [\ln\{(z - d/z_{0m})\} - \psi_h\{(z - d)/L\}]$$

$$r_2 = (1/ku^*) \ln(z_{0m}/z_{0t}) = (1/ku^*) \, KB^{-1}$$

where KB^{-1} represents the ratio between mechanical and thermal roughness, expressed in logarithmic form, L is the Monin-Obukhov length and u^* is called friction velocity.

Becker and Li (1995) gave the relationship between these two temperatures as:

$$T_{sa} = T_{sr} + \frac{b}{1+b} T_a - T_{sr}$$

$$b = KB^{-1} / \, ku^* ra \quad (9)$$

Differences between aerodynamic and radiative surface temperatures may vary from –8 K to 4 K depending on the values of the foliar index (actes de télédétection IRT, 1993).

■ Accuracy of data

The accuracy of surface temperatures thus determined from satellite data is difficult to evaluate without choosing homogeneous zones of sufficient size so as to include many pixels. It can be estimated only locally from measurements with thermal radiometers (which are often of wide band, 8–13 μm) and without covering exactly the same spectral region as that of the sensor. As thermal radiometers view surfaces of limited dimensions, the measurements are affected by interfield variability. Spatial sampling of measurements has to be large enough to be representative of the variability of the zone under study. When the segments are small and varied in land cover, the pixels are heterogeneous. An overall measurement at the pixel level is obtained from the composite fields. Different methods are thus possible depending on the objective of investigation: work at the level of the pixel and adjust the models to low resolution or employ methods of resolving the pixels to arrive at the level of the field (see Chap. 15). At present, little work is reported on spatial integration or disintegration of T_s. Moreover, tables of emissivity covering the range of possible surfaces and records of variation of the characteristics of a surface with time are required.

Various studies have reported comparisons of accuracy in surface temperature obtained by different methods for correction of thermal data (Kerr, 1991; Ottlé and Vidal-Madjar, 1992). Among these, the tables of Chanzy (1991), which are useful in evaluating the variability of measurements depending on the values of atmospheric radiance and emissivity employed, are given in Table 1.3. As values of emissivity are often unknown, these tables show that approximation in assuming $T_s = T_b$ introduces errors which are larger for smaller values of emissivity and atmospheric radiance.

Table 1.3: a. Errors in determination of surface temperature when $T_b = T_s$ for different values of atmospheric radiance and emissivity.
b. Errors in determination of surface temperature when emissivity is assumed constant at 0.95, for different values of atmospheric radiance and emissivity of soil (after Chanzy, 1991).

R_{atm}	Emissivities							
	a				b			
($W{\cdot}m^2$)	0.85	0.90	0.92	0.98	0.85	0.90	0.92	0.98
100	9.21	6.04	4.81	1.18	6.37	3.14	1.87	– 1.83
200	6.56	4.32	3.44	0.85	4.55	2.25	1.34	– 1.32
300	3.97	2.63	2.10	0.52	2.77	1.38	0.82	– 0.81
400	1.46	0.97	0.77	0.19	1.02	0.51	0.30	– 0.30

The accuracy reported in most studies is of the order of 1 to 2 K, which in absolute value is significant for point measurements but acceptable only when cumulative values over a year are analysed. Hence, one has to be cautious in using thermal infrared data.

The principal use of data of this spectral band is to evaluate the aqueous and energy states of surfaces. Before describing some examples of application in agriculture, it is necessary to review the relationships between surface temperature and evapotranspiration.

1.5.4 Relationship between surface temperature and evapotranspiration

■ Energy balance

Surface temperature is indirectly related to evapotranspiration of plant canopies through energy balance. In the case of a thin surface (bare soil or low canopy) which can be considered to be in instantaneous equilibrium, the energy balance is described as:

$$R_n = LE + H + G \tag{10}$$

where R_n is the net radiance representing the energy balance of various radiative fluxes of short and long wavelengths at the ground level. It is expressed as:

$$R_n = (1-a)R_g + \varepsilon R_{atm} - \varepsilon\sigma T_s^4 \tag{11}$$

where R_g is the total radiance, part of which is reflected by the surface (a being the albedo), R_{atm} the atmospheric radiance and $\varepsilon\sigma T_s^4$ the radiance emitted by the surface.

The various equations for each of the fluxes are not discussed here.

The conduction flux G of the ground can be neglected in the scale of a day for dense canopies. However, its instantaneous value can be significant in the case of bare and dry soils. For plant canopies, a conventional order of magnitude is 0.1 R_n to 0.2 R_n in diurnal conditions. G can be expressed in the form:

$$G = (K_c/\Delta z)\,(T_s - T_0) \tag{12}$$

where K_c is the thermal conductivity and T_0 the ground temperature at depth Δz

The sensible heat flux H, related to convective exchanges in air, depends on the difference of temperatures between the surface and air:

$$H = \rho C_p \frac{(T_{sa} - T_{air})}{ra} \tag{13}$$

where ρ and C_p are the density and specific heat of air, T_{sa} the aerodynamic temperature of the surface and ra the aerodynamic resistance, which varies depending on the wind velocity, surface roughness and stability conditions of the atmosphere (Guyot, 1997; Courault et al., 1996).

The latent heat flux LE (energy equivalent of evapotranspiration E expressed as mass flux) is written as:

$$LE = \rho C_p \frac{Te_{sat}(T_s) - e_a}{\gamma(ra + rs)} \tag{14}$$

where $e_{sat}(T_s)$ is the saturated vapour pressure at the temperature T_s, γ the psychrometric constant (0.66 $mb \cdot K^{-1}$), e_a the vapour pressure of air at the reference level (2 m) and rs the total resistance of the canopy. The parameter rs depends on climatic stresses to which plants are subjected and the moisture state of the soil.

Each of the fluxes can hence be expressed as a function of surface temperature. Evapotranspiration thus seems to be linearly related to T_s and, if flux G is neglected, can take the following form for a time interval of a day:

$$LE_d(T_s) = R_{nd}(T_s) - H_d(T_s) \tag{15}$$

T_s is the solution of the energy balance equation and attests exchanges between the surface and the atmosphere. In particular, it indicates the evaporation level of surfaces.

■ Simplified models

Various authors have proposed semiempirical models derived from energy balance on a daily scale. The most widely used model shows a linear relationship between the instantaneous temperature difference $(T_s - T_a)_d$ measured in daytime and the daily values of $(LE - R_n)_d$, expressed in mm of water per day (Jackson et al., 1977). Physical justification of this relationship is mainly based on conversion of instantaneous values into diurnal values (Seguin and Itier, 1983).

$$LE_d - R_{nd} = A + B(T_s - T_{air})_d \tag{16}$$

The coefficients A and B vary with the type of surface, in particular the roughness of the cover, as well as the wind velocity and conditions of thermal stability (Lagouarde and McAneney, 1992). They are obtained from ground measurements. Measurements by Seguin et al. (1982), for example, gave these values as: $A = 1$ $mm \cdot d^{-1}$ and $B = -0.25$ $mm \cdot d^{-1}$. They can also be determined from digital modelling (Lagouarde, 1991).

This relationship has been verified for various surface conditions and gives satisfactory results for clear days with an accuracy of about ±1 $mm \cdot d^{-1}$ for evapotranspiration. Examples of application are given in Chapter 26.

1.6 PRINCIPLES OF MICROWAVE REMOTE SENSING

Active sensors, viz., radars (acronym for Radio Detection And Ranging), are characterised by:

—a working range in which the atmosphere is transparent for most of the wavelengths used (from 3 mm to a little more than 30 cm);

—independence of the emission-detection system vis-à-vis the solar energy ('active' system);

—sensitivity for wavelengths at which the radiation-matter interactions correspond to totally different phenomena from those recorded by 'passive' systems.

Only the fundamentals required for understanding and utilising the microwave data are given in this book. The theory of radars can be found from more specialised books (for example, Paquet, 1997 and Polidori, 1997). Moreover, only the active systems are discussed, for which a large number of programmes of study and data for observation of the Earth are available.

1.6.1 Special laws at microwave frequencies

Radiant energies emitted by terrestrial surfaces in the wavelength range of about a cm (see Table 1.4), the most widely used, are very low compared to the thermal infrared band. A simplified Planck's relation (Rayleigh-Jeans law) can be employed in this case. The latter describes a linear relation between the temperature and the radiance of a blackbody at these wavelengths:

$$L_\lambda = 2kT / \lambda^2$$

where L_λ is the microwave radiance of the blackbody in $W \cdot m^{-2} \cdot Hz^{-1} \cdot sr^{-1}$, k the Boltzmann constant (1.38×10^{-23} $J \cdot K^{-1}$), T the temperature of the blackbody in K and λ the emission wavelength in cm.

The apparent radiant temperature (T_{app}) of an object is proportional to the energy emitted and hence to the emissivity of the object:

$$T_{app} = \tau(\varepsilon T_0 + \rho T_r) + (1-\tau)T_s$$

where τ is the transmittance of the atmosphere between the object and the sensor ($0 \leq \tau \leq 1$), ε the emissivity of the object ($0 \leq \varepsilon \leq 1$), T_0 the absolute temperature of the object in K, ρ the reflectance of

Table 1.4: Frequency bands (after Paquet, 1997).

Wavelength	OTAN nomenclature		Radar bands	
100 m	A	0 to 250 MHz		
10 m			HF	3 to 30 MHz
	B	250 to 500 MHz	VHF	30 to 300 MHz
1 m	C	0.5 to 1 GHz	UHF	0.3 to 1 GHz
30 cm	D	1 to 2 GHz	L	1 to 2 GHz
	E	2 to 3 GHz	S	2 to
10 cm	F	3 to 6 GHz		4 GHz
5 cm	G	4 to 6 GHz	C	4 to
	H	6 to 8 GHz		8 GHz
3 cm	I	8 to 10 GHz	X	8 to 12 GHz
	J	10 to	Ku	12 to 18 GHz
2 cm		20 GHz	K	18 to 27 GHz
1 cm	K	20 to 40 GHz	Ka	27 to 40 GHz
	L	40 to 60 GHz	V	40 to 70 GHz
	M	60 to 100 GHz	W	70 to 100 GHz

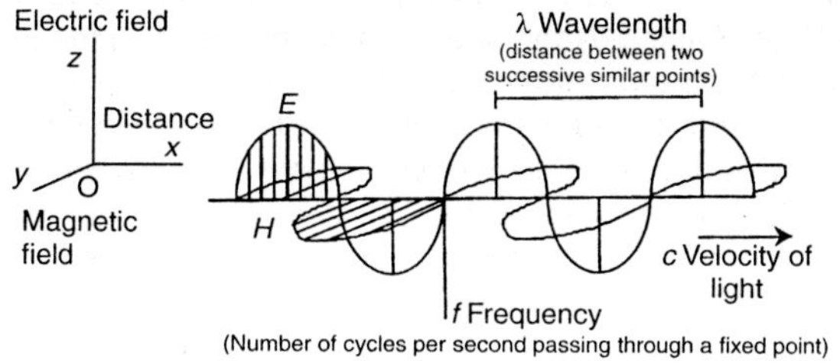

Fig. 1.12: Characteristics of a wave (after Bonn and Rochon, 1992, p. 26, reproduced with the permission of the editor. Extracted from: *Remote sensing abstracts*, vol. 1, *Principles and methods*, University of Quebec Press, Quebec).

the object ($0 \leq \rho \leq 1$), T_r the equivalent temperature of the radiation incident on the object (in K) and T_s the absolute temperature of the atmospheric layer situated between the object and the sensor (in K).

The value of emissivity ε depends on a number of factors such as surface roughness, polarisation and wavelength of the radiation, dielectric constant and temperature of the object etc. For example, the emissivity of water is close to 1 in the thermal infrared band and about 0.35 in the microwave band.

1.6.2 Frequency bands

The frequency bands employed in active and passive remote sensing are shown in Table 1.4. While the OTAN nomenclature applies to radar and radio bands, a radar nomenclature (unofficial) is also given because very commonly used in active remote sensing, especially in the frequencies between 1 and 40 GHz.

1.6.3 Polarisation

An electromagnetic wave can be represented in space by electric (*E*) and magnetic (*H*) field vectors. The vector *E*, which is perpendicular to the X-axis can rotate around this axis (Fig. 1.12).

Phase φ determines the manner in which *E* behaves in a plane parallel to *yOz*. If φ varies with time randomly, the wave is non-polarised and if φ remains constant, the wave is polarised. In the case of reflection of electromagnetic waves, two types of rectilinear polarisation can be distinguished:

—vertical polarisation, when the vector *E* lies in the plane of incidence;

—horizontal polarisation, when the vector *E* is perpendicular to the plane of incidence.

The polarisation phenomenon exists in various spectral domains used in remote sensing. However, it is particularly significant in microwave regions in which it carries information about the object under investigation, its surface roughness, for example.

1.6.4 Doppler effect

The Doppler effect produces a shift in frequency between the wave emitted by a moving source and a receiver, due to relative motion between the source and the receiver. It has important applications in microwave frequencies for the synthetic-antenna radar.

1.6.5 Backscatter signal

In a radar system, the source and the receiver are almost at the same place. An electromagnetic wave of frequency *f* and power P_s, is emitted by the source in the direction of the surface under observation. On reaching the surface, the wave undergoes an interaction and wave is scattered in all directions in space. A part of it is thus returned in the direction of the radar: this is the backscatter signal (σ_0) with a power P_{bs} (Fig. 1.13). From the thematic point of view, it is necessary to understand how the surface under investigation affects the backscatter signal.

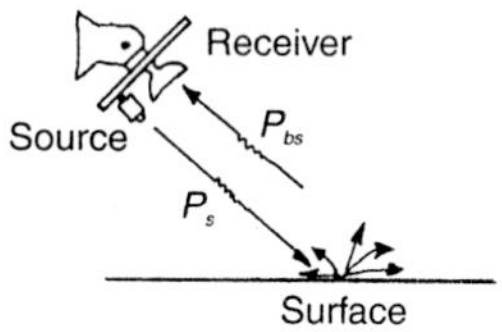

Fig. 1.13: Backscatter signal.

1.6.6 Radar equation

For a given wavelength and polarisation, the radar equation describes a relationship between P_s and P_{bs}, separating the effect of logistic parameters, that characterise the instruments used and the source, from the effect of the surface on backscattering of the signal:

$$\frac{P_{bs}}{P_s} = \frac{\lambda^2}{(4\pi)^3} \times \frac{(G_s - G_r)}{R^4} \times \sigma_0 S$$

Characteristic of instrument — Backscatter coefficient

where P_{bs} is the energy received, P_s the energy emitted, λ the wavelength, R the distance between the surface and the antenna, G_s the gain of the source antenna, G_r the gain of the receiver antenna, σ_0 the reflecting power per unit area of the surface irradiated and S the area of the trace on the ground.

It is therefore necessary to thoroughly understand the logistic parameters before identifying and interpreting σ_0 and its relationship with the properties of the surface under investigation.

1.6.7 Logistic parameters

The main features of logistics are very similar for all active systems whether they are point scatterometers (with fixed posts) or imaging radars (airborne or satellite mounted) and are divided into three groups.

■ Source-receiver system

The source-receiver system of a radar, by design, has its own characteristics defined by the following parameters:

—power of the source P_s,
—gain of the source antenna G_s,

—gain of the receiver antenna G_r,
—loss of energy p.

These parameters undergo variations which can be controlled and corrected or compensated by calibration coefficients depending on the system used.

■ Geometric parameters

The principal parameters in measuring σ_0 are (Fig. 1.14):

—Angle of incidence of the beam, determined by altitude and inclination of the antennas and the distance between the receiver and the surface viewed, on the ground. The depression angle corresponds to the inclination of the antenna on the platform relative to the horizontal as well as to the complement of the incidence angle of the wave relative to the surface when the latter is horizontal;

—Cone in which the energy backscattered by the irradiated surface (imprint) is measured, defined by the field of view θ_0 of the receiver antenna.

The angle of incidence is important in measurements since it affects detection of the roughness of the irradiated surface (see the section below). Depending on the type of receiver, the choice of incidence angles is limited due to constraints imposed by the platform, among others.

The field of view of the antenna and the geometry of its positioning directly influence the geometric resolution of the system. In the case of an airborne or satellite system, the concept of ground resolution is complicated due to movement of the platform (longitudinal resolution) (see Chap. 26).

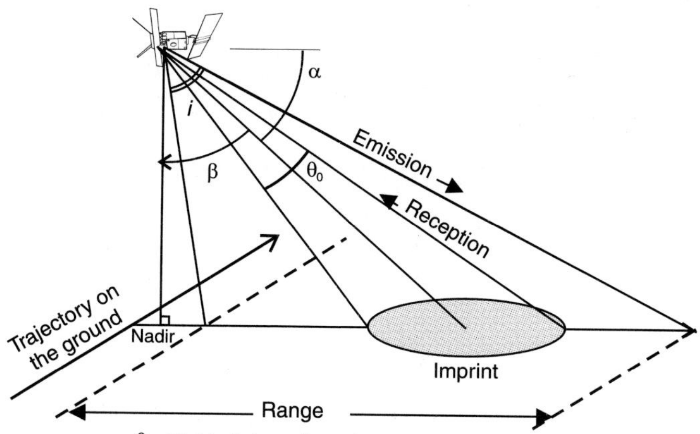

Fig. 1.14: Characteristics of an antenna.

■ Source characteristics

The characteristics of the source signal are the wavelength (λ), signal modulation, its trajectory and its coherence.

Signal modulation is a common characteristic of all active systems. The objective of modulation is to provide a means of comparing the source signal with the backscatter. This comparison ensures precise measurement of the distance covered by the backscatter signal and the ratio of the energies of the source signal and the backscatter signal. The latter gives information about the properties of the irradiated surface.

For observing surfaces situated at large distances, time variations in signal amplitude are used. An example is the pulse radar, which has an emission duration of a few microseconds (Fig. 1.15).

This type of modulation is not possible in small range (0 to 250 m) observations for fixed-post scatterometers, since the shift between the source signal and the backscatter signal is too small to be measurable. In such cases, another system of reference known as frequency modulation is employed: the source energy remains constant and the frequency varies from one cycle to another (Fig. 1.16).

In this case, variation of frequency with time takes place in a saw-toothed pattern (Fig. 1.17), with a frequency shift ΔF, of a few megahertz, around the central frequency F_0.

Whatever be the type of modulation used, the path of the signal in the source-receiver system can be schematically represented as follows (Fig. 1.18).

A microwave signal is produced by the source (1). Through a series of amplifiers and multipliers, the signal is modified to the chosen frequency. A fraction of the source signal (6) is directly sent to the mixer (5) and acts as the local oscillator. Another part of the signal (2) is routed through the transmission system (3) with a given polarisation. The transmission antenna irradiates the target surface. A part of this signal is backscattered towards the receiver antenna (4) and combines in the mixer with the fraction (6) of the source signal. This mixed signal is sent to the analyser (7) to determine the shift

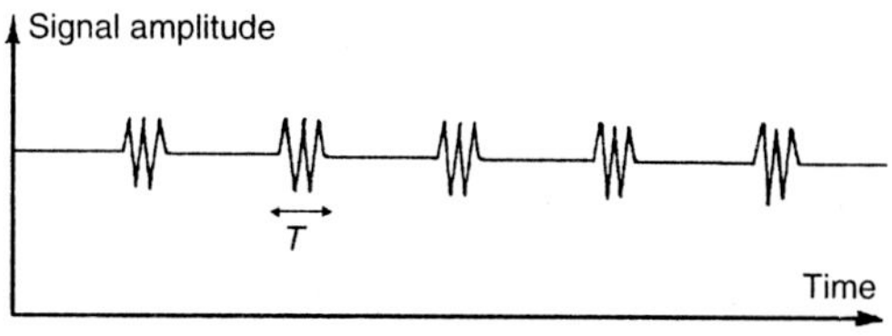

Fig. 1.15: Pulse modulation.

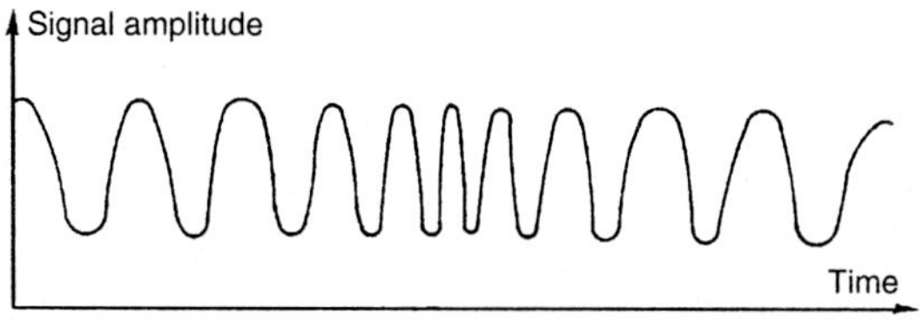

Fig. 1.16: Frequency modulation.

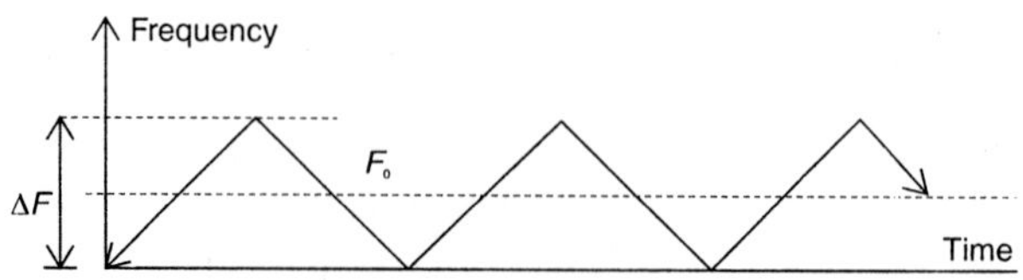

Fig. 1.17: Triangular modulation.

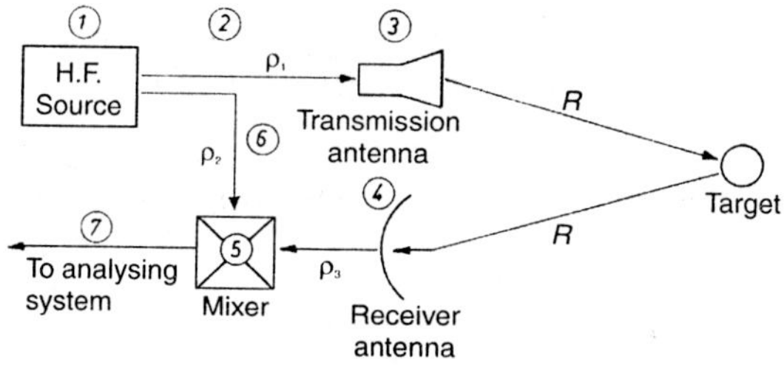

Fig. 1.18: Signal path.

between the two signals. This shift corresponds to the time t_0 taken by the signal to travel the distance $2R + \rho_1 + \rho_2 + \rho_3$. The value of σ_0 is obtained in this way.

The source wave, characterised by its wavelength, polarisation and modulation, is a coherent monochromatic wave. This wave has totally different physical effects compared to those of incoherent light waves used in passive systems.

In fact, interaction of coherent monochromatic radiation with the surface produces interferences during scattering in all directions in space. Very strong local variations in energy are observed while recording the backscatter signal P_{bs}: this is referred to as fading. To compensate for this major disadvantage of radars, averages of several measurements need to be taken:

—at the same frequency but over uncorrelated surfaces (i.e., independent samplings),

—at several frequencies but over the same surface.

Choosing the optimum between the time of integration of various measurements and multiplicity of less accurate but independent determinations of σ_0, in other words, choosing between enhancement in the accuracy of measurement (in decibels: dB) and improvement in the spatial resolution of the system, is a difficult task.

■ Conclusion

Logistic parameters affect the energy of the backscatter signal. These parameters should be available to the user so that he can concentrate on the information derivable from σ_0 about the object or medium under study. That is why research teams are at present engaged in two different aspects of investigation:

—analysis of elementary processes involved in the backscatter of the microwave signal. This refers to the study of interactions of radiation with matter on a large scale, either in the laboratory or in the field with scatterometers.

—use of systematically acquired image data.

This assumes knowledge of the manner in which the surfaces react to pulses received from a radar depending on the physical properties which per se depend on two types of parameters: geometric (angle of incidence, roughness) and dielectric parameters corresponding to the nature of the material constituting the surface.

1.6.8 Analysis of physical processes of backscatter

■ Geometric parameters

□ Angle of incidence

The angle of incidence in which σ_0 is measured has a very important role. As shown in Fig. 1.19, the backscatter intensity, which is very high at near vertical incidences (10°–20°), gradually decreases at steep incidence angles (60° and above).

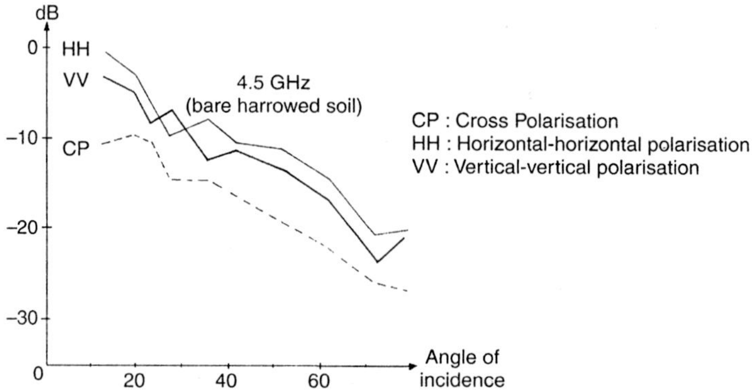

Fig. 1.19: Effect of incidence angle on backscatter signal (after King, 1979).

□ Roughness

Natural terrestrial objects, in particular plant canopies, often represent inhomogeneous media. The incident wave is transmitted into such media and scattered by various components. The backscatter signal is related to the characteristics of the volume concerned. However, the volume scattering is very low compared to surface scattering. Most often, surface scattering is studied in remote sensing.

Surface scattering depends on two parameters: angle of incidence of waves and roughness of the surface under observation.

It should be remembered that the geometric parameters of the objects affect the measured signal only if their dimensions are not too different from the wavelengths used. The same surface behaves differently depending on the condition of its surface and the wavelength of the incident radiation (Fig. 1.20).

Surface roughness originates from several sources. Its significance varies depending on the type of surface: bare soil (cropping activity), vegetation cover (roughness related to growth pattern of plants and their spatial arrangement) or water body (roughness of waves more or less related to the direction and force of wind). Surface roughness is usually expressed by two parameters, viz., Root Mean Square deviation (h) of surface irregularities and a measure of horizontal dimension of roughness. Such a formulation of surface roughness, widely used for bare-soil and water bodies, poses some problems in the case of vegetation. In fact, height and distribution of vegetation, orientation and dimensions of leaves, alignment of trees etc. are difficult to understand and more so when they are natural canopies. Theoretical models have been proposed for cultivated plants, approximating them to moist cylinders. However, morphological modifications undergone by plants during their growth and development should be integrated into such models. Further, 'mixed' surfaces, very common, introduce additional difficulties. Even a feeble coverage of bare soil by crop may suppress the effects due to soil. On the other hand, difficulty is often experienced in observing plants in the course of germination since the soil completely masks their contribution to the backscatter signal. This phenomenon is produced in the case of wheat until it covers at least 5% of the total surface.

Roughness is reported relative to the wavelength (λ) of the source radiation by the formula $h < \lambda/8 \sin \gamma$, where γ is the depression angle.

—A smooth surface, relative to the wavelength used ($h < \lambda/8 \sin \gamma$), reflects the incident radiation as a specular surface (see Figs. 1.3 and 1.4), and σ_0 is zero or very low.

—On a more or less rough surface ($h > \lambda/8 \sin \gamma$), the reflection properties of the surface gradually change from a specular reflection to diffuse reflection, which follows Lambert's law (see Figs. 1.3 and 1.4), and σ_0 increases with increasing roughness. The specular component proportionately decreases and only diffusive reflection remains.

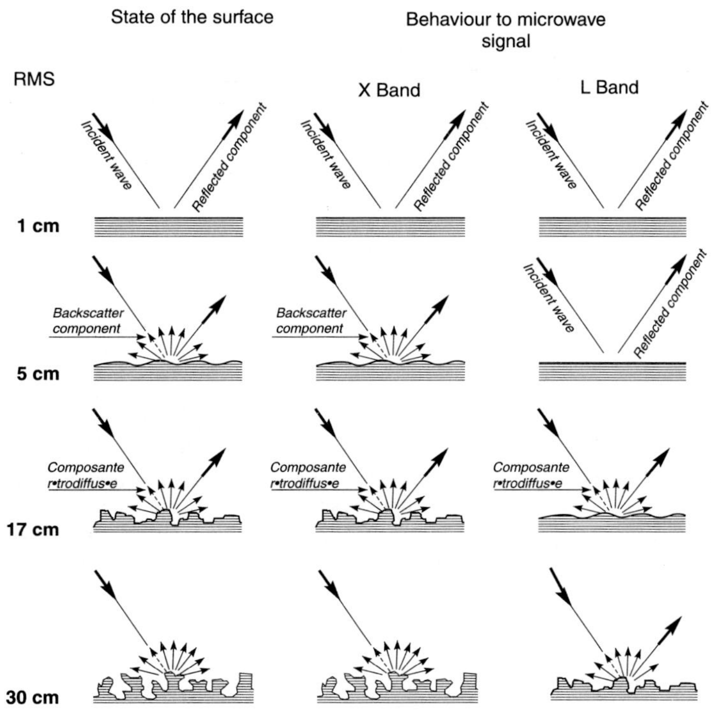

Fig. 1.20: Reflection behaviour of a surface depending on the wavelength of the incident radiation (after King, 1979).

The following example illustrates this effect. Measurements of σ_0 were carried out on a cultivated soil, successively subjected to three types of agricultural operations, viz., tilling, clawing and harrowing, using a scatterometer at wavelengths of 25 cm (1.5 GHz) and 6.6 cm (4.5 GHz). At 1.5 GHz, the effect of roughness was observed irrespective of the value of the angle of incidence and the tilled surfaces indicated σ_0 higher than the harrowed surfaces (Fig. 1.21). At 4.5 GHz, the effect of roughness is negligible, in particular between the incidence angles of 20° and 60° (Fig. 1.22).

Measurement of σ_0 gives information on roughness but the choice of wavelength relative to the degree of roughness studied is very important.

In littoral environments, on exposed tidal flats, sedimentary facies with large gradients in grain size and hence roughness (generally coarse on high zones and fine on low zones) and moisture content (low in high zones and high in low zones) are observed depending on hydrodynamic activity.

Studies of tidal flats observed by SAR of ERS-1 showed that radar measurements are sensitive to the grain size of the sediment. In fact, application of the formula ($h < \lambda/8 \sin \gamma$) to the SAR-ERS-1 data ($\lambda = 5.6$ cm, $\theta = 23$ at the centre of the scene) showed that roughness measurement is significant for $h = 0.8$ cm. This result indicates the possibility of discriminating sedimentary facies such as sands, gravel and pebbles. On the other hand, during very calm periods (sea with no wind), in the presence of vast open tidal flats comprising mud and sandy-mud facies, the land-sea contact cannot be clearly differentiated, as the two targets have almost equal coefficients of roughness.

■ Dielectric parameters

Dielectric constant (ε) and permittivity are the two components of dielectric properties of matter. They are related to the volume of a body and are dependent on its water content and the wavelength

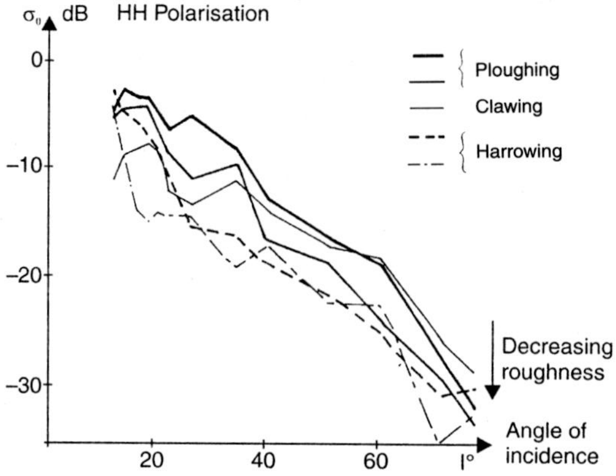

Fig. 1.21: Variation of a 1.5-GHz signal backscattered by a bare soil (moisture content by weight 12–16%) with surface roughness (after King, 1979).

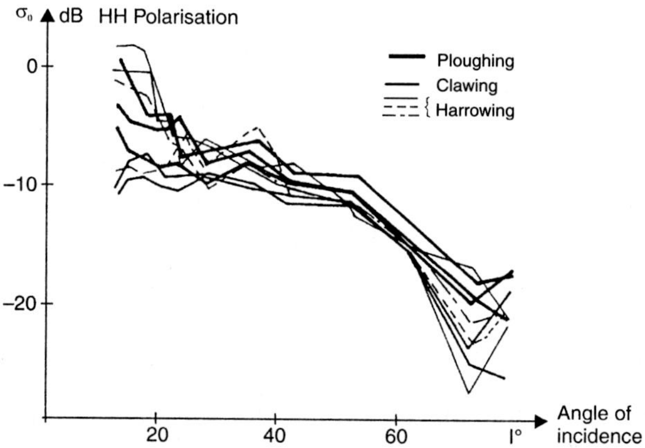

Fig. 1.22: Variation of a 4.5-GHz signal backscattered by a bare soil (moisture content by weight 12–16%) with surface roughness (after King, 1979).

used (Table 1.5). That is why, even small quantities of free water in any material, air and soil or vegetation and soil, produce considerable variations in its dielectric constant. The mineralogical and chemical properties of the objects observed also influence the dielectric constant, but their effects are masked by the predominant effect of water.

Table 1.5: Relative dielectric constant of soil and water for a frequency of 10 GHz ($j = \sqrt{-1}$) (after Ulaby et al., 1981).

Material	Relative dielectric constant
Dry loamy soil	$\varepsilon = 3.5 - 0.4\,j$
Humid loamy soil	$\varepsilon = 17.9 - 7.2\,j$
Water	$\varepsilon = 54.4 - 36.8\,j$

Influence of humidity of a soil on σ_0 was demonstrated by fixed-post experiments using a scatterometer. For example, at 1.5 GHz for a clawed soil, the backscatter signal increases with surface humidity (H = water content, by weight, of the top 2 cm) irrespective of the angle of incidence (Fig. 1.23).

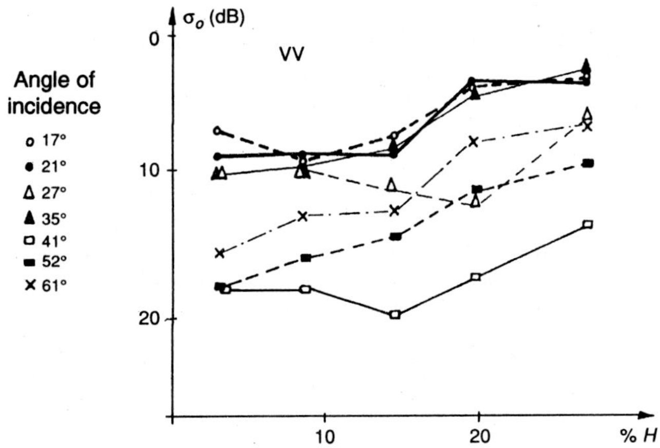

Fig. 1.23: Variation of backscatter signal (1.5 GHz) with surface moisture content for a clawed soil (after King, 1979).

To establish a general relationship between σ_0 and moisture content of a soil or plant canopy, the nature of moisture needs to be defined:

—for a soil, be it gravitational water, absorption water or bound water, and the soil thickness affecting the σ_0 measurement;

—for a vegetation, its water content varies with a number of parameters: vegetation growth stage, soil water content etc.

Having established the influence of moisture and roughness on σ_0, investigations were undertaken to determine the favourable range of frequencies for studying moisture alone. In the case of bare soils, correlation between moisture and σ_0 was found to be better than 70% for frequencies in the range of 4.5 GHZ to 6.0 GHz, with a maximum at 4.75 GHz. Although moisture and roughness are related, it is the water content, less prone to possible errors, which is estimated. Measurements should be carried out with an incidence angle of 10° for the results to uniquely correspond to the water content. This is possible for fixed-post scatterometer measurements but not for radars, since only a small fraction of images corresponds to this angle.

In the case of soils covered by vegetation, effect of water content in the underlying soil on the signal backscattered by plants was observed at 4.3 and 7.5 GHz in VV polarisation. Vegetation on a humid soil produces a higher σ_0 than the same vegetation on a dry soil. Plant cover increases the surface roughness but decreases the contribution of the soil to the backscatter signal. It is necessary to evaluate the effect of vegetation cover on scattering at microwave frequencies. In view of the difficulty in quantifying the geometry of a plant, Ulaby and Bush showed that knowledge of moisture content suffices to determine its stage of growth. They thus developed a model describing the backscatter effect as a function of only this parameter and considering the vegetation as a cloud of scatter elements having an intrinsic moisture.

Many studies were carried out on the effect of snow on microwave signals. Two types of snow are distinguished: dry (powdery) snow, which contains no water in liquid form, and wet snow containing a certain quantity of free water. Surface roughness little influences σ_0 in the case of dry snow and significantly affects σ_0 of wet snow. The backscatter effect of snow increases with increase in its thickness. For a given type of snow, it increases with frequency of waves, while the influence of the angle of incidence becomes negligible for 35.5 GHz (Stiles et al., 1981). This is the reason for studies on snow at frequencies close to the latter.

In the case of sea-water, the radar wave penetrates only a few millimetres from the surface due to the high dielectric constant of water. Hence, one of the predominant factors influencing backscatter is surface roughness. The roughness directly related to agitation of the sea surface can be observed and measured. It is the resultant of several phenomena, viz.:

—local action of wind on the surface (capillary waves);

—sea waves, whose direction and wavelength are modified by coastal morphology and bathymetric level;

—currents, whose directions and velocities are influenced by bathymetric level, especially in zones where the latter is small, i.e., close to the coast;

—presence of a surface pollutant (see section 25.3.4, the case of hydrocarbons).

These phenomena, interdependent, represent complex models, making interpretation of SAR images of coastal marine environment difficult. Nevertheless, several factors, in particular those related to waves and sea rollers, can be extracted from the marine data obtained by synthetic aperture radar such as those mounted on the satellites SEASAT, ERS-1 and ERS-2. Such measurements are independent of meteorological conditions (cloudiness, rainfall etc.) and accessible during the day as well as night.

Radar measurements can be used to determine a surface effect that may be related to the sea-floor topography. Thus, hydraulic dunes, which characterise the floor of the Calais Strait, were detected by radar image. The method requires a shallow zone under the effect of strong hydrodynamic activity. It may be emphasised that this approach is applicable only in special conditions.

Sea-ice is one of the important materials studied by microwave remote sensing in Arctic and Antarctic regions. The ice in such regions contains salt, unlike pure ice, but less than that of normal sea-water. Consequently, ions are distributed across the ice, and variations in their concentrations are represented as variations in the backscatter signal. The properties of sea-ice are dependent on the salinity of the original sea-water, and temperature, pressure and porosity of ice. The structure of ice varies according to the rate of its formation, its age and history. Melting of ice in spring and summer produces cavities inside the ice mass through which water may escape. An ice of first year (less than 1 year of age) melts faster than an older and less saline ice. The salinity and temperature of an ice vary with depth and its properties also change in the course of time, leading to a modification of its dielectric properties. The direction and velocity of wind and currents, number of freeze-defrost cycles, extent of snow cover on the surface etc. cause variations in thickness and surface roughness of ice and lead to variations in σ_0.

■ Penetration properties

Penetration properties of microwave signals were indicated in the preceding examples. These signals effectively penetrate a certain depth of the dielectric (soil or vegetation) before being completely absorbed. Depth of penetration α is defined as the depth at which the source wave is attenuated to about 1/3. In practice, this depth varies from a few cm to 1 m. Very few cases of large penetration, greater than a metre, were observed for highly homogeneous and dry desert sands.

■ Conclusion

In addition to understanding the elementary processes, field measurement campaigns have witnessed two kinds of development:

—defining the most suitable airborne or satellite sensors (with minimum influence of their operational parameters on σ_0 in the ratio P_s/P_{bs}),

—formulation of predictive models or simulation of responses of objects according to their state and characteristics. These models ought to be gradually generalised to larger areas than the test zones.

These investigations have thus brought together a number of signal physicists and thematic specialists engaged in description of the object of their study. In their respective domains, they adjust themselves to the two languages of the model and the reality.

1.7 GENERAL CONCLUSION

The physical basis described above enables us to understand the characteristics of different data acquisition systems in remote sensing as well as the interactions between radiation and natural surfaces. The latter, in the case of visible and reflective near- and middle-infrared, are illustrated in Chapter 4 dealing with spectral characteristics.

References

Actes de télédétection infrarouge thermique, 1993. Des échanges énergétiques et hydriques de la végétation en combinaison avec d'autres capteurs. La Londe les Maures, 20–23 sept. 1993. Cemagref, CETP, Pennstate, 330 pp.

Barton J. 1989. Comparison and optimisation of AVHRR Sea Surface Temperature algorithms. J. Atmospheric and Oceanic Technology, 6–89.

Becker F. 1978. Physique fondamentale de la télédétection, In: École d'été de Physique spatiale: Principes physiques et mathématiques de la télédétection. CNES, pp. 1–108.

Becker F. 1987. The impact of spectral emissivity on the measurement of land surface temperature from satellite. Int. J. Remote Sens., 8: 1509–1522.

Becker F, Li Z.L. 1995. Surface temperature and emissivity at various scales: definition, measurement and related problems. Remote Sensing Reviews, 12: 225–253.

Bonn F, Rochon G. 1992. Précis de télédétection, 1: Principes et méthodes. Presses de l'Université du Québec/ AUPELF, 485 pp.

Chanzy A. 1991. Modélisation simplifiée de l'évaporation d'un sol nu en utilisant l'humidité et la température de surface accessibles par télédétection. Thèse de Doc. Ingénieur INAPG, 210 pp. et annexes.

Courault D, Clastre P, Guinot J-P, Seguin B. 1994. Analyse des sécheresses de 1988 à 1990 en France à partir de l'analyse combinée de données satellitaires NOAA-AVHRR et d"un modèle agrométéorologique, Agronomie, 14: 41–56.

Deschamps P-Y, Duhaut P, Rouquet M-C, Tanre D. 1984. Mise en évidence, analyse et correction des effets atmosphériques, sur les données multispectrales de Landsat ou de SPOT. In: IIe Coll. Int. Signatures spectrales d'objets en télédétection. Bordeaux, 12–16 sept. 1983. Éd. Inra Publ, Les colloques de l'INRA, n°23, pp. 709–722.

Guyot G. 1997. Climatologie de l'environnement. De la plante aux écosystèmes. Masson, Paris, 505 pp.

Jackson RD, Reginato RJ, Idso SB. 1977. Wheat canopy temperature: a practical tool for evaluating water requirements. *Water Resour. Research*, 13 (3): 651–665.

Kerr Y. 1991. Corrections atmosphériques dans l'infrarouge thermique. Cas de l'AVHRR. 5e Coll. int. Mesures physiques et signatures en télédétection. Courchevel, ESA SP 319, pp. 29–34.

King Ch. 1979. Contribution à l'utilisation des micro-ondes dans l'étude des sols. Thèse INAPG, 122 pp.

Kneizys FX et al. 1983. Atmospheric transmittance/radiance: computer code Lowtran 6. Technical report AFGL-TR-83-0187, Optical Physics Division, US AIR Force geophysics laboratory, Hanscom air force base, MA, États-Unis.

Lagouarde J-P. 1991. Use of NOAA AVHRR data combined with an agrometeorological model for evaporation mapping. Internat. J. Remote Sensing, 12 (19): 1853–1864.

Lagouarde J-P, MacAneney KJ. 1992. Daily sensible heat flux estimation from a single measurement of surface temperature and maximum air temperature. Boundary, Meteorology, 59 (4): 341–362.

MODTRAN 3, 1995. User Instructions and Comments, version 1.3. États-Unis.

Norman JM, Becker F. 1995. Terminology in infrared remote sensing of natural surfaces. Remote Sensing Reviews, 12: 159–173.

Ottlé C, Vidal-Madjar D. 1992. Estimation of land surface temperature with NOAA9 data. Remote Sensing of Environment, 49 (1): 27–41.

Ottlé C, François C. 1993. Atmospheric corrections of ATSR-IR data. Actes de télédétection IRT. La Londe les Maures, 2–23 septembre CEMAGREF. CETP, Pennstate, Carlson et al. (eds), pp. 79–82.

Paquet G. 1997. Detection électromagnétique: fondements théoriques et applications radar. Masson, Paris, 320 pp.

Perrin de Brimchambaut C. 1985. Bilan thermique de la Terre. Encyclopedia universalis, vol. 3, pp. 612–613.

Polidori L. 1997. Cartographie radar, Gordon and Breach Science Publishers, Canada, 287 pp.

Privette J-L, Fowler C, Wick GA, Baldwin D, Emery WJ. 1995. Effects of orbital drift on advanced very high resolution radiometer products: normalized difference vegetation index and surface temperature. Remote Sensing of Environment, 53: 164–171.

Seguin B, Itier B. 1983. Using midday surface temperature to estimate daily evaporation from satellite thermal IR data. Internat. J. Remote Sensing, 4 (2): 371–383.

Seguin B, Baelz S, Monget J-M, Petit V. 1982. Utilisation de la thermographie IR pour l'estimation de l'évaporation régionale. II: Résultats obtenus à partir de données satellites. Agronomie, 2 (2): 113–118.

Stiles WH, Ulaby FT, Fung AK, Aslam A. 1981. Radar spectral observation of snow. IGARSS'81 Digest, Washington, pp. 654–668.

Stoll MP. 1988. Mesures de la température et de l'émissivité de surface par télédétection: modèles et méthodes. Télédétection spatiale, aspects physiques et modélisation. CNES, école d'été, pp. 845–904.

Ulaby FT, Moore RK, Fung AK. 1981–1982. Microwave Remote Sensing Active and Passive, vols. 1–2. Addison-Wesley Cy. 1064 pp.

2

Sensors and Platforms

2.1 SENSORS

In remote sensing, various sensors are used to measure in a given wavelength band the radiance of objects under study. Conventionally two types of sensors, viz., active and passive, are employed. Active sensors transmit a signal and receive a part of it returned by the objects. Examples of such sensors are RADAR (Radio Detection And Ranging), LIDAR (Light Detection And Ranging), LASER (Light Amplification by Stimulated Emitted Radiation), fluorometers etc. Passive sensors receive the energy emitted or reflected by the objects (radiometers, cameras, spectroradiometers); the main source of energy in this case is the Sun.

The principle of the sensors is as follows: radiance is measured within the field of view of a sensor over a given surface, i.e., over the pixels whose dimensions are determined by the solid angle of the sensor and its altitude. The altitude evidently depends on the platform employed, while the field of view, solid angle and wavelength bands in which measurements are made depend on the receiver.

Let us consider, as an example, a radiometer (Isco) (Fig. 2.1) mounted on a support at a height of 1.4 m with a solid angle of 90°. At this height, the area covered on the ground by the detector is a little more than 6 m^2, which represents the area of a pixel. As the receiver in this case is on a fixed mount that does not move, we define only a pixel (short form of picture element) but not a field of view. This radiometer has twenty wavelength bands. The EXOTECH radiometer has only four bands corresponding to the MSS sensor of the LANDSAT satellite. The CIMEL radiometer consists of three bands corresponding to the HRV sensor of SPOT satellite.

The photographic systems will not be specially discussed in this book even though they continue to be widely used. It has become common to digitise photographic data and process it as an image.

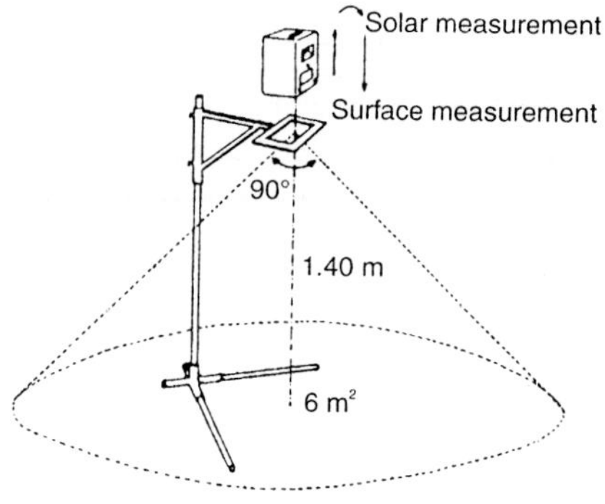

Fig. 2.1: Schematic diagram of the Isco radiometer.

However, care has to be taken to digitise the data in coherence with the grain of the film. The special problems posed by the geometry of measurement are different for aerial photographs and satellite photographs. The latter is easier to rectify and restore (Chapters 13 and 14).

2.1.1 General scheme

Various types of sensors are constructed based on a similar scheme. They comprise several components ranging from signal reception to signal storage (Fig. 2.2), viz.,

—system to receive radiation from the pixel and a telescope (objective),

—calibration source and spectrometer,

—amplifier and recording system.

The sensor thus represents an imaging radiometer, often called scanner, essentially consisting of a radiometer supported by a system of image acquisition, pixel by pixel.

The electromagnetic radiation from a pixel received on a primary mirror and transmitted to a secondary mirror, reaches the recording system through an optical fibre after being amplified and calibrated (Fig. 2.3). The area of a pixel is determined by the receiver system and the sensitivity of the detector. The angles of instantaneous field of view are very small and consequently the energy reaching the sensor is very low, about 10,000 times lower in a satellite system than in an airborne system. Another limiting factor is the duration of exposure, which varies inversely with the square of instantaneous field of view. The exposure times are of the order of a nanosecond to a millisecond depending on the receiver system.

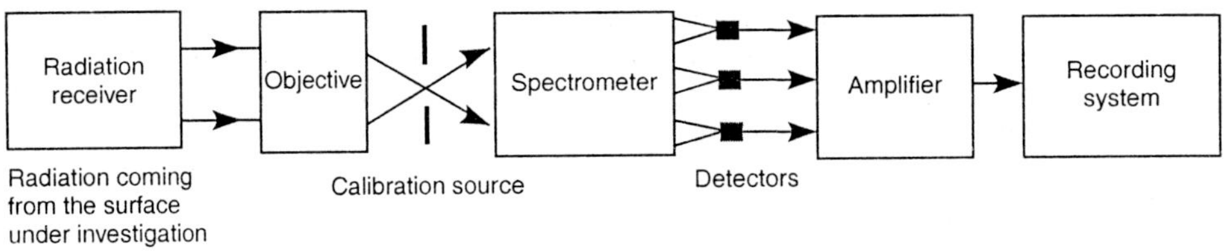

Fig. 2.2: General scheme of a digital sensor.

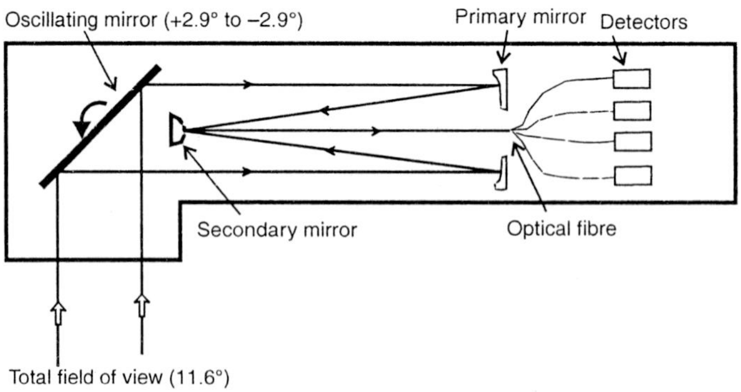

Fig. 2.3: Configuration of a multispectral scanner (LANDSAT MSS).

The amplifier is essential for enhancing the signal, which most often is very weak. The energy received is converted into a digital signal (with discrete values expressed as number of bits: between 6 and 12) by a recording system, which varies according to the detectors.

2.1.2 Radiation receiving systems

Two types of radiation receiving systems are presently in vogue: whiskbroom scanners and pushbroom scanners.

■ Whiskbroom scanners

The whiskbroom or across-track scanner system was first used in LANDSAT-1 satellite in 1972, as well as in several other scanners employed in airborne units (Fig. 2.4).

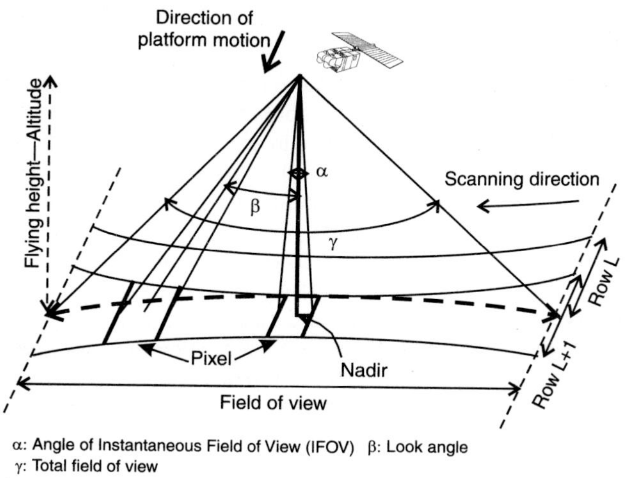

Fig. 2.4: Principle of whiskbroom scanning.

Scanning is carried out by means of a rotating or oscillating mirror, inclined at 45° on the vertical and situated perpendicular to the direction of motion of the platform. The energy coming from the Earth's surface is received in a solid angle, which determines the instantaneous field of view (IFOV). The IFOV is about 10^{-3} to 10^{-5} radian. This angle, which is constant for a given scanner, determines the area of the ground pixel for a given altitude of the platform that carries the scanner. This constitutes the geometric resolution of the sensor.

The image received consists of a series of strips limited in width by the field of view of the receiver and partly by the movements of the mirror. Successive scan rows are covered as the receiver and the platform on which it is installed move forward. In order to obtain a complete coverage without gaps, the velocity of the satellite, its altitude and the rate of rotation of the mirror need to be perfectly synchronised.

When a rotating mirror is used (for example, on an aircraft, the Daedalus scanner with a field of view of 120°), the ground signals are received during a part of its rotation and no signal is received during the other part. This time is utilised to calibrate the receivers by means of standard targets; in particular, a blackbody is employed to record information in thermal bands. When the FOV is small, as in the case of satellite-mounted sensors (for example, 11.6° for LANDSAT-MSS or 14.8° for TM), oscillating mirrors, whose standards are situated on either side of the zone of oscillation, are used.

The geometry of whiskbroom scanners should be fixed precisely since it affects the restoration and rectification of images. Depending on whether the pixel is situated at the nadir or laterally on a scan line, the value of the side of the square, i.e., the geometric resolution, varies (Fig. 2.4). This gives rise to image distortions; however, such distortions are quite small for satellite-borne sensors, which have very small solid angles. As for the radiometric aspect, the atmospheric effects are greater since the atmospheric thickness traversed is larger for an inclined view than for a vertical view (Fig. 1.6).

Geometric rectification of images eliminates such systematic errors. However, before undertaking geometric rectification of images, it is necessary to evaluate the relative magnitude of errors produced by geometric distortions and the errors that could be introduced by various processing techniques or by visual interpretation of images. For example, the interpretation accuracy for delineation of landscape units is less than 100 m. Thematic errors for scales between 1/1,000,000 and 1/100,000 are generally greater than geometric errors. Evidently, if thematic information is to be incorporated into a geographic information system (GIS), it is essential that the image be accurate enough to be superposed on the GIS map and geometric corrections need to be applied (Chapter 13).

■ Pushbroom scanners

The pushbroom or along-track scanner system is more recent and uses no scanning mirror. It consists of a linear array of 1728 to 12,000 CCD (charge-coupled device) detectors, which simultaneously receive information from 1728 or 6000 pixels aligned in a single row (Fig. 2.5). Each detector converts the detected radiance into an electric signal through storage of electrons in holes created on the surface of a semiconductor. The charge thus collected is transferred to an electric circuit wherein the current is amplified and stored on a magnetic base.

Each detector corresponds to a separate pixel. Since all the detectors are identical, no deformation occurs due to optics of the instrument, unlike in photographic units. Thus for every 1 1/2 millisecond, 6000 values are obtained corresponding to a ground swath 10 m wide and 60 km long in the case of SPOT. Each CCD detector measures only 13 μm × 13 μm. They are grouped into an array of 1728 detectors, which together form a bar of about 2.25 cm and do not occupy much space. All the detectors scan a row, which is generated by the advancement of the platform: consequently, this system is called 'pushbroom'.

Each detector is characterised by its own transfer function and has to be calibrated individually. As the detectors are identical, the solid angle that controls the instantaneous field of view is also

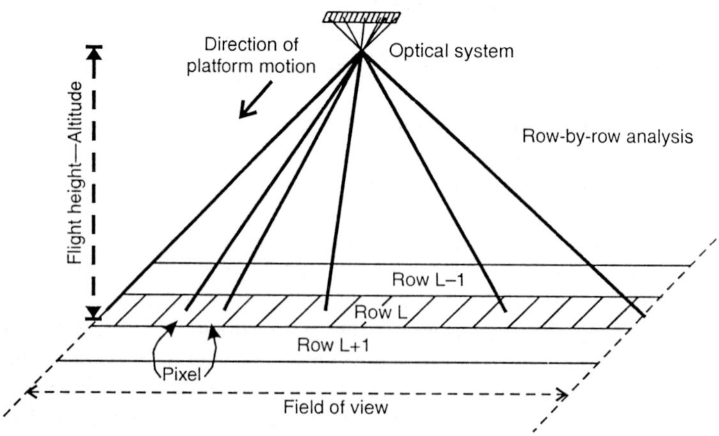

Fig. 2.5: Principle of a pushbroom scanner.

identical; the dimensions of the pixels are slightly larger on the edges of the row since the system has a conical projection (Chapter 13).

An advantage of the pushbroom scanner is the greater duration of exposure than in the scanning-mirror system for the same platform velocity. For the pushbroom scanner:

$$D = p/v$$

where D is the exposure time (in seconds), p the pixel size (in m) and v the velocity of the satellite (in m/s). Thus, for SPOT, D = 10 (m) / 6660 (m/s) = 1504 milliseconds.

In the scanning-mirror system:

$$D = p/nv$$

where D is the exposure time (in seconds), p the pixel size (in m), n the number of pixels per scan-line and v the velocity of the satellite (in m/s).

As the number of pixels on LANDSAT per line is 3000 with p = 30 m, the exposure time is 1000 times shorter. The pushbroom scanners perform better from the radiometric point of view. Each array of detectors corresponds to a spectral band. The pushbroom scanner is employed in HRV and MOS-1 sensors.

2.1.3 Spectral bands

It is important to arrange the detectors in different narrow spectral bands in order to detect specific phenomena of certain spectral domains.

A detector should receive a certain minimum energy to be active. Hence, if a wide band, such as the panchromatic from 0.5 to 0.7 μm, is used, the quantity of energy is greater than that for a narrow band and a pixel of smaller area can be taken. An inverse proportionality exists between geometric resolution and spectral resolution. If the geometry of a pixel is changed, the energy in each resolution element is changed by the same value. Consequently, a geometric modification leads to a modification of the radiance and vice-versa. It is, therefore, better to apply geometric corrections after processing than before so that the radiometric signal is not disturbed.

Different sensors have a variable number of spectral bands, generally three to ten (CD 2.1). Some sensors, employed on the ground or airborne, have more than 100 spectral bands. It is hence necessary to select bands that are most suitable for a given thematic application.

2.1.4 Airborne systems

At the time of launching of the first Earth resource satellites, it could have been thought that the era of airborne systems had ended. But it did not. Parallel to the pursuit of the conception and launching of satellites, a continuous development of airborne remote sensing missions with special sensors has been witnessed (Table 2.1).

There are three reasons for developing airborne remote sensing. Firstly, it is essential to acquire data at an intermediate level that enables a better understanding of certain features through the relationship between the level of acquisition and the level of organisation (Chapter 15). Secondly, in the case of regional studies covering a limited territory, it may be advantageous to investigate a specific problem through an airborne mission (more economical). Such is the case, for example, in evaluation of damages caused to agriculture by a drought or a devastating flood. Finally, the possibility of using narrow spectral bands, of a few nanometres to a few tens of nanometres wide, in the spectral domains for which high atmospheric absorption makes use of satellite data ineffective.

Table 2.1: Examples of airborne spectrometers

Characteristics	AVIRIS	Casi
Technology	CCD array measuring all the channels for 1 pixel; mechanical sweep of lines	CCD array. In spectral mode, selection of 39 columns of the array, information acquired for 288 bands. In spatial mode, selection of 15 bands, information acquired for 512 pixels.
Date of commencement of service	1987	
Spatial resolution	20 × 20 m at 20 km altitude	2.3 × 2.7 m in spatial mode, 2.3 × 9.5 m in spectral mode at 1860 m
Swath	12 km	FOV 35° 5
Number of spectral bands	224	288
Width of spectral bands	10 nm	1.8 nm
Wavelength range	410 to 2450 nm	430 to 870 nm

Absolute positioning of the aircraft in regions where no ground control points are available and high costs have long been the major handicaps in using airborne remote sensing. Application of GPS (Global Positioning System) in association with the American GLOSSNAR satellites has resolved this problem.

2.2 PLATFORMS

Any moving vehicle that can carry a sensor can be considered a platform for remote sensing. Thus, several types of platforms can be identified:

—those operating at a height of a few metres from the ground: cranes or other vehicles that support radiometric or photographic equipment;

—those operating at a height of about ten metres to ten kilometres: aeroplanes, helicopters and balloons;

—those operating between ten and hundred kilometres: stratospheric balloons;

—those operating between 200 and 40,000 km: satellites, manned or unmanned, subjected to terrestrial gravity. The latter are the most common for observation of the Earth.

Airborne platforms, such as aeroplanes or stratospheric balloons, are not discussed here, as these were described in our book of 1975 or in the book by Bonn and Rochon (1992). Only Earth observation satellites are described in this book. Orbital characteristics of the satellites launched to study other planets of the solar system are also not considered. Only some orbital data of satellites are reviewed in the following section. Illustrations and details regarding the geometry of remote sensing can be found in Lliboutry (1992).

2.2.1 General principles of orbital motion

Two masses E (Earth) and S (satellite) separated by a distance d attract each other with a force $F = g\,ES/d^2$, where g is the gravitational constant. Computed by Newton in 1687, this force controls the motion of the two bodies according to the three Kepler laws established in 1609.

■ First law

A satellite moves along an elliptical orbit around the Earth, which occupies one of the foci of this ellipse (Fig. 2.6). This orbit is described by its semimajor axis a and eccentricity $e = c/a$, where c is the distance of a focus of the ellipse to the perigee. Perigee is the position of the satellite when it is closest to the Earth (818 km for SPOT) and apogee the position when it is farthest from the Earth (833 km for SPOT). In order to avoid deceleration due to the atmosphere and falling on the Earth, the perigee of a satellite ought to be greater than 200 km above the ground.

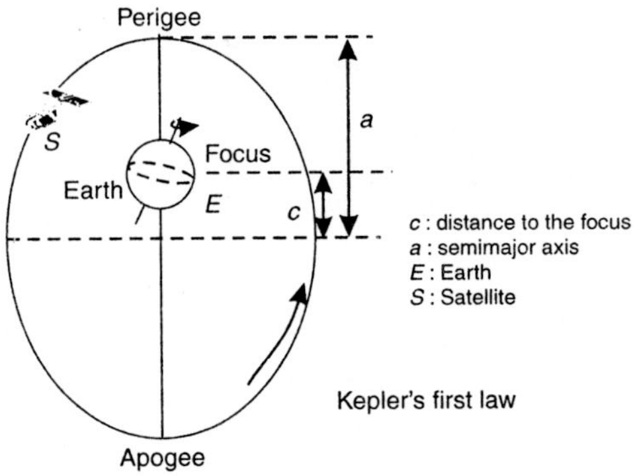

Fig. 2.6: Satellite orbit according to Kepler's first law (after Gilliot, 1994).

■ Second law

The satellite does not travel along the ellipse with a constant linear velocity but in such a manner that the radius vector joining the satellite to the Earth sweeps a constant area in a given time (Fig. 2.7). Hence when the satellite is at its perigee, its velocity is greater than when it is at the apogee.

■ Third law

The period P (in years) of a satellite is related to the semimajor axis a expressed in astronomical units (1 a.u. = Sun–Earth distance) and to the masses of the Earth (E) and satellite (S):

$$(E + S)\ P^2 = a^3$$

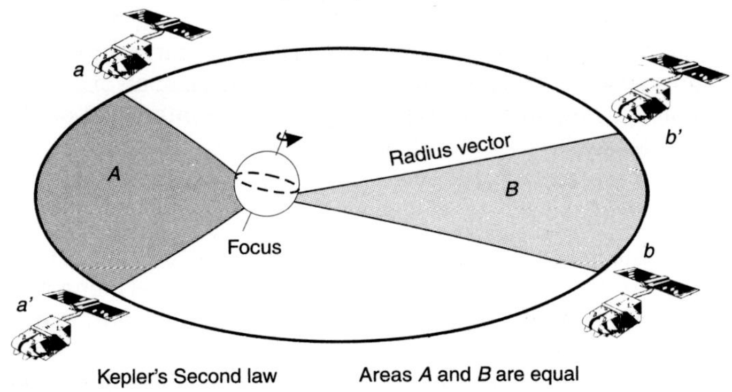

A is the area described by the radius vector between the points *a* and *a'* in a time t
B is the area covered by the radius vector between the points *b* and *b'* during the time t

Fig. 2.7: Satellite velocity according to Kepler's second law (after Gilliot, 1994).

2.2.2 Orbits

The plane of the elliptical orbit is known as the orbital plane of the satellite (Fig. 2.8). The angle between the orbital plane and the equatorial plane defines the angle of inclination of the satellite. The line of intersection of the two planes is the nodal axis (or line). A node hence corresponds to the intersection of the equator with the path of the satellite (orthogonal projection of the satellite orbit onto the Earth).

In Earth-observation remote sensing, circular, geostationary and sun-synchronous orbits are important.

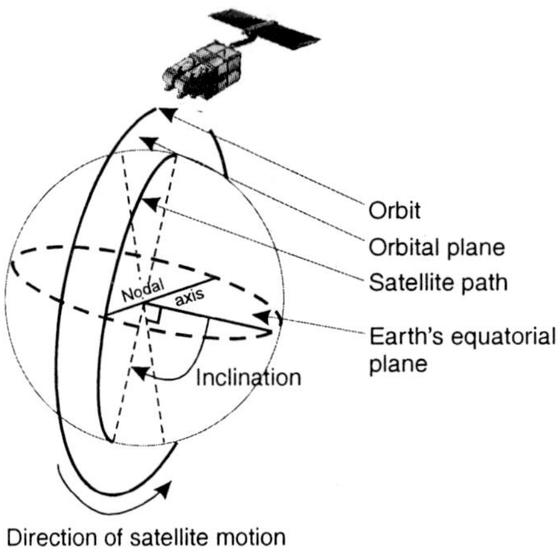

Fig. 2.8: Orbital parameters of a satellite (after Gilliot, 1994).

■ Any circular orbit

If the eccentricity of the ellipse becomes zero, the orbit becomes circular. The satellite revolves around the Earth at a constant altitude. The minimum velocity of the satellite at an altitude of 200 km, i.e., for $a = 6500$ km, would therefore be 7.77 km/s and the periodicity about 1 h 30 min.

At 200-km altitude, at which military observation satellites operate, the temperatures are of the order of 750° C plus or minus 300° C. These temperatures are essentially generated by the kinetic energy of the monoatomic molecule of oxygen, the most abundant element between 120 and 1000 km of altitude. The aerodynamic drag around 200 km limits the lifespan of a satellite to a few months. In order to remain in orbit, observation satellites need to be operated at about 800 km of altitude.

The orbits of the satellites TIROS, NOAA and ERS-1 are circular.

■ Geostationary orbit

A geostationary orbit is one in which the satellite is always situated at the zenith of a point above the Earth's equator. Hence the orbital plane coincides with the equatorial plane and the orbit is circular. The satellite revolves with the same angular velocity as that of the Earth and therefore is fixed in a terrestrial reference frame. Kepler's third law gives the radius R of the orbit as 42,164 km, or 35,786 km above the equator (the equatorial radius of the Earth being 6378 km). The satellite velocity v in its orbit is thus $v = 2\pi R/t$, where $t = 1$ $d = 86{,}164$ s and $v = 3.07$ km/s or 11,063 km/h.

The satellites METEOSAT (France), GOES (Geostationary Operational Environmental Satellites, USA), GMS (Japan) and INSAT (India) are geostationary.

■ Sun-synchronous orbit

A satellite in a sun-synchronous orbit always passes over the same point on the Earth at a given local solar time. The orbital plane of the satellite hence remains constant relative to the orbital plane of the Earth around the Sun. Consequently, the nodal line makes a constant angle with the axis joining the centres of the Earth and the Sun.

The solar day is defined as the duration in which a given point on the Earth returns to the same position relative to the Sun. By definition, the solar day is equal to 24 hours. The sidereal day is defined as the period of rotation of the Earth relative to a fixed point in space (vernal point). This period is equal to 23 h 56 min 4 s. Consequently, the Earth moves around the Sun in 365.2425 sidereal days. In order to be synchronous, therefore, the orbital plane of the satellite should rotate by 360° in 365.2425 days, i.e., by 0.9856° per day (Fig. 2.9). This corresponds to an angular velocity of retrograde precession. Hence, a sun-synchronous orbit can be defined as an orbit having an angular velocity of precession equal to the angular velocity of the Earth around the Sun.

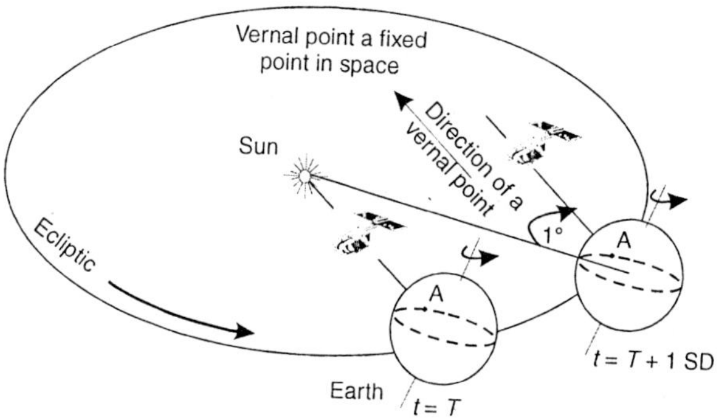

Fig. 2.9: Displacement of the orbital plane of a sun-synchronous satellite (adapted from Gilliot, 1994).

For a circular orbit at an altitude of 832 km, as in the case of SPOT, the inclination should be 98.7°. Consequently, the north and south poles are never covered. Hence for each pole there exists a circle of precession with its centre at the pole and a radius of 966.6 km for which no images can be obtained. Otherwise, the entire Earth can be covered by remote sensing images.

The relatively low altitude of sun-synchronous satellites facilitates a good resolution. However, as these satellites are subjected to deceleration by terrestrial atmosphere, the orbital parameters need to be corrected periodically. Similarly, the altitude of the satellite has to be permanently controlled.

In the case of sun-synchronous orbits, the azimuth of the Sun is constant throughout the year but the height of the Sun varies. This gives rise to shadows of varied lengths but in the same direction.

The satellites NIMBUS, LANDSAT and SPOT have a sun-synchronous orbit.

Thus satellites with any circular orbit or sun-synchronous orbit are distinguished from geostationary satellites (Fig. 2.10).

In the case of unmanned satellites, sensors are mounted on the platform once and for all. The conditions of image acquisition are determined by these two entities, viz., platform and sensors, which hence constitute a single system. To understand the interactions between sensors and platform, let us study the SPOT system (also see text on the CD giving details of the SPOT system and SPOT 4 in particular) and the data on other systems.

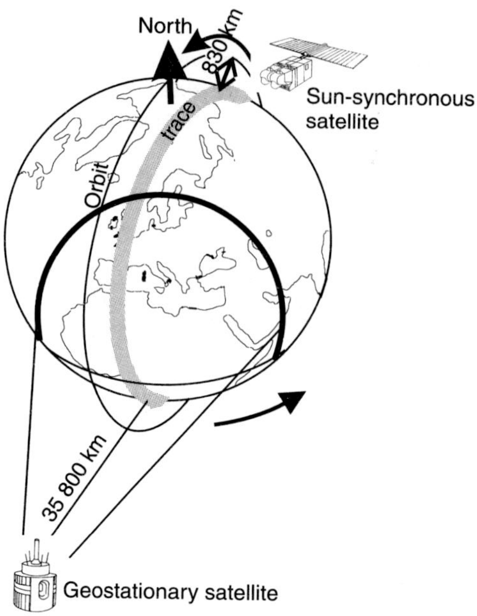

Fig. 2.10: Relative positions of sun-synchronous and geostationary satellites.

2.2.3 SPOT system

The SPOT satellite has a sun-synchronous, quasipolar (98.7°) and circular orbit with an eccentricity less than one per thousand. It acquires images in a descending path during the day, i.e., while going from the north pole to the south pole, and nominally passes through the equatorial node at 10 h 30 min on 15 June. The duration of passage above a given region is more or less constant at about 15 min.

The velocity of the satellite relative to the ground is 6.6 km/s or 23,760 km/h. The average duration of one revolution is 101.4 min. During this revolution, the Earth rotates by 25.35° or 2823 km at the equator (whose perimeter is 40,075 km). This constitutes the interval between chronologically successive paths. The orbits and the paths converge at the poles. The spacing between adjacent paths is 108.6 km at the equator, 76.7 km at 45° latitude and 16.5 km at the circle of precession.

The satellite thus makes 14 and 5/26 revolutions in 24 hours (Fig. 2.11). It covers the same point every 26 days after completing 369 orbits. The chronologically successive orbits are offset westwards because of the Earth's rotation. The paths are hence organised in the same way as those of a ball of thread from a skein: the Earth moves eastwards while the orbital plane of the satellite does not change.

The satellite passes over Europe in two adjacent paths every 5 days (Fig. 2.12).

Evidently, remote-sensing data in the visible domain, which requires solar irradiation, is acquired only during the descendant half-orbit. The satellites SPOT 1, 2 and 3 comprise 4 spectral bands, viz., panchromatic (P): 510 to 730 nm, green (B1): 500 to 590 nm, red (B2): 610 to 680 nm, and near infrared (B3): 790 to 890 nm.

■ Stereoscopy and revisit capability

Because of the repeated coverage (see CD) of the same zone in successive orbits, an important characteristic of SPOT is the acquisition of images, panchromatic or multispectral, at two different

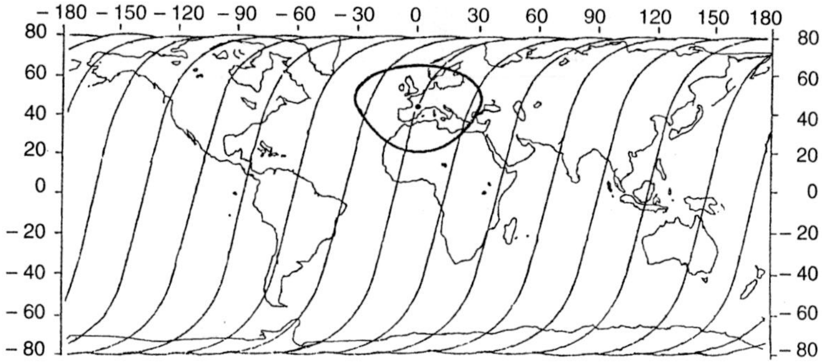

Fig. 2.11: Orbits in a day (after CNES illustration).

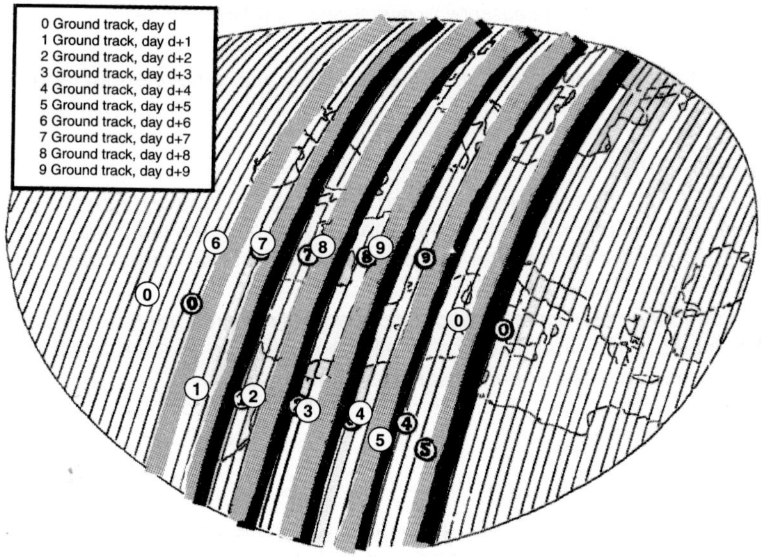

Fig. 2.12: Descending orbits over Europe (after CNES illustration).

angles (pre-processing stage 1A or 1B). This feature facilitates three-dimensional observation of a zone using a stereoscope (see CD and Fig. 2.13) and thus construction of digital elevation models (CD 2.2). Stereoscopy is very useful for numerous thematic studies pertaining to morphology, soil science, geology (ground slides), landscapes and evidently photogrammetric analysis.

The overlapping acquisition ensures a greater revisit capability and this enhances the probability of acquiring images, without clouds (CD 2.3), over a given region.

■ SPOT 4 (see CD)

The orbit of SPOT 4 is the same as that of the other three SPOT systems. It is hence possible to simultaneously use all these satellites not only for increasing the number of images acquired, but also for generating stereoscopic pairs the same day from two satellites. In fact, the two high-resolution visible and infrared (HRVIR) sensors have the same spectral bands as those of other SPOT satellites, to which a middle infrared (MIR) band of 1.58 to 1.75 nm, with a spatial resolution of 20 m, is added.

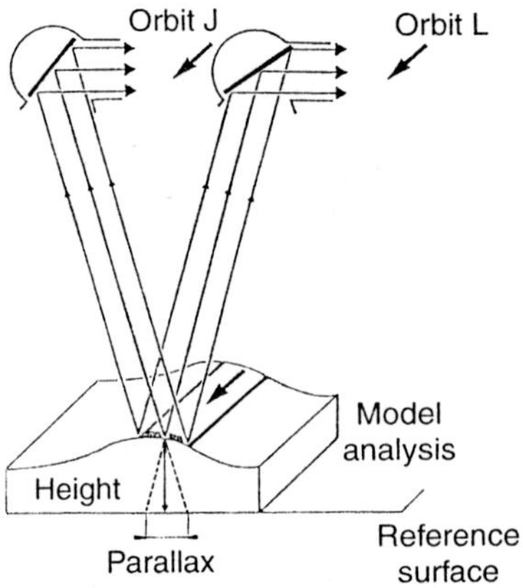

Fig. 2.13: Stereoscopy by lateral viewing (CNES illustration).

The panchromatic channel is replaced by a single band (b2) with a wavelength range of 610 to 680 nm and a resolution of 10 m.

□ Instruments on SPOT 4

PASTEL (*Passager SPOT de télécommunication laser*) transmits images to the satellite ARTEMIS through an optical connection or directly or through a replay of mass memory.

PASTEC (*Passager technologique*) enables study of orbital environment: radiation, ageing, vibrations, electric potential etc.

DORIS (*détermination d'orbite et radiopositionnement intégrés par satellite*), already existing on SPOT 2 and 3, enables determination of the satellite position with an accuracy better than 10 cm, as well as the ground position of the scanner with the same accuracy.

POAM III (Polar Ozone Aerosol Management), an American payload, measures the ozone and aerosol contents at the poles.

□ Vegetation instrument (VGT)

The vegetation scanning system on SPOT 4 is a novel tool vis-à-vis the previous SPOTs. With its very large field of view, medium spatial resolution and high radiometric resolution, it mainly enables a global monitoring of phenomena related to vegetation, as well as bare soils and water. Its spectral bands are b2, b3, MIR and b0 (430 to 470 nm).

■ SPOT 5

SPOT 5 has been successfully launched during the night of 3 to 4 May 2002. It provides service continuity to the previous SPOT 1 to 4 since the orbit is similar, i.e. a circular, quasipolar orbit at an altitude of 830 km and a pass over the equator at 10.30 a.m. (local time at descending node). The main improvements compared to SPOT 4 consist in: i) a dedicated instrument for along track stereo acquisition, ii) a higher ground resolution: 5 m and 2.5 m (instead of 10 m) in panchromatic mode, 10 m (instead of 20 m) in multispectral mode. Otherwise the spectral bands are the same as in SPOT 4 with the short wave infrared band maintained at a resolution of 20 m due to limitations imposed by the CCD sensors used in this band. The VGT (vegetation instrument) is maintained, as well as the field width (60 km) and the oblique viewing capacity. These choices answer the growing demand in cartography, agriculture, planning and environment.

2.3 OTHER SYSTEMS

Besides SPOT, other important Earth resource satellites presently in operation are (Table 2.2) METEOSAT, NOAA, LANDSAT. The satellites ERS, RADARSAT, JERS are used for studies in high frequencies. Specific thermal infrared sensors are given in Table 2.3 and high frequency sensors are indicated in Chapter 26. The CZCS (Coastal Zone Colour Scanner) of NIMBUS 7, which was operational between 1978 and 1986, will be mentioned in the study of water colour (Chapter 4).

2.3.1 METEOSAT

METEOSAT (Fig. 2.14) is a geostationary satellite and has an altitude of 35,800 km above the equator, at 0° latitude (Table 2.2).

Lateral swing is achieved by rotation of the satellite per se. An oscillating system ensures changeover from one row to another. The satellite makes 100 rotations per minute. As 2500 rotations are needed to cover the Earth, it takes 25 min to produce an image. Obviously, the geometry is altered by the rotundity of the Earth.

METEOSAT 5 has three bands, viz., 400 to 900 nm with a resolution of 2.5 km and 5.7 to 7.1 μm and 10.5 to 12.5 μm with a resolution of 5 km (CD 26.2). Ground receiving stations are at Lannion (France) and Kiruna (Sweden). The angle of view is 17°. The zone of observation extends from 60° North latitude (Sweden) to 60° South latitude (southern part of South Africa) and from 60° West longitude (centre of Brazil) to 60° East longitude (East of Saudi Arabia).

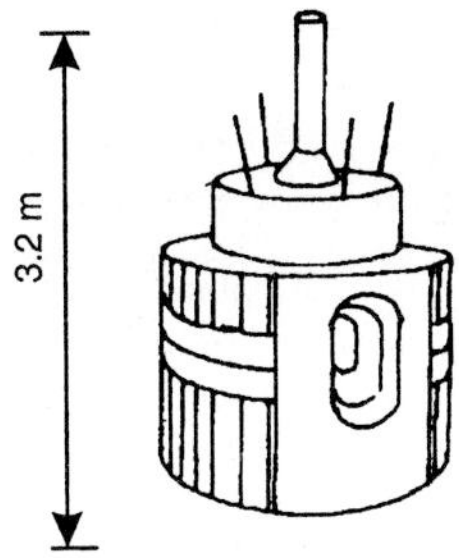

Fig. 2.14: METEOSAT satellite (after Lliboutry, 1992).

The first METEOSAT satellite was launched in 1977 followed by others, viz., METEOSAT 6 in 1993, and the seventh to be launched 20 years after the beginning of the programme.

The entire Earth, except the polar regions, is continuously observed by METEOSAT and four other geostationary satellites (Fig. 2.15): two American satellites GOES (Geostationary Operational Environmental Satellite) located at 75° West and 135° West, the Indian satellite INSAT situated at 75° East and the Japanese satellite GMS at 140° East.

2.3.2 NOAA

The National Oceanic and Atmospheric Administration (NOAA) operates several kinds of satellites successively, including TIROS (Television and InfraRed Observation Satellite) launched in 1960. TIROS-N launched in 1978, the NOAA-6 and the subsequent satellites (Fig. 2.16) are equipped with an AVHRR sensor (Tables 2.2 and 2.5).

Table 2.2: Characteristics of the main Earth observation satellite systems in visible and near infrared bands

System characteristics	METEOSAT	NOAA-AVHRR	LANDSAT MSS	LANDSAT TM
Orbit:				
Type of orbit:	Geostationary	Circular	Subpolar sun-synchronous	Subpolar sun-synchronous
Altitude (km)	35800	850	705	705
Revisit capability	25 min	12 h	16 d	16 d
Sensors:				
Scanning	Satellite rotation	Rotating mirror	Oscillating mirror	Oscillating mirror
Spatial resolution	2.5 km (S1); 5 km (S2, S3)	1.1 or 4 km	56 × 79 m	30 m
Spectral bands (μm):				
S1	0.4–1.1	0.58–0.68	0.5–0.6	0.45–0.52
S2	5.7–7.1	0.72–1.1	0.6–0.7	0.52–0.60
S3	10.5–12.5	3.55–3.93	0.7–0.8	0.63–0.69
S4		10.3–11.3	0.8–1.1	0.76–0.90
S5		11.5–12.5		1.55–1.75
S6				10.4–12.5
S7				2.0–2.35
Panchromatic				
Scene dimensions (km)	Globe	2400	185	185

System characteristics	JERS1-OPS	SPOT	Early Bird	IRS 1D	
Orbit:					
Type of orbit:	Sun-synchronous	Subpolar sun-synchronous	Polar sun-synchronous	Subpolar sun-synchronous	
Altitude (km)	570	830	470	905	
Revisit capability	44 d	26 (or 1–5) d	2–5 d according to latitude	22 (or 2–4) d	
Sensors:					
Scanning	CCD arrays	CCD arrays	CCD matrix	LISS: CCD	WIFS CCD
Spatial resolution	18 m	10 m (pan); 20 m (S)	3 m (pan); 15 m (S)	5 m (pan) 20 m (S)	190 m
Spectral bands (μm):					
S1	0.52–0.60	0.50–0.59		0.52–0.59	0.62–0.68
S2	0.63–0.69	0.61–0.68	0.49–0.60	0.62–0.68	0.77–0.75
S3	0.76–0.86	0.79–0.89	0.615–0.670	0.77–0,86	
S4	1.59–1.70		0.79–0.875	1.55–1.70	
S5	2.0–2.1				
S6	2.1–2.2				
S7	2.2–2.3				
Panchromatic		0.51–0.73	0.445 –0.65	0.50–0.75	
Scene dimensions (km)	75	60	3	131	774

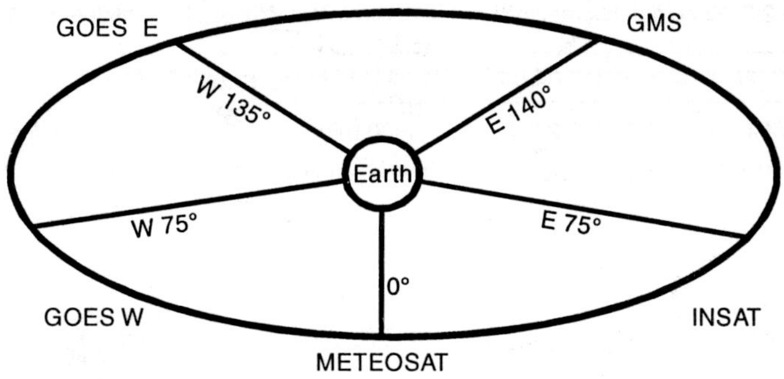

Fig. 2.15: Disposition of meteorological geostationary satellites.

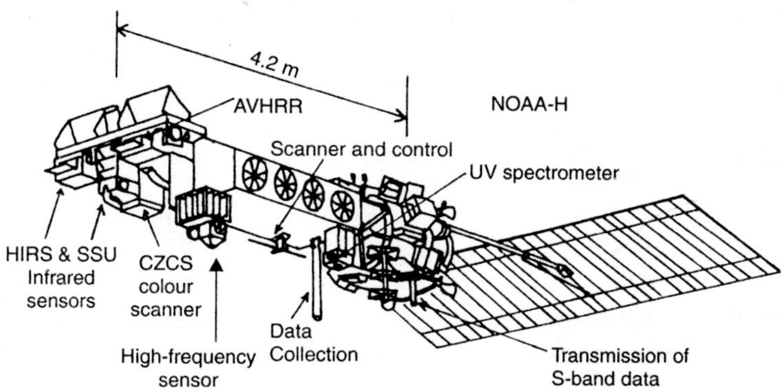

Fig. 2.16: The NOAA-H satellite (after Lliboutry, 1992).

NOAA-7 passes over the same region every 12 h, i.e., at 1400 h and 0200 h. It passes above the same zone every 19 days and completes 268 orbits to cover the globe. Its field of view is 2700 km by 2700 km.

The system consists of AVHRR (Advanced Very High Resolution Radiometer) with a rotating mirror and has the spectral bands of 580–680 nm, 720–1100 nm, 3.55–3.93 μm, 10.3–11.3 μm and 11.5–12.5 μm for NOAA-8 and 9. The resolution is 4 km for NOAA-10.

Examples of image interpretation are given in Chapter 26 (CDs 26.1 and 26.3; also see CD 4.3).

2.3.3 Thermal sensor systems

In view of the importance of applications of thermal infrared remote sensing, this band is installed in a number of satellites. Table 2.3 summarises the main systems that presently have (or are planned in the near future) channels in this spectral domain. The criteria of spatial resolution and temporal resolution are important for agricultural applications and, in particular, for monitoring water conditions of crops. It can be seen from the table that presently no system exists which is suitable for these objectives, simultaneously combining high spatial and temporal resolutions. Meteorological satellites (such as METEOSAT or NOAA) have the advantage of high repetition, which enables temporal surveillance of areas but have low spatial resolution. Consequently, various airborne sensors have been employed to

Table 2.3: Principal thermal infrared sensors (after Prevot et al., 1996, in INRA publ.).

Satellite	Sensor	Spectral band (μm)	Revisit capability	Resolution	Field of view	Date
METEOSAT	MVIRI	10.5–12.5	30 min	5 km	Geostationary	
NOAA	AVHRR2	3.55–3.93 c4: 10.3–11.3 c5: 11.5–12.5	Twice per day	1.1 km	3000 km	
ERS 1 ERS 2	ATSR ATSR2	1.6 3.7 10.8 12	3 to 35 days	1 km	50 km	1991 1995
LANDSAT7	ETM+	10.4–12.5	16 days	120 m	185 km	1998
EOS-AM	ASTER	5 8–12	16 days	90 m	60 km	1998
EOS-AM1	MODIS	0.4–14.4 (14c)	2 days	1 km	2300 km	1998
MSG 1	SERIVI	3.7 8.8 10.8 12	15 min	3 km	Geostationary	2000
Airborne sensors						
German Scanner	DAIS 7915	6 channels 8 to 14 μm		512 pixels/line	78° view	1993
Italian Scanner	MIVIS	10 channels		755 pixels/line	71°	1994
French Camera	Inframetrics 760	1/3 possibilities		256 pixels/line	7°–20°–80°	1992

acquire data of higher spatial resolution but, in this case, repetition of measurements is low. Moreover, airborne systems often have multiple view angles (Table 2.3); this feature is useful to test the spectral and spatial resolutions pertinent for a better monitoring of the state of soil and vegetation with the objective of proposing future satellite sensors.

2.3.4 LANDSAT

The Earth Resources Technological Satellites (ERTS) programme, using ERTS-1 satellites and renamed as LANDSAT (Land Satellite), is launched by NASA. The EROS (Earth Resources Observational Satellites) centres were established in 1966 to distribute images of the missions APOLLO, GEMINI and SKYLAB. From 1986 until recently the products have been commercialised by the EOSAT society, but now Landsat 7 products (launched in April 1999) are commercialised by the US Geological Survey (USGS).

There are two series of LANDSAT satellites: numbers 1 to 3 and numbers 4 to 7.

LANDSATs 1 to 3 (Fig. 2.17) contained a Return Beam Vidicon (RBV) camera (479–575 nm; 580–680 nm; 690–830 nm) and a Multispectral Scanner (MSS) with four spectral bands (Table 2.2). One pixel of MSS images corresponds to 79 m width over 56 m along the track. The four spectral

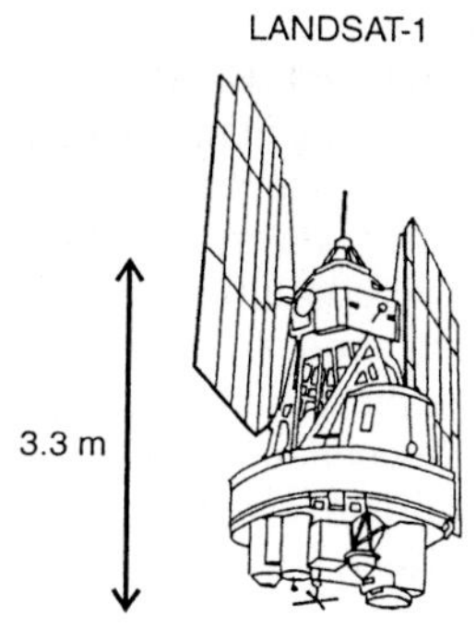

Fig. 2.17: LANDSAT 1 satellite (after Lliboutry, 1992).

bands are MSS4 = 500–600 nm, MSS5 = 600–700 nm, MSS6 = 700–800 nm and MSS7 = 800–1100 nm (CD 2.1). The total field of view is 11.56° and spatial resolution 56 m by 79 m.

MSS was the first system to procure multispectral images for scientific applications. Studies of these images showed that channel 6 was not very useful since bare soils and vegetation could not be distinguished. Although this band was not retained for subsequent LANDSAT satellites, this sensor has always existed for purposes of comparisons with the images acquired since 1972.

LANDSAT 5 is a sun-synchronous satellite in a subpolar orbit. Its altitude varies from 696 km at the equator to 741 km at the poles, with a standard altitude of 705.3 km (438 miles). Inclination is 98.22. The satellite passes over the equator at 9 h 37 min local time. With a period of 98.9 min per revolution, it makes 14 and 9/16 orbits per day. The periodicity of passage is 16 days. It completes 233 paths to cover the globe and acquires 248 scenes per path. Path 1 intersects the equator at 64.6° W. Overlapping between two scenes is 7.6 % at the equator and this value increases with latitude, reaching 54% at 60° latitude.

The satellite has no onboard recording system since it transmits the data in real time to receiving stations. When there is no direct contact, the data is transmitted through relays using communication satellites TDRS (Tracking and Data Relay Systems).

Every LANDSAT 5 scene measures 170 km in the north-south direction and 185.2 km (100 nautical miles) in the east-west. The territory of France is covered by 210–219 paths and 25–30 rows.

LANDSAT 4 and 5 carry an MSS scanner and a Thematic Mapper, which comprises 7 spectral bands (Table 2.2)—three in the visible with one closer to blue, one in the near infrared, two in the middle infrared and one in thermal infrared. The resolution is 30 m except for the thermal infrared band where it is 120 m (see the Thematic Mapper images of Brienne on the CD).

Sixteen photodetectors in parallel are employed to scan 16 rows simultaneously for all the channels, excluding channel 6 for which, evidently, only four detectors are available. Differences in sensitivity between the 16 photodetectors can produce parallel strips on the images (termed 'stripping') (CD 2.4).

LANDSAT 1 was launched on 22 July 1972 and functioned up to 6 January 1978. LANDSAT 2 was sent into orbit on 5 November 1975 and worked up to 27 July 1983, when some abnormalities affected its sensors subsequently. LANDSAT 3 was launched on 5 March 1978 and retired on 7 September 1983 following some trouble in the scanner system. LANDSAT 4, launched on 16 July 1982, also had technical problems such as breakdown in the TM source unit from February 1983 and trouble in two solar panels out of four. Launched on 1 March 1984, LANDSAT 5 functioned with no problem. LANDSAT 6 crashed into the sea during launching. LANDSAT 7 was sent into orbit in April 1999.

2.3.5 ERS-1 and 2

The ERS series of satellites (Fig. 2.18 and Tables 2.3 and 2.4) carry active sensors, viz., a radar altimeter, a backscatter reflector for laser telemetry from the ground and an active microwave instrument (AMI), designed and developed by the research centre for physics of environment. The latter consists of a scatterometer for measuring the height of waves, another scatterometer to record the force and direction of wind on the sea and a synthetic aperture radar (SAR) unit. In addition, a passive sensor with a scanning system, viz., ATSR, is also used. A microwave radiometer measures the total water-vapour content of the atmosphere, while another infrared radiometer measures two different thicknesses of the atmosphere, vertically and at 50°. Sea surface temperature is obtained with an accuracy of 0.3–0.5° C, with a resolution of 1 km, over a 500-km swath.

ERS-1 was placed in orbit on 25 July 1991. The IFREMER centre in Brest processes and archives the data of the altimeter and scatterometers.

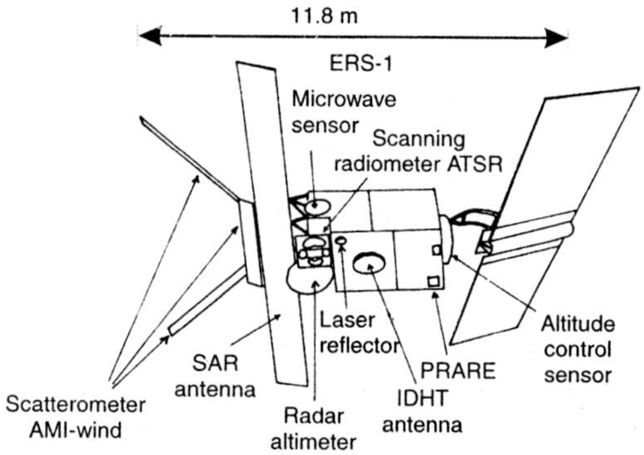

Fig. 2.18: ERS-1 satellite (after Lliboutry, 1992).

2.3.6 RADARSAT

RADARSAT is the only system which provides viewing at different angles (Table 2.4) and a choice between a high geometric resolution (10 m in fine mode) and a coarse resolution (100 m). These characteristics are programmable; this facility offers many advantages, provided the type of data required is clearly identified. These various possibilities ensure a better understanding of the phenomena under study but sometimes make acquisition difficult.

2.3.7 JERS

JERS is the only radar system operating in the L band (Table 2.4). The L band can be very useful, especially when data are available from JERS and ERS or RADARSAT for a given region, since the phenomena detected by these different sensors are not the same (see Chapter 26). The backscatter signal, for example, is much more sensitive to the row effects of plants, which may be an advantage or an inconvenience depending on the application under investigation.

2.3.8 Evolution of sensors

Instruments of various resolutions are currently under development.

Table 2.4: Examples of satellites with microwave sensors

	ERS	JERS	RADARSAT
Date of launching	1: 07/25/1991 2: 04/20/1995	02/11/1992	11/4/1995
Altitude	785 km	570 km	800 km
SAR (Synthetic Aperture Radar) sensors	Image mode		
Spectral band	C	L	C
Frequency	5.3 GHz	1275 MHz	5.3 GHz
Polarisation	V-V	H-H	H-H
Special characteristics	Possibility of a mode to study waves and wind velocities	44 days revisit capability	Various view angles, 24 d cycle, 1.5 d revisit capability at 45° North latitude possible
Angle of incidence in middle of swath	23°	35°	10° to 60°
Spatial resolution	25–30 m	18–18 m	10 to 100 m depending on view angle
Swath	100 km	75 km	50 to 500 km
Other sensors onboard	**ERS-1:** Radar altimeter, ATSR scanner (4 channels centred on 1.6, 3.7, 10.8 and 12 μm) **ERS-2:** The same plus GOME (Global Ozone Monitoring Equipment)	OPS (Optical sensor): 3 channels in visible and NIR, 4 channels in infrared	

Instruments of medium resolution, 1,000 to 100 m, provide increased opportunities for image acquisition due to their higher frequency of passes (4 times a day). Applications of these instruments include monitoring dynamic phenomena of the environment over vast areas. Instruments envisaged for launching at the beginning of this century, in addition to the SPOT 4 vegetation instrument already in operation, include:

—vegetation instrument of SPOT 5, France (5 bands from visible to near infrared and reflective medium band, resolution 1000 m, swath 2000 km);

—MSU-SK of RESURS-01 mission, Russia (5 bands from visible to thermal infrared, without MIR, resolutions of 170 or 600 m for a swath of 600 km);

—MODIS of EOS missions, USA, Proto-Flight Model (PFM) launched on 18 December 1999 (36 bands: 20 between 0.4 and 6.0 μm, 16 between 3 and 15 μm, with a resolution of 250, 500 or 1000 m and a swath of 600 km);

—MERIS of ENVISAT mission, Europe (15 programmable bands between 0.4 and 1.05 μm, with a resolution of 300 to 1200 m, and a swath of 1450 m).

Instruments of high spatial resolution, 15 to 8 m, enable acquisition of more accurate information about ground features, in particular detection and mapping of land-use, differentiation of plant species, moisture content of soils and plants etc. The HVIR of SPOT 4, already operational, belongs to this category as well as the following instruments:

—PAN/LISS-3 of IRS-1D mission, India (5 bands from visible to MIR, resolution 5.8 m in panchromatic, 23.5 m in multispectral, 70.5 m in MIR, and a swath of 148 km, probable revisit capability 5 days in panchromatic);

—AVNIR of ADEOS, Japan (5 bands from visible to NIR, resolution 8 m in panchromatic and 16 m in multispectral, and a swath of 80 km, probable revisit capability 3 days).

Instruments of very high spatial resolution, 8 to 1 m, may prove to be very important for applications in urban studies as well as in environmental investigations for detection and monitoring of narrow linear structures (such as roads, afforestation zones, silviculture strips etc.). Important among them are:

—HRG of SPOT 5 missions, France (5 bands from visible to MIR, 5 and 2.5 m resolution in panchromatic, 10 m in multispectral and 20 m in MIR, swath 60 km, estimated revisit capability 4 days);

—IRS-1C mission, India (5.8 m resolution in panchromatic, 16 m in the blue band);

—ADEOS 1 mission, Japan (8 m resolution in panchromatic);

—Earth Watch programme using the satellite Early Bird, launched on 24 December 1997; unfortunately, it has been out of order since 28 December. It was expected to acquire panchromatic images (3 m resolution) and multispectral images (15 m resolution) of 3 × 3 km, with a repetition of 2 to 5 days over the same site. The programme is to be continued with the launching of Quick Bird (resolution 1 m).

An example of future high-resolution satellites (American and French) as of September 1994 is given in Table 2.5 (after Mussio and Light, 1995).

Table 2.5: Characteristics of some satellites

System characteristics	Lockheed/E-Space Imaging Systems Inc.	Sci Orbital, Itek, GDE, Eyeglass	IKONOS	SPOT 5
Orbital characteristics				
Orbit type	Polar sun-synchronous	Polar sun-synchronous	Polar sun-synchronous	Polar sun-synchronous
Altitude km	680	700	681	832
Cycle (days)	14	16		26
Revisit capability	1–3	2	2,9	5
Sensor characteristics				
Sensor type	CCD arrays	CCD arrays	CCD matrix	CCD arrays
Spatial resolution:				
Panchromatic	1 m at nadir	1 m at nadir	1 m at nadir	5 m and 2.5 at nadir
Multispectral	4 m at nadir		4 m at nadir	10 m at nadir
Spectral bands (μm)				
Panchromatic	0.50–0.90	0.50–0.90		0.51–0.73
S1	0.45–0.52		0.45–0.52	
S2	0.52–0.60		0.52–0.60	0.50–0.59
S3	0.63–0.69		0.63–0.69	0.61–0.69
S4	0.76–0.90		0.76–0.90	0.79–0.89
S5	1.55–1.75			1.58–1.75
Stereoscopic coverage	Front-rear	Front-rear	on the track	on the track
GPS	Yes	Yes	Yes	No
Field of view (degrees)	1.3	1.2		0.34
Other characteristics				
Scene dimensions (km)	60 × 60	15 × 15	11 × 40	60 × 60
Ground stations	3 local + abroad	TBD	4	21
Date of launch	1998	1998	1999	2002

User application of such data is beset with two difficulties, viz., large computer capacities required for processing such large volumes of data and cost of data acquisition and processing.

While it is never certain that every proposed system will become operational, these forecasts indicate the major tendencies towards which remote sensing is progressing.

One of the approaches chosen by the European Space Agency (ESA) concerns the concept of small satellites that facilitate testing and validation of new instruments. Their smaller weights ought to enable their launching by lighter carriers or simultaneous launching of multiple satellites. The objective is to separate the industrial sector from the research sector so as to meet the needs of the users better. Users in particular demand a guarantee of continuity in service for a sufficient duration, which is possible only through involvement of public authorities supporting private initiatives.

2.3.9 Satellite photography

Acquisition of photographs of the Earth from satellites requires return of the films to the ground. For this,

—the flight has to be manned and the space vehicle returned to the ground, as in the case of GEMINI and APOLLO missions of the USA or Soyuz missions of Russia, or

—the films have to be in modules or satellites that can be recovered on the ground, as in the case of Big Birds, the American satellites flying at an altitude of about 200 km, and the Russian Cosmos. In some cases, photographs taken with remote cameras can be transmitted to the ground in digital form; resolutions achieved in such cases would be much better and in the range of decimetres and not in metres (Lliboutry, 1992).

The Earth has been photographed during manned satellite missions by SKYLAB and by space shuttles. Photographs were also taken with metric cameras onboard SPACELAB in 1983 and 1985. They were stereoscopic, colour infrared, with a resolution of about 10 m.

2.4 CONCLUSION

The possibilities of obtaining satellite or aerial photographs are becoming increasingly numerous and the sensors more diversified. It is now possible to choose the date of acquisition of images and multiply them by repetition. Several different fields of view can be used. Geometric and spectral resolutions have become finer. One can acquire analog data and digitise them, or directly obtain digital information. Collection of information is becoming increasingly easier through multiplication of receiving stations. More spectral bands are being used; sensors with tens of bands are envisaged. These developments lead to two important consequences:

—the choice ought to be made according to the thematic application of interest (see part E);

—the quantity of data becomes larger and larger.

A TM image with a 185-km field of view roughly corresponds to 9 SPOT images and 625 conventional aerial photographs of IGN. A multispectral pixel of TM (900 m^2) is equivalent to slightly more than 2 multispectral pixels of SPOT (2 × 400 m^2) and 9 panchromatic pixels. The volume of a TM image hence corresponds to 6 channels of 36 million pixels, plus 1 channel of 2.25 million pixels (for thermal infrared), each pixel being of 8-bit code, which amounts to 1746 megabits, or 218 megabytes. For an HRV image, the volume of information is 63 megabytes for four spectral bands.

The number of spectral bands can be increased considerably but it is to be noted that data processing using colour filters can be carried out only for three channels at a time (see Chapter 3). Hence, one has to choose the most important among the possible channels or devise new channels by combining a larger number of spectral bands. The resolution can be improved but, for a given field of view, the number of pixels increases by a power of two. It may be noted that in the near future spectral bands will be coded on 10 and more bits. Processing of such volumes of data demands computers of very large RAM memory. Effectively, this is the present-day dynamics of computer designers.

After processing, the data need to be interpreted based on physical and geometric models. This requires some time for developing mathematical and statistical models and new approaches to tackle the problem. Researches in the coming years ought to be focused on this aspect, if at all any predictions can be made on remote sensing, this young science whose usage has just commenced, although it is already thirty years old.

References

Bonn F, Rochon G. 1992. Précis de télédétection. 1: Principes et méthodes. Presses de l'Université du Québec/ AUPELF, 485 pp.

Gilliot J-M. 1994. Traitement et interprétation d'images satellitaires Spot: Application à l'analyse des voies de communication. Thèse de Doctorat, Université Paris V, 197 pp.

Lillesand T, M, Kiefer R, W. 1994. Remote Sensing and Image Interpretation. John Wiley & Sons Inc., NY, 3rd ed., 750 pp.

Lliboutry L. 1992. Sciences géométriques et télédétection. Masson, Paris, 289 pp.

Mussio L, Light D, L. 1995. Sensors, platforms, and imagery symposium. *Photogrammetric Engineering & Remote Sensing*, 61: 1339–1344.

Prevot L, Laville P, Xing-Fa GU. 1996. Les mesures de télédétection dans l'infrarouge thermique. Actes de l'école chercheurs INRA en bioclimatologie, Le Croisic, 25–29 mars 1996, Tome 2 du couvert végétal, pp. 69–80.

B

PHYSICAL INTERPRETATION OF DATA

3

Composition of Colours

Only the basic information required for understanding and use of colours in remote sensing is presented in this chapter. In fact, colorimetry per se is a very complex science. Books published by the International Commission for Colorimetry (1931) and that by Wyszecki and Stiles (1982) can be consulted for details. Among recent publications related to remote sensing, the papers by Escadafal et al. (1988, 1989, 1993) which are particularly concerned with applications in soil science are useful.

In remote sensing, a thorough knowledge about colours is necessary for:

—visual interpretation of images printed on paper,

—interpretation of colours on screen,

—preparing colour composites of processed data to make results readable,

—choosing the best possible printing of results on paper.

In order to establish comparisons between the objects that our eyes see and the colour of these same objects on images or photographs, it is also necessary to understand the systems of colour composition by our eyes, by computer monitors and by pigments.

Colours in light are called 'additive colours' and the colour impressions obtained by means of pigments are known as 'subtractive' colours. It is hence evident that all the colours that can be seen on a screen cannot be reproduced on paper. The latter depends on the pigments used for preparing various types of ink.

3.1 THE HUMAN EYE AND COLOUR

3.1.1 Vision

Light penetrating the eye induces chemical reactions which are converted into precoded electric signals. These signals are transmitted through the optical nerve in the brain, which perceives a colour. The process is complex since any excitation gives a colour perception even when there is no colour sensation coming from the eye, as in the case of dreams, for example.

The light received by the eye reaches nearly 120 million rod-like cells on the retina, which enable scotopic vision (vision in dim light), and another 6 to 7 million cone-shaped cells, which provide colour vision. The remaining light is absorbed by the melanin of chloride, preventing any diffusion. There are three types of pigments for cone cells. These are: erythrolabe, which has a maximum absorption at 560 nm and perceives red, yellow and white; chlorolabe, which has a maximum at 530 nm and perceives yellow, green and white; and cyanolabe which has a maximum absorption at 450 nm and perceives blue, red and white. The eye perceives a few thousand or tens of thousands of colours.

3.1.2 Sensitivity

The spectral sensitivity of the eye (Fig. 3.1) varies from approximately 400 nm to 700 nm, with a maximum between 540 and 560 nm, which corresponds to green-yellow. In this band, a variation $d\lambda$ in wavelength leads to a very small change in spectral sensitivity, which cannot be differentiated by the eye. Consequently, the eye does not perceive small nuances in yellow shades. On the other hand, the same variation $d\lambda$ in the red or blue-green band produces a significant change in spectral sensitivity and hence the eye perceives many variations in blue-green and red wavelength bands.

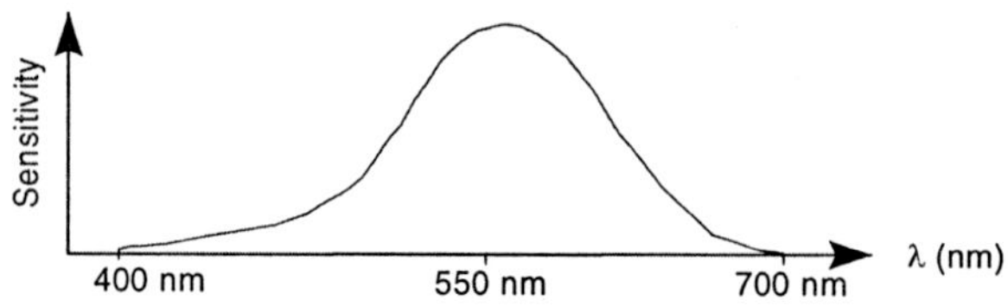

Fig. 3.1: Sensitivity of the eye (after Lliboutry, 1992).

3.1.3 Contrasts

The human eye exhibits contrast effects in the course of time. Thus, after observing an intense colour for a long time (for example, a red setting sun), by turning the face, we 'see' a complementary colour (green sun) during the time necessary for reconstitution of retinal pigments altered by light (red).

The eye also shows spatial contrast effects. The same grey colour appears less clear if surrounded by white and more distinct if surrounded by black. The effect is similar for complementary colours: orange looks more like yellow if surrounded by blue. To eliminate such effects, every colour needs to be surrounded by grey colour.

There is an edge effect as well. Near the boundary between dark grey and light grey zones, the dark grey would appear darker and the light grey lighter. This is the reason why it is believed that a large number of grey levels can be distinguished. But in fact, the eye perceives little more than a dozen grey levels, ranging from black to white, when these levels are not juxtaposed.

3.2 RED-GREEN-BLUE SYSTEM

Computer and television screens emit three colours, viz., red, green and blue, which are intermixed without this being shown on the screen. These are the primary colours. This trichromatic system is based on the principle that any colour can be produced by an appropriate mixture of these three primary colours. The International Commission on Illumination (ICI, 1931) adopted the following wavelengths for the primary colours: red, λ (R) = 700 nm, green, λ (G) = 546 nm and blue, λ (B) = 436 nm. These were chosen from empirical results which showed that this combination produces the largest range of colours.

The colour of an object can hence be computed as a vector C of co-ordinates *r*, *g* and *b* (Fig. 3.2):

$$\vec{C} = r\vec{R} + g\vec{G} + b\vec{B}$$

The units are chosen in such way that white colour is obtained when $b = g = r = 1$.

The curves of colour sensations of an average observer, or the effects of mixing, are generated based on a series of colorimetric measurements carried out with a field of view between 1° and 4°.

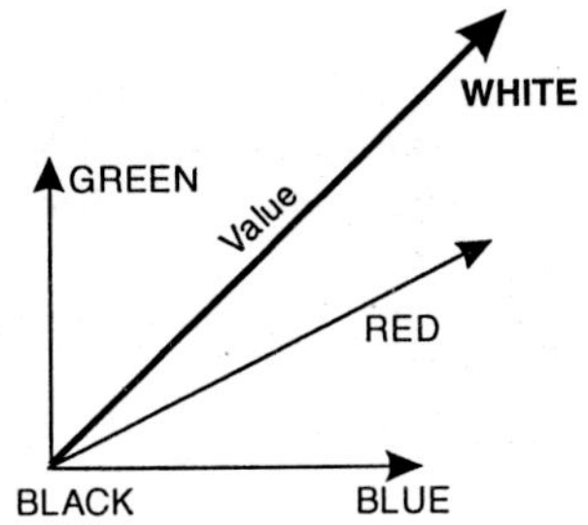

Fig. 3.2: Computation of co-ordinates *r*, *g* and *b* of a colour vector.

They define the characteristics of the reference observer (ICI, 1931) and represent the percentages of red, green and blue to be mixed for obtaining the sensation corresponding to each of the monochromatic radiations of the visible spectrum. Negative values of *r* can be clearly observed, with a minimum around 510 nm (Fig. 3.3). The existence of colour sensation curves also constitutes the basis for colorimetry.

These colours are hence called additive primary. Addition of primary colours defines the secondary colours (Fig. 3.4 A), as mentioned below:

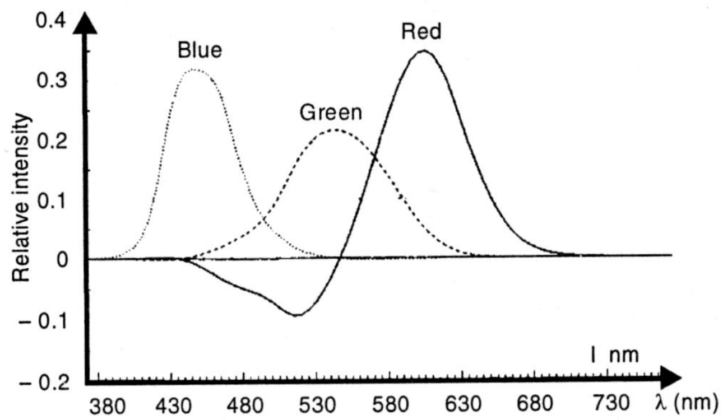

Fig. 3.3: Combination of monochromatic colours for the standard observer (ICI, 1931).

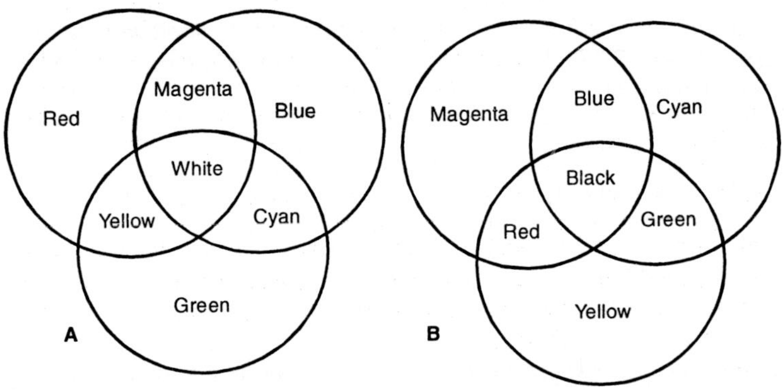

Fig. 3.4: Additive (A) and subtractive (B) colour compositions of R, G and B.

—blue + green = cyan
—green + red = yellow
—red + blue = magenta.

When the three colours are added, white is obtained, i.e., blue + green + red = white. Black represents absence of colour. Therefore, we can define an achromatic axis passing through black and white.

3.3 CUBIC REPRESENTATION

3.3.1 Additive system

The colour space can be represented by a cube whose three orthogonal axes constitute the proportions of red, green and blue (Fig. 3.5). At the origin, values of the three axes are equal to zero and the resultant is black. An identical maximum value for the three axes gives white. The corners of the cube represent white, magenta, red, yellow, green, black, blue and cyan. The diagonal connecting the two corners, black and white, constitutes the achromatic axis, which comprises all the possible grey levels. Other colours are produced by changing the proportions among the three principal colours.

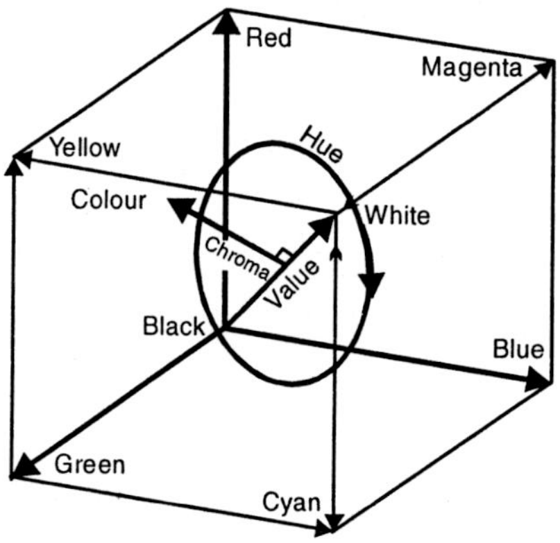

Fig. 3.5: Trichromatic cube.

A computer or television monitor operates on the principle of additive colours. Colours can be reconstructed in this manner (CD 3) if 256 values for each of the three primary colours are available. This number is much higher than what our eye can distinguish.

3.3.2 Subtractive system

Using the same cubic system, a similar colour space is obtained when the axes are defined by cyan, yellow and magenta (Figs. 3.4 B and 3.5) instead of red, green and blue. The origin of these three axes represents white and the maximum value for the three axes black. These three colours are hence called *subtractive*: the transmitted light is effectively lower in intensity than the light received. Any colour can be produced using white light, from which yellow, magenta and cyan colours are subtracted by means of filters as follows:

—the magenta filter transmits red and blue but stops green;
—the yellow filter transmits red and green but absorbs blue;
—the cyan filter transmits green and blue but stops red.

Consequently, when two filters are used, such as yellow and magenta for example, only one colour, i.e., red, can pass through. As a corollary, the magenta and cyan filters allow blue, and the cyan and yellow filters transmit green. Depending on the filters, the light intensity after passing through a filter is greater or smaller. Thus, any colour can be composed. When all three filters are used, all three colours are subtracted and no light exists, resulting in black.

The colour pigments are determined from subtraction of colours. The pigments absorb certain colours and, when they are mixed, colours are accumulated and shades absorbed. This is the principle involved in printing inks produced from yellow, magenta and cyan colours. In practice, in the offset process, the density of inks does not vary continuously. The different colours are obtained by the method of half-tints: colour points of different inks are distributed on grids of points, which are superposed at different angles; this results in overlaps. The points hence become large or small and the intensity of absorbed light is less or more, resulting in restoration of diverse colours.

In additive as well as subtractive systems, complementary colours can be defined as those whose sum gives grey. The latter corresponds to the two opposite points of the achromatic axis. The complementary colours are therefore red and cyan, yellow and blue, and green and magenta.

3.4 TRIANGLE OF THE INTERNATIONAL COMMISSION ON ILLUMINATION

With the objective of describing colours in a Cartesian system, the ICI makes use of the trichromatic co-ordinates X, Y and Z derived from R, G and B, which can be computed as follows:

$$X = 2.7659\,R + 1.7519\,G + 1.13020\,B$$

$$Y = R + 4.5909\,G + 0.06012\,B$$

$$Z = 0.0565\,G + 5.59440\,B$$

The colour of an object can thus be computed from its spectral reflectance curve ($C\lambda$) and the composition of light that illuminates it ($H\lambda$), using the respective mixing functions, viz., $r\lambda$, $g\lambda$ and $b\lambda$.

$$R = k \int_{380\,\mathrm{nm}}^{780\,\mathrm{nm}} C\lambda \cdot H\lambda \cdot r\lambda \cdot \mathrm{d}\lambda$$

$$G = k \int_{380\,\mathrm{nm}}^{780\,\mathrm{nm}} C\lambda \cdot H\lambda \cdot g\lambda \cdot \mathrm{d}\lambda$$

$$B = k \int_{380\,\mathrm{nm}}^{780\,\mathrm{nm}} C\lambda \cdot H\lambda \cdot b\lambda \cdot \mathrm{d}\lambda$$

Using the above, the trichromatic co-ordinates x, y, z can be derived from the relation:

$$x = \frac{X}{X+Y+Z} \text{ or } \frac{x}{X} = \frac{y}{Y} = \frac{z}{Z} = \frac{1}{x+y+z}$$

The co-ordinates are conventionally represented in a right-angled triangle (Fig. 3.6). The point named 'White' is the achromatic point and the dotted line gives the triangle RGB. Thus, every colour can be characterised by brightness (Y) and its chromatic co-ordinates x and y.

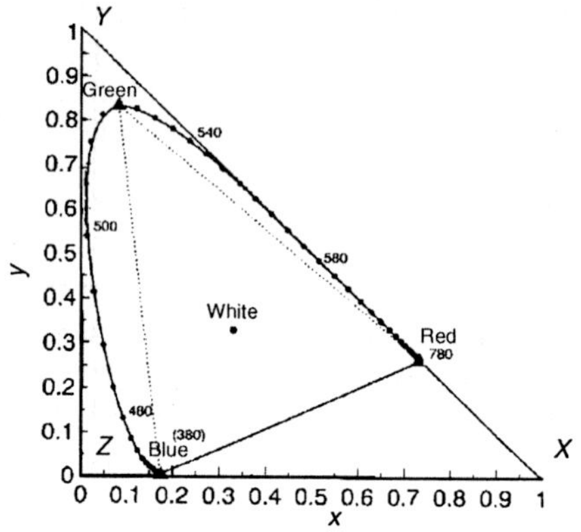

Fig. 3.6: The ICI chromatic triangle based on the co-ordinates X, Y and Z.

3.5 MUNSELL SYSTEM

While the Red-Green-Blue system is represented in cubic form, the Munsell system is based on cylindrical co-ordinates (Fig. 3.7).

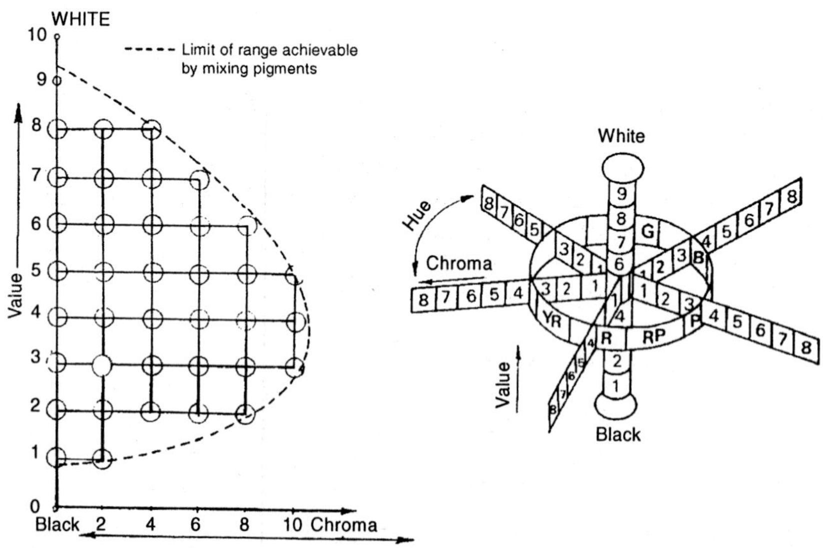

Fig. 3.7: Example of a Munsell page and cylindrical co-ordinates (after Escadafal, 1989).

The axis of the cylinder corresponds to the diagonal of the RGB cube: it is an achromatic axis which ranges from black to white, passing through all the grey points, and corresponds to a Munsell value. The square of the Munsell value gives brightness. For a given value of the achromatic axis, a perpendicular plane can be defined on which the angle of rotation corresponds to the hue. On this plane, from the achromatic axis a circle is defined whose radius characterises the value of chroma. Every colour is thus defined by three co-ordinates: value, hue and chroma (intensity (lightness), colour and saturation are also used).

A catalogue of Munsell colour codes exists. A digest chart that gives soil colour codes is used in soil science worldwide. However, all colours are not represented since the ranges of value and chroma do not exceed 8, whereas the theoretical ranges are 10 and 24 respectively (Fig. 3.7).

Colour pellets made from a mixture of pigments that produce a colour sensation are employed. The three pigments used correspond to the three relative maxima on the reflectance curve, viz., 430 nm, 530 nm and 590 nm (Fig. 3.8).

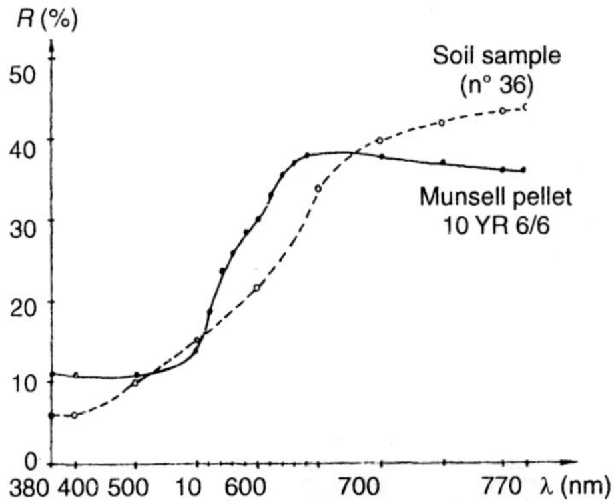

Fig. 3.8: Reflectance curves of a soil sample and the Munsell pellet giving the same colour print.

Conversion of the Munsell code to the RGB code is not simple and necessitates conversion tables or computer programs, since the geometry of the two colour spaces differs (Wyszecki and Stiles, 1982). As an example, one can see in Fig. 3.9 the projection of values of colour of hue 7.5 YR and 5 YR on the Red-Green plane, on the one hand, and on the Red-Blue plane, on the other. It is very clear that the colours are better separated on the Red-Blue plane than on the other plane. The influence of hue is small; chroma mainly affects the Red component; value affects the two components, and the distance between two values is greater in high values than in low values.

3.6 METAMERISM

The term metamerism denotes the phenomenon whereby objects of different spectral properties may produce the same colour sensation. Thus, an object with high reflectance in green and red bands may appear the same yellow as an object that reflects in a monochromatic band of wavelength corresponding to yellow. Hence, the same colour sensation may be obtained with very different reflectance curves (Fig. 3.10). As no bijective relationship exists between reflectance and colour, it is generally not possible to predict the reflectance curve of an object from its colour.

Two colours are generally metameric only under a given illumination. Thus, the colour of a soil and that of the Munsell pellet 10 YR 6/5 (Fig. 3.8), which give the same colour sensation, exhibit distinctly different reflectance curves that intersect one another thrice. It has been established that the

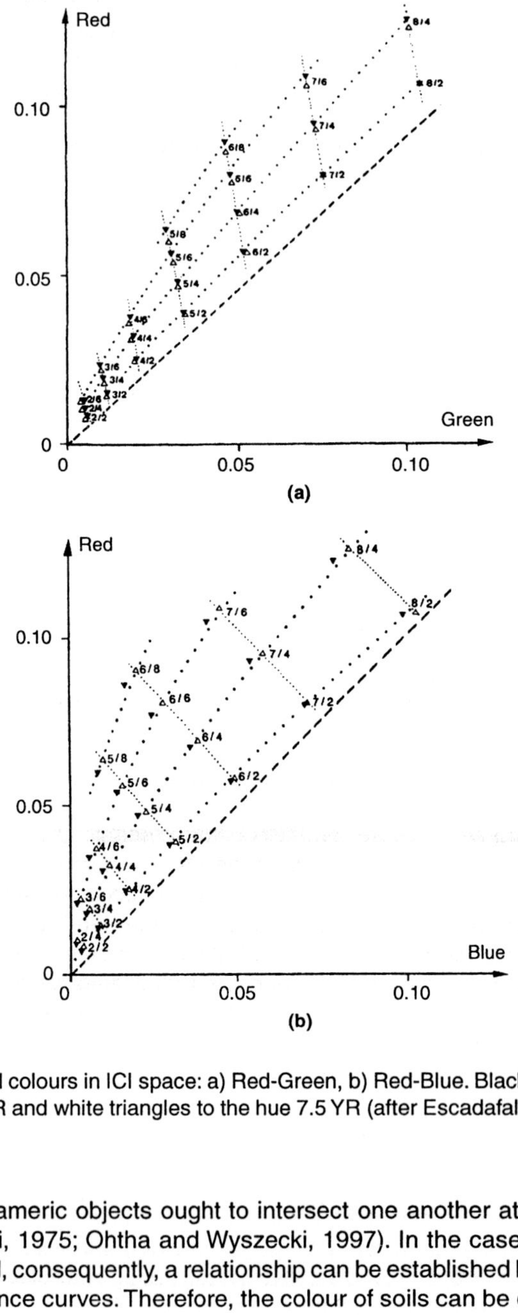

Fig. 3.9: Projection of Munsell colours in ICI space: a) Red-Green, b) Red-Blue. Black triangles correspond to the hue 5 YR and white triangles to the hue 7.5 YR (after Escadafal, 1989).

reflectance curves of metameric objects ought to intersect one another at least three to five times (Takahama and Nayantani, 1975; Ohtha and Wyszecki, 1997). In the case of soils, the reflectance curves do not intersect and, consequently, a relationship can be established between the colour of soil surfaces and their reflectance curves. Therefore, the colour of soils can be determined from satellite images.

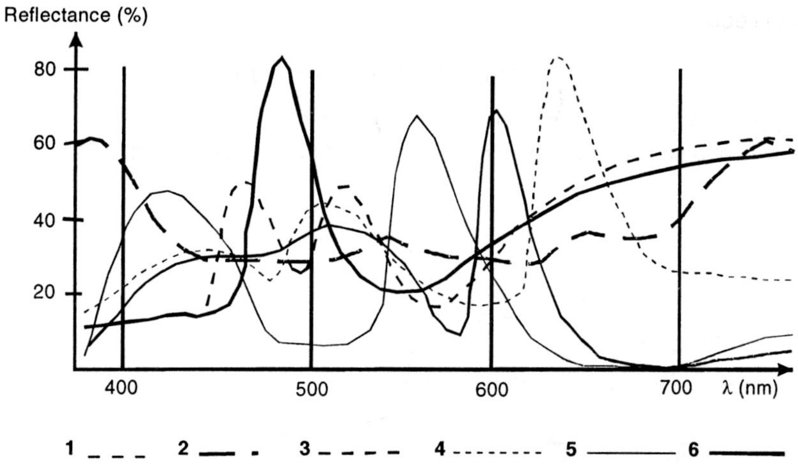

Fig. 3.10: Six metameric curves that give the same colour imprint.

3.7 COLOUR IN PHOTOGRAPHY

3.7.1 Principle of emulsions

Use of emulsions can be readily understood by considering a black and white emulsion. An emulsion consists of a thin layer of photosensitive silver halide grains coated on a base. When the emulsion is exposed to light, a chemical reaction takes place. During the process of development the silver salt is reduced to silver, which is black in a mono-atomic state. After some time, development is stopped by immersing the film in an acid solution. The emulsion is then placed in a fixer bath, which removes the unexposed salt grains and makes it chemically stable. The negative is then washed and dried. As a result, a grey level proportional to the energy received (E_i) is obtained. The latter is determined by the integral of the reflectance curve between the two limits of the spectral band (Fig. 3.11).

3.7.2 Panchromatic and infrared

Transmission T is defined as the ratio of light quantity transmitted to the emulsion, E_t, and the total incident light, E_i:

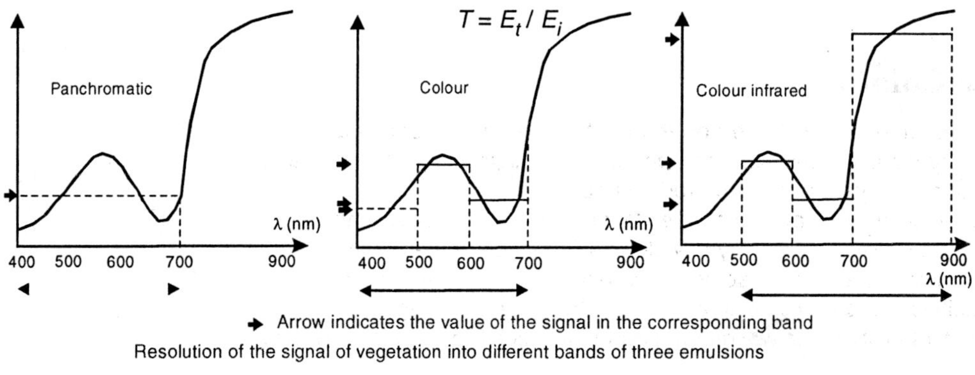

Fig. 3.11: Energy coming from vegetation and received by different emulsions.

Opacity is the inverse ratio: $Op = 1/T = E_i / E_t$

Density, D, is equal to the logarithm of opacity:

$$D = \log (Op) = \log 1/T = -\log T.$$

Thus, we get: $D = 0.3$ for $T = 100\%$ and $D = 1$ for $T = 50\%$.

The sensitivity of a film varies according to wavelength and density (Fig. 3.12).

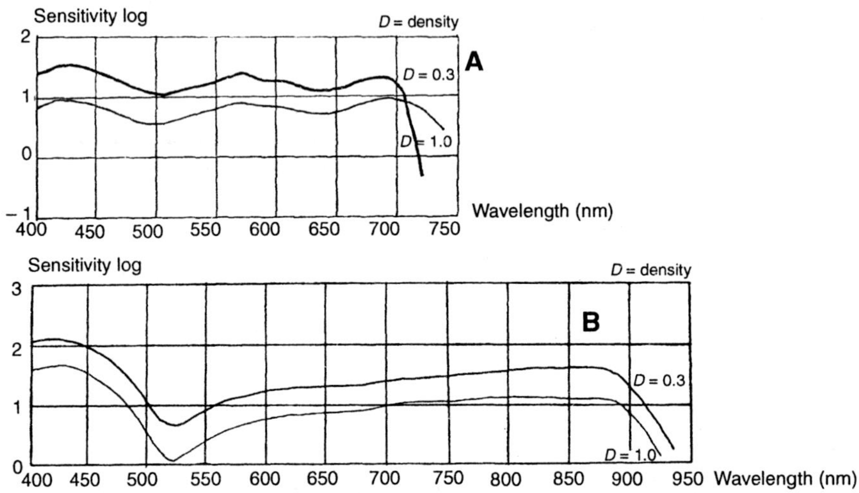

Fig. 3.12: Spectral sensitivities of panchromatic (A) and black and white infrared (B) (after Kodak).

Aerial photography conventionally uses panchromatic emulsions that are sensitive to visible light, 400 to 700 nm (Fig. 3.12 A). The emulsion is limited to 400 nm since this is the lower boundary of the visible spectrum but it is possible to make ultraviolet photographs as well. However, the photographic lens absorb wavelengths below 400 nm.

Infrared emulsion, 700 to 900 nm (Fig. 3.12 B), is also used in aerial photography by filtering the light below 700 nm. Beyond 900 nm the emulsion is photochemically unstable. The glass of the photographic lens absorbs wavelengths exceeding 900 to 1000 nm. Quartz lens can be used to extend the sensitive range to 1100 nm.

3.7.3 Colour

Colour photography is based on the subtractive system of colours, yellow, magenta and cyan. The yellow layer absorbs the blue component of white light, the magenta layer absorbs the green component and the cyan layer absorbs the red component.

Emulsions are made up of three layers, coated on a base, each layer being sensitive to a spectral band (Fig. 3.13A). The first layer, yellow, is sensitive to blue, while the second, magenta, is sensitive to blue and green. A filter is also used to absorb blue. This filter also serves as the third layer, cyan, which is sensitive to red and blue. In each layer, the quantity of recorded energy is inversely proportional to the light intensity of the wavelength band for which the layer is sensitive.

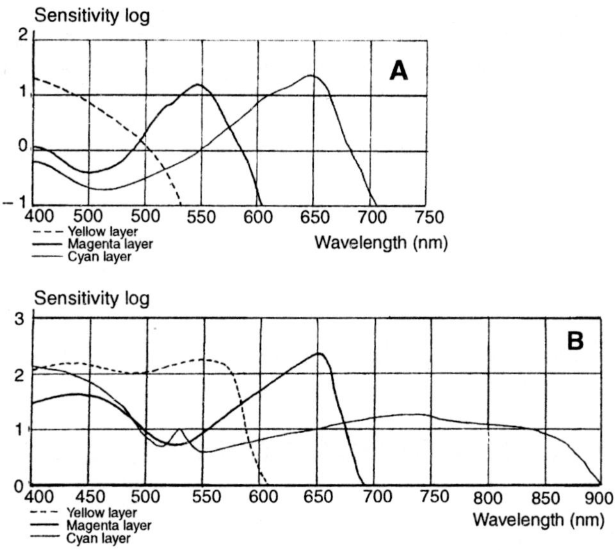

Fig. 3.13: Spectral sensitivities of colour (A) and colour-infrared (B) emulsions (after Kodak).

After development and printing, the initial colours of blue, green and red are obtained (Table 3.1). However, when the distance between the photographic camera and the Earth increases, it is necessary to use a yellow filter in order to avoid atmospheric disturbances, which occur mainly in the blue band.

3.7.4 Colour infrared

Colour-infrared emulsion (earlier called false colour) was developed forty years ago for detecting camouflages. It is useful in distinguishing live vegetation from cut tree branches. It covers the spectral domain of 500 nm to 900 nm, i.e., the green, red and near-infrared bands. Blue is filtered since this

Table 3.1: Principle of operation of colour and colour-infrared films

Spectral bands	Blue	Green	Red	Infrared
Colour film				
Normal sensitivity	Blue	Green	Red	
Colour of layers	**Yellow**	**Magenta**	**Cyan**	
Resultant colour	Blue	Green	Red	
CIR film				
Normal sensitivity	Blue	Green	Red	Infrared
After filtering		Green	Red	Infrared
Colour of layers		**Yellow**	**Magenta**	**Cyan**
Resultant colours		Blue	Green	Red

band, very much affected by atmospheric disturbances, is of no importance in aerial or satellite photography. The remaining part of the spectrum is covered by three bands (Fig. 3.13 B). As these three bands are sensitive to blue, a filter is used to suppress all the corresponding wavelengths.

The principle of this emulsion lies in shifting all wavelengths: when reflected, the green light appears blue, the red appears green and the near-infrared appears red (Table 3.1). Since the eye perceives a large number of shades in red, the possibility of detection of various features is increased as most elements of the natural environment (except water) have higher reflectance in the infrared than in the visible. This is particularly distinct for vegetation and hence this emulsion is important for agronomy, forestry and vegetation study in general.

Application of colour infrared developed with satellite remote sensing since images of Earth-observation satellites are almost always presented in this colour composition. Thus, the quick-look entries in the reference catalogue of SPOT images: SIRIUS (see the chapter SPOT on CD) are in colour infrared (Commonly known as IRC).

Colour as well as infrared aerial photographs are now available for France[1].

3.8 TREATMENT OF COLOURS ON COLOUR SCREEN

3.8.1 Colours on screen

It has been seen that colours can be composed from the three primary colours. For each pixel of the colour screen (monitor), additive mixing of the three (luminophors) primary colours, viz., blue, red and green, is carried out. The chromatic gamut that can be managed by a monitor is limited by the luminophors and hence all colours of the chromatic triangle (Fig. 3.6) are not reproducible on a monitor. The quality of visualisation depends on the number of rows and columns of the monitor, the number of colours available and the number of colours that can be displayed simultaneously (CD 3.1).

On the colour screens used for image processing of remote sensing data, 256 levels can be used for each of the three primary colours. Hence, 256^3, i.e., 16,777,216 different colours can be composed on the screen but not all of them can be seen since our vision is limited. Thus, the colours produced from the first 40 to 50 levels of digital values coded on 8 bits (256 levels) practically cannot be distinguished from black. In most cases, for low values the digital counts need to be enhanced at least by 10 values, such that two colours can be differentiated by the eye. If the number of levels for each primary colour is limited to 16, the number of colours that can be generated would be 4096, which per se is very high to be readily distinguished by the eye.

3.8.2 Colour display

Displaying colours on a screen consists in giving to each of the three basic components, viz., red, green and blue, a more or less strong intensity. For example, in order to generate red colour, the level of red is fixed at 256; however, to obtain pink, it is necessary to brighten the colour and hence increase the energy. This is achieved by enhancing the green and blue components by the same quantity.

It is interesting to note that if several persons are asked to construct a colour, orange for example, through composition of three colours, the results can be somewhat different. This is because the term 'orange' may correspond to several concepts, such as a mixture of red and yellow or the colour of the fruit. Depending on the idea of the person who composes the colour and his (her) visual capacities, the three colours, red, green and blue, are not the same. Thus, orange colour can be produced with varying digital values:

---

[1]They can be ordered from IGN-2, Pasteur Road-94 160, Saint-Mande, France. For website, see the list enclosed in the book.

For red:	217 to 255
For green:	81 to 160
For blue:	0 to 30.

The differences are still greater when the colour to be generated is denoted by terms such as dark brown or beige and, *a fortiori*, when qualitative terms such as café au lait, chocolate, apple green etc. are employed. This only proves the paucity of our vocabulary for denoting the thousands of colours we are able to distinguish.

It is nevertheless to be noted that the number of colours that can be produced on the screen is most often greater than what we can identify with our eye.

Black colour results from absence of colour emission, but on a colour screen what is perceived as black is often a green or a blue or a very dark brown. Similarly, white colour is a mixture of equal proportions of the three basic colours; however, it is common to see on a colour screen a 'white' when a very high intensity of pixels tends towards very bright green or red.

3.9 USE OF COLOURS IN IMAGE PROCESSING

Almost all image-processing systems are based on colour coding. However, two approaches can be distinguished.

3.9.1 Colour code of an image (8 bits)

The colour seen on the screen corresponds to a colour table known as Look Up Table (LUT), which assigns a colour to a value of a pixel, most often coded on 8 bits.

Any colour can be arbitrarily assigned to any value of a pixel, as in the case of classification. It is thus possible to assign to each class one of the 256 possible colours. Thus, a very large number of colour representations can be used in a single classification. These representations are important since they will be interpreted visually by the reader and, depending on the chosen colours, the felt impression may be different. Classes assigned to neighbouring shades of colours may be mixed up and be interpreted as a single group. If such classes are to be interpreted as separate, they should be allotted clearly distinguishable colours. For example, complementary colours ought to be used for zones that have common boundaries because they stand out better. The choice of colours also depends on the colour perception of the person viewing them! Lastly, it is to be remembered that colours have a symbolic significance, which varies with cultural background and context. For example, white is the festival colour in Europe versus red in the East, and white the colour of mourning in the East versus black in Europe—except for mourning the queen of France, which was observed in white. Leafy trees are green, but only for four or five months in temperate climate. Water is blue, but only when the sky is blue and not when in a glass. Soil is maroon, but this excludes white, red, black or brown, etc. soils.

3.9.2 Interpretation of a 3-by-8 bits code

A colour view on the screen corresponds to a look-up table coded in 3-by-8 bits. Such is the case when the RGB systems are used on a computer monitor. In different systems, it is possible to erase one or the other colour, and it is seen that for each pixel the intensity varies with each of the colours, blue, green and red. There are 256 possible intensities for each colour.

Colour composition is very commonly employed to create a colour or colour-infrared image from three channels of a satellite image. However, care should be taken in interpreting an image thus coloured when it is generated from different wavelength bands, such as TM3, TM4, TM7, etc. The same is true when a diachronic image is prepared assigning, for example, blue to visible red (675 nm) for a winter image, green for that of spring and red for a summer image:

—*Spring crops* appear dark cyan. In fact, in winter the area is a bare soil and hence appears blue. In spring, the crop is in growth and there is less green than blue. In summer, the crop is full of chlorophyll and there would be little red in the image (see Figs. 4.12 and 4.13).

—*Winter crops* appear medium red. The plot in winter is chlorophyllous and hence blue of low intensity in the image. In spring, the crop is full of chlorophyll and there is less green than blue. In summer, the crop is harvested, leaving stubble in the field, and hence much red in the image.

—*Mountain pastures* are covered by snow in winter and their hue is intense blue. In spring, bare organic soil is visible and shows as very pale green. In summer, the young vegetation is highly chlorophyllian and so very little red appears in the image.

Obviously, colours composed for such images change if other wavelength bands are used. Thus, for the same objects and the same colour combinations for the same seasons, but coding the near infrared channel instead of the visible red (675 nm), spring crops would be yellow, winter crops appear very light yellow, while mountain pastures would look magenta.

Interpretation of colours can become complex if different channels are used for different seasons and if positive channels are combined with the negative. In such a case, a look-up table should be compiled, indicating for each wavelength the values of digital numbers and the corresponding colour for each of the three components used, to obtain the resultant colour composite.

3.9.3 Choice of colours

For clearly distinguishing various classes during processing, it is recommended that colours be chosen, at least initially, in such a way that neighbouring pixels belonging to different classes on an image are given quite distinct colours of high contrast. This provides better detection of the boundaries between two groups and hence correct decisions regarding the digital value of the dividing line between them.

At the end of processing, colours should be chosen in such a way that the user can read the document with utmost ease. This depends not only on his preference of colours, but also the manner in which the document will be viewed. It may be preferable to use highly contrasting and bright colours if the document is to be only seen at a glance. On the other hand, if the material is to be studied for some time, more harmonious colours may be used.

The choice of colours in the legend should also correspond to the symbolism of colours. The succession of green, orange and red can evidently be used to indicate a gradation from good, inadequate to bad. Two extreme situations of the legend can be distinguished using contrasting 'hot' and 'cold' colours.

Colours are to be chosen taking into consideration the surface area of each unit in the final document, as well as the vicinity between colours. For example, red bordering green gives an impression that the red zone 'overflows' the green: the green area appears smaller than it actually is. Only the surrounding colours need to be changed in order to perceive the effect of various colours. Hence, the choice of colours has an immediate influence on what is perceived at first glance.

A document of successful colour composition ought to facilitate its reading at various levels. At first glance, one should be able to differentiate major units, estimate their relative areas, their dispersion in the zone under study and their organisation vis-à-vis other units. It is necessary that during an advanced stage of spatial analysis, units situated in secondary levels of a semantic hierarchy be differentiated. Several solutions exist for this. One is to indicate first-order units in colours representing different values: yellow is always very bright whereas maroon is always dark. Second-order units are represented in the same colours as those of the first-order but with more or less pure shades.

3.9.4 Colour printing of documents

A colour image prepared on screen may get distorted when transferred to a colour printer. It needs to be verified whether or not the computer code of the colouring system of the image-processing software is coherent with that of the printer. If it is not, the computer folder of the processed image should be routed through a graphic system that is per se compatible with the printer.

It should also be remembered that the inks used in a printer are made up of pigments, and all the colours possible cannot be produced with the pigments available. Such a situation is observed in pages of the Munsell code, wherein colours exist for which co-ordinates can be defined but which cannot be printed (Fig. 3.7). It is therefore possible to view colours on the screen which cannot be reproduced on a colour printer.

Lastly, it should be noted that when documents are exposed to daylight, different inks fade at different speeds: red disappears most rapidly and blue remains the longest. Hence, on documents that may be exposed to sunlight, features considered most important should not be represented in red since they will be the first to fade.

For more details about colours, refer to the various books on graphic semiology and, in particular, the one by Bertin (1967).

3.10 CONCLUSION

The minimum knowledge presented here on colour perception by the eye is necessary to interpret satellite and aerial images and photographs in colour or colour infrared. It should be remembered that visual interpretation is generally employed alone or combined with computer processing of images. In fact, the eye has the capacity of interpretation of forms of objects and their distribution, which is superior to what can now be achieved through modelling and structural processing of images.

References

AGFA. Introduction à la numérisation: Le prépresse couleur assisté par ordinateur, vol. 4, 41 pp.

Bertin J. 1967. Précis de sémiologie graphique, les diagrammes, les réseaux, les cartes. Mouton et Gauthier-Villars, Paris, 431 pp.

Escadafal R. 1989. Caractérisation de la surface des sols arides par observations de terrain et par télédétection. Application: exemple de la région de Tataouine (Tunisie). Études et Thèses. Orstom, Paris, 317 pp.

Escadafal R. 1993. Remote sensing of soil color: principles and applications. Remote Sensing Reviews, 7: 261–279.

Escadafal R, Girard M-C, Courault D. 1988. La couleur des sols: appréciation, mesure et relations avec les propriétiés spectrales. Agronomie, 8: 147–154.

Escadafal R, Girard M-C, Courault D. 1989. Munsell soil color and soil reflectance in the visible spectral bands of Landsat MSS and TM data. Remote Sensing of Environment, 27: 37–48.

Girad C-M, Girard M-C. 1975. Applications de la télédétection à l'étude de la biosphère. Masson, Paris, 186, pp.

ICI. 1932. ICI Proceedings 1931. Cambridge Univ. Press, Cambridge.

Lliboutry L. 1992. Sciences géométriques et télédétection. Masson, Paris, 289 pp.

Monget J-M. 1986. Cours de télédétection, 1. CTAMN, Octobre, Sophia-Antipolis.

Munsell Color Company. 1975. Munsell Soil Color Charts. Munsell Color, Kollmorgen Corp., Baltimore, MD.

Ohtha N, Wyszecki G. 1997. Location of the nodes of metameric color stimuli. Color Res. Appl., 2: 183–186.

Takahama K, Nayantani Y. 1975. New methods for generating metameric stimuli of object color. J. Opt. Soc. Amer., 62: 1516–1520.

Wyszecki G, Stiles WS. 1982. Color Science: Concept and Methods, Quantitative Data and Formulae. John Wiley & Sons, NY, 950 pp.

4

Spectral Characteristics

Interaction of radiation with matter constitutes the basis for interpretation of remote sensing images. An object situated in a given geographic position at a given moment, viewed under a given field of view and receiving a given radiation, exhibits a spectral behaviour that is specific to it. Hence, some authors use the term spectral signature. This term is inappropriate since a signature implies constancy whereas, in reality, the spectral behaviour of an object varies with time, place, mode of data acquisition and incident radiation.

The spectral behaviour of objects is an important means of analysis and interpretation of remote sensing images since it is based on general laws of physics. Therefore, as will be seen later, this approach facilitates formulation of an interpretation model that can be generalised for a large number of cases.

Although the objects under study may be many, they can be reduced to a few general cases such as vegetation (organic matter), soils (mineral matter), water, snow and ice. The first three constitute the major targets of remote sensing of the Earth.

4.1 VEGETATION

The term vegetation is commonly employed in a very general sense to refer to the existence of photosynthetic activity. Optical properties of vegetation covers depend on the plants they contain, their spatial arrangement and the underlying soils. Spectral behaviour of isolated leaves needs to be distinguished from that of an individual member of a given plant species and from that of a population belonging to the same species and the same variety. For the purposes of accurate investigations, it is therefore necessary to differentiate between different species and, for a given species, between various phenological stages and physiological states. Moreover, spectral behaviour of a single-species population needs to be distinguished from that of a multi-species population, for example alfalfa field from a permanent grassland. In fact, in such cases, a change is observed from a more or less regular repetition of spacing between individual plants, belonging to a single vegetation species, to an irregular distribution of individuals pertaining to various species that differ morphologically and physiologically.

The major types of spectral characteristics of vegetation are described in this section. It should be remembered, however, that like all living beings, there are as many forms as number of individuals. Hence, the spectral characteristics obtained from laboratory measurements under controlled conditions (in particular, in diffuse light) are discussed first, followed by results of in situ field measurements in direct illumination. It should be noted, however, that in the case of dense chlorophyllian vegetation (for which spectral contribution of soil is negligible), the spectral characteristics are similar irrespective of the precision level of data acquisition and the species considered, although the reflectance values may vary significantly.

The spectral characteristics presented here were obtained by different laboratories of the United States Department of Agriculture (USDA) who carried out the first investigations on cultivated species (Allen et al., 1969; Knipling, 1970).

4.1.1 Laboratory measurements

The spectra of reflectance, absorptance and transmittance determined in the laboratory (Fig. 4.1) are valid for all chlorophyllian leaves irrespective of the species. Three types of spectral behaviour, corresponding to the major spectral domains, can be identified:

—in the visible spectrum, vegetation exhibits a particular behaviour due to the existence of chlorophyll pigments;

—in the near infrared, spectral behaviour is related to the structure of tissues;

—in the reflective middle infrared, the water content of the tissues has the main influence (CD 4.1).

Reflectance and transmittance vary in approximately the same manner. The parameter investigated in remote sensing applications for vegetation is not actually reflectance, however. On the contrary, it is the efficiency of interception of radiation and hence absorptance which is normally used for yield assessments, as this controls the production of biomass. An indirect estimation is possible using linear relationships between the radiation reflected by vegetation (vegetation indices, see below) and yield. Other methods based on agro-meteorological models are mentioned in Chapter 22.

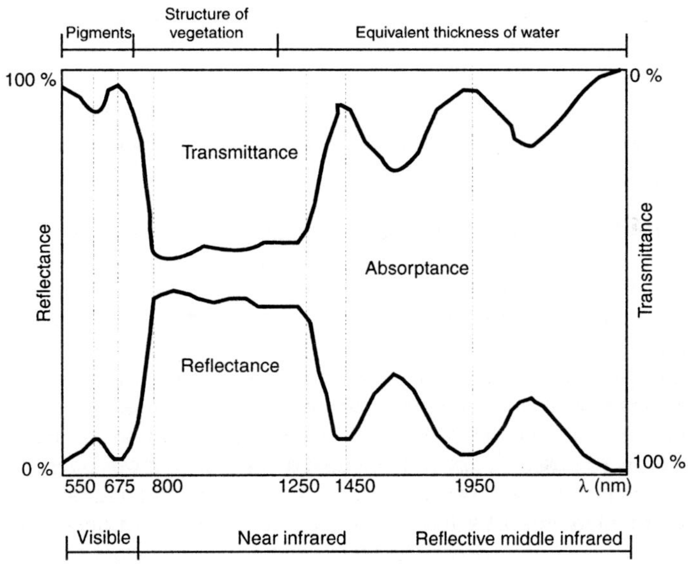

Fig. 4.1: Absorptance, reflectance and transmittance for chlorophyllian vegetation (after Guyot, 1997).

■ Visible domain (380–700 nm)

In the visible spectral domain, vegetation exhibits a low reflectance (maximum 15%), with a maximum at 550 nm. The latter is due to pigments, in particular chlorophyll a and b, which have two absorption bands in the blue (450 nm) and red (660 nm) domains. Other pigments such as carotene, xanthophylls, etc., also have absorption peaks in the visible band but the absorption peaks of chlorophyll usually mask them. Effects of other pigments on spectral characteristics are significant only for very young or, on the other hand, senescent vegetation, or one subjected to stress or deficiency or for plants with albino or variegated leaves. Laboratory measurements with DK2 Beckmann spectrophotometer on leaf samples of various species of permanent grasslands (*Plantago media* L., *Trifolium pratense* L., *Brachypodium pinnatum* L. (Beau.), *Festuca pratensis* Huds., *Prunella vulgaris* L., *Primula veris* L., *Salvia pratensis* L., *Sanguisorba minor* Scop.) for which the contents of different pigments [chlorophyll (a+b), chlorophyll (a) and carotene] were also determined, showed an inverse relationship between reflectance at 550 nm and chlorophyll a content (Fig. 4.2(a)). The greater the chlorophyll a content, the lower the reflectance at 550 nm.

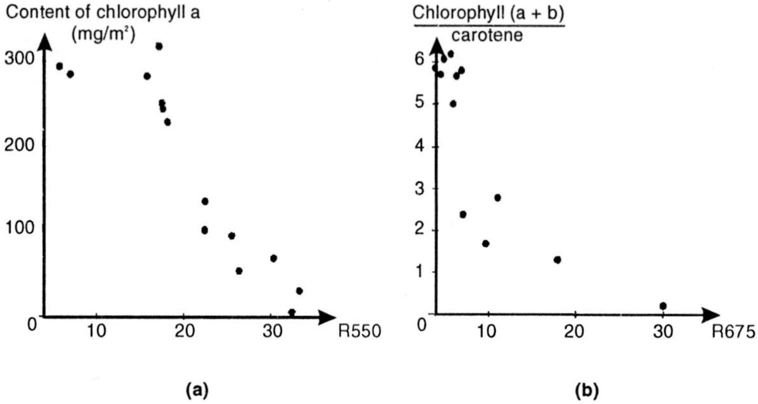

Fig. 4.2: a. Inverse relationship between chlorophyll a content and reflectance at 550 nm. b. Inverse relationship between reflectance at 675 nm and ratio of chlorophyll (a+b) / carotene.

Studies by Gaussman et al. on maize leaves (*Zea mays* L.) of different ages, in which the chlorophyll content decreases with age, showed that reflectance increases with age (Table 4.1). Similar observations were reported for cotton leaves (*Gossypium hirsutum* L.).

At 675 nm, an inverse relationship was observed between the ratio of chlorophyll/carotene and reflectance (Fig. 4.2(b)).

Chlorophyllian vegetation has a lower reflectance in green-yellow and orange spectral bands than does senescent vegetation in which the chlorophyll content is less.

Reflectance of vegetation is affected by chlorophyll content when the latter is greater than 3 $\mu g \cdot cm^{-2}$ and lower than 10 $\mu g \cdot cm^{-2}$. For values outside this limit, a lower but constant reflectance is observed due to the optical characteristics of vegetation tissues. In the case of mature leaves with similar chlorophyll content, differences in reflectance observed in laboratory measurements between various species are mainly due to a waxy cuticle (sclerophytes or succulent species subjected to xeric conditions) or dense pilosity on the leaves, which increases reflectance. These differences may not be

Table 4.1: Inverse relationship between age, chlorophyll content and 550 nm-reflectance of maize leaves

	Chlorophyll content	Reflectance at 550 nm
Young leaves	15.8 mg/L	13.7%
Old leaves	13.9 mg/L	14.4%

detected during field measurements of reflectance, partly because the radiometers often have very wide spectral bands and partly because the entire vegetation canopy, and not individual leaves, is measured.

The visible band spectral behaviour of senescent or dry vegetation totally differs from that of chlorophyllian vegetation. In the entire visible spectrum, for a given species the reflectance of green leaves is always lower than that of dry leaves (Fig. 4.3).

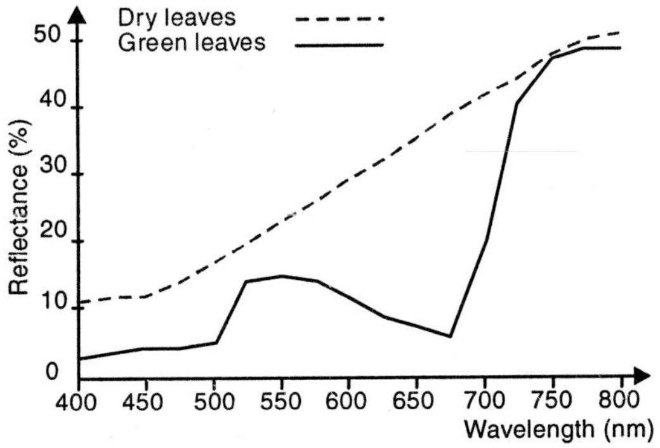

Fig. 4.3: Reflectance curves of green (of a year) and dry leaves (preceding year) of the chalk false-brome (collected from the same individuals).

Diseases, parasites, mineral deficiencies, etc. affect the chlorophyll content and can be detected using the visible band. However, the effects of such must be very severe to enable identification without ambiguity. Moreover, as pigment variations may be produced by extremely diverse sources, detection of particular spectral characteristics provides no information as to their causes.

Pigment contents may also vary in the presence of coloured flowers, as in the case of permanent grasslands. Such variations can be distinguished from laboratory spectrophotometric measurements of inflorescence (or parts of inflorescence) of various species of grass (Fig. 4.4).

White flowers—ray florets of daisy, inflorescence of white clover—exhibit strong reflectance for all wavelengths of the visible spectrum.

Yellow flowers—Ragwort, bulbous buttercup and horseshoe vetch—have very low reflectance in the blue band and maximum reflectance in the yellow.

Violet flowers—Red clover, Self-heal, knapweed—present very low reflectance in the green band and, depending on the colour shade, higher or lower reflectance in blue and red domains. Thus, the inflorescences of red clover have lower reflectance in the blue than do Self-heal and knapweed.

These differences are perceptible in field measurements on permanent grasslands containing abundant flowered species (Fig. 4.5). To facilitate a ready comparison of reflectance values in blue and red, these curves have been offset by 10% reflectance at the wavelength 550 nm.

The presence of many flowers, yellow (primrose), white (cuckoo flower) or orange (marsh marigold), is responsible for strong reflectance, particularly at 650 and 675 nm. This increase is greater for marsh marigold than for other species, since these large-size flowers contribute more to the spectral behaviour by partially masking the spectral contribution of the foliage. The marsh marigold grasslands can be spectrally differentiated from one another in places where bare soil is visible through the vegetation because of its low reflectance values at 400 to 500 nm. The effect of bare soils on spectral behaviour of low-density vegetation cover is discussed further in this chapter. On satellite images, only cultivated vegetation canopies (such as rapeseed, sunflower, linseed etc.) or fallow lands (*Phacelia tanacetifolia*) contain an abundance of flowers sufficient to modify spectral behaviour perceptibly.

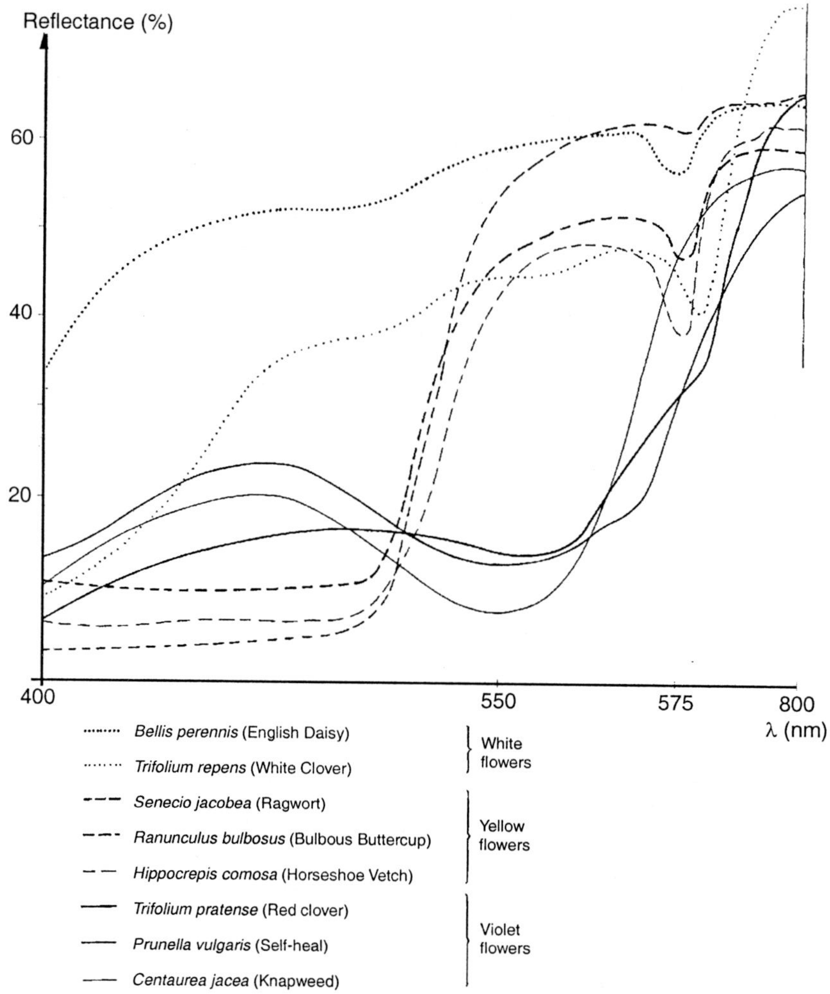

Fig. 4.4: Laboratory-measured curves of reflectance for different colours of inflorescence.

■ Near infrared (700 nm to 1.3 μm)

In the near infrared, pigments do not influence spectral behaviour since the quantity of radiation absorbed by leaves is very small. The internal structure of foliage, in particular various tissues and arrangements of cells and intercellular space, are principally responsible for differences in reflectance. A simplified transverse section of a leaf of a flowering plant consists of the following tissues, from the upper surface to the lower (Fig. 4.6):

—waxy cuticle of variable thickness,

—layer of cells forming an epidermis,

—more or less developed hypodermis,

—palisade parenchyma consisting of elongated and well-arranged cells, mainly containing chloroplasts,

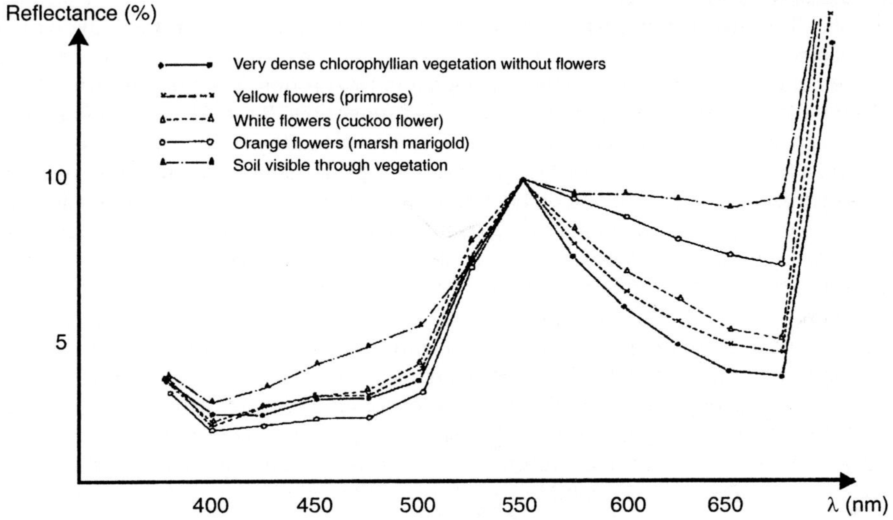

Fig. 4.5: Field-measured reflectance for various permanent grasslands.

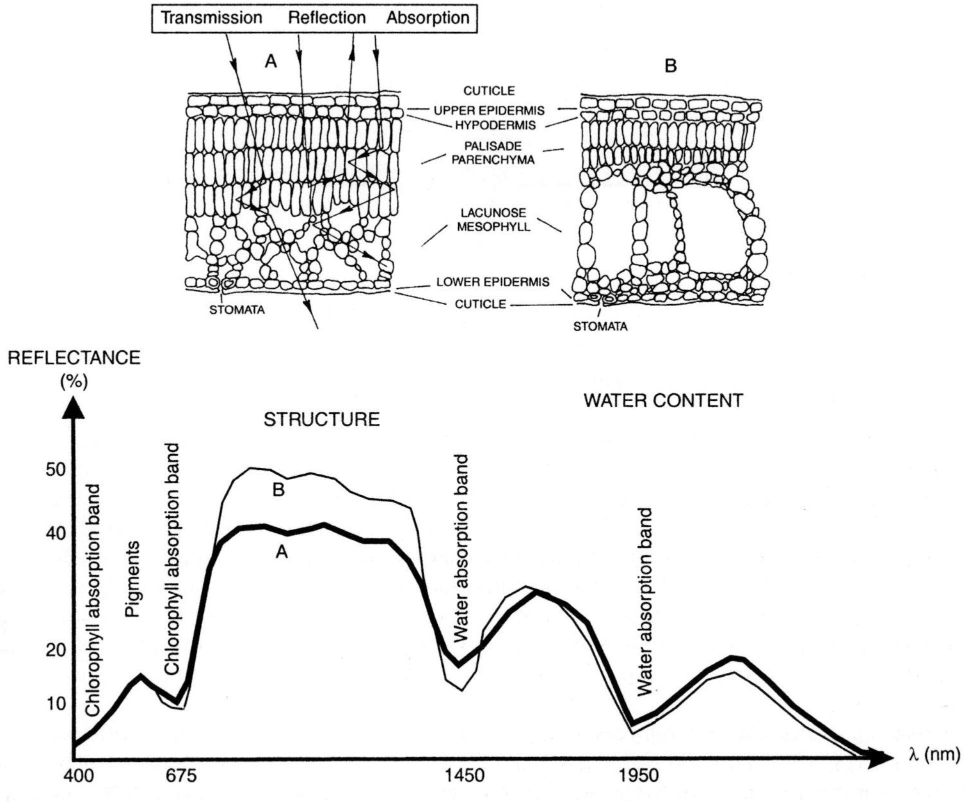

Fig. 4.6: Influence of internal structure of leaves on their spectral behaviour.

—mesophyll whose irregular-shaped cells form a loose network with a number of intercellular gaps filled with air or water vapour,

—layer of cells forming an epidermis in which stomata that ensure leaf transpiration are located,

—thin cuticle.

The leaves of plants can be arbitrarily classified into two groups. The first group contains a well-developed palisade parenchyma and a less-developed mesophyll (type A). The second group has a less-developed palisade parenchyma and a well-developed mesophyll (type B). A lower reflectance in the near-infrared band is observed for type A than for type B. Spectrophotometric measurements coupled with microscopic observations of thin sections of leaves have shown the significance of the internal structure of tissues in spectral characteristics of leaves. This is due to the discontinuities in refraction indices between cells, water and air of intercellular spaces and lacunae of the mesophyll.

In fact, the path of radiation differs in different tissues depending on their refractive indices. A light ray incident at an angle θ_i on the interface between two media of refractive indices n_1 and n_2 may be reflected at an angle θ_r or refracted at an angle θ_{rr}. In the case of specular reflection $\theta_i = \theta_r$. According to Descartes' law, the angle of incidence θ_i and the angle of refraction θ_{rr} are related by the equation: $n_1 \times \sin\theta_i = n_2 \times \sin\theta_{rr}$ (Fig. 4.7). The refractive indices for air and water are 1 and 1.33 respectively, and for the elements constituting the vegetation cells 1.55 approximately. When a light ray propagates from a high refractive medium to a low refractive one ($n_1 > n_2$), Descartes' law apply for angles for which $\sin\theta_i \leq 1$ or $\sin\theta_i = n_2/n_1$. Beyond this angle of incidence ($\theta_i > \arcsin(n_2/n_1)$), the ray is totally reflected.

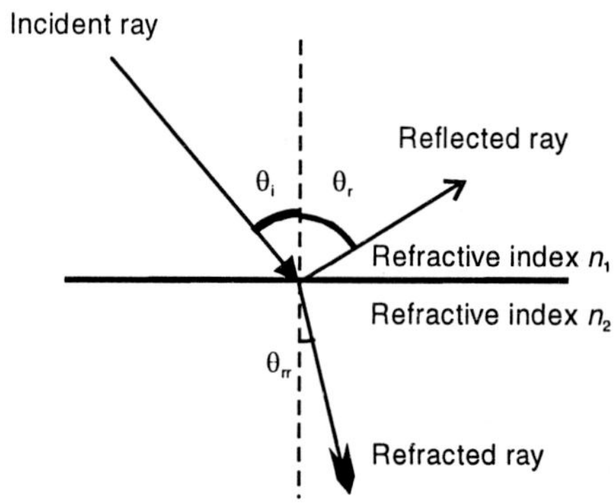

Fig. 4.7: Refraction of radiation.

A more or less strong reflection may occur on the cuticle. In palisade parenchyma, radiation is transmitted with a small path deviation. In fact, the cells in this parenchyma are roughly in the shape of parallelepiped and regularly aligned, without large lacunae between them and hence the refractive indices of the media traversed by light are similar. On the other hand, cells in mesophyll are more or less irregular spherical in shape and have many lacunae (sometimes of large size) between them containing air or water vapour. The presence of media of varied refractive indices gives rise to the situation of total internal reflection (for example during propagation from a cell to a lacuna) and hence a stronger possibility of reflection towards the upper surface of the leaf. This effect of intercellular lacunae and voids was detected from the decrease in reflectance at 800 nm during filling of these voids by water.

Physiological changes (pigment content, internal structure of tissues, water content) accompanying the processes of maturation and senescence produce significant changes in the spectral behaviour in visible and infrared bands. Senescent vegetation (yellow and dry) is generally characterised by higher reflectance in these two spectral domains than green and turgescent vegetation. In the near infrared, this phenomenon is due to the combined effect of changes in internal structure and water content. Senescence induces collapse of the cells of mesophyll, leading generally to horizontal layering of cell walls, accompanied by loss of water content. The decrease in reflectance in the near infrared, observed in the laboratory, for sick or nutrient-deficient plants (Fig. 4.8) corresponds to deterioration of cells and collapse or smaller tissue thickness.

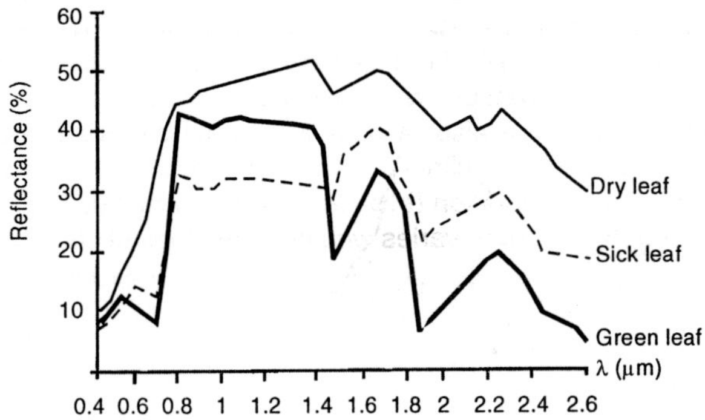

Fig. 4.8: Variation in reflectance of a leaf according to its physiological state.

Lastly, a small absorption occurs at 0.98 and 1.20 μm due to liquid water present in leaves. This effect is ordinarily masked by the very strong absorption of atmospheric water vapour. It can be detected in laboratory or field measurements but not in satellite data which are acquired in very broad bands. On the other hand, it has been noticed in the records of AVIRIS (Airborne Visible/Infrared Imaging Spectrometer).

■ Reflective middle infrared (1.3 to 2.5 μm)

The spectral characteristics of leaves are mainly affected by the water content of cells, as water has absorption bands at 1.45, 1.95 and 2.5 μm. A green, turgescent plant has low reflectance in the reflective middle infrared. If the water content decreases, either due to senescence and drying or following diseases and parasite attacks, its reflectance in this spectral band increases (Fig. 4.8). The bands 1.4 to 1.8 μm and 2.1 to 2.35 μm are more sensitive to variations in water content than the band 1.9 to 2.05 μm. Recent studies have shown that components such as lignin and cellulose also influence the reflectance of vegetation in this spectral range. Vegetation rich in lignin has high absorption at 1.72 μm, which increases with decrease in water content.

4.1.2 Field measurements

Spectral behaviour of vegetation inferred from laboratory measurements of leaves was described in the preceding section. What would it be when determined in the field for a vegetation cover? Several cases of vegetation are possible: total soil coverage (100%), low (< 20–30%) and intermediate. In the first case, spectral behaviour is similar to that of leaves measured in the laboratory, with variations induced by the structure of vegetation cover (growth and orientation of leaves, number of leaf layers, etc.). In the second case, spectral behaviour is close to that of bare soil while in the third case, all intermediate variations between the preceding two are possible.

In the case of more or less dense vegetation cover, in addition to the phenomena described in the preceding section, the processes associated with the structure of the canopy, mainly related to the disposition of leaves, need to be considered. Depending on the angle of inclination of the leaves to the horizontal, three types of vegetation can be distinguished. These are planophyll (angle close to 0°), erectophyll (angle close to 90°) and intermediate types, viz., plagiophyll (inclined leaves most frequent) and extremophyll (erect leaves most common). Dicotyledons are usually planophyll or plagiophyll whereas monocotyledons are erectophyll or extremophyll. Nevertheless, the same individual of a given species may exhibit significant angular variations in the course of its development. Such is the case in particular for grass in which the angle of leaf inclination decreases during maturation. Considering these differences in the disposition of leaves, the same leaf area index[1] leads to a smaller coverage for erectophyll species than for planophyll. The differences in the leaf attitude more or less affect the magnitude of shadow produced in the canopy. In the case of cultivated plants, the effect of leaf orientation is enhanced by the row effect. The latter is also sensitive in the near infrared for an erectophyll vegetation (leaf area index 10) for a field of view inclined at about 30° relative to the vertical, while it is not sensitive in the case of planophyll vegetation (even for low leaf area indices).

The reflectance of a vegetation cover varies with its phenological stage and physiological state (Fig. 4.9).

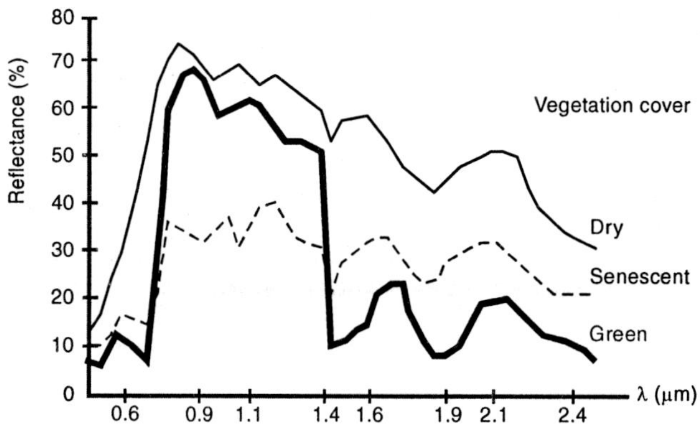

Fig. 4.9: Field-measured reflectance for vegetation cover in various physiological states.

In field measurements over senescent vegetation, reflectance in the near-infrared band is found to decrease. This decrease corresponds to the combined effect of changes in the internal structure of leaves and in the contribution of underlying soil to reflectance of vegetation, due to changes in leaf orientation as they become less dense. Consequently, the role of shadow produced becomes more important, which also contributes to reduction in reflectance.

In the reflective middle infrared band, for vegetation covers with similar water content, a greater absorption of radiation is observed in vegetation with the highest biomass. This band is sensitive to quantity of total water per unit area, i.e., the 'equivalent water thickness' defined by Allen et al. (1969).

In the case of total chlorophyllian vegetation cover (soil not seen and no spectral contribution from it), saturation of reflectance is observed starting from a leaf area index of 2 in the visible band and 8 in the near infrared. The saturation effect is reached at smaller leaf area indices for planophyll plants (such as beet) than for erectophyll plants (such as grass).

Moreover, multicoloured inflorescences (see supra) modify spectral behaviour in the visible band.

Non-chlorophyllian standing dry matter mixed with chlorophyllian aerial phytomass likewise modifies the spectral behaviour of the entire coverage to a varying degree, depending on the relative

[1]Leaf area index is the ratio of surface area developed by all the leaves to surface area occupied on the ground (Heller R. 1977. Abrégé de physiologie végétale. Masson, Paris).

proportions of the two mixing phytomasses. Such types of vegetation pose problems in estimating the fraction of chlorophyllian biomass.

In the case of low density of chlorophyllian vegetation, the spectral contribution of soil to the total spectral reflectance of bare soil and vegetation varies according to the spectral band used and the nature and state of soil. In fact, in the visible domain, bare soils most often have a higher reflectance than vegetation, while the reverse situation occurs in the near infrared.

When the coverage by soil equals that of vegetation, the more reflective the soil (very dry, low organic matter and iron content, smooth surface, etc.), the greater its spectral contribution in the visible band, and hence more readily detected. However, the spectral behaviour of the vegetation cover in the near infrared is not modified. Contrarily, a non-reflective soil contributes little to the spectral reflectance of vegetation in the visible, but significantly reduces it in the near infrared. This indicates that the proportion of vegetation cover at which the influence of soil on spectral behaviour becomes perceptible is highly variable. Our own investigations concerning Neoluvisols show that in the near infrared, the presence of green vegetation cannot be detected when the vegetation coverage is less than 20%, while a spectral behaviour characteristic of chlorophyllian cover is obtained when the coverage is more than 40%. Between these two values of coverage, the reflectance curves acquired in the field resemble neither those of the soil nor those of the chlorophyllian vegetation (Fig. 4.10). In-situ measurements at a height of 2 m above fields of *Panicum maximum* in different phenological stages and for different degrees of coverage illustrate this effect. The spectral behaviour of chlorophyllian *Panicum* with low coverage does not resemble that of vegetation.

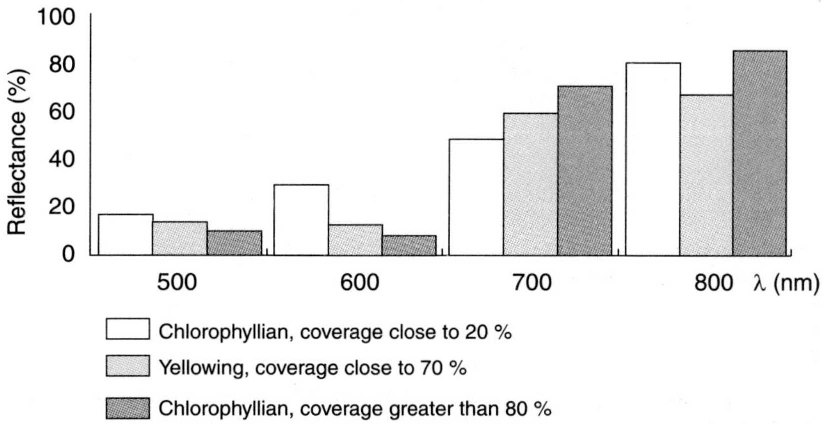

Fig. 4.10: In-situ measured reflectance graphs for *Panicum maximum*.

Similarly, a maize field, with a coverage of less than 20% (Fig. 4.11), exhibits a reflectance curve close to that of soil and does not indicate the presence of chlorophyllian vegetation.

The effects of spectral contribution of soil and different phenological stages for winter corn measured on different dates during the cultivation period are illustrated in Fig. 4.12. From October to June, a gradual variation of the reflectance curves from bare soil to chlorophyllian vegetation and back to senescence can be seen.

In the reflective middle infrared domain, for homogeneous vegetation cover such as annual crops, various authors obtained statistically significant correlation between reflectance values and water content of plants. Contrarily, in the case of heterogeneous covers of permanent grasslands (Orth, 1996), no such statistically significant correlation ($r^2 = 2.04 \times 10^{-5}$) was observed, except when different types of grasslands were separated according to height; in the latter case, significant correlation ($r^2 = 0.90$) was observed only for homogeneous and low canopies.

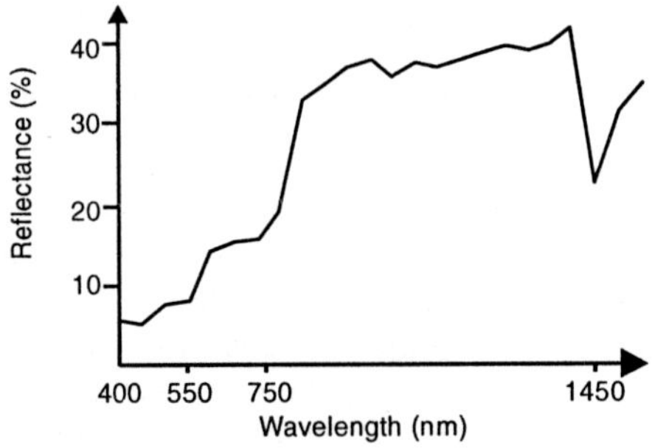

Fig. 4.11: Spectral behaviour of a maize field (coverage < 20%).

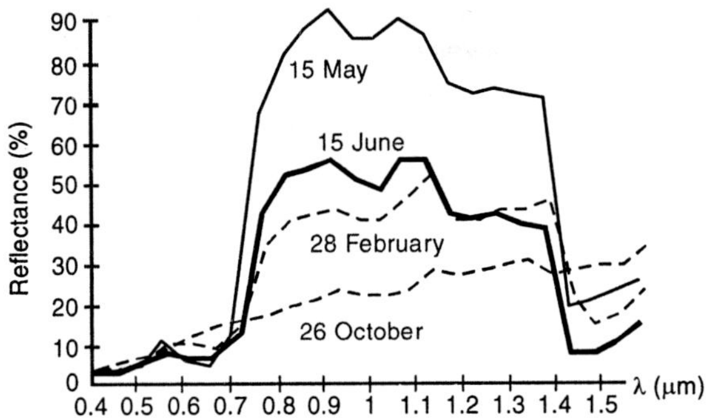

Fig. 4.12: In-situ measurements of reflectance on different dates for a winter corn field.

The foregoing discussion illustrates the influence of the structure of vegetation cover and indicates that results obtained for cultivated vegetation cannot be extended to natural or seminatural covers.

4.2 SOILS

Spectral characteristics of soils are discussed in detail in Chapter 23. In this chapter, the basics of spectral behaviour in the visible and reflective near- and middle-infrared bands are described, which constitute the minimum knowledge required for understanding vegetation indices.

Irrespective of whether the measurements are carried out in the laboratory or in the field, the reflectance of soils uniformly increases in the wavelength range from 400 to 1450 nm. It differs markedly from the reflectance of chlorophyllian vegetation: in the visible band, reflectance values of soil are usually higher than those of vegetation, hovering around 675 nm, whereas they are commonly lower than those of vegetation in near infrared.

Various soil horizons can be distinguished by means of a field radiometer (if care is taken to clear them in such a way that in-situ measurements on a horizontal plane can be obtained). For a Brunisol (Soil Reference Manual (SRM), 1995), the reflectance values can be interpreted when they are compared with the colours of horizons, organic matter content, total calcium content and moisture content (Fig. 4.13).

Horizons L1 and L2, which are very similar, cannot be differentiated (Table . 4.2). Horizons S1 and S2, which differ from the preceding horizons in colour and organic matter content, are distinguishable. Horizons C1 and C2 are identified from their total calcium content. Horizon C3, on the other hand, differs from horizon C2 by its lower moisture content.

Table 4.2: Results of analysis of a Brunisol (see reflectance variations in Fig. 4.13)

Horizon name (SRM)	Depth (cm)	Munsell colour	Organic matter (%)	Total calcium (%)	Moisture (%)
L1	0–5	10YR 3/3	5.1	3	28
L2	5–30	10YR 3/3	5.0	2	26
S1	30–60	10YR 4/4	1.2	0.5	20
S2	60–105	10YR 4/4	1.2	0.5	25
C1	105–130	10YR 5/6	0.8	32	26
C2	130–170	10YR 7/4	0.8	63	23
C3	170–220	10YR 8/3	Not analysed	66	19

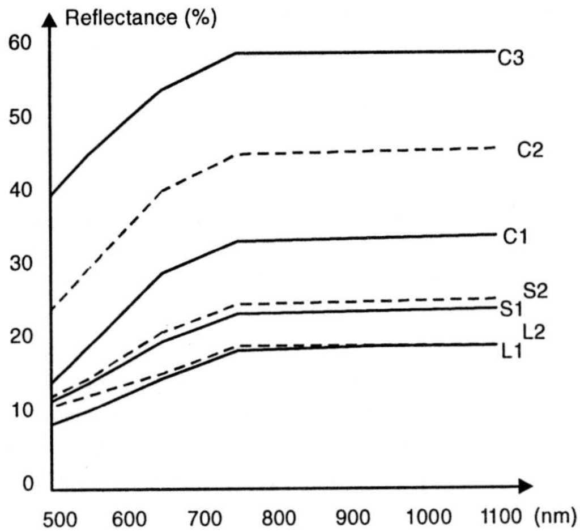

Fig. 4.13: Reflectance of various horizons of a Brunisol (measurements carried out with an EXOTEC radiometer using MSS spectral bands. For legend, see Table 4.2).

It can be seen that all the curves are approximately similar and reflectance increases with wavelength. Their major differences are expressed by the value of their integral. Spectral behaviour of soils can be interpreted by studying the shape of the curves (Fig. 4.13).

Lastly, while comparing reflectance curves of soils, the time of measurement should be taken into consideration, since the values change with time and moisture content, which varies with height of the Sun (Fig. 4.14).

In the reflective middle infrared band (1.6 to 2.2 μm), reflectance depends on water content. As soils always contain water, the absorption bands of water at 950, 1150, 1450, 1950 and 2450 μm can be observed on the curves obtained using a radiometer with narrow spectral bands.

Reflectance of soils is dependent on a number of internal and external factors, which can be classified as follows (see Chapter 23):

—*Surface roughness*, which depends on external factors, such as microtopography and crop (agricultural) activity, and on internal factors such as salt efflorescence, ferruginous encrustations,

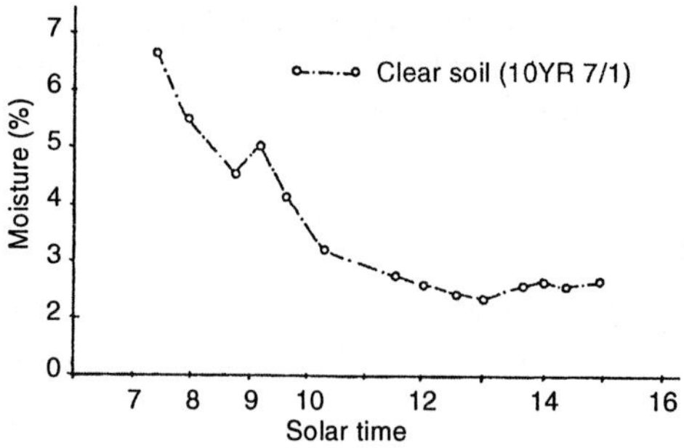

Fig. 4.14: Temporal variation of moisture content of a clear soil (10YR 7/1), 2 May 1989 (after Yongchalermchai, 1993).

cracks, gilgai, structure and porosity, slaking crusts, etc. Whatever be its origin, surface roughness modifies reflectance according to the amount of shadow viewed by the sensor. Reflectance is inversely proportional to the degree of roughness.

—*Physicochemical components*, such as organic matter, calcium, iron, water content, grain size, salts, etc., which increase or decrease soil reflectance.

Organic matter, iron and water content reduce soil reflectance in the visible and infrared bands. The higher the concentration of any (or several) of these elements, the smaller the reflectance. Contrarily, calcium and some salts enhance the reflectance of soils. Finally, the grain-size distribution either increases or decreases soil reflectance under different conditions. In fact, fine particles retain more water compared to coarse grains, leading to higher moisture content, which consequently reduces spectral reflectance.

A detailed description of the influence of physicochemical properties of soils on reflectance is given in Chapter 23. It may be remembered that in soil science, as in any biological discipline, factors involved are interdependent and variation of one leads to modification of another (or others). Interpretation of the spectral characteristics of soils hence necessitates a good understanding of soils and their functioning.

4.3 VEGETATION INDICES

Since the time satellite recording of spectral radiance of ground objects in visible and near-infrared bands became possible, many authors have developed various 'vegetation indices', based on a certain combination (sum, difference, ratio, linear addition) of intensities of these channels. These indices are used, on the one hand, to identify and monitor temporal variation of vegetation cover and, on the other, to estimate certain parameters such as aerial phytomass. Moreover, these combinations have the advantage of reducing the effect of factors external to vegetation, such as solar irradiance, atmospheric influence, spectral contribution of soils, etc.

4.3.1 Identification of vegetation cover

Identification of vegetation is based on differences between spectral characteristics of vegetation and soil (see supra). These differences are particularly distinct in the red band (around 675 nm) of the visible and near-infrared (800 to 1000 nm) channels, which are hence most often employed.

4.3.2 Estimation of vegetation parameters

Production of chlorophyllian aerial biomass by vegetation depends on the efficiency of interception ε_i, which is equal to the ratio of the photosynthetically active radiation absorbed (PAR_a) to the photosynthetically active incident radiation (PAR_i) (Fig. 4.15).

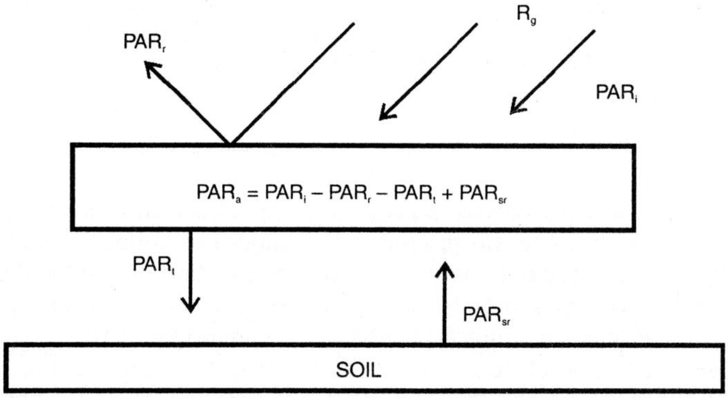

Fig. 4.15: Analysis of energy-absorbing efficiency of vegetation (after Varlet-Grancher, 1982). PAR_r: reflected PAR; PAR_t: transmitted PAR; PAR_{sr}: PAR of soil radiance; Climatic efficiency: $\varepsilon_c = PAR_i/R_g \approx 0.5$; Interception efficiency: $\varepsilon_i = PAR_a/PAR_i$ (dependent on vegetation, LAI, coefficient of extinction etc.); Biological efficiency: $\varepsilon_b = (c \times \Delta(MS))/ PAR_a$; $\Delta(MS)$ = increase in dry matter related to photosynthesis; c = energy equivalent = 12×10^6 J.kg^{-1}.

Interception efficiency evidently depends not only on the leaf area index, but also on the structure of the vegetation canopy, as illustrated in Fig. 4.16. Alfalfa and beans can be considered planophyllic (see supra), while maize and sugarcane are erectophyllic. It has been observed that for the first two types of plants, interception efficiency (ε_i) becomes asymptotic for leaf area indices between 2.5 and 3, whereas asymptote is reached for leaf area indices of 4.5 and 5 for the other two crops.

The structure of the vegetation cover should be taken into consideration while estimating ε_i from the leaf area index. Numerous studies (Baret and Guyot, 1991) showed that the spectral reflectance of vegetation cover is related to the total aerial phytomass at the time of measurement and that an asymptotic relationship exists between leaf area index of a vegetation and various combinations of reflectance in the red and near-infrared bands, i.e., the vegetation index. This relationship assumes

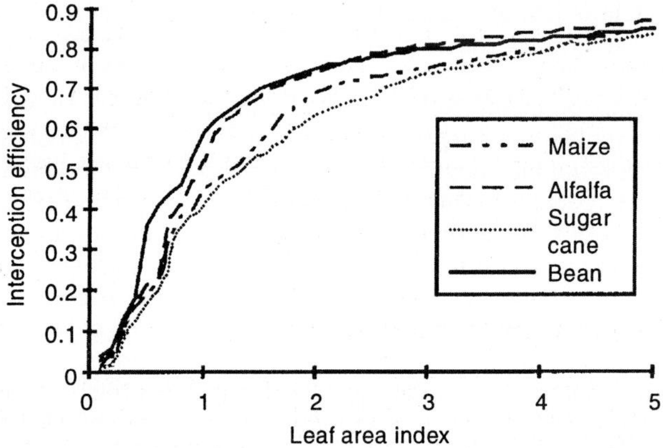

Fig. 4.16: Variation of interception efficiency with leaf area index (after Varlet-Grancher, 1982).

that the geometry of illumination and field of view (see Fig. 1.5) as well as inclination of leaves remain constant. However, this is not so in reality and the semiempirical relations between vegetation index and aerial phytomass are valid only locally and for a given instant.

Some vegetation indices are applicable only for vegetation covers that are dense (no soil visible) but not too dense (otherwise, the saturation effect mentioned at the beginning of the chapter becomes apparent) and that are chlorophyllian, with no mixing of standing dry matter with green. Other formulae ought to be used when coverage of chlorophyllian vegetation is low (seedling or harvesting stages of agriculture and annual steppes in arid zones etc.) or when standing dry matter accumulates, such as in natural herbaceous stands (Rondeaux et al., 1996).

4.3.3 Use of 'soil clusters'

The concept of soil clusters is based on knowledge of the spectral behaviour of soils (see supra and Chapter 23). Spectral characteristics are related to the degree of surface roughness (presence of slaking crusts, hardpans, etc.), organic matter content, composition of chemical elements ($CaCO_3$, Fe, etc.) and water content. If only one of these parameters varies, a linear relationship between the radiant intensities of visible and near infrared bands can be defined for a soil. This relationship is the 'soil line' (Huete, 1988) used as a reference line for studying sparse vegetation covers.

Strictly speaking, there is no single 'soil line' but a cluster (see Fig. 23.16) depending on various factors that influence the reflectance. This makes the correction for soil reflectance in vegetation studies difficult.

4.3.4 Various vegetation indices

Many authors have used vegetation indices computed from reflectance measured in the field or from digital values furnished by satellite data for identification of chlorophyllian vegetation and estimation of leaf area index, canopy development, stress effects, APAR (Absorbed Photosynthetically Active Radiation), evapotranspiration and efficiency. These studies have resulted in formulation of several equations (Table 4.3), with varied coefficients of balance depending on the data and subject under analysis (CD 7.8).

The vegetation indices can be grouped into two categories: those characterised by a slope (RVI, NDVI, SAVI etc.) and those characterised by a distance (PVI and Tassel Cap index).

The various indices, in fact, correspond to an empirical approach in estimating the vegetation parameters. They are more or less sensitive to the spectral contribution from soil (and to its variations depending on soil characteristics) as well as from standing dry matter mixed with green. Errors in estimation of vegetation parameters will be high if field spectral data on the components of the pixels are not available.

The normalised vegetation index (NDVI) should be used cautiously in view of its sensitivity to atmospheric effects and angular variations (as in large angle scanners such as NOAA-AVHRR). In fact, the flux reflected by a vegetation cover is higher in the visible red band than in near infrared if the system is in a back (hot spot) position relative to the sun, rather than facing the sun (against the light). Consequently, the NDVI values corresponding to the first configuration are systematically smaller than those corresponding to the second. The most commonly used correction to overcome this effect consists of taking the highest value of NDVI over a long temporal sequence (about a month, for example). Neglecting this effect may lead to erroneous interpretations of NDVI in studying variations in vegetation cover. Hence, satellite remote sensing, notwithstanding its synoptic view and ability to observe a given region at regular time intervals, may not be able to furnish all the information desired. In fact, in subjects of global importance, such as temporal monitoring of vegetation cover variations over long periods and over vast areas (at regional or continental level) or drought phenomena, etc.,

Table 4.3: Most commonly used vegetation indices
(NIR: Near infrared; R: Red; G: Green; B: Blue)

Index	Formula	Characteristics	Authors
Difference	R-NIR	Very sensitive to atmospheric variations	Monget, 1980
Ratio	RVI = NIR/R or of other channels; Pigment index: G/R	Saturation at high indices; sensitive to spectral contribution of soils and atmospheric effects	Knipling, 1970; Viollier et al., 1985
Normalised vegetation index	NDVI = (NIR-R)/(NIR+R)	Sensitive to atmospheric effects; smaller range of variation than preceding; sensitive to variations in view angle, according to position vis-à-vis Sun/hot spot*	Rouse et al., 1974; Tucker, 1979
Transformed vegetation index	TVI = $\sqrt{(NDVI+0.5)}$	Attempt to eliminate negative values; stabilisation of variance	Deering et al., 1975
Perpendicular vegetation index	PVI = a_1NIR – a_2R + constant	Decrease in spectral contribution of soils; sensitive to various soil parameters	Richardson and Wiegand, 1977
Tassel Cap	General formula a_1G + a_2R + a_3NIR + a_4NIR	Orthogonal transformation of 4 channels for minimising sensitivity to spectral	Kauth and Thomas, 1976
Greenness index (derived from the preceding)	GR4 = – b_1G – b_2R + b_3NIR + b_4NIR for MSS channels	contribution of soils; complete elimination not possible	Jackson, 1983
Soil-adjusted vegetation index	SAVI = (1+L) (NIR–R) / (NIR+R+L), where L = 0.5 to reduce soil effect	Several indices are derived from this to minimise soil effect (TSAVI, MSAVI, etc.)	Huete, 1988
Atmospheric reduced vegetation index	ARVI = (NIR–RB) / (NIR+RB); RB = R – γ (B–R) γ varies with aerosol type	Effect of aerosols in atmosphere on NDVI reduced but sensitive to spectral effects of soil	Kaufman and Tanre, 1992

*Hot spot corresponds to a view angle identical to incident solar radiation (backing sun). No shadows in this configuration and all areas viewed are illuminated.

large errors may accumulate due to atmospheric perturbations as well as vegetation characteristics, making the results ineffective.

It is evident that remote sensing data ought to be integrated into models, often very complex, together with other parameters of vegetation and environment (climate, soil, hydrological regime, etc.) for estimating yields (see Chapter 22), water stress, etc. The empirical approach of vegetation indices should be considered with great care.

4.4 WATER

4.4.1 Response in visible and near- and middle-infrared bands

The spectral behaviour of water is dependent not only on the water molecules but also on dissolved or suspended constituents (such as particles, algae, organic matter, etc.), as well as the state of surface roughness. Incident radiation is partly reflected specularly, more so when the water is calm and smooth, and partly refracted and transmitted into the volume. Specular reflection can be observed in aerial photographs as well as in satellite images. The phenomenon can be readily detected on aerial photographs since the same surface appears very white in one photograph and dark in the next. In satellite images, unless diachronic data are available for the same region, the specular surfaces may be confused with highly reflective mineral surfaces (such as quarries, alluvial-sand mines, etc.). Pure water reflects very little in the red and infrared bands. It appears black in black-and-white infrared as well as colour infrared photographs. On satellite images the digital values are very low for the corresponding bands (CD 4.2) and especially in reflective middle infrared.

The spectral response of the sea is directly related to interaction of several processes operating on its surface and inside the water mass. The important ones are:

—agitation of the surface under the effect of wind and surges,

—presence of a floating pollutant (hydrocarbons, wastes etc.), or of pollution from a ship,

—mixing of waters of different densities, buoyancies and temperatures (fresh and salt waters, rise of cold waters),

—currents,

—suspended sediment load,

—presence of phytoplankton and chlorophyll pigments in waters,

—presence of dissolved substances, etc.

The response of a layer of water in the optical domain, a phenomenon concerned with the land/sea boundary and the actual water mass, is discussed in relation to two types of application.

4.4.2 Approximate method of bathymetry in clear water

Under the conditions of observation of coastal environments with clear waters, the light signal in the visible band penetrates a layer of water where it is absorbed. In shallow seas, it may reach the sea-floor and traverse the water layer in return, thus providing information on depth to the bottom and its nature. The signal measured over a shallow sea depends on the reflection properties of the floor and the influence of the water column standing on it.

The spectral behaviour of the sea-floor can be described by a model of exponential decrease of the signal with depth at which the floor is situated. Several formulations of this model are reported (O'Neill and Miller, 1989; and others). All these models are similar and can be represented by the equation:

$$R(0, z) = R(0, \infty) + [A - R(0, \infty)] \exp(k_i z)$$

where $R(0, z)$ is the measured reflectance above a sea-floor of depth z, $R(0, \infty)$ is the reflectance measured above a sea-floor of infinite depth, A the albedo of the floor, z the depth of the sea-floor and k_i the coefficient of attenuation.

Thus, the reflectance of a water column bounded at the bottom by a reflecting surface at a depth z, can be described as a sum of the reflectance $R(0, \infty)$, of an optically identical water column with no bottom and the contrast of the bottom relative to $R(0, \infty)$ modulated by the attenuation in a two-way path between the surface and the floor. The parameter k_i characterises the rate of decrease of the incident solar energy with depth. It represents the radiant intensity of the water in a given spectral band and is related to the depth to sea-bottom. Spectra of diffuse attenuation coefficients for water with different concentrations of particulate and dissolved material are shown in Fig. 4.17.

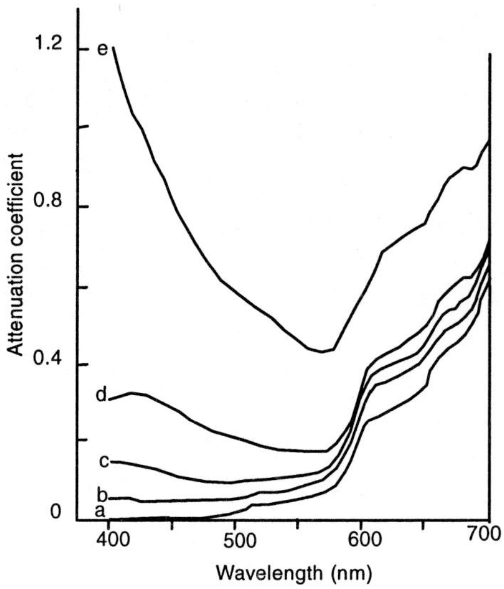

Fig. 4.17: Spectra of diffuse attenuation coefficients for descendant illumination for (a) pure sea water, (b) oligotrophic oceanic water, (c) clear coastal water, (d) eutrophic oceanic water (4–5 mg chlorophyll a per m^3) and (e) coastal water with high concentrations of optically active particulate and dissolved matter (Baltic Sea) (after Maritorena, 1993).

The attenuation coefficient k_i varies with wavelength. For clear waters in particular, the diffuse attenuation coefficient is small at short wavelengths and exhibits higher values beyond 570 nm. Thus, for clear waters penetration of radiation is high for channels in the wavelength range of 400–500 nm, whereas in waters with high concentrations of particulate and dissolved matter, largest penetration depths are obtained in the range 500–600 nm. Considering the values of attenuation coefficient, only a few bands are useful for determining the depth and nature of shallow sea-floor. These are TM1 (450–520 nm), TM2 (520–600 nm) and to a lesser extent TM3 (630–690 nm) of LANDSAT, and the channels b1 (500–590 nm) and less so channel b2 (610–690 nm) and P (510–730 nm) of SPOT. For extracting bathymetric information from remote sensing data of the visible domain, it is necessary to determine the best possible attenuation for the zone under study and the concerned spectral bands.

This complex process of bathymetric estimation can be simplified by employing a data-processing method. For a given sea-floor, the transformation $X_i = \ln (L_i - L_{\infty i})$, where L_i is the radiance for the wavelength i and $L_{\infty i}$ the radiance for an infinite floor, is approximately equivalent to linearising the signals corresponding to different depths (Lyzenga, 1978). In one graphic presentation in which $\ln (L_i - L_{\infty i})$ is plotted on the abscissa and $\ln (L_j - L_{\infty j})$ on the ordinate, the points associated with a given type of sea-floor form a cluster of points scattered along a straight line with a slope equal to k_i/k_j. The spread of the cluster of points is linearly dependent on the bathymetric level and the deviation of this cluster is a measure of the variation in the type of sea-floor. A rotation of the system of axes by an angle equal to arctan (k_i/k_j) gives a new reference framework (Yz, $Yf1$, $Yf2$, ..., $Yfn-1$) such that Yz is theoretically independent of the nature of the sea-floor while Yfi is independent of depth. Thus, we get:

$$Z = aYz + b$$

In practice, several methods exist to determine the real absolute *Z*. Use of calibration points facilitates determination of the coefficients *a* and *b* which are assumed constant for all the pixels of an image. An application of the method proposed above (Loubersac et al., 1989) by inversion of the model of exponential decrease of the signal with depth, linearisation, change of axes, calculation of a pair of coefficients (*a*, *b*) for each of the major types of floor detected by a threshold of *Yf*1 and calibration by introducing the in-situ measurements into the model, shows a relative error of 10% for the bathymetric model for the 0–10 m layer of water. Beyond 20 m the results of the model have no significance due to loss of the signal and presence of noise of the HRV sensor used. Andrefouet et al. (1998) described the various approaches and solutions, a critical analysis of the results obtained and the errors.

4.4.3 Measurement of water colour

When the radiance due to the sea-floor does not interfere with the signal received by the sensor, either because waters are saturated with particles or because the layer of water is sufficiently thick, information on the water colour can be extracted from the data of the visible band. In fact, the reflected flux measured above the surface is directly related to the concentrations of the constituents present in the water, pigments associated with plankton, degraded organic matter, suspended material and dissolved matter. However, as the contribution of the atmosphere to the measured signal is very high, up to 90%, in the narrow spectral bands used (see later), development of algorithms for atmospheric corrections is necessary.

Significant advances have taken place in the remote sensing of the colour of water thanks to the data of the CZCS (Coastal Zone Colour Scanner) sensor of NIMBUS-7 satellite, which became operational between 1978 and 1986. This satellite has six spectral bands, four in the visible, one in near infrared and one in the thermal range (Cassanet, 1990). The four bands of the visible domain positioned respectively at 433–453 nm, 510–530 nm, 540–560 nm and 660–680 nm facilitate measurement of water colour, especially in the presence of chlorophyll. These bands also enable atmospheric corrections.

In general, the algorithms developed by means of CZCS experiments for estimating the optical parameters of sea-water comprise three stages:

—Transformation of the digital counts furnished by the satellite sensor into luminance values and calibration in view of the gradual drift in the sensitivity of the sensor (Sturm, 1983; Singh et al., 1985).

—Atmospheric correction of the total luminance obtained for isolating the effect of the water layer under investigation.

—Calculation of the concentrations of pigments and suspended matter, and computation of the coefficient of absorption by means of specific algorithms based on the analysis of the relationships between channels, calibrated by in-situ measurements of pigment concentration (Bricaud et al., 1981; Sturm, 1981), or by inversion of a radiative transfer model (Fischer and Doerffer, 1987).

Results of such processing are used for applications in marine biology and ecology, such as monitoring of water quality, sedimentary transport and dynamic processes.

As the spectral bands of CZCS sensor are not optimally placed vis-à-vis causes of variation in the optical quality of sea-waters, we cannot obtain sufficiently reliable qualitative and quantitative parameters about the colour of water that are useful for problems of littoral management such as water pollution, for example. Hence, a strong expectation stands before the scientific community to generate data from new sensors, such as SeaWiFS launched in August 1997 and Envisat/Meris launched on 28 March 2002 for which spectral resolution is optimised relative to CZCS.

4.5 SNOW AND ICE

Snow, which is made up of fine crystals of ice varying from 50 µm to 1 mm in size, constitutes a highly diffusive medium due to its granular structure. Reflectance of snow in the visible band mainly depends on the content of pollutants, while in near infrared, it varies according to the geometry of the crystals, i.e., their shape and dimensions. Reflectance of snow in this range of wavelengths is inversely proportional to the size of the crystals (Fig. 4.18).

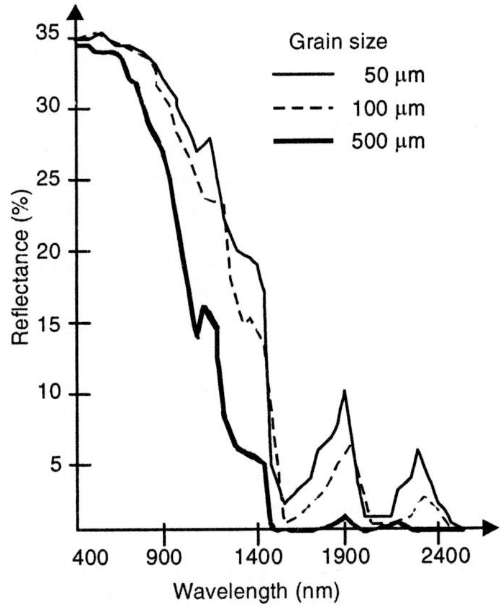

Fig. 4.18: Variation of the bidirectional reflectance of snow (computed from a model) with wavelength for various grain sizes. Angle of solar incidence 40° relative to the nadir (after Fily et al., 1997, p. 455, with permission from Elsevier Science Publishers).

This Figure explains the stronger reflectance of fresh snow compared to that of older snow (with coarser crystals) or the reflectance of firn which per se is more reflective than ice. Ice has optical properties close to those of water, except in the range 1.55 to 1.75 µm, where it absorbs more radiation than water.

References

Allen WA, Gaussman AJ, Richardson AJ, Thomas JW. 1969. Interaction of isotopic light with a compact plant leaf. J. Opt. Soc. Amer., 59: 1376–1379.

Andrefouet S, Loubersac L, Maritorena S, Morel Y. 1998. Mesure de la bathymétrie des zones côtières par télédétection passive dans le domaine visible. In: Manuel de télédétection océanique. Pêche et Océans. Gordon, Breach (eds.). Institut Maurice Lamontagne, Canada.

Bardinet C, Monget J-M. 1980. LANDCHAD. Télédétection et géographie appliquée en zone sahélienne du Tchad. Collection de l'École Normale Supérieure de Jeunes Filles, Paris, no. 12, 133 pp.

Baret F, Guyot G. 1991. Potentials and limits of vegetation indices for LAI and APAR assessment. Remote Sensing of Environment, 35: 161–173.

Ben Moussa H. 1987. Contribution de la télédétection satellitaire à la cartographie des végétaux marins: archipel de Molène (Bretagne, France) Thèse de doctorat, Univ. Aix-Marseille II, 122 pp.

Bricaud A, Morel A, Prieur L. 1981. Absorption by dissolved organic matter of the sea (yellow substance) in the UV and visible domains. Limnology and Oceanography, 26: 43–53.

Cassanet J. 1990. Satellites et capteurs. In: Télédétection satellitaire. Paradigme, 141 pp.
Deering DW, Rouse JW, Haas RH, Schell JA. 1975. Measuring forage production of grazing units from Landsat MSS data. Proc. 10th Int. Symp. Remote Sensing Environment, vol. II, pp. 1169–1178.
Fily M, Bourdelles B, Dedieu JP, Sergent C. 1997. Comparison of in situ and Landsat Thematic Mapper derived snow grain characteristics in the Alps. Remote Sensing of Environment, 59: 452–460.
Fischer J, Doerffer R. 1987. An inverse technique for remote detection of suspended matter, phytoplankton and yellow substances from CZCS measurements. Advances in Space Research, 7 (2): 21–26.
Guyot 1997. Climatologie de l'environnement. De la plante aux écosystèmes. Masson, Paris, 505 pp.
Heller R. 1977. Abrégé de physiologie végétale. Masson, Paris.
Huete AR. 1988. A soil-adjusted vegetation index (SAVI). Remote Sensing of Environment, 25: 295–309.
Jackson RD. 1983. Spectral indices in *n*-space. Remote Sensing of Environment, 13: 409–421.
Kaufman YJ, Tanre D. 1992. Atmospherically resistant vegetation index (ARVI) for EOS-MODIS. IEEE Trans. Geoscience Remote Sensing, 30: 261–270.
Kauth RJ, Thomas G. 1976. The Tassel Cap, a graphic description of the spectral-temporal development of agricultural crops as seen by Landsat. Proc. Symp. Machine Processing of Remotely Sensed Data. IEEE Catalogue, no. 76, ch. 1103-1 MPRSD, LARS. Purdue Univ., West Lafayette, IN (USA).
Knipling EB. 1970. Physical and physiological bases for the reflectance of visible and near infrared radiation from vegetation. Remote Sensing of Environment, 1: 155–159.
Loubersac L, Burban PY, Lemaire O, Chenon F, Varet H. 1989. Nature des fonds et bathymétrie du lagon de l'atoll d'Aitukai (Iles Cook) d'après des domnées SPOT 1. Photo Interprétation 89-5 et 6, fasc. 4.
Lyzenga DR. 1978. Passive remote sensing techniques for mapping water depth and bottom features. Applied Optics, 17 (3): 379–383.
Maritorena S. 1993. Étude spectroradiométrique de la colonne d'eau et des fonds en milieu lagonaire récifal. Implications sur l'imagerie télédétectée à haute résolution dans le visible. Thèse Univ. Française du Pacifique, Océanologie, Tahiti, 195 pp.
Monget J-M. 1980. cf. Bardinet & Monget 1980 Monget is the author of the index.
Myneni RB, Asrar G. 1994. Atmospheric effects and spectral vegetation indices. Remote Sensing of Environemnt, 47: 390–402.
O'Neill NT, Miller JR. 1989. On calibration of passive optical bathymetry through depth soundings. Analysis and treatment of errors resulting from the spatial variation of environmental parameters. Int. J. Remote Sensing, 10 (9): 1481–1501.
Orth D. 1996. Typologies et caractérisation des prairies permanentes des marais du Cotentin, en vue de leur cartographie, par télédétection satellitaire, pour une aide à leur gestion, Thèse INA PG, 149 pp. et annexes.
Richardson AJ, Wiegand CL. 1977. Distinguishing vegetation from soil background information. Photogrammetric Engineering & Remote Sensing, 43: 1541–1552.
Rondeaux G, Steven M, Baret F. 1996. Optimization of soil-adjusted vegetation indices. Remote Sensing of Environment, 55: 95–107.
Rouse JW, Haas RH, Schell JA, Deering DW, Harlan JC. 1974. Monitoring the Vernal Advancement of Natural Vegetation. NASA/GSFC Final Report. Greenbelt, MD, 371 pp.
Singh SM, Cracknell AP, Spitzer D. 1985. Evaluation of Sensitivity decay of CZCS detectors by comparison with in situ near-surface radiance measurements. Int. J. Remote Sensing, 6: 749–758.
Sturm B. 1981. Ocean color remote sensing and quantitative retrieval of surface chlorophyll in coastal waters using CZCS data. Marine Science Series, 13: 267–279.
Sturm B. 1983. Selected topics of CZCS data evaluation. In: Marine Science and Technology. Reidel Publ. Comp., Dordrecht, pp. 137–167.
Tucker CJ. 1979. Red and photographic infrared linear combinations for monitoring vegetation. Remote Sensing of Environment, 8: 127–150.
Varlet-Grancher C. 1982. Analyse du rendement de la conversion de l'énergie solaire par un couvert végétal. Thèse de Docteur ingénieur ès Sciences Naturelles, Univ. de Paris-Orsay, 144 pp.
Viollier M, Belsher T, Loubersac I. 1985. Signatures spectrales des objects du littoral. Proc. 3rd Int. Coll. on Spectral Signatures of Objects in Rem. Sens. Les Arcs, France. ESA SP 247, pp. 253–257.
Wiegand CI, Richardson AJ. 1990. Use of spectral vegetation indices to infer leaf area, evapotranspiration and yield. 1: Rationale. Agron. J., 82: 623–629.

C

PROCESSING AND INTERPRETATION

5

Visual Interpretation of Photographs and Images

5.1 VISUAL INTERPRETATION

5.1.1 The eye

Visual perception differs according to the colour of an image. The eye perceives up to 16 grey levels in black and white and about 10,000 colours. These figures correspond to the number of levels that can be recognised absolutely, i.e., when they are not intermixed but are isolated from one another. The concept of contrast is not taken into consideration in this count. When contrast is involved, the eye exhibits better performance. In fact, a large number of shades are perceived when intermixed grey levels are compared, but it is not certain that the eye can reliably measure the same grey levels when the two levels compared are not close to one another.

The light signal perceived by the eye, by the order of its spatial resolution, is about 0.1 mm. If the pixel of an image or a photograph observed is greater than 0.1 mm in size, it is visible, otherwise it cannot be perceived. For a SPOT (1 to 4) image with a resolution of 20 m and in the scale of 1:200,000, the resolution perceived by the eye of 0.1 mm would be 20 m. Thus, we do not get the impression of seeing the pixel. At larger scales, the pixel of the image is perceived. Thus, at 1:20,000 scale the pixel size on the image is 1 mm and hence discernible. Therefore, images in the range of scales of 1:100,000 to 1:50,000 can be interpreted visually without being disturbed by the pixel, which plays the same role as that of the grain in a photograph. Obviously, visual interpretation on monitors does not present these inconveniences since image resolution can be enhanced or diminished as desired by using the zoom.

5.1.2 The brain

Interpretation of the signal perceived by the eye is done by the brain, where all forms of processing take place. Five types of processing by the brain can be identified and grouped as:

—pre-processing of the signal (1),
—textural processing (2 to 4),
—structural processing (5).

1. The first form comprises perception of colours, contrasts and shapes, in particular linear fields and angles between them.
2. The second is identification, carried out by recollection of objects or shapes seen previously.
3. The third is classification of objects into types or reference groups.

4. The fourth form of processing corresponds to spatial analysis of the distribution of objects in two-dimensional space. For example, a type of land cover, viz., hill slopes with vineyards and orchards in the Yonne district, is observed in a single part of the image (in the centre) and absent elsewhere (Fig. 5.1). On the other hand, another type, hardwood forest bordering a talweg, is seen throughout the image (Fig. 5.2). Thus spatial analysis can be carried out after textural processing of images (CD 5.1).
5. The fifth form is analysis of organisation of objects in three-dimensional space (if stereoscopy is feasible) or at least in two-dimensional space (images on monitor). This involves defining textural groups of objects whose spatial organisation is specific and visually recognisable from more or less defined patterns, often empirically or through chorological laws, involving studies of relationships between characteristics of semantic units (intrinsic factors) observed and their distribution in three-dimensional landscape (extrinsic factors).

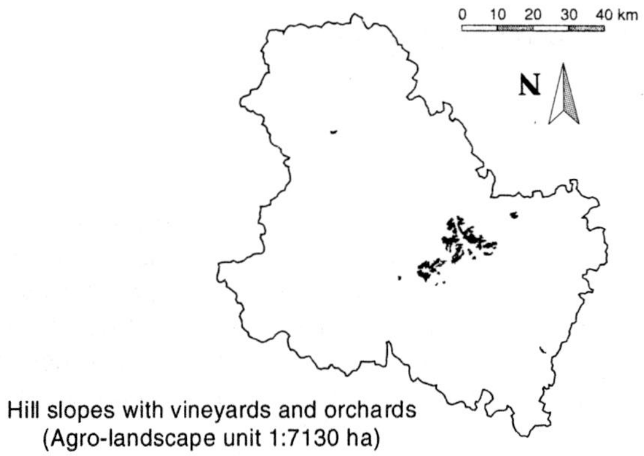

Fig. 5.1: Distribution of hill slopes with vineyards and orchards in the Yonne district, interpreted from SPOT images (after Bertrand, 1994 and Mollet, 1994).

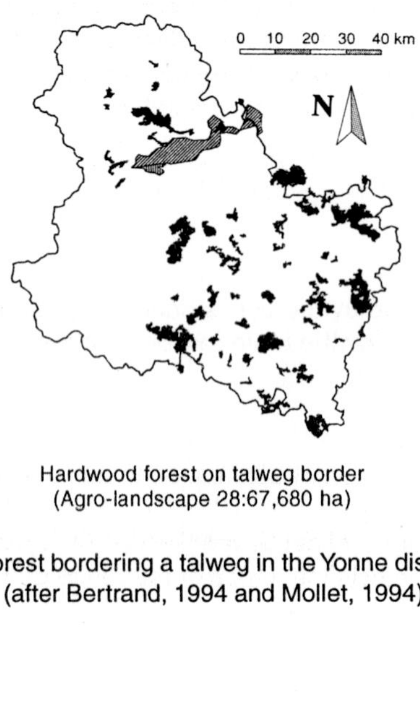

Fig. 5.2: Distribution of hardwood forest bordering a talweg in the Yonne district, interpreted from SPOT images (after Bertrand, 1994 and Mollet, 1994).

5.1.3 Interpretation procedure

An aerial photograph or an image represents aggregate data received at a given time in a given field of view and with a resolution defined by the characteristics of the source-receiver pair. It is raw data for which an interpretation has to be made. Aerial photographs or images thus differ from a map for which a legend corresponding to a given theme exists. *A map is the result of classification of the objects depicted on it; it represents spatial distribution and organisation of these objects as well as their conceptualisation.* An aerial photograph or an image is the result of a series of measurements carried out for each element of resolution in a given field of view; it does not define objects. Hence it is necessary to recognise in the aerial photograph or image the objects present, to classify them and to study their distribution and organisation before comparing the information from a map with that of remote sensing data.

In a photograph or an image, the eye detects the most contrasting features. These contrasts may be represented by lines which, in most cases, are open. Hence they do not delineate entities. It is therefore not simple to define features generated and determined by what is enclosed inside an envelope. For example, just by applying isodensity analysis on an aerial photograph or an image it can be seen that objects earlier defined are not correctly delineated.

Two major types of visual interpretation are commonly used. The analytical method is described below and the method of synthesis or landscape is given in Chapter 18.

5.2 PHOTO-INTERPRETATION

Interpretation of aerial photographs has been done for several decades now for many thematic subjects concerned with the investigation of natural sciences, such as geology, geography, agronomy, soil science, botany and urban studies (Agache, 1970; Chevallier, 1971; CRU, 1969; Girard and Girard, 1970, Girard et al., 1996; Guy, 1969; Lillesand and Kiefer, 1994; Mulders, 1987; Smith, 1968, and others). Evidently, interpretation varies with the object viewed, which determines the semantic and graphic precision to be achieved, the theme concerned and the methods of operation. However, in almost all cases we are interested in analysing aerial photographs for diverse thematic problems, retaining only those that would be important at the end of study. In fact, like a satellite image, an aerial photograph is a general and non-specific document on which all types of thematic data are overlapping and recorded at a given date and time.

The basic approach to interpretation of an aerial photograph is to proceed from the simplest to the most complicated. Thus, the brain interprets the various features appearing on the image in the order of legibility and stores them in memory. It is then easier to choose features of interest from a thematic point of view.

Let us consider, for example, a 1:17,000 colour infrared photograph of the Vitteaux region (Côte d'Or) in Burgundy. While the method is general, the tables presented here comprise only outlines pertinent to the example.

5.2.1 Lines and points

■ Communication routes

Features most readily discernible in a photograph or an image are lines and straightline aspects such as roads, boundaries of agricultural or forest zones, etc., which in most cases are produced by the effect of human activity on the environment. It is often important to draw or extract these features by filtering onto a layer of information and to establish a legend (Fig. 5.3).

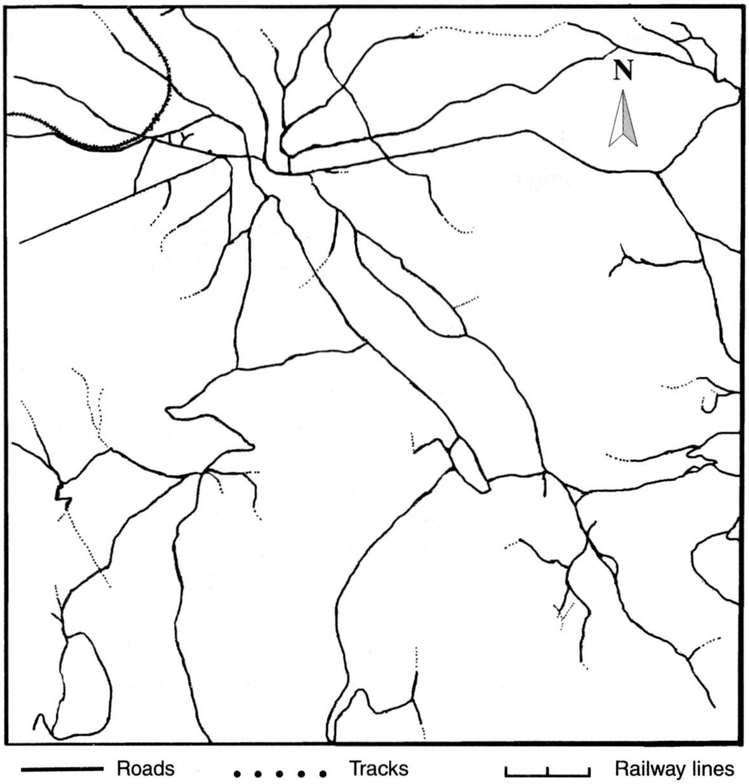

Fig. 5.3: Interpretation of communication lines in the Vitteaux region. Approximate scale 1:40,000.

These features can be analysed by studying:

—general directions of straightline segments using rose diagrams,
—density of lines per unit area,
—length of lines as a function of their shape (rectilinear, curvilinear, etc.),
—intersections of lines,
—open or closed drainage patterns.

A curved communication line in the north-western part of Fig. 5.3 can be very clearly distinguished. It represents an old railway track.

Roads are usually indicated as long lines covering the entire image. They are numerous and exhibit a radial disposition all around Vitteaux. Two roads, merging into one, are directed south-east, encircling a river. It can be inferred from this that the river overflows from time to time between the two roads and that caution is taken to travel roads in this zone without risk of floods. Another series of roads run eastwards and join at the village Massingy. Some areas without roads are also evident, which correspond to inclined zones (see geomorphology in a subsequent section). A road with a hairpin turn is also seen in this slopy region.

■ Parcels

Parcels are analysed on the basis of criteria of size, shape, abundance and mode of closure.

☐ Size

The sizes of a parcel are expressed in metres. This is possible when the scale in the case of a photograph and resolution in the case of satellite image are known. The following notations are used:

- *L* : the longest side,
- *l*: the shortest side,
- *p*: sides of equal dimensions,
- *r*: curved sides.

☐ Shapes

Six definite shapes are mentioned in Table 5.1 but other forms can also be described depending on the region and mode of cultivation.

Table 5.1: Shapes of plots

Shape	Formula	Description
Strip-form	$L \times l \, (L \geq 3l)$	Highly elongated rectangles
Rectangular	$L \times l \, (3l > L)$	Four perpendicular sides, trapezium, parallelepiped
Draught-board	$L \times L$ or $l \times l$	Four equal sides, or squares
Polygonal	np	More than four sides of equal dimensions
Curved	$L \times r$	One or two curved sides
Rounded	$r \times r$	Complete circles, or more than two curved sides

☐ Modes of closure

The following types of closure can be identified:

—open plots such as open fields,
—plots enclosed by hedges, with bushes or trees
—plots enclosed by dry stone walls (in particular in clearing zones),
—plots surrounded by trenches or embankments.

☐ Abundance

Abundance refers to giving an estimation of the sum of areas of each type of plot, relative to a wider group such as landscape, landscape unit or landscape element. A code of 0 to 1 or 1 to 5 (see Table 5.2 below) can be used.

☐ Pattern of parcels

When studies are carried out over a large spatial field[1] , the general pattern of the plot compartments enables a better understanding of the relationships between various components of the landscape and the environment, or between the environment and its exploitation by man. It is preferable to integrate the plots of land and forest with those of agricultural plots. Six types of plot compartments can be defined:

[1]Spatial field or field of view: Largest than the area analysed and comprising all the sites under study.

Circular

The shapes of curved or approximately circular features are due to constraints of the physical medium, such as hills or depressions (common in karst zones), from which they get their configuration, or due to technological constraints, such as circular plots prepared according to a sprinkler irrigation system. This pattern, little favoured in Europe, is more common in North America or in semidesert regions (Saudi Arabia, Australia, Southern Algeria, for example) where modern cultivation operations have modelled the land use characteristics according to available technology.

Parallel

Striplike plots are often non-uniform along their longer side particularly in alluvial plains or on slopes. Their general direction may be rectilinear or curved.

Radial

Plots that are most often rectangular or striplike converge by their longer side towards a centre (a village, for example). This pattern combines the circular and parallel patterns. It is characteristic of circular zones which are recognised from the boundaries of plots organised as circles or crescents and extend over hundreds of metres and kilometres. Such patterns are quite common in regions of ancient culture (Europe, Syria etc.). Several studies have interpreted such patterns (Soyer, 1970). They can also be detected in satellite images (Girard, 1995).

Perpendicular

Some plot compartments clearly show two preferential alignments for the boundaries of plots, such as newly cultivated zones or areas recently reclaimed from former wet or aquatic zones.

Disseminated

Various types of scattered plot compartments are observed in forests, lands and grass plots. They indicate low human activity or a zone difficult to cultivate or old abandoned agricultural attempts.

Random

In some cases, neither a spatial order nor a structural pattern is observed in the distribution of plots. Such is the situation in many zones. It may also arise when the field of study is not large enough for a pattern extending over a wider zone to be detected.

Infrastructures and superstructures may be observed within such patterns. For example, in the Loire valley near Longué, squared features are noticed, often indicated by communication tracks, inside which parallel patterns comprising strips or rectangles or random structures can be delineated (Girard and Girard, 1989).

It is also possible to detect roads cutting across old patterns of the area or, inversely, zones adapted to Roman roads (Agache, 1970; Chevallier, 1997). In some cases, the pattern of the feature is dictated by an environmental component. Thus, in narrow flat-bottomed valleys, parallel features are frequently observed due to the insertion of roads and plots between slopes.

Three types of features can be identified on the image shown in Fig. 5.4, which are briefly summarised below:

1. Zones with no plot boundaries. Firstly, towns and villages are recognised by zones bounded by polygons with numerous entries (A). Secondly, zones of large area with tightly undulating and circular boundaries are evident. These are grasslands, fallowlands on slopes or in some cases, forest blocks (B).
2. Groups of very small plots, most often striplike, densely packed one against the other (C). East of Vitteaux, these features are larger and more rectangular (D).

Fig. 5.4: Interpretation of plot compartments in the Vitteaux region.

3. Other plots are medium to large in size and more or less polygonal in shape (E). Grasslands of rounded shape are also distinguishable near the river (F).

■ Habitat

Another layer of information can be established from remote sensing data by identifying every point-type element of the image such as dwellings and constructions. For this theme, analysis of point density as well as their spatial distribution often provides useful information. Thus, various farm and residential constructions, sections of a town and diverse urban zones of an agglomeration can be distinguished (see the satellite images of towns on the CD). Areas of human impact on the environment such as development zones, drainage or irrigation zones, etc. are also discernible.

The pattern of dwellings in the global analysis can be coded in two ways:

—Coding the habitat units with values of 0 to 3:

0. No habitat,
1. Scattered habitat,
2. Grouped habitat,
3. Mixed (scattered and grouped) habitat.

—Coding each mode of habitat by 0 or 1. The value 0 in each of these three groups, viz., scattered, grouped and mixed, indicates absence of habitat.

It is also very useful to superpose this layer on the preceding one for analysing relationships between habitation or urban zones and roads, as well as between open lands and habitation zones. A new interpretation can thus be made by excluding topographic and land cover data. Not only are conventional relationships invariably confirmed, but a new method of detecting new specific relationships between the media under investigation is developed. One of the basic premises of visual interpretation is to be prepared to be astonished and to discover a priori unknown relationships or to accept "new" discovery of them.

It can be seen from Fig. 5.5 that habitation is strongly grouped in several suburbs and in Vitteaux proper. It should be noted that the railway track does not pass through Vitteaux but some habitations have developed between the city and its suburbs. Between Boussey and Saffres habitation is developed along the road. Some isolated hamlets exist on the plateau west of Vesvres and in the east between Saffres and Massingy.

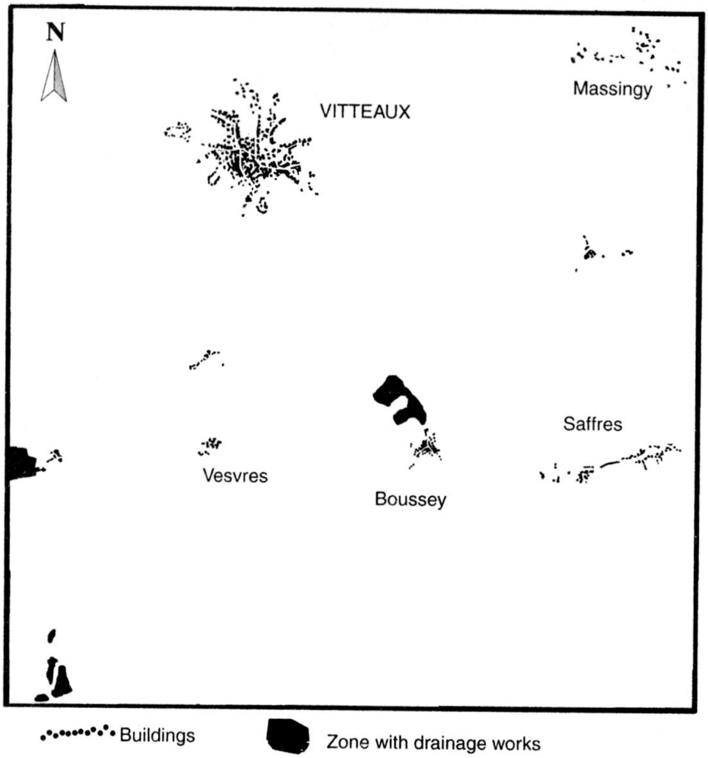

Fig. 5.5: Interpretation of construction zones (and drainage works) of the Vitteaux region.

Processing of all such information (scanning, for example) is facilitated by digitisation of interpretation and its incorporation in a geographic information system.

■ Hydrology

All information pertaining to water can be extracted from remote sensing images. In fact, most common forms associated with water are either punctuate (ponds, lakes etc.) or linear (rivers). Moreover, if the photograph is in the near infrared band, the contrast between the water body and all neighbouring objects is very sharp since reflectance of free water in this wavelength band is zero (see Chapter 4).

The hydrology of the Vitteaux region (Fig. 5.6) is mainly represented by the river Brenne, a very serpentine watercourse with numerous meanders. Several secondary watercourses with dendritic stream systems exist, some of which have only temporary flows.

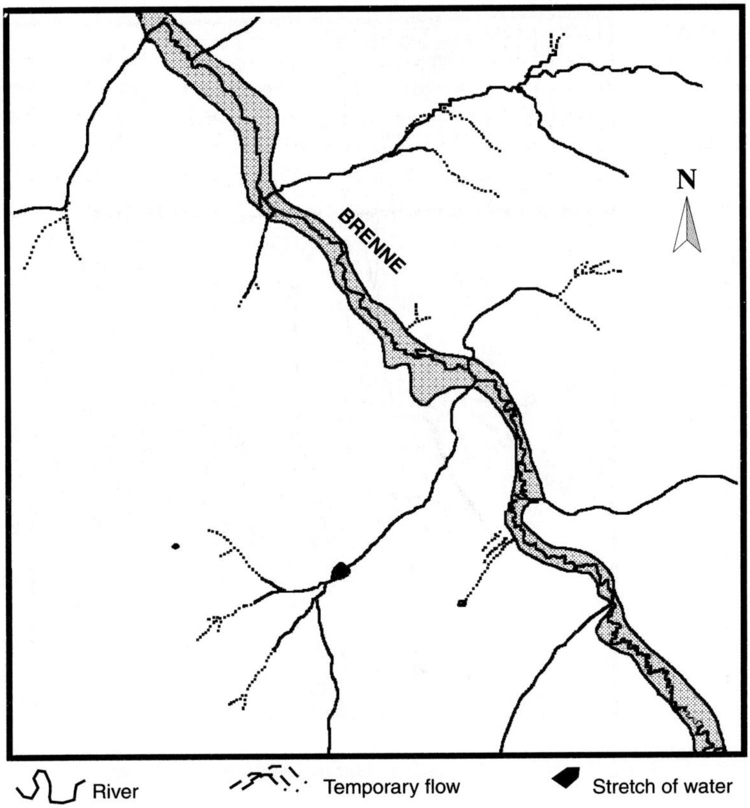

Fig. 5.6: Hydrological interpretation of images of the Vitteaux region.

In this case also, each preceding layer can be compared with the next. The existing hypothesis may be confirmed or new chorological laws may be defined. Integration of two interpreted layers of information for the habitat and hydrology constitutes a good reference for subsequent interpretations. In particular, the number, shape, spatial distribution of watercourses (permanent or temporary), provide valuable information about geology and geomorphology (Girard and Girard, 1970).

■ Geomorphology

Geomorphology may be represented by lines or partly closed contours. Geomorphological maps indicate various types of information and the conventional signs denote different morphological forms such as talwegs, alluvial cones etc., as in the given example. Slopes are represented by most expressive and most readily interpretable signs such as hatches drawn on the greatest slope and spaced at one-fourth their length. Colour presentation of maps enhances their legibility (Fig. 5.7).

5.2.2 Closed areas

Another phase of interpretation of a photograph or an image consists of delineating closed areas, map units[2], with which a ground feature is associated. Hence, it becomes necessary to define the

[2]Map unit: a graphic unit with a closed contour having a local semantic content and assumed homogeneous at a definite probability level.

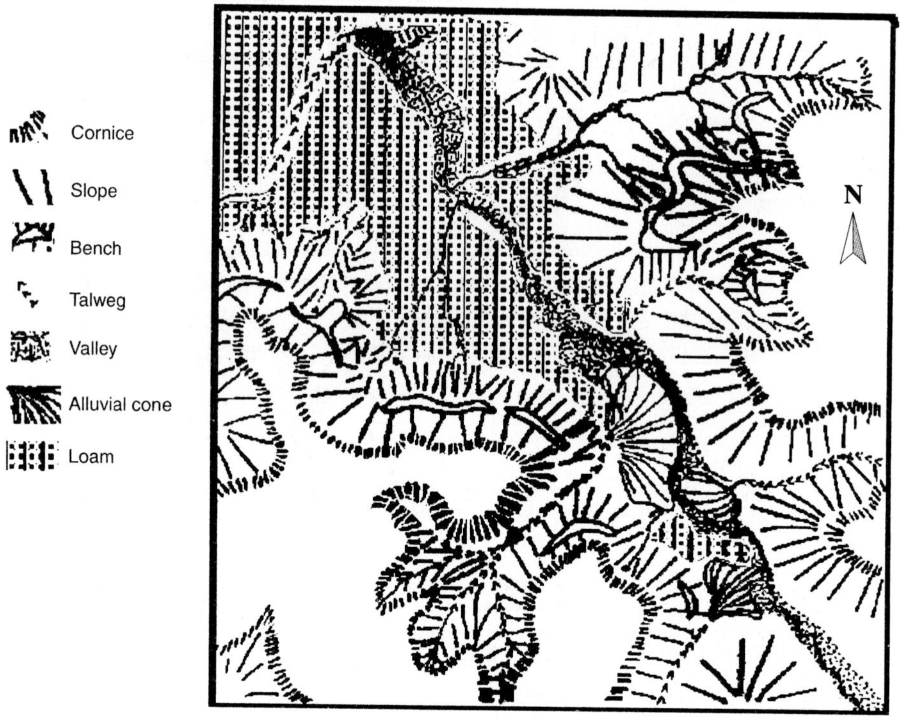

Fig. 5.7: Interpretation of morphology of Vitteaux environs.

objects we wish to recognise, either a priori or while drawing the map units. This phase differs from the preceding ones since closed areas are drawn and especially since thematic maps are prepared through iteration between graphic and semantic data. Drawing boundaries leads to defining an object, i.e., we go from the container to the content. On the other hand, the object permits defining such boundaries, i.e., we go from the content to the container. The capacities of the interpreter gradually increase and play an important role. The interpreter ought to come out of the confines of his specific ability and be capable of defining all semantic and spatial models that he uses.

■ Land cover

In the study of the natural medium, the first information layer often is the land cover map. In order to process this theme correctly, it is necessary to start with relationships that tend to be distinct: for example, that a land cover is delineated by the boundaries of a particular plot. As these boundaries would have been drawn previously, all such plots are identified as having the same land cover, even if a track or a road separates them (Fig. 5.8).

Depending on the importance of the theme and the experience of the interpreter, analysis of land cover data can be carried out in detail. For example, several types of forest cover can be delineated taking into consideration the tree species, their density and their height. In some cases, only one class such as 'forest vegetation' can be used but contrarily bare soils can be classified into many groups by their grey levels, hues or brightness. If it is desired to obtain a maximum number of classes, it is often convenient to initially differentiate tree and scrub vegetation, then herbaceous vegetation followed by analysis of soils, for example. Thus, various information layers are generated, which can be readily

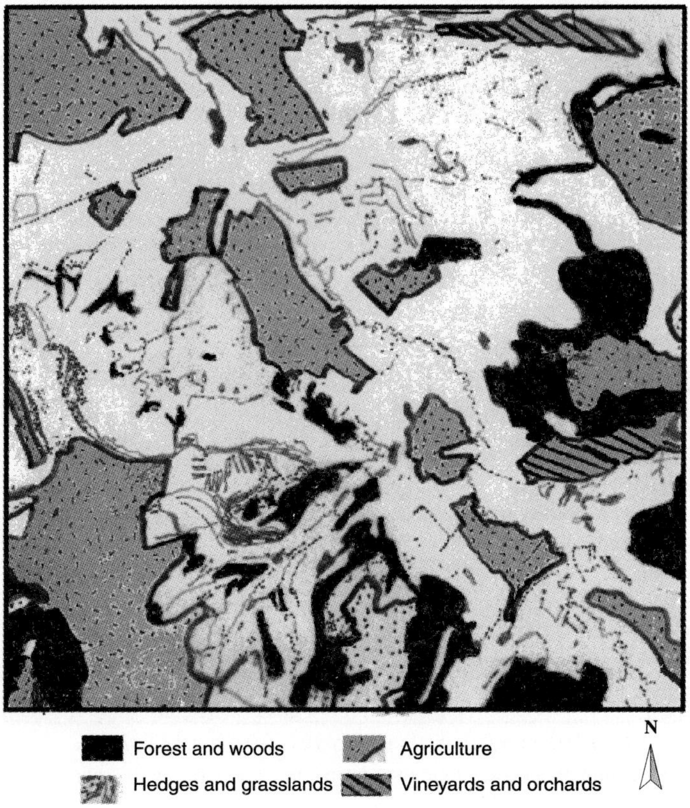

Fig. 5.8: Interpretation of land cover in Vitteaux environs.

compared with the functions of GIS. In fact, as the graphic base of all these documents is the same—either a photograph or an image—superposing them one over the other poses no problems.

■ Soil science and geology

Every subtheme can be compared with the preceding layers with the objective of better defining the chorological laws that enable construction of a reconnaissance map.

Such a procedure was followed in the example of Vitteaux with regard to soil mapping (Fig. 5.9). A geological map could also be prepared from the same data (Fig. 5.10).

5.2.3 Identification of thematic objects

Certain constituent elements of the natural medium, such as deep soil horizons, groundwater table, geological layers, etc., are not directly detectable on images or aerial photos (Fig. 5.10). In such cases, aerial photographs or images can be used differently from the manner described above. In this approach, we look for features that are detectable on aerial photos or images and which are directly related to the theme being investigated. Thematic objects are thus delineated using boundaries of image features visible in the remote-sensing data. This enables recognition of chorological laws that

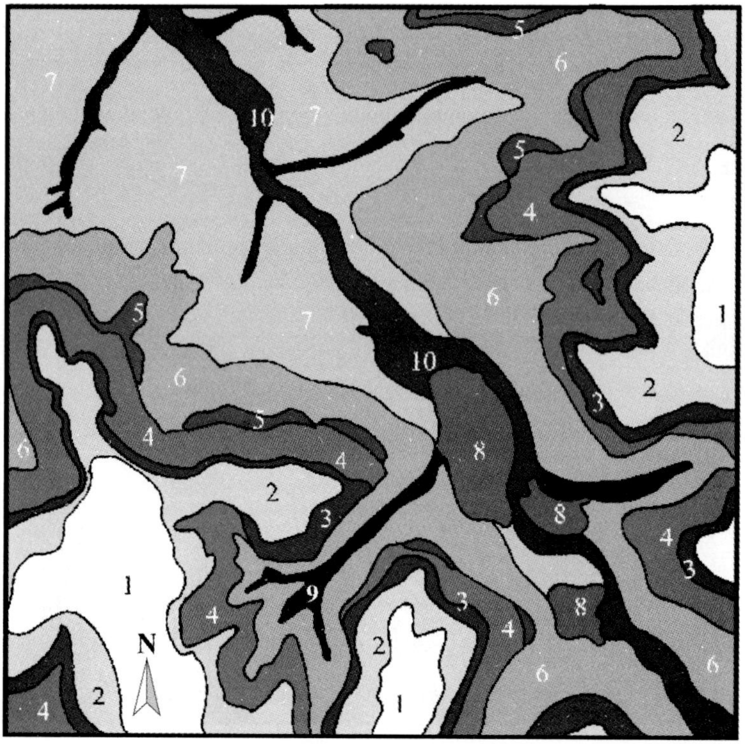

Fig. 5.9: Preparation of a reconnaissance soil type map from interpretation of aerial photographs of the Vitteaux region.
1. CALCISOLS, thick and stony; 2. LITHOSOLS, limestone, on slopes; 3. Humic CALCISOLS, very thin; 4. Humic RENDOSOLS, on marls; 5. RENDOSOLS on limestone, on benches; 6. Thick CALCISOLS on marls; 7. NEOLUVISOLS of loams on limestone; 8. BRUNISOLS with coarse material of alluvial cones; 9. COLLUVIOSOLS of secondary valleys; 10. Hydromorphic, loamy and carbonaceous FLUVIOSOLS.

prevailed in making the thematic map. This new information layer can subsequently be integrated with others because the characteristics of geographic projection are the same since they are derived from the same data.

Aerial photos and, to the extent possible, satellite photos are obviously analysed in stereoscopic vision.

5.2.4 Vinicultural "terroirs"

This term has been variously defined by different specialists (Vaudour, 1997). It can be briefly defined as a geographic space of limited size affected by interactions between the natural environment and human activity. Remote sensing facilitates detection of these spatial features from pertinent indicators related to agronomic potentialities. If the latter are integrated with the structures associated with human activity, such as community appurtenance and skill, mode of rural life and typical and specific productions," terroirs" can be defined by means of remote sensing.

Interpretation of aerial photos is often utilised for studying terroirs and demands a very fine geographic resolution and synoptic view.

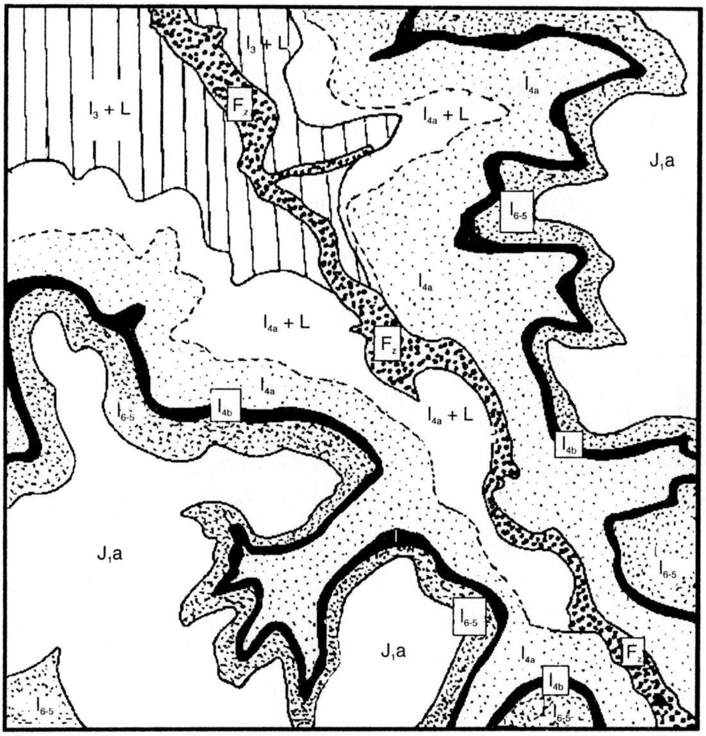

Fig. 5.10: Preparation of a geologic map of the Vitteaux region based on interpretation of aerial photographs; map 1:50,000.
J_{1a}: Hard limestones of the Lower Bajocian; l_{6-5}: Marls of the Upper Liassic; l_{4b}: Limestone with giant Gryphaea of the Middle Liassic; l_{4a}: Domerian marls; l_3: Sinemurian limestone; L: Loams; Fz: Recent alluvium.

The terroir mapping of Côtes-du-Rhône area in the Nyons-Valreas region can be cited as an example. This map of 21 soil-landscapes (Chapter 18) grouped into 5 morphological units enabled determination of 8 vinicultural terroirs (see CD 5.2). These terroirs were validated by data on grapes and were subsequently used for classification from the SPOT image based on the maximum likelihood method (see Chapter 9).

5.2.5 Conclusion

Analysing every phase and every theme is the best method for extracting all the information contained in an aerial photo or image. Comparisons are then made between various layers with a view to reconstructing a spatial model relative to a given approach to the natural environment.

Each zone can hence be described by a group of analytical variables defined earlier (Table 5.2). In the given example, a generalised description key is obtained for all map units. For each map unit, the following modes are indicated.

—either absence (0) or presence (1);

Table 5.2: Format for description of analysis

Format for description of aerial photos			
Communication	**Mode**	**Land cover**	**Mode**
Road		Agriculture	
Track		Vineyards and orchards	
Railway track		Grasslands	
Shape of plots		Woods, forests and isolated trees	
Strip-form		Conifers	
Rectangular		**Geomorphology**	
Draught-board		Loam	
Polygonal		Alluvium	
Curved		Flat-bottomed talweg	
Rounded		V-shaped talweg	
Plot compartments		Alluvial cone	
Circular		Cornice	
Parallel		Slope	
Radial		Bench	
Perpendicular		**Hydrology**	
Disseminated		Spring	
Arbitrary		Temporary flow	
Habitat		**River**	
Scattered		Free water	
Grouped			
Mixed			
Drained zones			

—relative abundance, which to be unaffected by variations generally possible from different interpreters and considering the duration required for interpretation, can be given by the following values:

0—absence,
1—0 to 5%,
2—5 to 35%,
3—35 to 65%,
4—65 to 95%,
5—95 to 100%.

This code has been tested for the last twenty years. It has undergone a slight change compared to the one proposed in 1975 (Girard and Girard, 1975). It proved to be quite sufficient in defining various landscape units by subsequent statistical analysis.

The GIS greatly facilitates comparisons. It is also possible to establish interpretation models and, using them, simultaneously analyse the results in spatial as well as semantic domains.

5.3 VISUAL INTERPRETATION OF SATELLITE IMAGES

5.3.1 Interpretation of photographic prints

In order to extract maximum information from these images, it is possible to enlarge them in such a way that the resolution of the image corresponds to the resolution of the eye. Thus, for a panchromatic SPOT image, the 10-m resolution is represented by 0.1 mm, which then corresponds to an image of scale 1:100,000. To facilitate drawing, a scale two to four times larger can be used, i.e., 1:25,000 for a resolution of 10 m, or 1:50,000 for a resolution of 20 m.

One of the main advantages of satellite images is a synoptic view that enables comparison of colours or grey levels from one end to the other of the scene, i.e., over thousands of square kilometres. On the other hand, common aerial photographs provide such comparisons for only a few tens of square kilometres. It is therefore necessary to preserve the entire spatial field. For a 10-m resolution image, such as that of SPOT in the scale 1:25,000, the entire field corresponds to a photograph of 2.4 × 2.4 m. This is not convenient to work with. Contrarily, interpretation becomes very easy with 60 cm × 60 cm photographs, which correspond to a scale of 1:50,000 for a quarter of the SPOT multispectral scene.

■ Points and lines

On a satellite image it is difficult to mark all the points or lines that often do not cover a pixel:
—either because their size is insufficient and they do not cover an entire pixel; in such a case, a composite response, a mixel is obtained (Chapter 15);
—or because they lack sufficient contrast with adjoining objects.

Such a situation is observed in the case of constructions and a large number of roads. The boundaries of plots appear only when the land cover feature has a strong contrast. Water bodies can be delineated if they are sufficiently large (more than four times the size of the pixel). Streams can be detected even if their width is of the order of the size of a pixel since most often a strong contrast exists with neighbouring zones; however, the relationship between pixels of the same stream cannot always be identified. Contrarily, they are readily confused with shadows and hence caution must be exercised.

■ Designing a legend

Interpretation can be carried out object by object in the manner described above and validity of interpretation rapidly verified. To prepare the legend for the entire field under investigation, interpretation in different zones is recommended so as not to forget the modalities in diverse variables. The legend has to be defined as interpretation progresses for all the thematic aspects of the zone—the earlier the better.

In order to ensure uniformity of processing for the entire visual field and good reproducibility between interpreters, when there are several, or between procedures followed by the same interpreter on different days, a description format is prepared (Table 18.1), which should be completely filled in for each map feature.

5.3.2 Interpretation on computer monitor

Interpretation can be carried out directly on monitors when an image-processing software package is available in the computer system. In this case, boundaries of various map units are directly drawn on the monitor. Two aspects need to be harmonised, viz., high precision of the pixel for tracing the boundaries and maintenance of a sufficiently large field of view to facilitate decision making. For this, it is necessary to have a largest possible monitor and instantaneous access to forward (enlargement) and backward (reduction) zoom. One has to verify whether this function is practicable and immediate in the software and whether it can be used directly with the mouse employed to trace the boundaries.

The major advantage of computer interpretation on a monitor for tracing boundaries lies in the ability to process:

—several images of the same scene,
—multiple channels of the same image,
—several colour combinations of various channels,
—images produced from diverse treatments and classifications, and
—images representing topography generated from digital terrain models (CD 3.1).

In fact, since tracing the boundaries is primarily dependent on contrasts, it would be easier when using these various combinations to choose the pixels through which a boundary is to be drawn. But then it is essential to understand the information included in the image displayed. This is not so simple, as will be shown later in the study of digital image processing.

Contrarily, regions corresponding to images acquired on different dates can be delineated and the temporal variation of objects thus analysed. This is based on the assumption that geometric corrections are applied (Chapter 13) so that the images can be superposed one over the other. Radiometric distortions would still remain and have to be corrected to the extent possible.

For visual interpretation on the screen, there should be a facility for instantaneous display of the output image and for repeatedly accessing it. This imposes constraints on the image processing software used. It is important to have a series of interpretations, all of which have the same projection and can be compared with one another. Any boundary of a map unit can then be readily modified when different themes are compared and when it is perceived that differences in delineation of some pixels, between different thematic features, raise the semantic significance above random variation.

All the thematic data thus interpreted are digitised in raster form. It is possible to directly process them in geographic information systems. They can also be represented as vector data (several image processing programs incorporate this function).

When a combination of an image processing software and GIS software is employed, it is necessary to georeference the images to be used. Then the boundaries can be traced directly in vector mode on the GIS, while retaining the required images at the bottom of the screen. The importance of this approach is that by the end of visual interpretation, the geometric base of the GIS is established and the topology of various map units traced from the images is automatically obtained.

5.3.3 Stereoscopic vision with satellite images

Stereoscopic vision can be acquired when two images covering the same ground objects at two different angles are available (Chapter 14). Spot images provide stereoscopic vision (see Chapter 2 and the CD SPOT system). In fact, by its capability to acquire images at various angles, it is possible to observe a region stereoscopically using images acquired at one or several days' interval.

The larger the difference between view angles, the better the perception of topography. It is hence recommended that images with a large positive angle for one and a large negative angle for the other of the stereoscopic pair be obtained. It should be remembered, however, that in this case distortions will be maximal and consequently interpretation on GIS becomes more difficult. It may therefore be recommended that, whenever possible, one image be acquired with maximum angle and the other with a smaller angle in another direction or with zero angle. In this case, the boundaries are traced on the second image since image distortions would be minimal and topographic information more accurately extracted.

The stereoscopic vision of a satellite image differs significantly from that of an aerial photograph since the relief is less accentuated and since the image is acquired through a more synoptic view. Consequently, it is more difficult, for example, to differentiate various terraces in large valleys or alluvial cones. Contrarily, level differences that extend over tens of kilometres can be readily traced and determination of large zones becomes easier since large features are distinctly identifiable and readily classified. Similarly, drainage networks are easy to interpret given their hierarchy and structure. Talwegs can be readily identified by analysing winter images in which they are observed below deciduous forests and on bare soils.

In some cases, images required for stereoscopic vision have been acquired at an interval of several months. It should be remembered that over time the solar angle changes and inevitably some shadows will silhouette the apparent relief. Consequently slopes of strongly incised valleys are wholly shadowed and little information about them ascertainable. Such a situation can lead to many errors in interpretation in as much as shadows and water can have very close radiometric characteristics.

5.4 CONCLUSION

Visual interpretation of satellite images has been a standard procedure for many years. One might have thought in the 1980s that it would displace expert interpretation of aerial photographs. As a matter of fact no such displacement has occurred. On the contrary, interpretation of aerial photographs has taken on greater importance for several reasons.

Given the lower resolution of satellite images, aerial photographs offer many advantages in answering questions that require high resolution, such as quantum of vehicles, persons and trees, accurate areal computation, land-use management, etc. Further, with the advent and ready availability of GPS, demands for accuracy are on the rise. Present-day Earth observation satellites do not yet acquire data accurate enough to meet these demands but that capability may not be far off.

On the other hand, since photographs can be digitised and processed digitally in the same manner as images, a revival of interest is seen. Further, the high resolution of aerial photos can be combined with the power of computer processing. A model of a transformed landscape can be developed by modifying one or several factors of the physical or human setting and display the new model in the same scenario. This enables visualisation of changes that might be effected through a directorial variation in territorial management.

Lastly, with the advent of GIS and digitised mapping data such as digital terrain models, which give altitudes and all the derived variables, viz., slopes, exposure, crests and talwegs, etc., it has become possible to integrate visual interpretations with the data obtained from computer processing. It is possible to process the information at multiple levels of resolution, although problems concerned with change of scale (Chapter 15) have yet to be satisfactorily solved.

Visual interpretation is obviously simplified since it can be directly carried out on a computer monitor:

—with integration of topography by digital processing if needed,

—with integration of aerial photos and satellite images,

—with the possibility of synthesising a multiband image providing resolution of aerial photographs.

Visual interpretation today is the fastest method for integrating structural information. This will perhaps change in the near future. However, to answer the present-day demand of professionals in a time frame compatible with their requirements, visual interpretation is often one of the components to be integrated into such an approach.

Introduction of the stringent analysis necessary for computerisation of data and their digital processing has led to profiling visual interpretation methods. Thus, formats for description of images and photographs have been designed and computerised, which have to be filled in systematically. Consequently, visual interpretation procedures are freed from the word 'approximately' so often used by experts. Laws of interpretation are now emerging from this need for absolute accuracy. It will hence become possible to formulate a set of chorological laws that will be usable at least in a given geographic region, and which will be re-usable for the same region at a subsequent date. It may even be estimated that when this set of laws is no longer applicable in one region, it may be used in another. Delineation of regions may be based on this fact. Similarly, it may be considered that there has been a temporal change in a region when a set of chorological laws is no longer applicable.

References

Agache R. 1970. Détection aérienne de vestiges protohistoriques gallo-romains et médiévaux. Bull. Soc. de Préhistoire Nord, 7: Musée d'Amiens.

Bertrand P. 1994. Élaboration d'une base de données localisées sur les agropaysages à partir d'images satellitaires. Application à l'étude des organisations spatiales et à la segmentation du département de l'Yonne. Mémoire de Mastère 'Système d'informations localisées pour l'aménagement des territoires'. Institut national agronomique, Paris-Grignon, 46 pp.

Chevallier R. 1971. La photographie aérienne. Armand Colin, 227 pp.

Chevallier R. 1997. Les voies romaines. Picard, Paris, 343 pp.

CRU. 1969. Photographie aérienne et urbanisme. Centre de recherche d'urbanisme, Paris.

Girard CM. 1995. Persistance de terroirs circulaires dans le Gâtinais occidental et relations avec les îlots boisés. Photointerprétation. 4/95.

Girard M-C, Girard C-M. 1970. Cours de photo-interprétation. Polycopié, INA PG, Grignon, 208 pp.

Girard C-M, Girard M-C. 1975, Applications de la télédétection à l'étude de la biosphère. Masson, Paris, 186 pp.

Girard M-C, Girard C-M. 1989. Télédétection appliquée. Zones tempérées et intertropicales. Masson, Paris, 260 pp.

Girard M-C, Girard C-M, Bertrand P, Orth D, Gilliot J-M. 1986. Analyse de la structure des paysages ruraux par télédétection. C.R. Acad. Agri. Fr., **82** (4): 11–25.

Guy M. 1969. La photo-interprétation. Encyclopedia Universalis, Paris.

Lillesand TM, Kiefer RW. 1994. Remote Sensing and Image Interpretation. John Wiley & Sons, Inc., NY, 3rd ed., 750 pp.

Mollet S. 1994, Élaboration d'une base de données des agropaysages du département de l'Yonne—application à l'étude des dynamiques financières agricoles. Mémoire Diplôme d'Agronomie approfondie, Institut national agronomique, Paris-Grignon, 55 pp.

Mulders MA. 1987. Remote sensing in soil science. Elsevier, Amsterdam, 379 pp.

Smith JT Jr. 1968. Manual of color Aerial Photography. Amer. Soc. Photogrammetry.

Soyer J. 1970. La conservation de la forme circulaire dans le parcellaire français, Mémoire de photo-interprétation de L'EPHE, vol. VI. SEVPEN, Paris.

Vaudour E. 1997, Analyse spatiale et caractérisation des terroirs du bassin viticole de Noyons-Valréas (AOC Côtes-du-Rhône). Memoire de DEA, Institut national agronomique, Paris-Grignon, 34 pp.

6

Image Processing—General Features

6.1 INTRODUCTION

The following chapters are devoted to description of various methods of image processing. In this book, special attention is given to processing satellite images or digitised aerial photographs. However, most of these methods are applicable to any digitised data such as photographs obtained from as different domains as archaeology (site detection as well as pottery analysis), soil science (micromorphology), medicine, artificial vision in robotics, industrial zone surveillance, etc.

After a presentation of generalities (Chapter 6), the following topics are covered in successive chapters: preliminary image processing of single or multichannel data (Chapter 7), unsupervised classification (Chapter 8) and supervised classification (Chapter 9). This part is concluded by methodology of image processing in remote sensing (Chapter 10); bibliographic references for these five chapters are listed in the last one.

The next chapters include structural processing of images (Chapter 11) by the software OASIS, digital filtering of images (Chapter 12) and, lastly, geometric transformation of images, which is necessary for superposition of various sources of geographic information (Chapter 13).

Names of the methods may vary according to different authors since they are mostly derived from older mathematical methods, revived for application to image processing. The description of these methods is not the same in general works on image processing, but we have used the most common terms in the various advanced software programs in remote sensing.

Image processing methods can be broadly grouped under six principal types: 1) measurement-space-guided spatial clustering; 2) single-linkage region growing; 3) variant of the preceding which takes into consideration not only the value of the pixel, but also its neighbourhood and is known as hybrid-linkage region growing; 4) spatial clustering; 5) centroid-linkage region growing; and 6) split and merge methods which use contour reduction.

Attempting to organise these different methods into rigid groups is of little use for applied remote sensing. Greater importance should be given instead to the possibilities and limitations of each type of processing. However, it is always a combination of various methods that leads from the information available at the beginning of the study to the best possible qualitative response to the problem posed by the user.

As it is often difficult to find examples of application of the various methods, we have applied all of them to a single image. Since colour figures cannot adequately express all the details of processing, all the results have been given on a CD. The simple programs of the TeraVue software (La Boyère publishers) on the CD facilitate visualisation of the numerous images presented and their partial modification. To derive maximum benefit from the text access to a PC with a CD-ROM drive is necessary. It is hoped that a large number of readers interested in satellite image processing have this facility. References to the CD are indicated in parentheses in the text with a number following the letters CD.

An image of Earth observation satellites can be used in several ways. It can serve as a basic plan that suffices to mark the objects constituting the references needed for the theme under study. Most often the reference points are villages, roads, rivers, forests, seashores, agricultural lands, etc. These markers are not visible on the image in the same manner. Their visibility depends on the quality of the image, date of acquisition of the scene, etc. Even for such basic usage, it is necessary to carry out a preliminary visual interpretation or image processing to obtain a readily understandable document.

Use of satellite images through visual interpretation (see Chapter 5) always requires a minimum amount of processing such as pre-processing of the signal (see Chapter 2) and processing abutting in colour composites. Sometimes this stage of processing is more elaborate and visual interpretation is carried out on colour composites priorly subjected to digital classification.

Use of satellite images almost invariably necessitates statistical processing of data by means of information technology.

6.2 IMAGE PROCESSING METHODS

Before undertaking processing of images, a judicious choice of the possible methods to be employed should be made. Thus two different types of methods can be identified:

—Point analysis is based on the laws of spectral behaviour of objects, for which one should have at least an interpretation model of objects (see Fig. 7.4). A better solution would be to conduct field measurements of reflectance and take into consideration the most possible transformations of the signal by the atmosphere as well as due to satellite conditions of acquisition. This group of analyses is known as the *radiometric method.*

—Spatial analysis is based on characterisation of objects by their geographic position and by their relative spatial position. Identification of objects based on field studies and in association with already acquired information (mainly from existing maps) constitutes the basis of processing. This type of analysis is called the *geographic method.*

Generally both methods are used depending on the phase of processing, objectives and data available in addition to the satellite images (see Fig. 10.1).

In both cases the results obtained ought to be verified by comparing them with data other than that of remote sensing. It is hence necessary to identify the method of validation to be employed in order to choose the type of processing that would be coherent.

Several approaches exist for image analysis, depending on whether one is a mathematician, computer scientist, statistician or thematic expert. These different approaches are complementary and become enriched during analysis of satellite images. However, since each has its own specific language, it is necessary to precisely define the latter when working with different scientific disciplines, all the more so because the same terms may have different significance in different disciplines. For this reason the terms and methods essential in image processing are described below.

6.2.1 Texture

Texture of an image is defined as a combination of textural elements. A textural element is a group of resolution elements (whose area is determined by the characteristics of the sensor) which have the same value of radiance (or of a function of radiance) and which are connected (spatial dimension).

A textural element in the case of a raw satellite image is a group of connected pixels having the same digital value. For a classified image, it may also represent a group of pixels belonging to a map unit and hence ranked in the same class. Texture is the characterisation of the entirety of these units.

One of the expressions of texture can be represented as a diagram on a Cartesian system of co-ordinates, with the perimeter of the map unit as abscissa and its area as ordinate (Fig. 11.4). Analysis of such a graph facilitates characterisation of shapes of various zones. Characterisation of shapes of textural elements hence forms a part of texture.

6.2.2 Structure

Structure of an image is a combination of structural elements. A structural element is defined by repeated relationships existing between the textural elements.

These relationships may be, for example, relative positions between the textural elements of the same shape or of different shapes, distances between textural elements, contrasts, etc. Obviously, characterisation of textural elements by means of chorological laws, or spatial laws such as convergence of shape, forms an integral part of structure.

From a didactic point of view, image processing in remote sensing can be classified depending on whether it is based on structural or textural data. However, most of the programs presently available in image processing software pertain to textural analysis. Hence we start with the description of textural processing methods in the next chapters, followed by structural analysis in Chapter 11, devoted to VOISIN and OASIS.

6.3 CLASSIFICATION

6.3.1 Multiple languages

Image processing involves several aspects. It is based on mathematical and statistical methods, often quite old. However, their development could not have been possible without computers since they demand very long and tedious computations. With the advent of remote sensing data, these methods progressed as large amounts of data requiring use of these tools became available. Hence a processing language has developed based on mathematical, statistical and computer languages.

Thematic experts obviously use image processing every day since it constitutes a statistical extension of the visual interpretation they used to carry out on non-digital images and still carry out on digital images. The presently existing image processing techniques are not capable of replacing interpretations by the brain. The language of image processing is impregnated with that of interpreters of various thematic fields. This language may differ according to whether the theme of study is closer to a scientific or literary approach, or to the theme of a geologist, a soil scientist, a botanist, an agricultural scientist or a geographer, and so forth. One of the most patent examples is the definition of 'structure' and 'texture', terms which according to different thematic experts or mathematicians acquire opposite meanings (see Glossary).

6.3.2 Segmentation and classification

If the approach of mathematicians working on data processing is followed, segmentation has to be differentiated from classification.

Segmentation is the action of dividing an image into classes without knowing what these classes thematically represent. This most often corresponds to unsupervised classification in remote sensing data processing.

Classification is the process of extrapolation to the entire image using the previously chosen test zones or nuclei for which the thematic significance and relationship with an object (or a group of objects) have been established. This generally corresponds to supervised classification in remote sensing data processing.

6.3.3 Classification and grading

The terms 'classification' and 'grading' need to be distinctly differentiated.

Classification consists of determining the laws (most often statistical, arithmetical or logical in remote sensing) which lead to the choice of classes or categories to be retained and to the mode of grading. Classification may or may not be hierarchic, supervised or unsupervised.

Grading consists of distributing the pixels of an image into groups or categories pre-established by classification. Similarities exist within a category or class and the characteristics are common to all the pixels included in it.

However, in the common parlance of image processing in remote sensing (see the discussion of various programs), the term classification is generally used without differentiating between the two.

To avoid any confusion about the so-called automatic classification, it should be remembered that an external intervention in processing always exists. In a number of cases, a rather too fast use of a processing method may give the impression that the thematic specialist is not involved. This is never the case. If the thematic specialist is not involved and wishes a result to be displayed on the monitor, it means that he/she accepted *by default* the options inserted by the author of the software in the algorithm. Some examples of the various types of external intervention in processing (discussed later) are:

—division of a dendrogram and the method used for it in a hierarchical ascendant classification,

—number of classes retained in a hierarchical ascendant classification or in the classification by mobile centres,

—choice of number of iterations in the classification by mobile centres,

—choice of the probability threshold in the maximum-likelihood classification,

—choice of nuclei in classification such as maximum-likelihood or structural classification, viz., OASIS,

—choice of the window size in structural classification (OASIS or VOISIN methods).

'Nucleus' is a group of pixels chosen a priori to characterise an object for which a classification is to be designed. A *thematic group is the population of pixels which as a result of classification and hence a posteriori, is combined into a single group.* At the end of classification, a thematic group *carries the same name as that of the nucleus used for classification.*

6.3.4 Ascendant and descendant methods

Descendant methods are based on selection of geographic or radiometric criteria for dividing an image into various parts. This division is then iterated for each of the preceding parts with new criteria that differ from the previous. The procedure *starts* with the entire *image* and its division into parts that gradually become increasingly smaller and may ultimately reduce to the size of a *pixel.* The main question in this regard would be: what is the basis for the criteria for division in every iteration and are these criteria coherent with one another? It is necessary to verify the stability of the entire division operation when the criteria chosen change slightly. One has also to look into the question of heterogeneous pixels, viz., mixels, that encompass several objects (Fig. 15.3).

These methods quite often correspond to visual interpretation of a synthetic nature, which is based on structural analysis. They are also used when the objectives of classification are well defined and when only geographic distribution of objects for which digital values are clearly established is analysed. These methods are little developed for satellite image processing.

Ascendant methods consist of aggregating a group of pixels that exhibit pre-defined characteristics. The latter may refer to their radiometric properties or relative geographic positions. These are generally the same criteria as used in successive iterations that *finally* lead to grouping all the pixels into a single category. Ascendant methods hence *start* with a *pixel* and arrive at the entire *image* through successive additions. The question to be considered here would be: on what basis are the criteria of grouping (distance, probability, etc.) defined for a given iteration (metric) and between different iterations (ultrametric). The validity of the resultant groupings needs to be verified through assessment of the quality of classification by means of the distance to a given reference point or probability of appurtenance to a given category (see below).

These methods are most commonly employed in textural processing of satellite images.

6.3.5 Concept of mathematical distance

The statistical methods of classification used in image processing are mainly based on the concepts of probability and mathematical distance.

The concept of probability is readily understood. The greater the probability of a pixel joining a group, the better its grading if it is included in this group. This approach is used in maximum-likelihood classification.

The concept of mathematical distance is less common but equally easy to interpret. The greater the distance between a pixel and a group, the less likely its classification in this group.

Mathematical distance is computed in a multivariable space, i.e., multiband and/or multidate. Several procedures exist for computing mathematical distance between two objects. An example is useful in understanding computation of mathematical distance.

On the image of Brienne (CD 6.1) three points were identified (Table 6.1), one characteristic of dark bare soils (S), the second a mixture (a mixel, M) of water and soil and the third characteristic of chlorophyllian vegetation (V).

Table 6.1: Mathematical distances of three pixels

Object	Row	Column	*b*1	*b*2	*b*3	Distance between objects	Euclidean distance	Manhattan distance
Soil (S)	796	847	75	56	66	S and V	43	61
Water and soil (M)	779	838	91	84	86	S and M	40	64
Vegetation (V)	795	839	66	46	106	M and V	50	83

The mathematical distance *D* between pixels S and V is designated as D_{SV}. For each channel (b_i), a difference between the digital values of each pixel S and V is computed: $[S_{bi} - V_{bi}]$.

The distance is equal to the sum of these differences and is given by:

$$D_{SV} = [S_{b1} - V_{b1}] + [S_{b2} - V_{b2}] + [S_{b3} - V_{b3}]$$

Different types of distances are defined. The χ^2-distance is related to frequencies and Mahalanobis distance is related to probabilities. Two most commonly used distances are the Euclidean and Manhattan.

Since in principle the distance V to S is equal to the distance S to V, each difference $[S_{bi} - V_{bi}]$ should be equal to $[V_{bi} - S_{bi}]$. Thus, the following two distances are defined:

—Euclidean distance is obtained as the square root of the sum of squares of all the differences:

$$D_{SV} = [(S_{b1} - V_{b1})^2] + (S_{b2} - V_{b2})^2 + (S_{b3} - V_{b3})^2]^{1/2}$$

—Manhattan distance (or *L*1 or block-city distance) takes the absolute value of each difference:

$$D_{SV} = |S_{b1} - V_{b1}| + |S_{b2} - V_{b2}| + |S_{b3} - V_{b3}|$$

The computed distances for the three points mentioned above are given in Table 6.1. The Manhattan distance is greater between soil and water-soil mixel (64) than between soil and vegetation (61). An inverse situation is observed in the case of the Euclidean distance: the latter is smaller between soil and water-soil mixel (40) than between soil and vegetation (43). This is due to the fact that the differences channel by channel between the soil and the mixel are of the same order of magnitude (16, 28 and 20); consequently squares of these differences are not very high: 256 to 784. Contrarily, between soil and vegetation, the channel-to-channel differences are high (11, 10 and 40); hence squares of these differences are much higher—from 100 to 1600. Thus, the corresponding distance is greater.

Euclidean distance is much more sensitive to a large difference for a channel compared to the Manhattan distance. Consequently Euclidean distance is greater between two pixels that have a single large difference for a channel than in the case of two pixels that have smaller differences over several channels.

The choice of distance to be used, termed *metric* distance, is very wide. However, this choice is much reduced in the satellite image processing programs available and so unfortunately the user rarely has much choice. Software designers use a particular metric distance. It is better to know which one. Since at present satellite data are mostly coded in 8 bits, it is considered that quantitative information is processed and Euclidean distance used as default. Some program propose several distances but then it is necessary to search in specialised books for the incidence use of a given distance for classification.

6.3.6 Utilisation of distances for classification

Distances provide a simple way of expressing the *postulate* on which the classification is based. A classification is a grouping of pixels of a region in such a way that they are at the closest possible distance from one another within a group and at the farthest possible distance from one another in different groups.

Distances constitute the basis of classification. *Intragroup distance* is defined as the distance characterising not two pixels, but a group of pixels. *Intergroup distance* characterises the separation between two groups. The rules of utilisation of these two distances for classification and their computation correspond to *ultrametric* criteria.

■ Intragroup distance

Intragroup distance can be determined by computing the weighted average of the distances of all the pixels belonging to a group channel by channel. If there are *G* pixels in a group, there exist $G(G-1)/2$ distances to be computed. It is also possible to choose the intragroup distance as the longest distance between pixels of the same group.

The longer the intragroup distance, the more scattered is the group, the smaller the intragroup distance the more compact is the group.

A mean value for the group for each channel can also be defined, which determines the spectral characteristic of a 'mean pixel'. However, it should be remembered that this is an approximation, since all values in image processing ought to be integers in order to be displayed on the monitor, which is rarely the case at the end of computation of a mean or median. Then the standard deviation or the median deviation is computed. However, we may also compute the sum of distances of the pixels of the group relative to the central value (median, mode or mean).

■ Intergroup distance

Intergroup distance expresses the differentiation between groups retained in a phase or at the end of classification.

It can be determined by computing the mean of all the distances existing between the pixels of the two groups under consideration, taking 2 × 2 pixels at a time. Between two groups containing G and K pixels respectively, there would be ($G \times K$) distances to be computed. Another central tendency such as the median can also be used. The minimum distance between G and K is also important since it poses the greatest risk that the two groups are not different but a singular.

A large intergroup distance indicates that the two groups are totally different or well separated. Contrarily, a small intergroup distance indicates that the two groups are not separate and that they may be ultimately combined at a higher hierarchical level.

Another solution consists of defining a statistical population for each of the two groups G and K and investigating by conventional statistical tests whether the two populations (assumed Gaussian) are similar or not and with what probability. For this, generalised variance analysis can be done. One can also compute the distance between two fictitious individuals (mean, mode or median) representing the two groups.

■ Ultrametric

The ultrametric parameter corresponds to the choice made for comparing a pixel with a group and a group with another group. This is essential for classification. In fact, when two pixels are compared with each other, it is considered that they have a zero intrapixel distance. But this is no longer true if groups are involved. Thus replacing a group of G pixels with a fictitious pixel (mean, median, etc.) introduces distortions upon comparison with another pixel if its intragroup distance is not taken into consideration during decision making. The same would hold true in a comparison of two entire groups.

Decision rules for determining the group to which a pixel is to be attached are relatively simple. First, it is necessary to determine the distance between the pixel and the group.

The distance between a pixel and a group is determined in most cases by computing the distance from a pixel to the fictitious pixel representing the group. In this case we do not consider the fact that a single pixel (for which the intradistance by definition is zero) is compared with a group of pixels (whose intragroup distance is rarely zero). The same procedure is followed for all groups.

The decision rule for attaching a pixel to a group is as follows. The group for which the pixel-to-group distance is the smallest is identified and thereby the group to which the pixel is to be attached is determined. This is the method of minimum sorting distance: DIMITRI (Girard, 1983).

Other more accurate methods exist for determining the distance of a pixel to a group G; they are based on intragroup distances (Van Den Driessche, 1965). For this, the distance of the pixel to each of the pixels in a group is computed. The mean of these distances is used as the criterion for defining the position of the pixel relative to the group. The same procedure is followed for all the groups and finally the pixel is allotted to the group for which the distance is the shortest. These methods are rarely used in remote sensing.

Several possibilities exist for defining the decision rules for combining two groups.

Intergroup distances can be considered as the distances between pixels and the groups showing minimum distances can be combined. If the intragroup distances, far from negligible at the end of classification, are not taken into consideration, the errors would obviously be large.

Groups can be combined on the basis of mean intergroup distances. This is the common practice.

It is possible to take the smallest of the distances existing between two pixels of two different groups. The pixels aggregate one with another . Consequently, if several distinct groups are formed, it is certain that they are well separated since the ultrametric parameter does not favour separation.

It is possible to take the longest of the distances existing between two pixels of two different groups. This expresses the condition to establish connections only when the pixels of the two groups are very close to one another; thus, formation of numerous groups is favoured. Since the ultrametric criterion favours separation in this case, it is necessary to verify whether they are truly different at the thematic level.

■ Quality of results

Upon completion of classification, the quality of results must be verified. When classification is based on mathematical distances, it follows that each pixel is placed in a group. As the distance of each pixel to the group can be determined, a display showing these distances on the monitor provides information on the quality of the results (CD 11.4). The smaller the distance of the pixel to the group, the better it is classified.

Depending on the assumption made at the beginning of classification, if the intragroup distance is smaller, the group is more likely to be made up of pixels having close spectral characteristics (Fig. 6.1, groups A and D). A long intragroup distance (Fig. 6.1, group C) indicates that the group may be subdivided.

The greater the intergroup distances, the better separated they are (Fig. 6.1, A and B, A and C, B and C). A smaller intergroup distance shows that the two groups can be considered as a single group (Fig. 6.1, B and D).

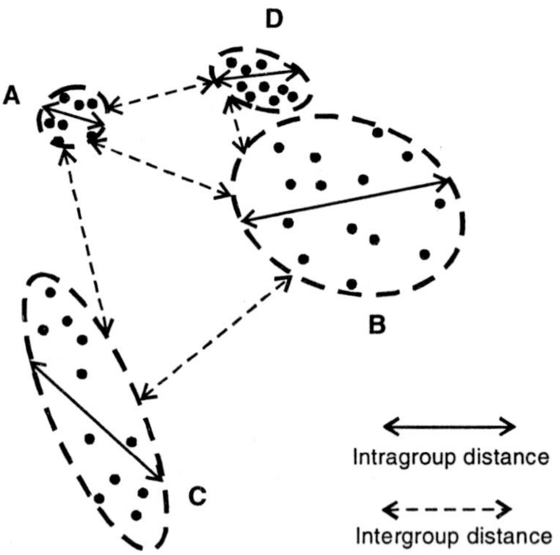

Fig. 6.1: Relative positions of groups as a function of their intragroup and intergroup distances.

6.4 CONCLUSION

In commercial softwares, classifications are often imposed with little choice. This is not a serious limitation if their merits and demerits are known. Some program, much more expensive, enable choice of a specific method of classification. Clear understanding of the fundamentals of various classification techniques is necessary for using them effectively, however.

7

Preliminary Processing

Before undertaking any sophisticated processing of images, it is necessary to fully understand the data contained in each band. For this, the statistical composition of the image has to be analysed to generate contrast stretched and colour images.

In the case of multispectral data, after processing each band as mentioned previously, various arithmetic treatments are carried out and the data of different bands are combined. Principal component analysis and masking are often useful before commencing classification.

7.1 PROCESSING OF SINGLE-BAND DATA

The following treatments may be carried out on digital data of a single band, which can be:

—a panchromatic or a multispectral band of a satellite image,
—the result of a processing such as an axis of PCA,
—the result of a preceding classification, or
—a combination of multispectral data, such as a vegetation index, brightness, etc.

7.1.1 Histogram analysis

At present, satellite image data are coded in 8 bits, i.e., in 256 levels. Various data processing programs can be used to obtain histograms of the digital numbers of a band. Histogram analysis is important in several respects and should be carried out before any processing.

Firstly, when an image is displayed on the monitor, it is possible that nothing can be inferred, and in such cases the histogram aids in identifying the information contained in the image. If no information exists, all the values are 0 (black) or 255 (white) or there may be no value at all. On the other hand, it is possible that all the pixels on an image are dark and the eye cannot distinguish them. For example, such a situation occurs in the case of the result of a classification into a small number of groups, say 10 to 30, which are displayed with intensity values between 10 and 30. Moreover, hues (tints) separated by a level difference of 1/256 cannot be visually differentiated, especially in the range of low values (Chapter 3). It may hence be thought that the image comprises no different classes, since the eye cannot discern them, when de facto it does.

An histogram is primarily characterised by its shape. Rarely is the distribution of the digital numbers of an image Gaussian. This can be verified by standard statistical tests but these may not be directly available in a single program. It is hence necessary to verify that the available image processing software allows storage of all the histogram-related information in a file whose format is compatible with the formats needed for conventional statistical programs.

In the case of satellite images, Gaussian type histograms may be observed occasionally. However, most often the distributions (Fig. 7.1) are bimodal or trimodal (b1 band) or of two separate populations

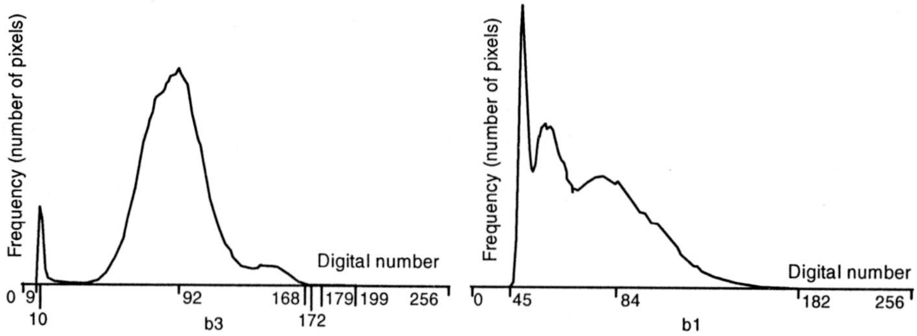

Statistical data of histograms:

Band	Minimum value	Maximum value	Mean	Standard deviation
b3 (Infrared)	9	199	91.68	24.95
b1 (Green)	42	254	83.87	24.53

Fig. 7.1: Histograms of bands b3 (infrared) and b1 (green) of the image of Brienne region (also see Table 7.1).

(b3 band). In most cases, histograms of unprocessed (raw) data are highly skewed to the right, with very low frequencies of large digital numbers.

The exact significance of the histograms has to be studied. Some programs give the values of mean, median, standard deviation, etc. along with the histograms (Fig. 7.1). These values should be interpreted with every precaution, especially when the histograms are not Gaussian. In any case, these values, which correspond to estimates, are not very important since exhaustive information is available in a satellite image.

Radiometric analysis provides a definite method for interpretation of the digital numbers of an histogram.

■ Radiometric method

Models can be used to convert the digital numbers into reflectance values taking into consideration the atmospheric transformations of the signal (Deschamps et al., 1984) (see Chapter 1). Models such as LOWTRAN (LOW Resolution TRANsmittivity program) are included in some softwares such as TeraVue. Using these models, the digital numbers can be evaluated on the basis of information about the spectral characteristics of the objects, viz., soil, vegetation, water, snow, roads, etc. For this, among others, co-ordinates of the satellite scene, date and time of acquisition need to be known.

It is also possible to calibrate the digital numbers using the reflectance values measured in the field. Presumably field measurements are obtained on the same day as acquisition of the satellite image or at least on a date very close to the latter so that the characteristics of the objects under investigation do not change.

■ Geographic method

In some cases, locations of an object clearly known on the ground can be readily identified on the image. By ascribing a given colour to all occurrences of this object, the spectrum of digital numbers characterising a geographically defined object can be established (CD 7.1). For example, water bodies are identified in the southern part of the infrared colour image (band b3) of Brienne region. The digital numbers for water vary from 10 to 19 and include the values 22, 25 and 33. Thus an object whose geographic position is known can be characterised in terms of radiometric values but not by reflectance.

It is then possible to thematically interpret the digital numbers of the histogram of each band. This process is essential for optimal utilisation of the various software tools of the image processing technique.

Visualisation of an image on the monitor passes from the raw image through two phases, viz., a function of transformation of raw digital numbers and Look Up Table (LUT). The pixel values are not changed; rather their visualisation on the monitor is modified depending on the function applied.

7.1.2 Transformations of digital numbers

■ Dynamic range enhancement: thresholds

One of the primary methods of transformation of an image is enhancement of the dynamic range. The raw data of an image hardly ever covers the entire range of 256 values available for display. It is hence recommended that this interval be utilised to the maximum extent possible.

One method of image enhancement involves stretching the dynamic range of the digital numbers by giving 0 to the minimum value and 255 to the maximum value, i.e., by using the *real maximum interval*. Thus we obtain an image that exploits the entire capacity of the monitor. This procedure is very simple and conserves all the information of the image; however, dynamic range of the image on the monitor is not always improved (CD 7.2).

The second method is based on determining the limits of statistical population of the digital numbers by leaving on the left (minimum values) and on the right (maximum values) 1 per 10,000 values: 1-per-10,000 integral (Table 7.1 and CD 7.2). A reduced interval of digital numbers is thus obtained, which enhances the dynamic range of the image on the monitor. Consequently pixels with digital numbers less than 1 per 10,000 (45 for b1) or more than 9999 per 10,000 (214 for b1) are perceived as black or white respectively. If this transformation is used for subsequent processing, these pixels will no longer be differentiated. It is therefore necessary to ensure that important objects to be classified are not in these intervals of digital numbers. As in the preceding case, the minimum and maximum values are changed to 0 and 255. This procedure is readily applicable when a function that gives the integral of pixels below a chosen threshold is available. One can also choose the 1 per 1000 integral instead of 1 per 10,000; however, most often the objects corresponding to the edges of the histogram, indicated as black or white, become visible on the monitor and may ultimately disturb the processing. Lastly, it may be difficult to differentiate between water and shadows, both of which appear as black since their digital numbers are small.

The third method consists of fixing the thresholds from the first or the last value of a digital number for which the pixel frequency is greater than or equal to 1 per 10,000 (1 per 10,000 class)

Table 7.1: Various threshold values (in digital numbers) for dynamic range enhancement of an image. Image of Brienne

Band	Real maximum interval	1 per 10,000 integral	1 per 1000 integral	1 per 10,000 class	Mean	Standard deviation
b1 (green)	42 to 254	45 to 214	49 to 184	45 to 182	83.87	24.53
b2 (red)	22 to 254	23 to 203	25 to 176	23 to 175	68.88	31.10
b3 (infrared)	9 to 199	11 to 179	11 to 168	10 to 172	91.68	24.95

(Table 7.1, CD 7.3). This is the simplest solution and was used for processing the image of Brienne. Precautions to be taken are the same as in the previous case.

Once the limits are chosen, it is possible to apply several transformation functions to the digital numbers.

■ Transformation functions

The objective of a transformation function is to obtain a majority of pixels in the grey levels that can be readily detected by the eye, for example between the digital numbers 50 and 200. It is hence possible to choose a transformation function such that most pixels of the band under study lie between these values.

The simplest transformation function is the *linear function*. After fixing the two limits of threshold (Table 7.1), the raw signal is transformed linearly to present the image on the monitor (Fig. 7.2a and CD 7.3).

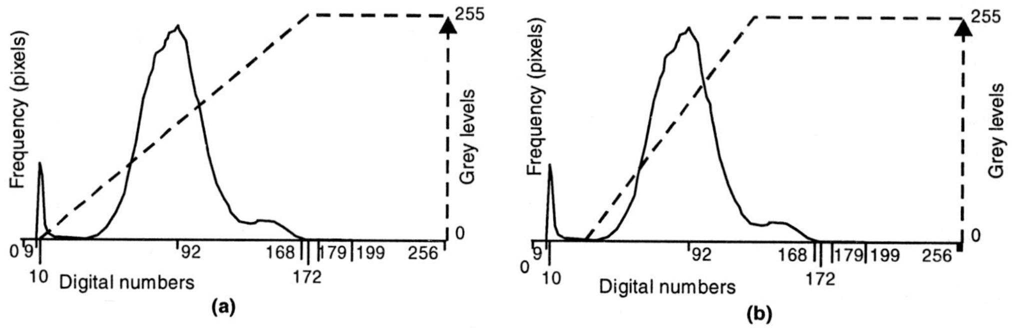

Fig. 7.2: Linear transformation of band b3 (image of Brienne). a. 1-per-10,000 class; b. suppression of values corresponding to water and cultivated plots.

This method has the advantage of maintaining the relative positions of objects in the range of digital numbers. The dynamic range of the image is enhanced but not necessarily optimally. It can be used to make only a portion of the image visible. This portion corresponds to the interval of digital numbers chosen for defining the linear range of transformation (Fig. 7.2b).

The *square root* of the transformation function gives a satisfactory contrast if the digital numbers of the band are relatively small. The minimum values, if not too close to zero, are thinned out and the values close to 255 are saturated to very bright values. Consequently the image often has no contrast (CD 7.4) and its dynamic range is not optimal.

Logarithmic transformation gives an effect similar to the preceding one, and further reduces the dynamic range. The image is generally bright (CD 7.4).

The *exponential* function, when applied to an image without defining the limits of dynamic range, gives a very dark image for the more common medium and low values. On the other hand, high and very high values appear very bright (CD 7.4). This function can be employed when the receiver gain is high and saturation occurs at high values.

The last three transformation functions are mainly useful for immediate application on the original bands. The bright values and the dark zones can be directly marked in the image.

The *equal-population* transformation, or continuous anamorphosis, optimises the dynamic range of the band (CD 7.4). The image is divided in such a way that each class of grey levels on the monitor comprises an equal number of pixels. This function is preferred when the image is processed by the geographic method since it provides a better perception of objects in the image. Contrarily, transformations related to radiometric values vary for each wavelength band and consequently application of radiometric models is rarely possible.

These methods of transformation are of the continuous type. Consistency requires their application to raw data obtained from a continuous signal (even if it is digitised in the satellite during data acquisition). On the other hand, when indices that result from classification are analysed, the data are interpreted as discrete classes and not as a continuous parameter. In this case, other transformation functions are employed.

The transformation function enables us to allot any grey level to any class, point by point or *class* by *class* on the monitor. In this way, the value of a digital number can be modified in its representation on the monitor. Hence, the result of a classification can be transformed into a band comprising as many grey levels as the classes, assigning the desired grey level to each class.

The *polygonal* function is useful in constructing for any interval defined by the user a straight line of the desired slope, viz., small or large, positive or negative. It is thus possible to represent the digital numbers of low value in white and those of high value in black; a negative slope is thereby obtained. As the interpreter cannot discern the variations in the dark and bright zones with the same degree of acuity, some objects can be more readily identified on a negative slope than on a positive one. This is often the case with long linear features such as rivers, roads, railway tracks, etc.

The polygonal function also facilitates putting a greater or smaller range into a class, allotting an entire interval to the same grey level, etc. All variations within a limited spectral range can be evaluated. For example, all 256 grey levels can be used solely for pixels corresponding to water bodies (CD 7.4). Thus, this tool is useful for carrying out:

—*Binary segmentation* of image, which consists of placing all the pixels below a given grey-level threshold in black and all the pixels above it in white (this constitutes a kind of masking).

—*Multithresholding*, which consists of creating various classes assigned to different grey levels (Fig. 7.3). The thresholds are determined from modelling of the spectral characteristics of various

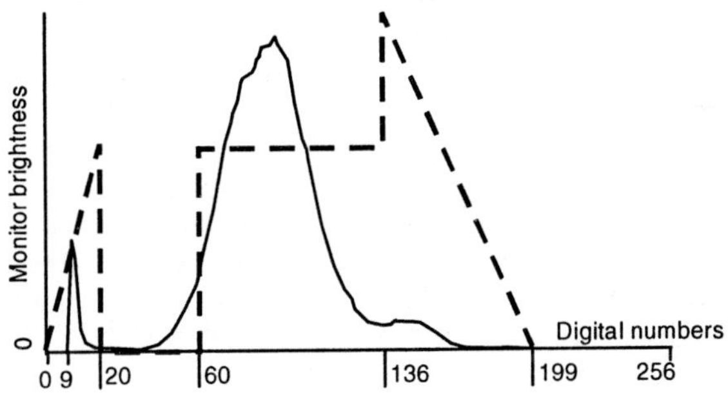

Objects	Water	Water side and dark soils	Crops, grasslands, forests, soils	White soils and green plots
Digital numbers	0 to 20	20 to 60	60 to 136	136 to 199
Image intensity	0 to 150	0	150	255 to 0
Transformation functions	**Positive threshold**	**Binary threshold**	**Binary threshold**	**Negative threshold**

Fig. 7.3: Multithresholding of band b3 of the image of Brienne.

features or from field measurements. The representative classes in such a case correspond to known objects. This procedure belongs to the radiometric method.

A special transformation function corresponds to radiometric masking (see below). If the spectral characteristics of the image feature are used, it is possible to prepare a band in terms of final interpretation of a desired theme. For example, if natural vegetation is studied, the following aspects can be taken into consideration:

—Free water bodies, which correspond to the lowest values of digital numbers in the infrared band; these are assigned 0 value;

—Cultivated zones with intense chlorophyll activity with very high values of digital numbers in the infrared band; these are assigned the value of 255.

All the other values of digital numbers belonging to themes assigned between 0 and 255 are then spread out (Fig. 7.2b). Water bodies appear as black and green vegetation as white; all other features are distributed between black and white over the 255 values (CD 7.5).

■ Look Up Table (LUT)

Once a transformation function is defined, a colour is assigned to each of the 256 digital numbers and a LUT is prepared. Several tables are currently in use (CD 7.6). One table codes the digital numbers from black for 0 to white for 256. There are also tables that proceed from black (0 value) to white (value 128) and return to black (value 256). A monochromatic table can also be used which allots the same colour to the entire image, from dark for 0 to bright for 256. Monochromatic tables can likewise be prepared with less than 256 classes; this facilitates visual interpretation since the eye cannot detect so many classes. Tables of 16 colour classes are also common.

Further, one can build one's own LUT by assigning any colour to any class. For display during processing phases, it is important to choose colours of high contrast and not dull ones. In fact, the role of colours is to facilitate recognition of a given image feature, which requires its distinctive differentiation from all other features.

Choice of colours is particularly useful in making good hard copies of the results of processing. The palette chosen for printing an image is obviously a matter of taste, but it should be such that the results to be evaluated are enhanced and not diminished by the colours selected. Moreover, legibility of an image largely depends on the quality (make) of monitor used.

The possibility of assigning a given colour to a particular value of a digital number provides a kind of sorting that represents a preliminary stage of data processing.

Thus, a colour can be given to a pixel on the grey level image and all the occurrences of the same value, coloured the same way, can be obtained (CD 7.1). This aids in determining the number, geographic distribution and location of all the pixels having the same value of digital number. The connection between radiometric and geographic methods is thus established.

■ Contrast analysis

Contrast is the ratio of the typological distance and the geographic distance between two sites. It thus represents the ratio of their semantic distance to geographic distance.

In the case of satellite images, contrast is defined as the ratio of the mathematical distance between the values of digital numbers of two pixels and their geographic distance.

Several types of mathematical distances can be used. The simplest, in the case of one and the same band, is to take the difference between the absolute values of two digital numbers (Manhattan distance). The geographic distance can be expressed in metres or in units of resolution for a given satellite.

On a black and white image, the expression of this contrast appears to the eye as the difference in intensity between two connected pixels. However, if colours are assigned to various pixels, the contrast is then expressed by the difference that the eye establishes between the colours of two connected pixels. But, this difference manifests differently for each observer. In colour, the effect of brightness is greater than that of hue; thus, yellow appears brighter than red or violet. It is necessary to give attention to the colours chosen for finalising a given processing since visual interpretation, indispensable for the cartographic method, depends as much on the choice of colours as on the result of processing. In all cases, it should facilitate interpretation of the image by the user.

The contrast on a computer monitor depends on the image per se, the transformation function used for contrast stretching, the LUT, the colours assigned to various units, as well as on the contrast regulation of the monitor. The contrast is important since it aids in evaluating an image feature or zone.

The book *Graphical Semiology* (Bertin, 1974) may be consulted for preparation of a final document. Lastly, it should be remembered that it is not possible to obtain exactly the same colours on a colour print as on the monitor, since the print uses pigments that do not support all colour combinations observed on a monitor (see Chapter 3).

■ Classification

Most programs of satellite-image processing provide classification using three bands and it is generally not possible to process a single band. This is inconvenient because one may desire to process a panchromatic band, for example derived from a satellite image or a digitised aerial photograph. This difficulty may be overcome by taking the same band three times and applying three-band classification. Obviously, the image remains black and white. The neighbourhood method of image processing, OASIS (see Chapter 11), can be used for single-band processing.

A possible method of processing panchromatic data is to initially carry out contrast enhancement of the image. Segmentation can be done on the histogram, but this is often insufficient for making a significant interpretation. Subsequently, for the image exhibiting different zones, if necessary, masking can be applied (see below) and classification carried out on each unmasked part of the image using the program OASIS. All the masked parts can then be grouped into a single image, the result of classification displayed in colour.

7.2 PROCESSING OF MULTISPECTRAL DATA

Most often, three bands are used for image processing since they can be displayed simultaneously on a monitor operating on a three-fold 8-bit code. This provides an immediate interpretation of colours.

The three channels may be three multispectral bands of the same scene, three channels of the same spectral band acquired on different dates, or one of the principal components (over an image of multispectral bands) chosen for the same scene of three different dates or even three classifications of the same scene, etc.

In most programs, it is possible to assign any one colour to one, two or three planes and display the composite of the planes (see Chapter 3). As 256 grey levels are available for each principal colour, i.e., blue, green and red, 256 × 256 × 256 or 16,777,216 different colours can be composed and displayed on the monitor. In fact, the eye cannot perceive all of them.

7.2.1 Comparison between colours and spectral characteristics of objects

The colours can be interpreted when the significance of each of the basic colours for the objects represented on the image is known. With regard to spectral behaviour of objects, the following points should be recognised:

1. There is no direct relation between reflectance of an object in a spectral band and the value of digital number corresponding to the band since it is the reflected energy that is measured by satellites and not reflectance;
2. The calibrations made between the energy received by satellite sensors and the 8-bit digitisation of the signal (256 levels) are not the same for different bands (see gain of the SPOT system on the CD);
3. The dynamic-range enhancements of different bands are not the same.

There are at least three models available for converting the spectral characteristics into colours. More models can be used if colours are compared with field measured reflectance data since in this case other factors such as atmospheric modifications, calibration of diverse field targets, drift of satellite sensors, etc. need to be considered.

Consequently, the same value of digital number in different bands does not correspond to the same value of radiance. As it is possible to obtain three values of digital numbers for each pixel, a multiple line of 'digital characteristic' of a pixel, corresponding to a given object can be constructed. However, we should not search for an absolute relation with the spectral characteristics curves obtained from field measurements of reflectance for an object of the same nature. Depending on the situation, the digital characteristic may be very close to the spectral characteristic, but often it is quite different.

The relative positions of digital characteristics of objects for a given band are comparable to the relative variations of reflectance for the same objects. Consequently, linear dynamic-range enhancements for each band would facilitate such an analysis.

General models can also be applied but one cannot go too far in interpretation unless much trouble is taken for inversion of the three transformation models.

7.2.2 Choice of band combinations

Various types of colour composites and in particular infrared colour (IRC) images can be prepared based on the models mentioned above. For this, red is assigned to the channel corresponding to the near infrared band that varies between 750 and 1100 nm depending on the sensor, green to the channel corresponding to the band between 600 and 700 nm and blue to the band between 500 and 600 nm. In such a colour composite, water appears dark blue, soils more or less bright blue or cyan, vegetation more or less intense red and shadows very dark.

We can thus produce another colour composite similar to the colours that our eyes can normally detect by assigning green to the near infrared band, red to the 600–700 nm band and blue to the 500–600 nm band. Vegetation then appears green, soils brownish yellow and water violet-blue. This is a pseudo-true colour composite.

Each spectral band can be made positive or negative and it may be given one of the three basic colours. Then there would be 2 × 3 ways of colouring the first band, 2 × 2 ways of colouring the second and finally 2 for colouring the last one. Thus, there are 48 ways of preparing colour composites from the raw data (CD 7.7). If it is possible to prepare colour composites with only two bands or to assign different colours to the same band or to modify the transformation functions, the number of combinations further increases. It is to be noted that all the images thus obtained comprise exactly the same information but our eye perceives different organisations of the ground features on different composites. This is due to colour contrasts which the eye cannot interpret identically.

When more than three spectral bands are available, the best possible combination can be studied. The latter evidently depends on what one desires to detect.

In the case of LANDSAT TM, a large number of combinations, as well as detection of varied themes are possible, with the seven bands ranging from visible to thermal infrared.

For example, the higher reflectance of water in the visible than in the infrared leads to preferential usage of bands 1, 2 and 3 when differences in depth or turbidity of water are to be determined.

On the other hand, to study vegetation bands 2, 3 and 4 or bands 3, 4 and 5 (or 7) could be used for infrared colour composites. In fact, reflective middle infrared holds important information on water content of vegetation covers, complementary to that provided by visible and near infrared bands. The thermal infrared band can be used but it is beset with problems due to its lower geometric resolution compared to other bands (120 × 120 m instead of 30 × 30 m) and difficult to interpret. In fact, for France acquisition of scenes early morning makes interpretation of thermal phenomena difficult.

Depending on the visual effect sought, the three colours can be assigned to three bands variedly (Table 7.2). It is hence possible to identify certain themes or phenomena by using on the one hand, the differences in spectral characteristics of objects in various spectral bands and, on the other, the differences in sensitivity of the eye to colours and colour contrasts.

The visual effects and possibilities of differentiation between various themes for the codes given in Table 7.2 are summarised in Table 7.3.

Table 7.2: Examples of colour composites prepared using LANDSAT TM bands

Code	Colours/ Bands			Characteristics of major land cover themes			
	Blue	Green	Red	Water	Vegetation	Bare soils	Buildings
A	1	2	3	Blue-green (brightness increases with turbidity)	—Chlorophyllian: green —Dry: maroon, more or less greenish	More or less bright beige	White
B	2	3	4	Dark to light blue (brightness increases with turbidity)	—Chlorophyllian: magenta—Dry: brownish green	More or less bright blue or cyan	Bright blue to white
C	5	3	4	Green (brightness increases with turbidity); non-turbid: black	—Chlorophyllian: magenta—Dry: violet	More or less bright mauve	Bright blue
D	3	4	5	Black to sea blue (for a low to high turbidity)	—Chlorophyllian: bright green to dark green according to nature of plants—Dry: pale pink	Mauve	Mauve
E	4	5	7	Black (turbidity differences cannot be distinguished)	—Chlorophyllian: bright to dark blue according to nature of plants—Dry: green to yellow	Yellow to orange	Orange to pink
F	6	5	4	More or less dark blue depending on currents (no turbidity differences)	—Chlorophyllian: various hues of colours: pink, orange, purple—Dry: almond-green to yellowish-green	Different shades of green to yellow	Blue

Table 7.3: Detection of themes using different colour composites

Code	Comments
A	Risks of confusion between water and vegetation, differences in water, differentiation between green vegetation and bare soils but no difference within these themes. Confusion between some bare soils and buildings.
B	Clear distinction between water and other themes, differences within the water, differentiation between vegetation and bare soils as well as within each of these themes. Confusion between some bare soils and buildings.
C	Clear distinction between water and other themes, differences within the water, risk of confusion between some classes of vegetation and bare soils, differences within each of these themes. Confusion between some bare soils and buildings.
D	Clear distinction between water and other themes, differences within the water, good separation between some classes of vegetation and bare soils, differences within each of these themes. Confusion between some bare soils and buildings.
E	Clear distinction between water and other themes, differences within the water, good separation between vegetation and bare soils, differences within each of these themes. Confusion between some bare soils and buildings.
F	Possible confusions between water and some elements of constructions, very little differences within water, risk of confusion between dry vegetation and bare soils, differences within each of these themes. Good separation between bare soils and constructions, detection of dip and orientation of slopes.

7.2.3 Two-dimensional or three-dimensional histograms

A plane defined by two perpendicular axes, such as the red (675 nm) and infrared bands for example, on which all the pixels of a satellite image are located, is called a two-dimensional histogram. For each point of the plane marked by the co-ordinates of the two bands, the frequency of pixels corresponding to these two specific values is given. The frequency may be coded by a colour, for example for increasing frequencies ranging from black to blue, green, yellow, red and white, or by dots of size proportional to the frequency.

It is interesting to analyse various shapes of two-dimensional histograms. In fact, it is possible to thus identify the relationships existing between the bands. In most cases, strong correlation exists between successive bands, which may not always be the case, in particular between visible bands and the near-infrared band.

Correlation between bands can be observed when the points are more or less aligned in a two-dimensional histogram. For two highly correlated bands two new channels can be created formed by the sum and difference of the two correlated bands. In this case, differences can be clearly distinguished and the two-dimensional histogram is much more stretched but this poses difficulty in interpreting objects using digital numbers. This difficulty occurs in any arithmetic combination between two different bands.

When the three conventional spectral bands, viz., green, red and near infrared of the Earth Observation Satellites are used and if the zone under investigation comprises green vegetation, more or less bare soils, water and shadows, often a very large deviation is observed in the 'red/infrared' two-dimensional histogram. However, this evidently varies with the zones under study and does not constitute a norm. If the area consists of only one object such as water, soil or vegetation, the results are different.

The image of Brienne was divided into nine subimages and correlation between the three spectral bands computed for each. The most important results (Table 7.4) show that:

—the strongest correlation is between b1 and b2;

—the lowest correlation is observed between b2 and b3 (the entire image and the ninth situated in the south-east) or between b1 and b3 (for the ninth in the south);

Table 7.4: Correlation between various spectral bands of the SPOT image of Brienne

Situation	Correlation coefficient: b1–b2	Correlation coefficient: b1–b3	Correlation coefficient: b2–b3
Entire image	**0.981**	0.147	0.079
South-east	0.988	0.198	0.147
South	0.963	**0.089**	0.225
East	0.975	**– 0.00**	**– 0.137**

—there may also be no correlation between b3 and b1 (ninth in the east where no water or almost no forest is present);

—sometimes, correlation may be negative (b2 and b3 for the ninth situated in the east).

The values given above correspond to the most common cases: they clearly indicate the difference between the near infrared and the visible bands and the strong correlation between the two visible bands.

When a greater number of bands are analysed as in the Thematic Mapper (TM) (Table 7.5), similar very strong correlation is observed between the three visible bands. Correlation between the near infrared and the visible bands is low. Correlation between bands 5 and 7 of the mid-infrared is high and these two bands also show strong correlation with visible bands but weak correlation with the near infrared band TM4. Lastly, the TM6 band of thermal infrared exhibits a special behaviour. It should be taken into consideration that the spatial resolution of this band is not the same as that of other bands.

Table 7.5: Coefficients of correlation between the TM bands for the image of Brienne

	c1	c2	c3	c4	c5	c7
c1	1.000					
c2	0.974	1.000				
c3	0.957	0.981	1.000			
c4	0.098	0.113	0.012	1.000		
c5	0.877	0.900	0.916	0.176	1.000	
c7	0.889	0.918	0.954	– 0.024	0.961	1.000
c6	0.340	0.381	0.453	– 0.031	0.549	0.543

■ Model of spectral behaviour for interpretation of digital numbers

It is interesting to note that if the red band data are plotted on the abscissa and the near infrared data on the ordinate, the form of the histogram is always almost of the same type (Fig. 7.4).

The points on such a diagram are distributed inside a triangle for which the three vertices are represented by:

—water or shadow (W),

—highly reflecting soils: bright or very smooth (S),

—most chlorophyllian vegetation (V).

Going from W to S, various states of the soil surface are observed, constituting a cluster of soils ranging from dark or rough to very bright-coloured or smooth. From S to V, soils partly covered

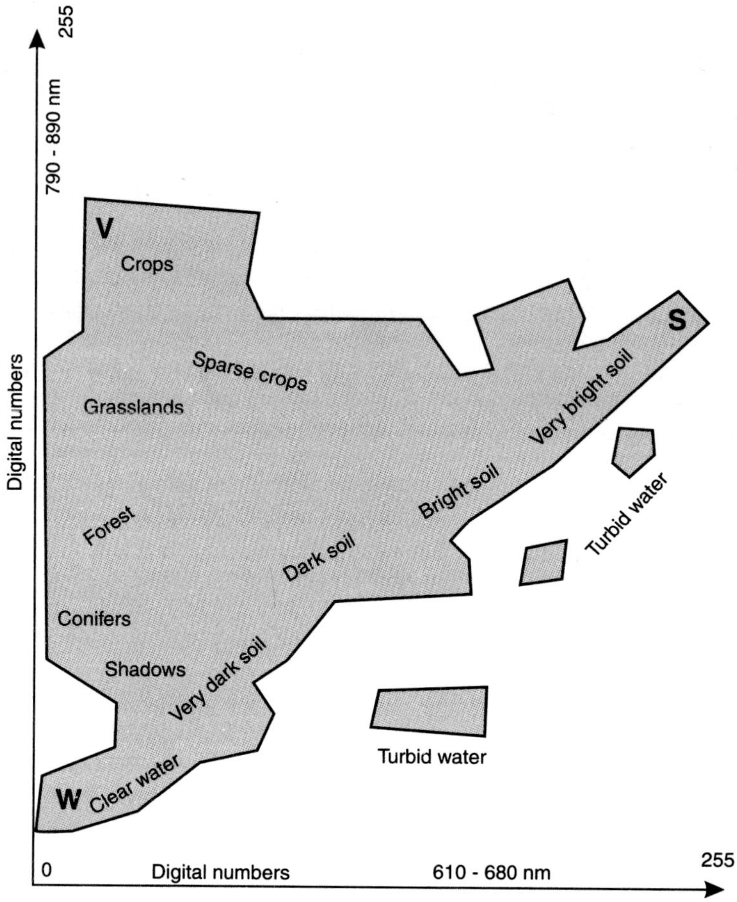

Fig. 7.4: Digital model of interpretation of soils, water and vegetation on a R/IR plane.

with increasing chlorophyllian vegetation are observed. From V to W, we find vegetation of decreasing chlorophyll content, with decreasing coverage and with changing nature of vegetative structure. For example, conifers are situated between W and V while hardwoods occur towards the pole V.

The features are thus distributed in the same way as on a graph with reflectance values as co-ordinates instead of digital numbers. Hence, objects can be detected from their relative positions in an histogram. However, this does not always suffice. It is therefore necessary to display three two-dimensional histograms, i.e., orthogonal projections of three planes of the cube representing three dimensions of the semantic information in a Cartesian co-ordinate system. Some image processing programs provide this facility. Others such as TeraVue are more advanced and enable projection of three-dimensional composition on any arbitrary plane. Thus, the cube can be rotated and the points can be projected on a Cartesian plane for which the equation of the axes is automatically given. This offers the advantage of defining any arithmetic combination of two axes and determining the position of points in three dimensions. This may be termed as three-dimensional analysis of histograms.

■ Study of spectral characteristics of image features

The next operation emerges from the analysis of histograms. It seeks to establish a relationship between the semantic information (digital numbers) seen on the histogram and the graphic information (positions of the corresponding pixels) observed on the image.

It is essential to 'navigate' through the image, i.e., to move onto any pixel and identify the digital numbers of this pixel in the three dimensions. This enables estimation of the type of object the pixel represents. For deeper exploitation of the histogram, it is necessary to determine the position of every pixel in the histogram and from it, the significance of the histograms on which the pixels are represented by a point of the same colour as that on the image. Frequencies in this case are indicated by circles of proportionate diameter. With such a function, it becomes possible to identify the digital numbers of a small zone on the image. Pixels pertaining to the small zone studied are marked successively and the values of each as well as its position in the general histogram of the image are noted.

This investigation of radiometric spectra of image features together with their geographic distribution aids in understanding the spectral characteristics of objects and in choosing the boundaries of classes. This constitutes the basis of any classification using chorological laws. *These laws determine the relationships existing between semantic or thematic units (defined by intrinsic factors) and their geographic distribution in three-dimensional landscape (defined by extrinsic factors).*

7.2.4 Segmentation of an histogram

Another method of analysing an histogram consists of defining a class in the histogram, assigning a given colour (preferably bright) and representing all the pixels belonging to this histogram class by the same colour on the image. The spatial distribution and organisation of all the pixels having the same value of digital numbers on the image can thus be analysed. The same procedure can be followed for a neighbouring class or for a group of classes, which aids in relating the geographic position of objects with their spectral characteristics on the image.

This phase constitutes the theoretical basis of classification to be ultimately carried out since it enables establishing the relationship between spectral and geographic aspects and hence defining the chorological laws. This would facilitate choice of radiometric boundaries for classifications or tracing of boundaries of training zones, which serve as nuclei for classifications.

All these operations can be carried out rapidly using image-processing software and information can be acquired without conserving it. When it is desired to protect this information on a geographic plane, a *new band* is constituted. The latter carries over the same geographic space the value of every pixel after its transformation. This new band can be used again in a subsequent stage of processing.

7.2.5 Arithmetic combination of bands

As the values of digital numbers of various bands (between 0 and 255) are geographically attached to a pixel, it is mathematically possible to make any arithmetic combination of bands and geographically assign it to the same pixel. However, it must be ensured that it corresponds to something. It is commonly used for several objectives:

—to construct indices by associating a theme with the value (on the basis of radiometry or diachronism),

—to improve the representation of information and to reduce the amount of information,

—to superpose or integrate the results of various classifications.

Computation of indices

Vegetation indices were developed from investigations on reflectance. Caution needs to be exercised when these are to be used on images, i.e., no longer from reflectance data but from digital numbers. In fact, unlike reflectance which when acquired in normal experimental conditions varies from 0 to 100% in each band, the same is not true for digital numbers. The dynamic range of each band is independent and consequently the same value of digital number does not necessarily have the same significance in each band. Thus, before interpreting the result of an index, it is necessary to spectrally calibrate the two bands relative to each other.

In 8-bit image processing, all the results of algebraic operations should be between 0 and 255 to enable display on the monitor. If the two bands are not spectrally calibrated, it is quite possible for example that the difference IR–R would not be positive. When negative values are obtained, depending on the program, they are either set to zero, in which case they appear on the histogram as a class with a high frequency, or the negative values such as –1, –2, –3, etc., are re-coded as 255, 254, 253, etc. To avoid this inconvenience, an arbitrary value such as 128 for example can be added. This aids in adjusting the result of computation of the index. If the results contain decimal values, they are rounded off. In such cases, the result is multiplied by a coefficient such as 10, 20, 100, etc., if they are to be interpreted.

After every computation of arithmetic functions, it is necessary to view the histogram for evaluating the distribution obtained. The image obtained is in black and white and can be used to make classes and hence to determine the limits. This phase is most difficult. It necessitates good field knowledge for obtaining reliable information. We can proceed as in the preceding case for identifying some classes by composing a Look Up Table.

Vegetation indices

Various vegetation indices (see Chap. 4) are most often computed based on an arithmetic combination of digital numbers of the near infrared (NIR) and the 675 nm (R) bands (CD 7.8). The difference NIR-R was computed for the image of Brienne. As this difference was small, a constant equal to 100 was added so as to obtain distinct grey levels on the monitor. Six classes were determined which could be coloured as indicated in Table 7.6. Using this colour composite, large forest blocks and riverine forests were identified as green, bare soils as black and orange, soils bearing chlorophyllian crops as yellow and mixed zones between soil and vegetation as blue and magenta.

The NIR/R index was also computed. If the digital numbers are high, vegetation exhibits a greater chlorophyllian activity. This index is not applicable for low digital numbers as interpretation is difficult, except when field-training data are available. Another index or another method has to be used.

The Normalised Difference Vegetation Index (NDVI) is computed from the same NIR and R data as (NIR – R)/(NIR + R). This index can also be described as [(NIR/R)–1]/[(NIR/R)+1]. Then it is similar

Table 7.6: Interpretation of NIR-R index for image of Brienne

Digital numbers of classes	Colour on image	Legend
100	Black	Very bright to dark soils and water
101 to 121	Orange	Very dark soils
122 to 133	Magenta	Soils with vegetation regrowth or seedlings
134 to 152	Blue	Chlorophyllian vegetation and shadows
153 to 186	Bright green	Chlorophyllian vegetation (trees)
187 to 255	Yellow	Very chlorophyllian vegetation

to the first and is based on the same type of interpretation. However, as it lies between 0 and 1, its dynamic range is much smaller and its interpretation is still difficult.

These three indices show good correlation (Table 7.7).

Table 7.7: Correlation between three vegetation indices

Index	NIR – R	NIR/R
NIR/R	0.961	1
NDVI	0.941	0.962

Three units, viz., grasslands, non-chlorophyllian vegetation and zones of sparse vegetation cover, were identified on the image of Brienne using Bayesian classification (see Chap. 9). The NDVI indices are practically the same. They vary respectively in the ranges 128 to 144, 131 to 140 and 125 to 135. Hence, it is impossible to differentiate between these three units on the basis of index, whereas it is possible using a two-dimensional red/infrared histogram. These indices do not provide finer discriminations. On the other hand, it is possible to distinguish between water, bare soils and vegetation (Fig. 7.5), for which the NDVI indices for the Brienne image are 108 to 120, 121 to 134 and 135 to 151 respectively.

■ Brightness index

Some authors have defined a brightness index as the square root of the sum of squares of the two bands, NIR and R:

$$BI = \sqrt{NIR^2 + R^2}$$

This amounts to calculating a value equivalent to albedo, limited to a wavelength interval of 500 nm to 1000 nm. Interpretation of this index is not always easy but it clearly differentiates shadow zones from illuminated zones. It gives an image that differs little from a panchromatic image.

Various soils could be distinguished on the Brienne image using this index; however, cultivated soils could not be identified (CD 7.9). Rivers are differentiated better on this image than with the NIR band and much better than the NDVI image.

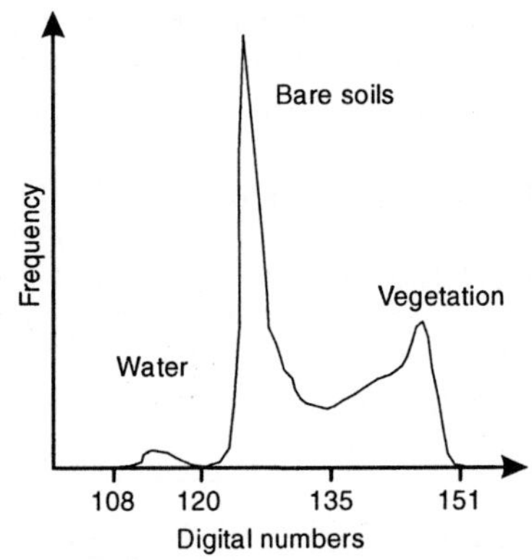

Fig. 7.5: NDVI index histogram for image of Brienne. Water (108 to 120), bare soils (121 to 134) and vegetation (135 to 151) can be distinguished.

■ Modification of representation of information

One can readily compute the sum IR+R and the difference IR–R of the two bands. The colour composite of these two new bands often gives a greater dynamic range than that observed in the case of the initial bands and the differentiation is much better. In a number of cases, the visual effect is hardly different from the more elegant solution, which consists of computing linear equations corresponding to the principal components (see below).

■ Integration of results of various classifications

When several classifications that are mutually exclusive are made in different parts of the image (see masks below), arithmetic combinations can be used to superpose them one over the other. Thus, let us assume that a classification is made by dividing the forests into five classes (F1, F2, ..., F5) and another classification by separating the cultivated soils into 12 classes (C1, C2, ..., C12). In order to obtain on a single document (I) interpretation of forest (F) and cultivated zones (C) from the entire scene, the band comprising cultivation data may be transformed by multiplying it by 10. The classes then vary from 10 to 120. The classification of forests is next combined with the latter and we obtain on a new band the entire information divided into 17 classes (see Chap. 8): Image = 10 × cultivation + forest.

In this application, arithmetic combination forms only an intermediate tool for preparing a final representation of information.

7.2.6 Statistical analysis of bands

Bands can be analysed pixel by pixel but nowadays we process images containing 1000 × 1000 pixels to 3000 × 3000 pixels or more, i.e., from one to ten million pixels. It is hence imperative to study the statistics of these images for characterising their principal components before undertaking classification.

Almost all image processing software includes statistical programs but in a more or less precise manner. These programs provide the shape of histogram, mean and standard deviation, but they are not necessarily the best parameters to define the distribution curves since the populations in image data are not Gaussian. Nevertheless, when diachronic bands are compared, these parameters may be initially useful to study radiometric variations of objects between two different dates.

Thus it was observed for TM images of Brienne (*© ESA (1990 and 1992) Acquired by Fucino Station, Distributed by Eurimage-Geosys, by courtesy of Eurimage*) that the means and standard deviations were greater by 2 to 3 values of digital numbers for the image of May 1992 (CM) than for the image of April 1990 (Table 7.8). However, this difference is much greater for band 4 since it changes from 58.9 to 91. It was verified that the vegetation has a higher chlorophyllian activity in the month of May (CD 7.10).

It is essential to study the histograms closely and preferably directly using the digital numbers.

It is difficult to give rules applicable to all cases for statistical interpretation of bands since several factors interfere. In fact, the values of digital numbers are not directly obtained from radiance received by the satellite sensor (see Chapter 2). In the case of SPOT images, gain may significantly modify the position of the histogram between its limits 0 and 255 (see CD: SPOT Satellite).

The respective positions of the histograms of various images are not necessarily the same as those obtained from reflectance curves. The histogram of the green band (b1) often shows a higher mean than that of the red band (b2). Values corresponding to the infrared band are usually higher (see Table 7.1). Thus, for classifying images of a given date the dynamic range of each band for each image has to be reduced to a common dynamic range for all. Reciprocally, when an image is divided into image segments, the histograms change.

Table 7.8: Mean, standard deviation and correlation between the bands of two Thematic Mapper images. M: Mean; SD: Standard deviation; c1 to c7: TM bands for the month of April; CM1 to CM5: TM bands for the month of May. Correlation coefficients between bands of the same image are shown in bold.

	M	SD	c1	c2	c3	c4	c5	c7	CM1	CM2	CM3	CM4	CM5
c1	70.51	10.04	1.000										
c2	31.89	7.63	**0.974**	1.000									
c3	34.52	13.54	**0.957**	**0.981**	1.000								
c4	58.90	17.22	**0.098**	**0.113**	**0.012**	1.000							
c5	63.37	23.86	**0.877**	**0.900**	**0.916**	**0.176**	1.000						
c7	30.14	17.12	**0.889**	**0.918**	**0.954**	**−0.024**	**0.961**	1.000					
CM1	73.70	12.84	0.371	0.361	0.333	0.177	0.297	0.291	1.000				
CM2	34.97	9.66	0.343	0.342	0.321	0.200	0.305	0.290	**0.968**	1.000			
CM3	35.75	16.67	0.323	0.326	0.312	0.190	0.301	0.289	**0.960**	**0.982**	1.000		
CM4	91.00	23.70	0.144	0.151	0.129	0.514	0.262	0.142	**−0.184**	**−0.140**	**−0.196**	1.000	
CM5	66.65	27.26	0.304	0.309	0.301	0.302	0.367	0.312	**0.837**	**0.848**	**0.875**	**0.025**	1.000
CM7	29.47	19.93	0.289	0.296	0.290	0.193	0.309	0.291	**0.894**	**0.900**	**0.940**	**−0.193**	**0.954**

A method of image segmentation is based on this principle. An image is divided into *n* (4 or 9 for example) segments of equal area. Histograms of each segment are compared with those of the complete image (see Table 7.4). If the difference is greater than a given threshold, the segments are distributed into different units. Then, classification is continued for each unit and the function is iterated.

Using the histograms in a similar manner, we can also compare two images, one complete and the other subsampled, and determine whether information is lost or not. This enables us to know whether noise at the spatial level is high or low: if the histograms are similar, this indicates that nothing is lost by subsampling and hence there is little useful information from the spatial point of view and thus 'noise' exists.

It is also advisable to examine correlations between bands of the same image. Some constants exist when the visible, near infrared and middle infrared bands are studied. In most cases, visible bands correlate very well with one another (Table 7.8) and fairly well with those of middle infrared. For the SPOT image of Brienne the correlation coefficient between the two bands b1 and b2 is 0.981 (CD 7.11). The near infrared band usually correlates less well and minimum correlation is observed for red band, even with the nearer blue band. These remarks evidently apply for an image comprising areas of water, soil and vegetation.

In the case of diachronic images (Table 7.8), when bands of images acquired in different seasons are compared, since the areas covered by vegetation change considerably the correlation coefficients also change but in most cases the separation mentioned above between visible and infrared bands persists.

7.2.7 Principal component analysis

As seen above, satellite image data exhibit different degrees of correlation. It is interesting to investigate a representation of the same information in another configuration, for example by projecting all the pixels on axes that do not correlate. The principal axes of principal component analysis derived from linear combinations of initial bands do not correlate. Thus, when three SPOT bands, b1, b2 and b3 are used, we obtain three principal components PCA1, PCA2 and PCA3 whose equations as a function of the initial bands are as follows:

General case	Example for SPOT image of Brienne
PCA1 = *a* b1 + *b* b2 + *c* b3	PCA1 = 0.701 b1 + 0.696 b2 + 0.157 b3
PCA2 = *d* b1 + *e*b2 + *f* b3	PCA2 = – 0.076 b1 – 0.146 b2 + 0.986 b3
PCA3 = *g*b1 + *h*b2 + *i*b3	PCA3 = 0.709 b1 – 0.703 b2 – 0.049 b3

These axes also exhibit the following interesting property: information on axis 1 is most widely stretched and its value is the highest. Hence it is the most important axis since it carries maximum information. Axes 2 and 3 follow successively.

If the information carried by each of the bands, b1, b2 and b3, is considered equal to 1, the total information would be 3 for a SPOT image. After a principal component analysis, it is common to obtain the information (actual values) distributed in a manner similar to that of Brienne region:

—for PCA1: 2.007;

—for PCA2: 0.977;

—for PCA3: 0.016.

This may be interpreted as follows:

—if the PCA1 axis alone is retained, nearly two-thirds of the information contained in the entire image is preserved;

—if the first two axes PCA1 and PCA2 are retained, almost the entire information is preserved and hardly a few parts per thousand are lost.

It follows from the above that two axes are often sufficient to represent the essential content of an image in three dimensions. The three-dimensional representation of all the pixels of an image resembles more a cutlet than a balloon!

It is hence necessary to analyse what the three axes of the principal component analysis signify. This is done by studying the linear equations that define the three PCA axes (vectors). In most cases, the following orders of magnitude are observed.

For PCA1, *a* and *b* are of the same sign and high, while *c* is of the same sign but small. Consequently, the first component is quite close to the sum of digital numbers for a given pixel. This is similar to the result that would be obtained for a panchromatic band. In other words, two-thirds of the information is contained in the sum of three bands.

For PCA2, *f* is very high and opposite in sign to *d* and *e*, which are smaller. As a consequence, the second component indicates the contrast between the visible and infrared bands. This is quite similar to a vegetation index, which could be roughly formulated as IR–R. Hence, diverse types of vegetation are detected better on this band. One-third of the information is contained in the difference between visible and near infrared bands.

In the case of PCA3, *g* and *h* are quite high and of opposite sign, while *i* is most often very small. This component indicates differences between the two visible bands since the infrared band plays almost no part. Noise is observed in this component, which represents only one- or two-hundredths or less of information. This band aids in detecting linear features of the image. The latter are visible in particular over water, which is understandable since it is on this component that the information relative to visible bands is maximised, in which water exhibits maximum dynamic range. It is also in this component that some elements not noticed in the other components are observed, viz., very bright zones of small extent such as open quarries. However, shadow zones are detected from this component. Hence the band derived from this component can be utilised to make arithmetic combinations that tend to highlight a particular point, such as a shadow zone, or contrarily suppress the noise.

When a PCA is carried out using the six bands of a TM image, with visible, near infrared and reflective middle infrared bands, the results are as follows. The first axis corresponds to the sum of all bands except the near infrared (TM4). The second corresponds to the near infrared band (TM4). The third axis represents the contrast between the visible bands (TM1, TM2, TM3) and those of the middle infrared band (TM5, TM7). The fourth shows the contrast of the blue band (TM1) against the green and red bands (TM2, TM3). The fifth axis indicates the contrast between the two middle infrared bands TM5 and TM7. The sixth mainly compares green (TM2) against red (TM3). Depending on the images, the composition of the axes varies, but often the same type of composition is identified.

Once the behaviour of the axes of principal component analysis is understood, it is useful to visualise the information of these axes in its spatial form by creating a new band for each PCA axis. Each band is constructed in black and white and used to study the geographic distribution of the various objects to be identified. In fact, the value of each pixel cannot be readily interpreted since a linear combination of the initial bands has to be analysed. The dynamic range of the signal of each PCA axis differs considerably from that of bands of the satellite image. It extends over the entire 256 possible values (Fig. 7.6). In some cases, the contrast of each new band can be improved by studying every histogram, as indicated earlier. If the three bands are represented on a three-dimensional histogram, it can be seen that point distribution is the widest possible. This is due to the fact that the bands are uncorrelated.

Using the first PCA axis, a new band of PCA1 can be constructed (CD 7.11), which can be interpreted visually or digitally in the same way as done for a panchromatic image. All geographic details (including roads) become apparent on this new image and hence identification of features becomes easy if the 'square-root' function is used for enhancement, as in the case of the Brienne image (Fig. 7.6).

In the case of PCA2, major land cover zones are clearly identified since the information on this axis corresponds to a contrast between visible and infrared bands. Hence it highlights various types of chlorophyllian vegetation, bare soils and water bodies.

On the image of Brienne (CD 7.11) cultivated zones can be seen as white hues and forests and riverine forests as grey. Bare soils are identified as dark grey areas and water as black. The histogram of PCA2 (Fig. 7.7) is similar in its distribution to that of band b3 (Fig. 7.1). The difference lies in the fact that the histogram of PCA2 is stretched over 256 levels instead of 190 levels for band 3.

In the case of PCA3, if there is no particular feature and if the noise is distributed over the entire image, everything appears a medium grey. Lineage effects are observed in this band. In some cases, this image that contains only a few per cents of information, can be visually interpreted quite well. In the image of Brienne, even the small amount of information enabled a correct interpretation and no

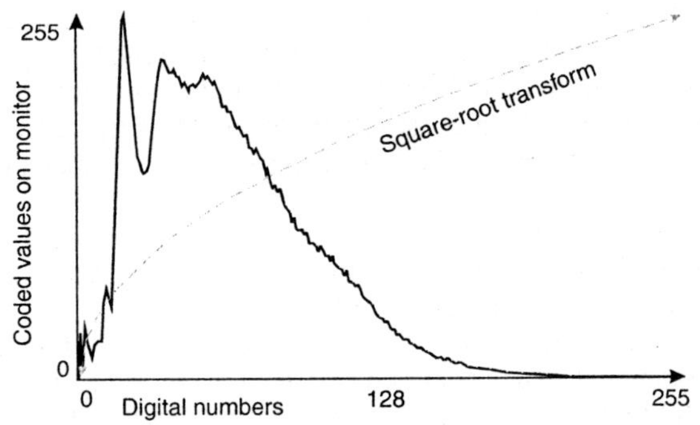

Fig. 7.6: Histogram of image of Brienne projected on PCA1.

evidence of noise was found. If high digital numbers of the histogram (Fig. 7.8) are carefully observed, a specific form of noise can be noticed between 128 and 180. This is spatially discernible in areas of reservoir waters (CD 7.11).

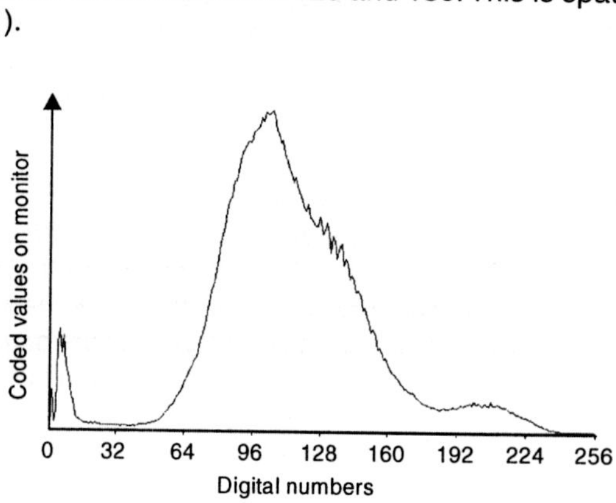

Fig. 7.7: Histogram of image of Brienne projected on PCA2 axis.

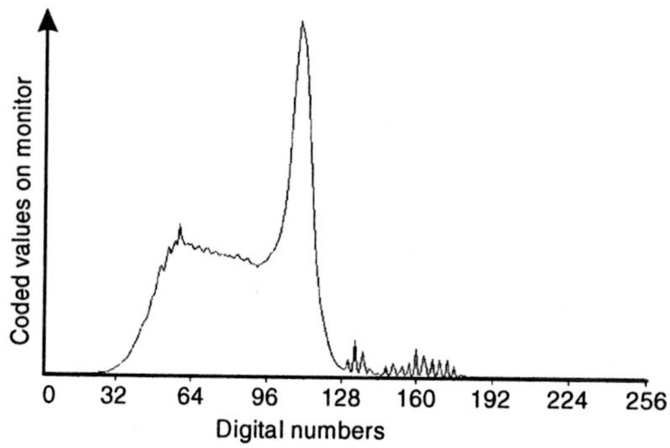

Fig. 7.8: Histogram of image of Brienne projected on PCA3 axis.

■ Colour composites

The importance of using the new bands constructed from the principal component analysis lies in the fact that the three components are independent, which is recommended for some classification methods.

Once the visual significance that can be drawn from an image derived from projection of each axis of principal component analysis is understood, a colour image can be constructed using two or three bands. The choice of colours for the final visualisation is important since the quantity of information differs for each component.

If such a component is to be evaluated, it is better to assign it a red colour, which offers the advantage of several shades. On the other hand, very few shades can be differentiated in blue; hence it can be used for a band whose visual impact is to be reduced.

A colour composite can also be constructed by choosing colours of each component in such a way that a given object is displayed in a predefined colour. In such a case, it is necessary to determine the characteristics of this object in each of the three components, which is not simple, and to apply rules of colour composition.

■ Comparison of new bands

The first component can be used to represent the entire image. This allows use of results of classification or arithmetic combinations (for example, an index) for the second or third band. By assigning a different colour to each of these three bands, the results obtained can be compared by analysing the colours on the monitor.

■ Diachronic comparison

When images of the same region acquired on different dates are to be compared by means of colour composites for a diachronic study, analysis should be limited to three bands. Consequently, one solution is to take for each date one of the principal components, the one considered as best representing the entire image of the given date. Once constructed, a study of colours on the monitor provides a diachronic analysis (CD 7.12).

■ Comparison of multiple processing methods

The results of several treatments, each of which constitutes a new band, can be compared by interpreting correlation coefficients, intrinsic values and vectors of principal component analysis. For this purpose, we identify the processing methods that play the same role (these are correlated with one another) from those that are dissimilar (these are not correlated) and determine the weight of each. Obviously, the comparison, statistical and textural, is carried out pixel by pixel and does not integrate the geographic approach (for the method of neighbourhood, see Chapter 11).

7.3 MASKS

Masking consists of hiding part of the image and preserving the other part intact. A mask may be geometric or radiometric.

Commonly, only part of the image is of interest for a given thematic application, such as land or marine region, plateaus, slopes or valleys, swamps or well-drained zones, etc. Moreover, in order to avoid confusion in classification of certain features, it is important to preliminarily segment the image into major land cover classes. Thus, we can avoid confusion between diverse bare agricultural soils and urban zones, or between crops and permanent grasslands. In montane zones, it is desirable to separate northern exposed slopes from southern; this facilitates land cover classification without application of radiometric corrections required for compensating solar illumination differences. Lastly, when a cloud cover shadows part of a scene, it is preferable to apply a mask so that pixels situated in the shadow zone can be classified separately from those outside this zone.

In one case, only part of the image is normally analysed and consequently it is preferable to suppress the geographic area that is not processed. For this purpose, a geometric type of segmentation is carried out. The boundaries of the zone of interest are traced and the remaining area is eliminated. This is known as geometric masking.

In the other case, it may be desired to eliminate a theme that does not form part of the study and which corresponds to a radiometric group. This is referred to as radiometric masking.

7.3.1 Radiometric masking

Radiometric masking is designed to suppress part of the objects in one or several bands based on digital numbers. In this case, the spectral significance of the image is analysed. As mentioned earlier, it is possible to prepare a band according to the ultimate theme of interest by suppressing all other objects that do not correspond to the theme of investigation. These transformations should be taken into consideration during the final processing of images. Suppression of some digital numbers is significant only when it is ensured that a two-way connection exists between the latter and the object under study. Very often, we may eliminate other pixels that belong to the object of study but in the conditions of image acquisition, have the same digital numbers as those of the pixels outside the theme. In such a case, geographic or logical masking should be applied.

An ascendant hierarchic classification (see below) is used for digital numbers of pixels. This classification provides a grouping of pixels on a unique statistical basis that ensures distinction between classes; on the other hand, grouping of objects of different nature but with identical digital numbers should be controlled. One way of doing this is to use the two- or three-dimensional histogram and ensure the nature of pixels by controlling them on the image. By this method we could separate permanent grasslands from other land cover classes on an end-winter scene of Lorraine, using visible and near infrared bands. After necessary geometric corrections, this classification served as a mask for scenes acquired in different seasons covering the same region.

7.3.2 Geographic masking

Two types of geographic masks may be distinguished, viz., manual and thematic.

■ Manual masks

These masks are obtained by manual delineation of the boundary of zones not included in the study.

When the region of study is not identical with that of the satellite image, it is essential to mask from the beginning of processing the part of the image not involved in the study. When a digital information layer corresponding to the boundaries of study is not available (otherwise a logical mask is applied, see below), boundaries of the region of study can be drawn manually. This can be defined as geographic masking. Once the boundaries are drawn, it operates like a logical mask.

In an investigation concerned only with agricultural areas, the boundary of the urban zones was determined by computer-aided visual interpretation on a SPOT scene in the west of Paris. The work of the interpreter was aided by preparation of a plane (LP), corresponding to a Laplacian filtering technique (see Chapter 12), of the near infrared band (b3) of the SPOT scene. Tricolour combination of this new plane (LP) with bands b3 and b2 enabled us to more readily identify urban expansion zones. In fact, the plane resulting from filtering (LP) showed a distinct contrast between urban zones, characterised by a very high density of sinuous curves (similar to spaghetti), and agricultural or forest zones of less marked contrast, represented by polygons of varied dimensions.

■ Thematic masks

These masks are normally developed from exogenous data and require a prior geometric calibration between these data and remote sensing data.

A digital terrain model for example, enables extraction of some slopes for isolating hill tops or valley floors. A slope map can be used to select the slopes included in the range of interest. If the variable of exposure is added to this map, we get a method of classifying land cover units exposed to different illumination conditions without applying radiometric corrections. Similarly, any thematic layer of a geographic information system emerging from a map or an earlier classification can be used to create a mask.

7.3.3 Logical masking

Depending on the software, several techniques of logical masking exist. In all cases, we start with an initial image I (characterised by digital numbers), make use of an image M to mask the unwanted part (characterised by the value 0) and by a logical operation generate the final image F. In this image, all the pixels that are not to be considered for the study are represented by the same value, for example zero (Table 7.9).

The logical expression used in masking is of the form:

IF (mask image M = 1) THEN put the pixel value of the initial image

IF NOT put the pixel value 0

A mask can also be prepared (Table 7.10) by considering that class 3, for example, of a classified image C constitutes the mask for the initial image I, and by obtaining the value of 50 for the pixels eliminated.

Table 7.9: Design of a logical mask, starting with an initial image I, to obtain the final image F using an intermediate image of masking M. The values indicated in the image are digital numbers.

Initial image I	Mask M	Final image F
12 24 25 17 58 69 64	0 0 0 0 1 1 1	0 0 0 0 58 69 64
11 28 31 67 98 85 79	0 0 0 1 1 1 1	0 0 0 67 98 85 79
17 21 35 62 78 29 27	0 0 1 1 1 0 0	0 0 35 62 78 0 0

Table 7.10: Design of a logical mask, starting with an initial image I, to obtain the final image F using an intermediate classified image C. Values indicated in the images are digital numbers.

Initial image I	Classified image C	Final image F
12 24 25 17 58 69 64	3 3 3 3 1 2 4	50 50 50 50 58 69 64
11 28 31 67 98 85 79	3 3 3 1 2 2 4	50 50 50 67 98 85 79
17 21 35 62 78 29 27	3 3 2 2 1 3 3	50 50 35 62 78 50 50

The mask for this case can be written as follows:

(C = 3) *take the class 3 of image C*

True: 50 *if it is true that pixel p of image C is 3 then put 50*

False: I *if it is false that pixel p of image C is 3, then put the value of the pixel of the initial image I*

In this case, the final image F generated comprises all the values of pixels of the initial image irrespective of the classes to which they belonged in the image *C*, except for class 3 in which the pixels take the value of 50.

Masks offer the advantage of dividing the image into several segments each of which can be processed differently (see CD 8.4). Subsequently the entire image is reconstructed by arithmetic or logical operations (see sec. 7.2.5). This is used for detailed analysis of dark soils and vegetation in studies by ascendant hierarchic classification.

A geographic zone corresponding to a civil parish for example can be delineated at the beginning of processing and the remaining area masked. In such a case, the statistics of different groups of classification of the civil parish can be defined.

7.4 CONCLUSION

The various techniques of preliminary processing enable construction of images suitable for visual interpretation (see Chapter 5). Even at this level of processing, some questions can be answered. Finally, it is necessary to understand well all these preliminary processing techniques since they are used not only before, but also after classification of images. The latter will be discussed in the following chapters.

8

Unsupervised Classification

Image processing methods can be categorised in two groups:

—*Unsupervised classification methods* in which the data are classified according to their structure. The interpreter gives no a priori information about the objects to be determined. However, as mentioned in Chapter 6, he (she) interacts by choosing the number of groups, thresholds, etc.

—*Supervised classification methods* based on searching for features that are similar to reference objects. The latter can be defined radiometrically on a multidimensional histogram or marked geographically on an image (nuclei, reference zones, etc.).

8.1 ASCENDANT HIERARCHIC CLASSIFICATION

8.1.1 Principles

Ascendant hierarchic classification is based on the measure of a distance, such as Euclidean, computed for each of the N pixels having different spectral characteristics. On an image comprising ten million pixels, the order of magnitude of N is only about ten thousand.

We search in a table of $N(N-1)$ distances two pixels that have the shortest distance between them and decide to group them together. These two pixels are then converted into a single fictitious entity, viz., a mean pixel.

A dendrogram is drawn representing various phases of classification, placing pixels on the abscissa and a value similar to an intergroup distance on the ordinate. The two pixels are joined on this diagram by a segment equivalent to their distance.

Distances between $N-1$ pixels (real or fictitious) are recomputed and pixels that have the shortest distance between them identified. These may be two real pixels or one real and one fictitious, representing a group. These two new units are combined. The procedure is thus continued until all the pixels are grouped. This involves $N-1$ phases of grouping of all the pixels or groups of pixels into a single large group (CD 8.1, Fig. 8.1, Tables 8.1 and 8.2).

Classifying all pixels into a single group amounts to obtaining a single group for the entire image. Such a classification is certainly correct since there is only one group and the intragroup distance is the shortest possible, since it corresponds to that of the entire image, but this does not constitute the objective of classification. Similarly, it is possible to classify every pixel into a separate group. Thus the intragroup (intrapixel) distance is the minimum, namely zero, but this likewise is not the aim of classification. Moreover, this information is always available since it is strictly comparable to the initial histogram.

8.1.2 Groups and legend

A proposed classification of all pixels is represented in a dendrogram. The number of groups desired must necessarily be selected, which constitutes external intervention.

This classification is based only on a mathematical model of classification and hence provides no information on the nature of the groups. One has to guess what the groups represent in order to establish a meaningful legend. Also, the choice of groups ought to be determined by the possibility of thematically defining each group identified.

■ Choice of groups

Choice of groups primarily depends on the four types of objectives pursued.

□ Reconnaissance

A primary aim could be preparation of a preliminary map of a region for which no information on the theme under study is available. In such a case, the main objective could be establishing a segmentation of the region from processing of satellite images. It follows that a priori all the pixels in the entire zone under investigation should be classified. It is not necessary to define a precise legend for each group since this would be done subsequently during field studies (CD 8.1).

□ General classification of units

A second objective could be to give a general classification of the most characteristic units of the region in a scale that does not indicate too small spatial units. As very small zones cannot be represented, they need to be combined with others so as to obtain an areal extent that is compatible with the minimum acceptable area for representation on the map or greater than a given threshold. Thus for example, the minimum size of objects represented on the CORINE Land Cover databank of 1/100,000 scale is 25 ha. In the multispectral SPOT image this corresponds to 625 pixels. Consequently, objects that are too small are combined with those of the semantically nearest group.

□ Definition of major thematic zones

A third objective could be to establish just a very general segmentation of the region under investigation to identify only a few major zones for a given thematic study. Unsupervised classification is very well suited for this objective. Often we are not interested in all the possible themes in a region and sort only the important zones. If, for example, only forests are being investigated, all bare soils need to be eliminated. This very rapid method of classification enables us to select on the dendrogram a minimum number of groups relevant to the theme 'bare soils'. This is done based on a radiometric model. Then these groups are masked and an image is thereby obtained on which all pixels corresponding to forests are preserved. In this process, it is better to retain pixels that might ultimately prove useless rather than eliminate those that might form part of the theme.

□ Precise thematic study

A fourth objective could be to study a precise theme without considering its size or spatial distribution. Areal extent of the zones retained does not constitute a criterion. As it is very fast, unsupervised classification may be very useful in such cases if it is observed that the unit under investigation appears in one of the proposed groups. Then only this group is retained and through masking an image obtained wherein the sizes and spatial distribution of this unit can be readily analysed, as was done for lakes on the image of Brienne.

■ Legend

The purpose of a legend is to give meaning to the classification created. Each group is assigned a name with reference to a semantic or a geographic model.

Irrespective of the objective of image segmentation by this method, a legend ought to be given for each group identified. The legend cannot be beyond this mode of classification since no a priori hypothesis is made about radiometric values. In order to define themes that might cover different groups, a histogram of digital numbers is determined for each group so as to characterise the digital behaviour of the theme relative to the group. For this purpose the interpretation model of digital numbers on a three-dimensional histogram is employed (Fig. 7.4).

Analysis of spatial distribution of pixels belonging to a single group may also give interpretable geographic information. In such a case, a geographic approach is adopted and the positions and forms of zones of each group obtained are analysed, which may give information about the objects combined into a given group. Rivers and talwegs are readily recognised by their linear and curved forms and by their hierarchic relationships such as joining of a watercourse with another to form a higher level of flow. Unlike roads, they do not criss-cross one another. However, there can always be exceptions as in the case of channels or ditches that more or less follow valleys or cut across hills or even may in some cases intersect rivers.

■ Analysis of a dendrogram

A dendrogram can be analysed systematically. Two groups are chosen and the geographic spatial distribution of each is observed on the image. As a matter of fact, the dendrogram gives only semantic information and the number of pixels pertaining to a group is not ascertainable. It varies from a few tens to several hundred thousand. This aids in judging the aptness of preserving a particular group: a group cannot be preserved because it represents a very small number of pixels or because the pixels it contains are scattered over the entire image and do not constitute sufficiently large zones.

The same analysis is repeated by successively increasing the number of groups and thereby the optimum number of groups for the objective of investigation is determined (CD 8.1).

■ Interpretation of a dendrogram

A dendrogram is provided in some programs such as TeraVue (see Fig. 8.1). This is essential for understanding the statistical approach of segmentation into groups.

A dendrogram is an expression of the results of classification. For interpretation, it needs to be considered similar to a mobile system. This means that the order of pixels and regroupings on the dendrogram constitute only a projection of a group, which in fact is a three-dimensional unit, onto a plane. Thus the following two representations (Table 8.1) are identical. It would be a serious error to consider that according to the first scheme group 18 is closer to group 9 than group 15, or that groups 5 and 15 are farther apart than groups 27 and 15.

Table 8.1: Two forms of representation of a dendrogram derived from an ascendant hierarchic classification

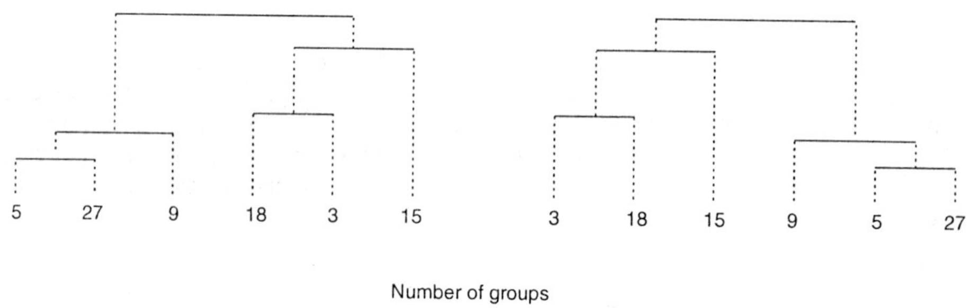

The dendrogram given as an example indicates that the two nearest groups are 5 and 27. They constitute a very homogeneous unit because they are separated by a small distance (grey level threshold expressed on the dendrogram). These two groups are connected to group 9 also by a small intergroup distance. On the other hand, groups 3 and 18 are also connected but with a larger distance than the group (5, 27, 9) since they necessitate three levels on the dendrogram. Group 15 is very different from the groups (5, 27, 9) and (3, 18). Nevertheless, it combines with group (3, 18).

Mathematically, division of a dendrogram is carried out at the same level. On proceeding through it the following segmentations will be obtained (Table 8.2):

1. Two groups: One a part of group A: 3, 18, 15 and the other part of group B: 5, 27, 9. The intragroup distance of A is greater than that of B; group A is thus more heterogeneous than group B.
2. Three groups: These are groups C: 15, D: 3, 18, and E: 9, 27, 5. Group C is thereby isolated and hence becomes very homogeneous since its intragroup distance is zero. Group D becomes more homogeneous than the preceding group A, since group 15, which significantly contributes to the high value of the intragroup distance of A, is detached from it. However, the intragroup distance of D is greater than that of E. Group E is similar to group B and its intergroup distance is the same.

   It should be noted that when we move to the lowest level in the dendrogram, the same three groups are found, and the same holds to true when we go down to three levels. This indicates that division into three levels is steadfast.
3. Four groups: These are groups F: 15, G: 3, H: 18, K: 9, 27, 5. Two groups are not affected by this segmentation. Group F retains the same characteristics as found in group C and group K those of groups E and B. Thus the former group D is divided into two groups, G and H, which become homogeneous since each intragroup distance is zero.
4. Five or more groups: When there are five groups, segmentation results in division of group K into two: on one side L: 9 and on the other M: 5, 27. At this level, intragroup distances are all very small or zero and on the other hand, the number of groups has considerably increased.

Table 8.2: Analysis of segmentation of a single dendrogram according to the choice of number of groups

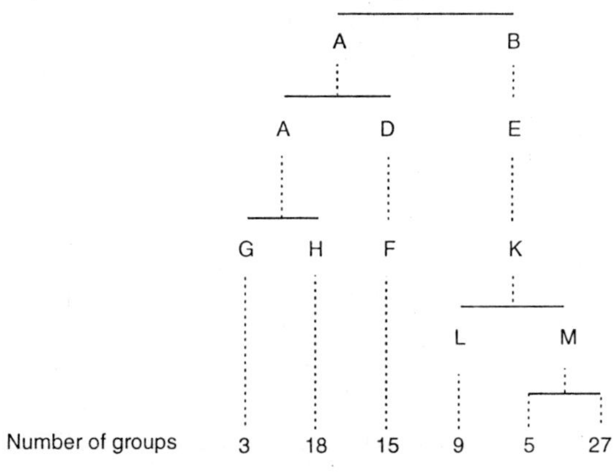

It is also possible to define 6 groups; but this amounts to further dividing each group of pixels: each intragroup distance is zero and the intergroup distances are maximal. Theoretically speaking, such a classification is the best; however, since no regrouping was carried out it is no longer a classification!

■ Interpretation of classification

We cannot determine on a dendrogram the area corresponding to each group nor the geographic distribution of pixels that occur in the same group. The possibility offered by image processing to immediately and interactively see the dendrogram, the number of groups chosen and the corresponding spatial result is of great significance in interpretation of classification (CD 8.1). It is thus possible to choose the desired number of groups best suited to the objective pursued.

When the cross-section of the dendrogram that is most suitable to the objective of interest is chosen, the groups can be combined based not only on statistical criteria, but also on thematic considerations. It suffices to allot a single colour and a single attribute to various groups that we desire to combine. Thus groups corresponding to different types of water (CD 8.1) or to different vegetation units can be combined. A new image is thereby obtained, which may serve as a mask in subsequent processing.

A new ascendant hierarchic classification can be applied to this image to be processed as done earlier. As it contains fewer pixels, a different and more diversified dendrogram would be obtained for themes that are not masked and would be retained.

8.2 EXAMPLE: IMAGE OF BRIENNE REGION

8.2.1 Statistical interpretation

■ Analysis

Various images obtained from division of a dendrogram at different levels that define different groups (Fig. 8.1) can be analysed. For each group, the displayed colour for each band is the mean of digital numbers of pixels belonging to this group (in TeraVue program).

The first division (Table 8.3) gives two groups, viz., very bright bare soils (group code No. 31) and all other themes. The second division displays water bodies. The third division indicates chlorophyllian crops which in the month of September could be beetroot and potato. The groups thus formed are derived from a division of groups previously detected.

Thus groups 5 and 6 were obtained from division of unit 4 which comprised slightly dark bare soils. Group 7 is segmented into many groups, viz., 8, 10, 11 and 13 (Table 8.3 and Fig. 8.1).

The first 11 groups are well separated from those that follow and hence the tenth division is robust. A new robust division can be made into 17 groups (Fig. 8.1).

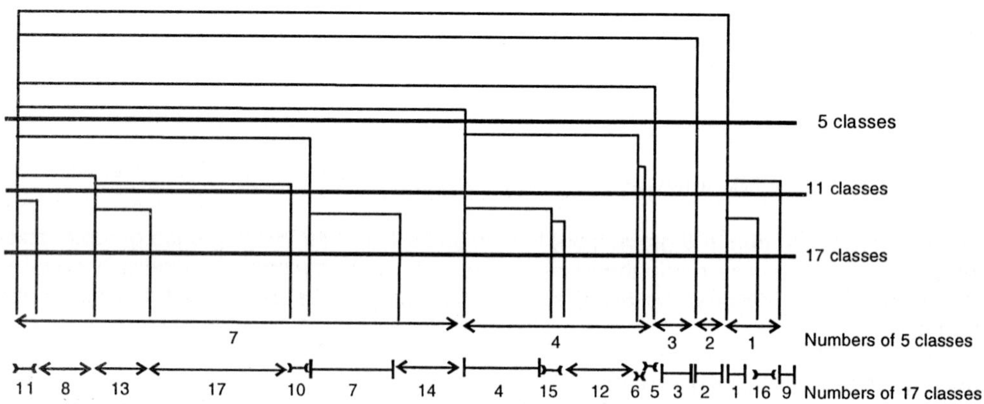

Fig. 8.1: Seventeen-group dendrogram obtained by ascendant hierarchic classification.

Table 8.3: Interpretation of dendrogram of image of Brienne

Division no.	Identification of groups appearing in division	Group code when 32 of them are required
0	Entire image	
1	Very bright bare soils	31
2	Water	25
3	Chlorophyllian crops	24
4	Slightly dark bare soils	16
5	Slightly clear water	23
6	Unclear water	22
7	Fairly chlorophyllian vegetation (including forests)	7
8	Vegetation less chlorophyllian than 7	5
9	Sparse chlorophyllian vegetation on very bright soils	32
10	Vegetation and shadows or conifers	11
11	Highly chlorophyllian vegetation	1
12	Bright bare soils	19
13	Vegetation more chlorophyllian than 11	6
14	Very dark bare soils	14
15	Borders of ponds: soil and water	18
16	Almost white soils	30

When all the 32 groups are completed, the following groups can be differentiated:

—12 groups for vegetation, the first of which (Table 8.3, No. 24) is separated very early;

—6 groups for water, 2 pertaining to turbid water;

—3 groups constituting a mixture of water and soil;

—9 groups for bare soils, 4 of which are very bright to white soils (in the Munsell code the value is greater or equal to 5) and five darker soils;

—2 groups represent soils sparsely covered by vegetation (winter crops already sown).

Clouds, being white, are confused with bright soils and their shadows with clear water. Only their shape is useful in differentiating them.

■ Characterisation of groups

The groups need to be characterised by their effective pixel number which is directly related to their readability on the image. As an approximate guide, the mean number of pixels can be taken as 36,000 (or 31.25 per thousand). Below this value, a group is under-represented.

Here, five units can be distinguished (Table 8.4) by taking the threshold limits as:

—twice the mean value or 72,000 pixels, division between A and B,

—mean value or 36,000 pixels, division between B and C,

—one per cent of the pixel population or 11,520 pixels, division between C and D,

—one per thousand or 1152 pixels, division between D and E.

The groups differ in effective number of pixels. In fact, the nine groups of units A and B, which have a mean frequency of more than 100,000 pixels, represent nearly 85% of the image processed.

Table 8.4: Regrouping of 32 groups of ascendant hierarchic classification (AHC)

Unit	AHC groups	Sum of pixels for the unit	Value per thousand	Mean number of pixels per group
A	12, 16, 7, 10, 8, 6	823,904	715.9	137,317
B	19, 9, 24	151,601	130.7	50,533
C	14, 25, 5, 27, 20, 11, 15	147,048	127.6	21,000
D	4, 30, 26, 13, 31	25,273	21.9	5,054
E	1, 3, 29, 18, 2, 23, 28, 17, 22, 21	4,174	3.6	379

■ Statistical interpretation according to objectives

If the objectives enumerated at the beginning of the study of ascendant hierarchic classification (sec. 8.1.2) are recalled, the following decisions can be taken:

For the first objective, viz., 'segmentation of the image prior to field verification', only groups with a sufficiently high frequency of pixels are taken to enable ready detection in the field. Hence, only groups A, B, C and D are used. Groups of unit E are too small as their mean area is only 15 ha. If pixels are not combined into a single zone, it is difficult to mark them in the field and establish sufficiently wide zones for making measurements. This also depends on the form and scattering of various map zones contained in each of the groups.

For the second objective, viz., 'general classification of the region', statistical analysis alone does not suffice; radiometric analysis is also necessary since the pixel frequency of a group is not the only criterion for retaining a given segmentation. If it is decided to take only groups of more than 625 pixels for a single zone, which corresponds to a resolution of 25 ha (as in the case of CORINE Land Cover), only groups A, B, C and D can be retained. It then becomes necessary to combine all other groups into the units retained. This necessitates radiometric analysis of the groups in order to give a name, at least approximate, to each group.

For the other two objectives, viz., 'thematic study of large zones' and 'precise study of objects', obviously the thematic significance of each previously defined group should be available.

8.2.2 Digital interpretation

One solution would be to define the groups from a field study. However, this involves a supervised method of classification. Here, the objective is to make a semantic analysis of groups from their digital characteristics.

Each group is defined by a statistical population for each band. Their spectral characteristics can hence be approximately described by means of digital numbers recorded in different bands of the image (three in the present case). These may be termed digital characteristics.

While comparing digital numbers it should be remembered that the pixel frequencies of various groups are extremely different. According to the radiometric model adapted to the values of digital numbers of an image (see Chap. 4 and Chap. 7, Fig. 7.4), the groups can be categorised into four themes (CD 8.1), viz., water, soil, vegetation and soil and vegetation.

For precise classification of objects from their digital numbers, it should be remembered that this approach is not effective unless the position of objects is studied band by band (see Chap. 9).

The following ten features may be defined based on their digital characteristics in the three spectral bands of SPOT (Fig. 8.2):

—clear water (group 25)	31,802 pixels,
—turbid water (group 13)	2,649 pixels,
—dark soil (group 15)	11,809 pixels,
—slightly dark soil (group 16)	198,786 pixels,
—bright soil (group 27)	17,194 pixels,
—white soil (group 31)	2,395 pixels,
—vegetation and shadows or conifers (group 11)	11,880 pixels,
—forest vegetation (group 10)	107,044 pixels,
—cultivated vegetation (group 24)	41,118 pixels,
—vegetation and soil (group 5)	26,983 pixels.

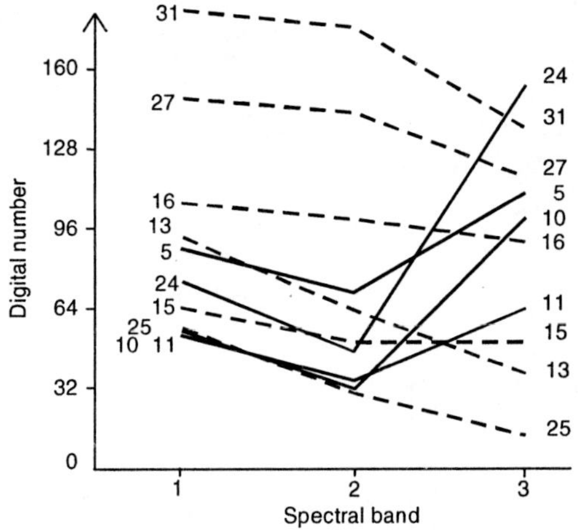

Fig. 8.2: Digital characteristics of ten objects (image of Brienne).
Numbers given to groups are defined in the text.

Other groups are combined into four themes defined by their digital characteristics, as shown in Fig. 8.3 (CD 8.1). For clarity of reading, the groups corresponding to vegetation have been distributed in two figures.

■ Water

The digital characteristics of water can be recognised from the fact that digital numbers are the smallest in band 3, but not in 1 or 2. The clearer the water (groups 25 and 26), the smaller the digital numbers. Moreover, the greater the concentration of suspended particles and hence more turbid (groups 23, 22, 21, 17, 28), the higher the values (Fig. 8.3, water).

■ Soils

Digital characteristics of soils exhibit little variation from band to band. However, digital values increase with increasing brightness of soils (Fig. 8.3, soils). In the case of white soils (groups 19, 27, 30), the colour due mainly to chalk or gravel material (groups 31, 29), a convexity is observed in the digital characteristic curve, which is distinct in band 2. For soils of medium brightness (group 16) the digital characteristic can be approximated to a straight line. Lastly, in the case of darker soils (groups 12, 14, 15) a concavity is observed in the digital characteristic, which is marked for band 2.

■ Chlorophyllian vegetation

In the example presented above, the digital characteristics of various types of vegetation are marked by low digital numbers in band 2, lower than those in band 1 and band 3, the latter DN values being very high. Two units can be identified. One consists of vegetation cover with the lowest values in band 2 (Fig. 8.3, forests and others), which corresponds to forests and riverine forests (groups 9 and 10) as well as much darker vegetation (group 11). The other unit comprises vegetation with the lowest values in band 2 and the highest values in band 3 (Fig. 8.3, crops and others). This unit represents plants for which chlorophyllian activity is the highest (group 24).

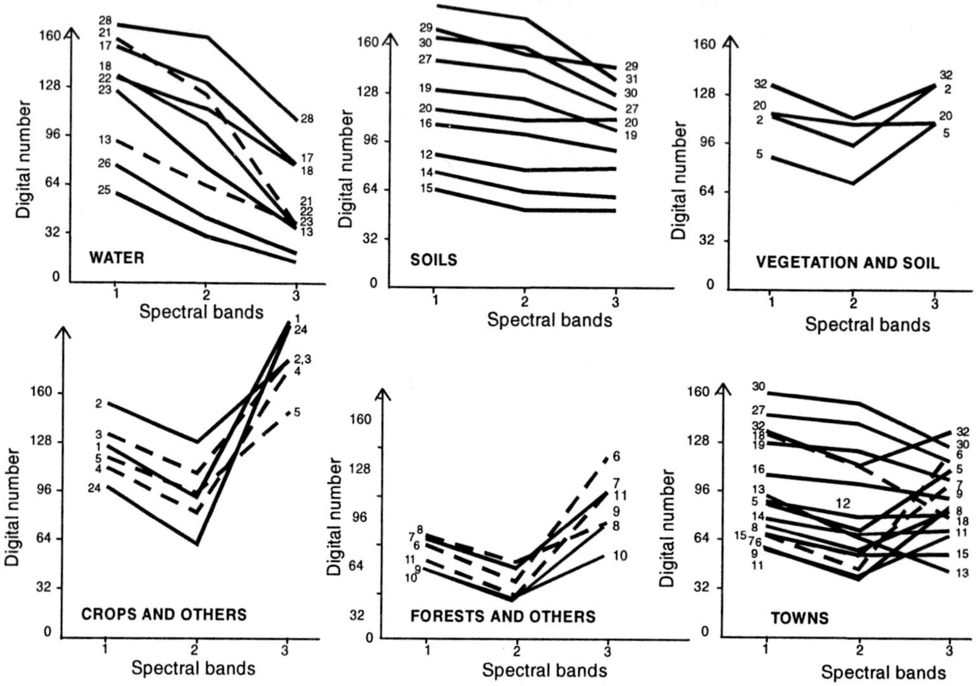

Fig. 8.3: Digital characteristics of 32 groups derived from AHC of image of Brienne.

■ Mix of soil and vegetation

After isolating major groups of objects, it was observed that areas exist which exhibit digital characteristics of vegetation (Fig. 8.3, vegetation and soils) but have a small concavity (groups 32, 2 and 5). Evidently these correspond to a mix of vegetation and soil, which has the effect of relatively decreasing the concavity of the curve of band 2. Groups 2 and 32 have very high values and on an IRC image a pink colour is perceived, indicating zones of sparse vegetation that may correspond to regrowth after harvesting or to germination after sowing.

■ Shadows and clouds

The IRC image shows some clouds (for example, row 725, column 605; row 728, column 450) as well as their shadows (row 660, column 580; row 660, column 430). Two clouds are represented by pixels belonging to groups 31, 29, 27, 20, 19 and 16. Hence, they are spectrally homogeneous and their digital characteristic is close to that of soils.

Shadows of clouds have very low values of digital numbers pertaining to groups 25, 26 and 15, interpreted as water or very dark soils.

■ Towns

The town of Brienne does not exhibit a specific digital characteristic but a mixed one (Fig. 8.3, town), as described below:

—digital numbers of bright zones (groups 18, 19, 27 and 30) which correspond to roofs of large industrial and commercial complexes are similar to those of bright soils;

—digital numbers of darker zones (groups 13, 14, 15 and 16) such as roads, shadows, roofs (tiles, slates) are similar to those of dark soils;

—zones of dense vegetation (groups 5, 6, 7, 8, 9 and 11) such as 'green belts' of the town (gardens, row trees etc.) exhibit digital characteristics of vegetation, partly mixed with shadows or soil.

The digital behaviour is characteristic of most urban agglomerations which, depending on their organisation, combine some pixels having a characteristic of vegetation and a large majority of pixels with the characteristics of soil. This heterogeneity characterises the digital behaviour of towns (CD 7.13). If the pixels are examined not by ascendant hierarchic classification, but directly in an IRC band, a still greater diversity of pixels will be obtained.

An urban agglomeration can hence be identified by its position in the landscape and its shape; digital heterogeneity confirms this assumption. It is therefore necessary to carry out spatial analysis of images.

8.2.3 Spatial interpretation

For spatial interpretation of groups derived from ascendant hierarchic classification there should be several images, each of which displays a single group. If all the groups are seen together the eye cannot perceive the spatial organisation of each image segment. The most useful tool for spatial analysis is an image in which the group under study appears in colour and all others appear black.

Visual spatial analysis of these groups is based on several factors which aid in analysing different image zones pertaining to a single group.

■ Number of pixels in a group

The number of pixels in a group determines the organisation of various groups. When the groups have a very small number of pixels, three cases are possible. A fourth case exists when contrarily the number of pixels is very high.

1. This is a group little represented on the image but which might include many more pixels if a slightly wider spatial field was investigated. Variation in the pixel frequency of this group may be very large if the zone of study is changed. Delineation of such a group is fully justified. It is recommended to expand the field of study for better characterising the group by a larger population to facilitate better evaluation.

   This may be the case of group 29 in the image of Brienne, which is always associated with group 30 or 31. If the image field of view comprises more bare soils on chalk (Chalky Champagne), it may be considered that this group has a higher pixel frequency.

   Such a situation occurs when image borders artificially intersect a unit such as a water body or when a new landscape is observed on the border.
2. A group may have a small number of pixels since it represents by nature a small surface area. Such is the case for example, of various types of water in ponds and lakes in the image of

Brienne. These can be combined into a single thematic unit if that is the objective (see third objective defined earlier). However, such a unit would evidently have a higher standard deviation that would affect subsequent processing. Groups 21, 22 or 23 correspond to this situation.

3. A group may have a small pixel frequency since the pixels combined are in fact a combination of different objects; they represent mixels. These only constitute a variant of a larger group (mixels of combination). On the other hand, the group may have spatial significance. It corresponds to a transition and its analysis provides an estimate of contrast between two spatial units (boundary mixels, see Chap. 15).
   In the image of Brienne, the pixels of group 3, which are contiguous to those of group 24, represent this case (the latter corresponds to chlorophyllian crops developed in fields of regular geometric form). When the zone is surrounded by bare soils, pixels of group 3 are observed on either side of the boundary between the two. This leads to their interpretation as mixels comprising a mix of bare soil and highly chlorophyllian vegetation. Groups 17 and 18 are mixels of a mix of water and very bright soil.
   This may also indicate a variation in land use of soils such as starting growth of some crops which on a given date have not yet sufficiently covered the soil. The same scene acquired 15 days earlier would not have shown this group since vegetation coverage was not yet significant (for example, less than 20%). The same scene acquired 15 days later would show a group of more numerous pixels corresponding to an ensemble of fields containing the same crop, since the coverage would be more than 40%. This group may be either combined with the group having the closest value of digital number or retained as the indicator of a temporal variation. Groups 2 and 32 in the Brienne image are of this type.
4. Contrarily, when a group has too high a pixel number it may be considered that the number of groups established by classification is insufficient. Groups 12 and 16 which represent bare soils of medium brightness are of this type. In such a case the number of groups needs to be increased. For this purpose all other groups are masked and a new ascendant hierarchic classification is started on the new image.

■ Form of image zones

Shape is one of the elements to which the human eye and brain are most sensitive. Analysis of forms constitutes an integral part of image processing. Unfortunately, it is not included in most image processing programs for remote sensing. It is still in the domain of research. It is true that the human brain is still much faster than processing algorithms that are now in full development and are based on fuzzy logic and on concepts (neural networks, etc.) and techniques of artificial intelligence.

In interpretation, form of image zones substitutes for the inadequacy of radiometric information. Forms can be readily analysed from simple values of perimeter and area, which can be directly obtained from a geographic information system. A greater or smaller consistency of image zones can be derived from it which, together with size, provides a typology of forms (see Chap. 11, Fig. 11.4).

The results of an ascendant hierarchic classification can be readily interpreted by separating image zones that are more or less large, are more or less compact and elongated in shape.

One of the common difficulties in interpretation of classification results is that a very large number of image zones of a single group covers a very small area. On the image of Brienne the 56 pixels of gravel pit water (group 28) are distributed in more than 30 zones! The same is the case with very bright soils (group 30) comprising more than 500 zones for 5626 pixels. A group (No. 21) with a small number of pixels (26 pixels) may also consist of only a single zone (pond in southern Brienne). Except in the last case, it is hardly possible to interpret the forms of very small image zones.

When image zones are large, it is possible to determine their shapes. The ponds and lakes in southern Brienne (group 25) correspond to compact and large forms. It is rare to observe this type of zone on an image after a precise interpretation.

Very large zones of the order of 1000 pixels (40 ha) for example are observed for group 19 (CD 8.1) but they are highly divided and distorted. As these are included in agricultural plots, they are inferred to be due to heterogeneities in soil and not to cultivation practices. Hence, similar forms ought to be observed on images acquired in different years. They can also be interpreted as erosional tracks or indices related to soil depth.

In some cases highly geometric forms of almost all image zones clearly indicate that they belong to agricultural plots, as in the case of group 24 (CD 8.1).

In some other cases highly tapered, elongated or distorted forms are noticed, as in the case of group 9 (CD 8.1) which follows the Aube River. It can be concluded that this group is associated with the river. This may be a land cover type related to alluvial zones.

■ Localisation of image zones

Localisation of image features can be studied by various criteria, viz., geographic position, dispersion, synthesis of forms and neighbourhood.

□ Geographic position

Geographic position forms a source of information about image features of a group under study. First of all, image regions where features are observed and not observed are determined. Thus bright bare soils are found only in the north-west, which corresponds to the chalky Champagne region. In the south-east all the soils are more or less dark (humid Champagne). Position of image features of a given group can be combined with other sources of information to facilitate interpretation of the group. For example if clear water is spectrally confused with shadows, then the topographic data indicate that the zone is on a northern exposed slope, prompting the conclusion that the group represents shadows.

□ Dispersion

Another analysis is based on dispersion of image zones of a group. Group 5 comprises a large number of small image features that are distributed throughout the cultivated zone (CD 8.1) and hence it ought to correspond to a particular vegetation cover. On the other hand, group 19 a soil of medium brightness is not scattered. It is almost absent in valley and forest zones and is mostly situated in the north-western half of the image (CD 8.1).

□ Synthesis of forms

The eye is able to identify on the monitor associations of image zones which indicate a preliminary hypothesis of interpretation. Piles, patchiness due to uncontrolled urban sprawling and alignments are determined in this way. Piles are combinations of image features close to one another giving an impression of compactness.

When an image is enlarged, an image zone earlier considered to be compact may reveal a large number of pixels pertaining to other groups, giving the impression of patchiness like a moth-eaten fabric. A somewhat similar situation is observed in the forest of Brienne (groups 9 and 10) where the effect is more apparent when all the groups other than 9 and 10 are depicted in black (CD 8.1).

Alignments not made up of a single digital zone but of many straightline segments are also determined in this way since the eye interprets them through integration as a straight line. Various image features in group 11 are not contiguous and correspond to vegetation with shadows, giving the impression of the Aube course (CD 8.1).

Image shapes of urban agglomerations are interesting. They appear as a unit of piled up pixels if taken group by group. However, they also exhibit alignments due to roads and patchiness associated with gardens (CD 7.3).

□ Neighbourhood

Neighbourhood of image features are very important for interpretation of objects. Evidently, groups 9 and 10 (CD 8.1, band 38) can be interpreted one knowing the other. Both mainly represent riverine forests developing in the Aube valley. In the case of crops, groups 1, 2, 4 and 5 commonly form the edges of cultivated plots of group 24.

It is also on the basis of unassociated and more distant features that a shadow is interpreted as that of a cloud. It was computed (see above) that shadows for the two clouds are offset roughly by the same distance, viz., 85 rows or 1.6 km northwards and 20 columns or 400 m westwards. This evidently depends on the time (here, 10 h 51 min 13 s) and date (16 September 1996) of acquisition of images, which determine the solar angle (43.1° solar height with an azimuth of 164°, the angle of incidence being 0.9°).

When more detailed analysis of image features is desired, structural classification methods need to be employed (see Chap. 11).

8.2.4 General interpretation

A general interpretation of results of ascendant hierarchic classification can be carried out using these elements. Groups are examined one after the other using three sources of information, viz., statistical, spectral and spatial. They are presented through major thematic groups.

The units obtained from this preliminary ascendant hierarchic classification can be categorised into the following groups: crops, forest and riverine forests, water, water and soil mix, dark soils, medium soils, bright soils and vegetation and soil mix.

■ Crops

Chlorophyllian crops are primarily represented by group 24 which corresponds to agricultural plots clearly identifiable by their linear boundaries. Considering the date of image acquisition, these crops may be beetroot or potato for example. This group comprises pixels corresponding to the chlorophyll pole on the two-dimensional histogram (Fig. 8.4). Having a medium pixel frequency, it represents a typical group that can be clearly identified in the image and on the ground.

This group may be combined with groups 1, 2 and 3 which are characterised by low frequencies and whose pixels are close to those of group 24. On the 'red-infrared' two-dimensional histogram

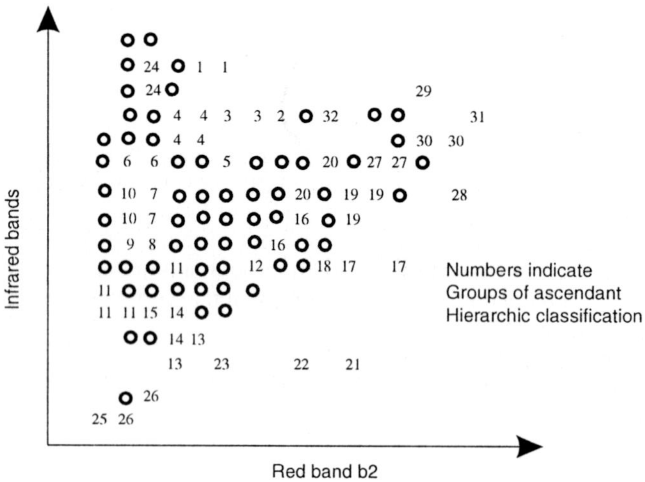

Fig. 8.4: Distribution of 32 groups defined by AHC on a red-infrared plane.

these groups are situated in the proximity of soil cluster and chlorophyll pole and hence are intermediate (Fig. 8.4).

This group can also be associated with groups 20 and 32 corresponding to germinating crops.

Group 2, represented by higher values of digital numbers in the red band can be further differentiated. It is probable that vegetation cover is less and hence this group comprises mixels pertaining to a bare soil and a cultivated plot.

The pixels of group 4, with a greater frequency than the preceding, correspond to heterogeneities in the agricultural plots of group 24. Group 4 can hence be interpreted as a crop that is relatively less chlorophyllian than that of group 24, due to earlier ripening for example.

Group 5 sometimes represents mixels surrounding the agricultural plots of group 24, sometimes heterogeneities in agricultural plots and sometimes a crop when it occupies several agricultural plots. This is explained by its position on the two-dimensional histogram (Fig. 8.4). Groups of this type are difficult to classify without ground information. They add to the difficulties of classifying all the pixels from a single image.

The groups of this unit are situated in the north-west half of the image in chalky Champagne and are absent in valleys. They are also observed a little north and south of Brienne. The unit of 'crop' groups is practically absent in the forest zone in the south.

■ Forests, riverine forests and grasslands

Other chlorophyllian units of the image belong to groups 6, 7, 8, 9, 10 and 11 which are very close to one another on the 'red-infrared' two-dimensional histogram where they are situated in the centre and towards the base of the chlorophyllian zone (Fig. 8.4). The frequencies are high in every group.

Group 6 is the most chlorophyllian of this unit and is observed in valleys. Interestingly, this group marks an ancient meander of the Aube valley (in a square situated between the points defined by 'row 105-column 225' and 'row 125-column 245'). It is also found in large forests in a disseminated manner except in the south-west where it constitutes compact forest plots. It is also observed in some agricultural plots. Hence this group does not correspond exclusively to forest zones but also to chlorophyllian herbaceous communities.

Group 7 is most often located in valleys or in the south-east part in the vicinity of darker soils. It ought to represent grasslands.

Group 8 is geographically closely associated with group 7. Based on its digital characteristics it can be considered as belonging to another group of grasslands.

Group 10 constitutes the major part of forest cover. In the image, it is recognised from the effect of relief due to different heights of the summits that generate shadows. This is the most chlorophyllian group constituting forests. It covers a major part of the valleys.

Group 9 is highly scattered but occurs only in valleys and forests. It is situated between groups 10 and 11 on the red-infrared two-dimensional histogram.

Group 11 has the lowest values in the infrared band. It can hence be concluded that it comprises either chlorophyllian vegetation and shadows or conifers for which the digital numbers in band 3 of SPOT satellite are always less than those of broad leaves. In the Brienne forest a compact zone is interpreted as a conifer plot. Image features that follow streams in the midst of riverine forests are interpreted as trees and shadows.

The unit of these groups corresponds to riverine forests developing in secondary valleys and around the Aube. Its full extension is observed in the south-east part of the image in humid Champagne which comprises grasslands and forests.

■ Water

The various types of water can be quickly separated while analysing the classification dendrogram. They are recognised from their digital characteristics represented by low values in the infrared band

(band 3). Their position in the red-infrared histogram is distinct (Fig. 8.4). Clear water has the lowest values (25 and 26) and more or less turbid water (21 and 22) or shallow water (17 and 18) occurs below the soil cluster.

Very shallow waters in which the chalky bottom can be seen are characterised by high digital numbers (28). It should also be noted that all the groups, except numbers 25 and 26 have very low frequencies. These correspond to various ponds.

Ponds or gravel pits in north-western Brienne are associated with various groups, which indicates that the waters contain highly variable quantities of suspended particles. Groups with low frequencies cannot be subjected to detailed spatial analysis. At best, one can state which pond contains what type of water, as in the case of groups 17, 18, 21 and 22.

Group 23 is essentially observed in a gravel pit in north-eastern Brienne. This gravel pit extends over an area of 20 ha (500 pixels) and exhibits a large variation since it comprises pixels belonging to 6 groups of water and 6 groups of soil.

The other three groups, viz., 13, 26 and 25, constitute water types that are increasingly transparent (Fig. 8.4) and exhibit a spatial organisation. Group 25 corresponds to two reservoirs of the Aube south-south-west of Brienne. Group 26 is situated on the borders of the latter and in particular in the concave edges where water is shallow and where suspended matter could also be higher. Lastly, pixels of groups 13, 16 and 12 are observed on the shores of reservoirs.

Surprisingly, no difference exists between waters of the two reservoirs.

It is observed that these groups do not facilitate tracing the Aube River. To do this, group 15 should be combined with group 11 but still there would be no satisfactory continuity.

When all these groups are combined into a single unit and displayed by a single colour, the envelope encompassing the water bodies fairly forms the boundary of the humid Champagne region.

Consequently, for a total interpretation of the theme 'water' in this image groups 13, 15, 17, 18, 22, 23, 25 and 26 ought to be preserved and others masked, and a new ascendant hierarchic classification attempted on the new image.

As mentioned above, a shadow of clouds is found to be associated with group 25. If this group was not connected to clouds, it would have been readily mistaken for a pond because of its shape. Hence, the possibility of confusion between water bodies and cloud shadows should be anticipated during classification.

■ Soils

Soils are represented by separate groups, except some which correspond to a mix of soil and vegetation or clouds. Three major classes can be identified in this soil cluster depending on their brightness.

Four groups show convexity in digital characteristic and hence represent bright soils. These are groups 31 and 30 whose frequency is several thousand pixels, and group 27 which consists of 17,000 pixels. Group 19 comprising more than 60,000 pixels also belongs to this class.

The classic chorological law is observed in this image relative to soils (see parallelepiped classification, Chap. 9): brightest soils (31, 30) are surrounded by darker soils (27, 19) (CD 8.1).

Groups 14 and 15, dark soils, are at the bottom of the soil cluster (Fig. 8.4). They are mainly situated on the boundary between humid Champagne and chalky Champagne. They are fairly scattered in small and compact zones constituting agricultural plots and boundaries of cloud shadows. Group 15 is also observed on the reservoir shore. It may represent moist soil.

Groups 12 and 16 are located between these two extremities. Their frequency is very high and these two together represent nearly 38% of the image area. Geographically, these are found everywhere, in chalky Champagne (where group 12 is predominant) as well as in humid Champagne (where group 16 is predominant). They do not have specific shapes and occur as scattered zones as well as compact agricultural plots.

These two groups situated in the central zone of the two-dimensional histogram (Fig. 8.4) have no neighbours and ensure 'filling' of the image. All the pixels that were not classified satisfactorily in the preceding groups are observed in these two groups. For a better understanding of the content of these groups, only these two should be retained in the colour infrared image by masking all the others, and a new ascendant hierarchic classification carried out.

8.2.5 Classification quality assessment

The quality of any method of classification, including the ascendant hierarchic, ought to be verified. The principle of this method is to classify all the pixels in an image. A pixel belongs to only one group and consequently division between groups is perfect. However, all pixels belonging to a single group do not have a single digital characteristic and it is not always possible to identify chorological laws that explain their spatial organisation.

The quality of ascendant hierarchic classification can be assessed by computing the distances of all the pixels in the centres of groups taken as reference (called nuclei) (CD 8.2). At the end of this computation a three-dimensional histogram is obtained from which the probability of each pixel belonging to the group into which it is classified can be ascertained (Fig. 8.5). Groups 12 and 16 contain a very large number of poorly classified pixels.

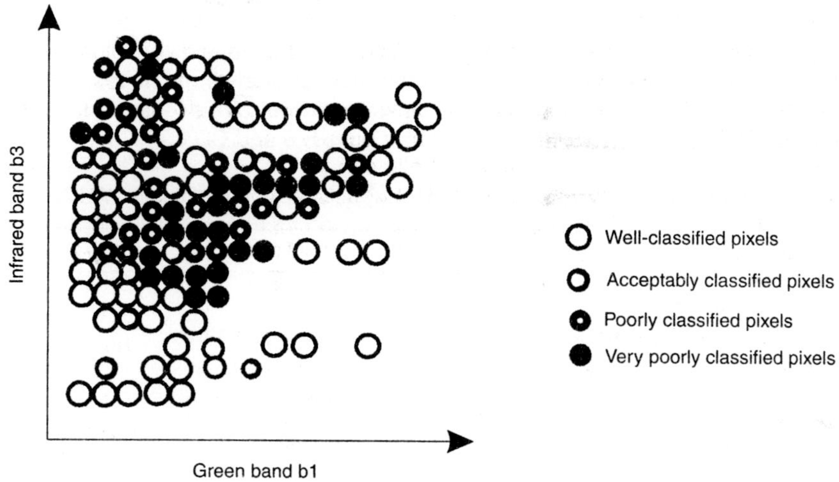

Fig. 8.5: Distribution of 32 groups defined by AHC on a green-infrared plane, indicating quality of classification.

8.2.6 Response to identified objectives

■ General classification of units

To achieve the second objective, viz., general classification of most characteristic units of the region, the following eight units (CD 8.3) are retained by reorganising the 32 initial groups (Table 8.5).

The radiometric regroupings carried out during the creation of these units can be seen from Fig. 8.4.

Table 8.5: Reorganisation of 32 AHC groups into 8 units

Unit No.	Legend	Groups combined	Colour	Area (km^2)
10	Cultivated plots	1, 2, 3, 4, 5, 24	Yellow	37
20	Forest and riverine forest	6, 9, 10, 11	Dark green	98
30	Grasslands	7, 8	Bright green	80
40	Soils and vegetation	12, 16	Bright maroon	175
50	Clear water	13, 25, 26	Dark blue	15.5
60	Very dark soils	14, 15	Dark maroon	18.5
70	Turbid water	17, 18, 21, 22, 23, 28	Bright blue	0.2
80	Bright soils	19, 27, 29, 30, 31	Beige	36

The groups obtained are of sufficient areal extent to be used for regional analysis. If necessary the two types of water can be combined.

■ Reconnaissance

To achieve the first objective, viz., preparation of a reconnaissance map from the preliminary segmentation of the region, it is imperative to proceed up to 32 groups and regroup them as mentioned above (CD 8.3). Groups 12 and 16 constituting unit 40 (Table 8.5) have a very high frequency—more than 400,000 pixels (38% of the image)—and hence further analysis is necessary. For this, after masking (CD 8.4), a new ascendant hierarchic classification of unit 40 alone is carried out in 32 groups. These are then reorganised into 5 new units (Table 8.6 and CD 8.4).

Table 8.6: Legend of second AHC of poorly defined groups of the first AHC

Unit number	Legend	Band colour	Area (km^2)
1	Very bright soils	Beige	44
2	Soils of medium brightness	Bright brown	55
3	Sparse chlorophyllian vegetation	Green-blue	45
4	Very dark soils	Maroon	31
5	Masked groups	Black	286

Group 5 corresponds to pixels already classified earlier and masked (units 10, 20, 30, 50, 60, 70, 80).

The results of the two classifications should then be integrated into a single image (CD 8.4) by an appropriate arithmetic combination. Groups of the first classification carry numbers in steps of ten from 10 to 80. By combining the two bands, unit numbers are obtained as 15, 25, 35, 41, 42, 43, 44, 55, 65, 75 and 85. All the pixels pertaining to units 1 to 4 of the second classification correspond in fact to only unit 40 of the first classification and hence groups 41, 42, 43 and 44 are obtained by addition. All pixels of unit 5 of the second classification belong to units 10, 20, 30, 50, 60, 70 and 80 of the first.

■ Delineation of major thematic zones

The third objective of classification was to determine major zones of a given theme. Our interest was riverine forests, recognisable from their shape that follows streams. The image classified into 32 groups was geographically investigated for all groups situated in riverine areas (CD 8.5). The following groups were thus identified:

—4, 5 and 24 corresponding to crops;

—6, 7, 8, 9, 10 and 11 corresponding to riverine forests and forests (which include practically all the groups in which chlorophyllian biomass is found);

—group 13 (which represents water) and groups 12, 14, 15, 16 and 19 corresponding to bright to dark soils.

Thus 15 of the 32 groups determined are situated in the riverine zone, of which 9 are major. This indicates the heterogeneity of this group and the diversity of these media, which constitutes a preliminary conclusion. If a new image was generated comprising all these groups, three-quarters of the image would be covered. Hence there is no unambiguous relationship. All these groups are necessary to delineate riverine but combinations of these groups clearly describe units other than riverine. The latter corresponds to a specific spatial organisation of these groups. Other types of classification such as OASIS are employed to determine these spatial organisations (see Chap. 11).

A similar approach can be followed for the theme 'chalky soil'. In this case, 11 groups would be needed, of which the most important 6 practically correspond to all bare soils and the others to a mix of chlorophyllian vegetation and bare soil and to fields in which crops still exist.

It is hence possible to characterise various landscapes with the same groups of objects. Differentiation occurs by the degree of abundance of the groups retained and their relative geographic positions. This leads to structural investigation of remote sensing images (see Chap. 11).

■ Precise thematic study

The fourth objective is the precise investigation of a feature such as water. It was seen in the second classification that clear water is readily separated from the rest of the image. It also comprises cloud, detected only when more groups are identified. A fifth classification is necessary for detecting a new group corresponding to turbid water. In the sixth classification the latter is subdivided into two. However, an eighteenth division is required for obtaining all the groups of water. A large number of subdivisions is necessary for precise identification of an object. This is achieved by a combination of 9 groups followed by masking and a new classification.

This detailed analysis led to identification of 21 groups illustrating the large diversity of pond waters situated in the Brienne region at the border between chalky and humid Champagne. Three groups characterise the three reservoirs (CD 8.6). The banks of reservoirs are represented by a large number of pixel groups, which also attests to their wide heterogeneity and diversity. Cloud shadow is still mixed with some types of water.

We can proceed further in this way by making more detailed groups. However, we then have to use bands b1, b2 and b3 directly, employing parallelepiped classification and limiting the analysis only to the reservoir zone on the image.

8.2.7 Conclusion

It follows from this study that ascendant hierarchic classification is very useful when no field spectral data or measurements are available. Computationally this is a very fast method. However, if the software provides choice, the distance and ultrametric parameter should be chosen carefully since results may be relatively different. This choice is not very simple if the objective is not clearly defined and if sufficient information on statistical methods is not available. Unfortunately, this is frequently the case in programs in which descriptions of statistical methods are often discrete.

While this method is rapid for computations, involving a few seconds, much time is needed for a correct interpretation. In fact, the statistical data, spatial (geographic) data and reference spectral data in the interpretation model (established on the basis of reflectance measurements) should be studied together. It should be remembered though that a serious and minute study ensures satisfactory results for all the four objectives proposed earlier.

8.3 CLASSIFICATION BY MOBILE CENTRES (OR "K-MEANS")

8.3.1 Principle

The method of classification by mobile centres is based on measurement of distance and an initial choice of the number of groups or 'nuclei'. For each, a certain number of pixels are randomly picked to statistically determine the radiometric centre of gravity of each nucleus (Fig. 8.6A). Then the distance of every pixel of the image to each of the nuclei is computed. Each pixel is attributed to the nucleus closest to it. From this, a kind of segmentation is obtained, constituting as many groups as there are nuclei. However, in spite of the fact that all the pixels are classified, they are not necessarily well classified. In fact, there is no certainty that the nuclei drawn at random are the most discriminant.

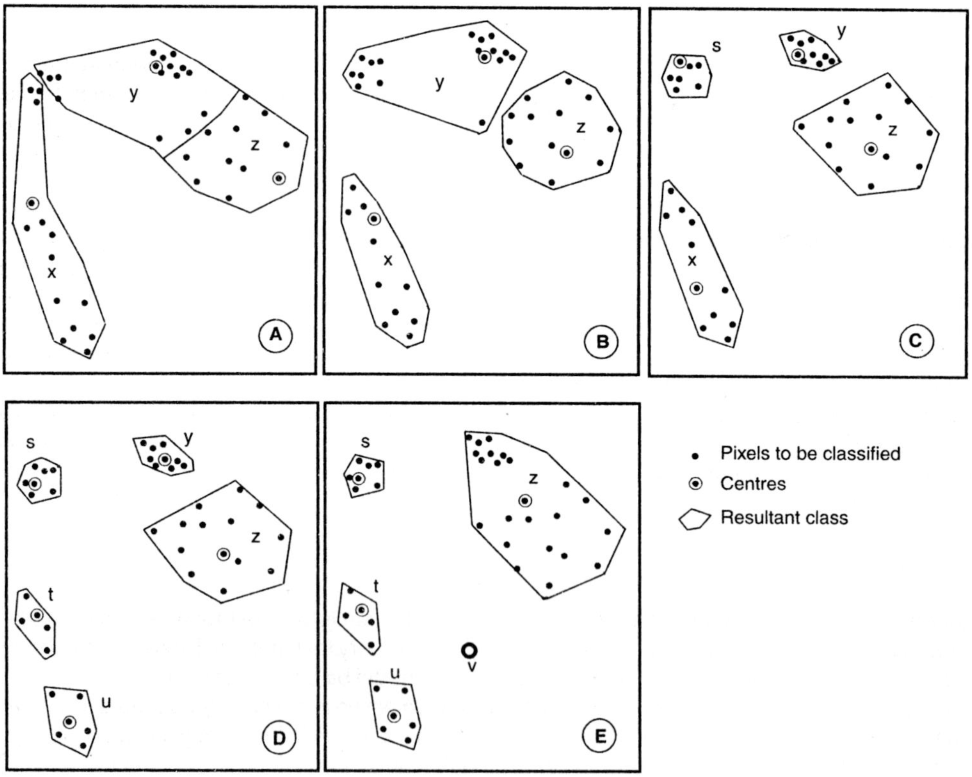

Fig. 8.6: Scheme of classification by mobile centres. Each of the five iterations (A to E) shows pixels to be classified, centres of nuclei, results obtained at the end of iteration (zone around the pixels placed in a single group).

Hence several iterations are carried out. After the first iteration, the centre of gravity is computed for each nucleus using all the pixels grouped in it, resulting in the definition of a new nucleus (Fig. 8.6B). If the nuclei of the first iteration include pixels far from the centre of a nucleus, the new centre will be quite different (nucleus x) and hence moves in the second iteration. This method is therefore known as classification by mobile centres. Other criteria can also be used to characterise new nuclei. Some nuclei may be removed if their frequency is too small, for example 1 per 10,000. A nucleus may be split into two if its variance is very high (Fig. 8.6C: nucleus y was subdivided into two, s and y; similarly, in the next iteration x was divided into t and u). Lastly, nuclei whose distance is very small can be combined into one (Fig. 8.6E: nuclei y and z). New nuclei can also be added (Fig. 8.6E, nucleus v).

All the pixels from these new nuclei are reclassified and the result of the next iteration obtained. The result of *n*th iteration is then compared with that of *n*+1. If they are still different, iteration is continued; if they are similar, the process is stopped. Various criteria are employed to decide whether to continue the process or not; most common is the criterion of convergence of the sum of distances of pixels to the nuclei to which they are allotted or the criterion of the ratio of mean intragroup variance to the intergroup variance.

It is possible that during an iteration no pixel is classified into one of the nuclei (Fig. 8.6E, nucleus v). Thus if originally *N* groups are required, at the end of classification only *N*–1 or even less number of groups may be obtained.

The choice of number of groups is very important. This is more difficult to decide in this method than in ascendant hierarchic classification if a precise objective is lacking. One solution is to examine the results obtained when a successively different number of groups is required.

8.3.2 Method of interpretation

■ Number of iterations

When an image is classified into a given number of groups, it is necessary to ensure that the number of iterations applied to the system is greater than the number of iterations required for converging. Otherwise, there would be risk of an incorrect result since convergence might not be reached. The effects of various iterations can be verified by analysing as earlier the consequences in spatial and spectral domains using a three-dimensional histogram.

Thus in the image of Brienne, if a single iteration is employed for two nuclei the result differentiates bright soils and shallow waters (G2) from the rest of the image (G1). The separation corresponds to a more or less high value of digital number in band b2 (Fig. 8.7A and CD 8.7). These groups do not represent the same area (Table 8.7). With two iterations, separation occurs at lower values (Fig. 8.7B and CD 8.7). With three iterations (Fig. 8.7C and CD 8.7), the boundary between the two groups becomes less distinct. This represents the case of mobile centres which is difficult to interpret thematically. A similar situation exists in the case of 4 and 5 iterations in which adjustment modifications occur. In iteration 6 (Fig. 8.7D and CD 8.7), pixels with low values in the red band (chlorophyllian vegetation and clear water, very dark soils) are combined into a group. Between iterations 7 and 10 (last) separations occur between soils more or less covered by vegetation (situated at the centre of the two-dimensional histogram) and more or less clear waters (Fig. 8.7E and CD 8.7).

Variation of nuclei with the number of iterations can hence be exactly traced in digital space (Fig. 8.7) as well as in geographic space (Table 8.7 and CD 8.7).

If the number of iterations is small, this means convergence is fast. The resultant nuclei have better a chance of being well-separated radiometrically relative to one another.

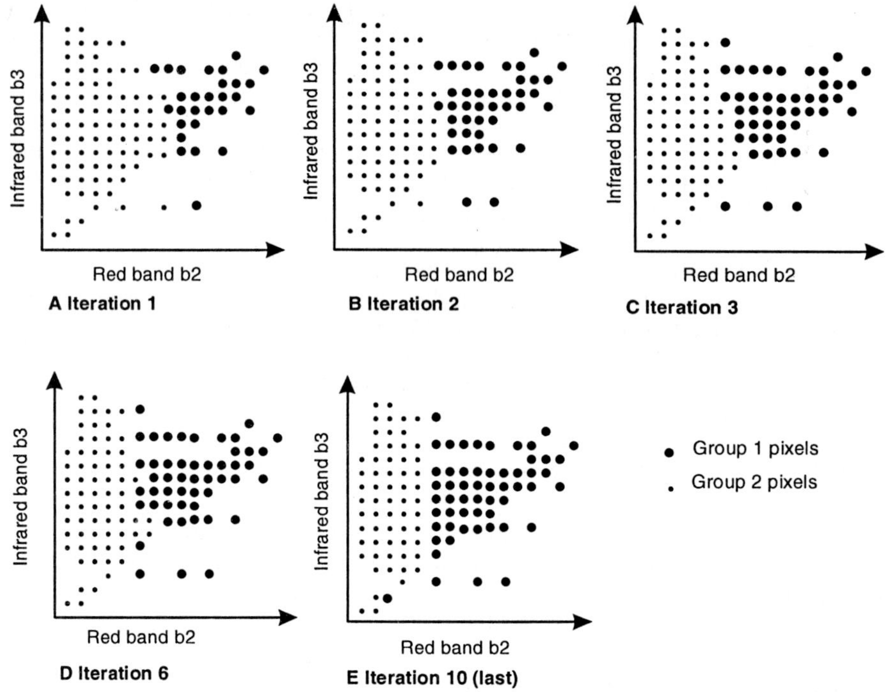

Fig. 8.7: Radiometric representation of two groups for different iterations.

Table 8.7: Variation in areas of two groups with number of iterations

Iteration	Group 2 area (ha)	Percentage	Group 1 area (ha)	Percentage
1	40,775	88.5	5,305	11.5
2	37,076	80.5	9,004	19.5
3	34,257	74.3	11,823	25.7
6	30,266	65.7	15,814	24.3
10	28,387	61.6	17,693	38.4

■ Number of groups

The number of iterations necessary for convergence and the number of groups actually identified can be analysed by demanding an increasing number of nuclei (Fig. 8.8).

Obviously, a single group is obtained in a single iteration. This represents the entire image whose characteristics correspond to the mean of all the pixels.

From 2 to 4 nuclei, 2 to 4 groups are obtained. As the number of groups increases, vegetation, water bodies, dark soils and bright soils are gradually differentiated. The number of iterations is only 5 for 4 groups and they are distinctly isolated. A satisfactory segmentation is achieved with 4 groups.

When 5 groups are required, 11 iterations are necessary, which is too high for such a small number of groups. When 6 nuclei are demanded, only 5 groups are identified. With 8 nuclei, 24 iterations

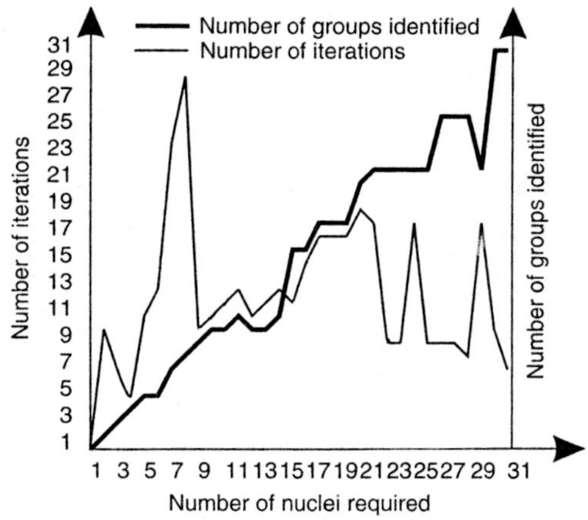

Fig. 8.8: Relationship between number of iterations and number of groups actually identified for a number of nuclei varying from 1 to 32.

are needed for convergence. Further on, a new stability is observed for 10 groups with 11 iterations when 10 to 14 nuclei are demanded. With 10 groups the following features are distinguished:

—cultivated plots,
—riverine forests and forests, not separated,
—grasslands,
—four surface states of bare soils,
—three types of water.

Subsequently, if the number of nuclei is increased the number of iterations increases with the number of groups. At 22 nuclei the number of iterations remains at 9 wherein a new stability is observed. One more step occurs at 26 groups and the next stage at 31 groups. Thus several points of stability are observed in the number of iterations for certain values of the number of groups.

Using this model, the optimum number of groups to be retained can be chosen. In the present case, we can take 4, 10, 22 or 31 groups (CD 8.7). Choosing an intermediate number of groups does not lead to vigorous interpretation.

As in the case of ascendant hierarchic classification, a legend has to be given to every group. If some groups have very low frequency, they can be combined with others; this depends on the objective pursued, as seen earlier. The 31 groups of the last classification have been regrouped into 11, similar to the case of ascendant hierarchic classification and with the same colours (CD 8.7).

Groups situated in the outer part of the two-dimensional histogram are classified quite fast whereas at the centre of the histogram with similar values between IR and R the pixels are poorly differentiated. Hence it is necessary to apply the same method as that used in ascendant hierarchic classification, viz., masking the classified and stable pixels, recomposing an IRC image based on the centre of three-dimensional histogram and making a new classification.

If the quality is assessed by estimating the probability of correct classification (see maximum-likelihood classification), it may be concluded that this method of classification is less accurate than the preceding.

8.3.3 Comparison with ascendant hierarchic classification

Validity of a classification cannot be determined without field data. However, the results of the two classifications, viz., ascendant hierarchic (AHC into 11 groups) and mobile centres (MCC into 31 groups, regrouped into 11 units), can be compared. The result (CD 8.7) can be broadly analysed by comparing the areas of each of the 11 groups (Table 8.8).

Table 8.8: Comparison of results of ascendant hierarchic (AHC) and mobile centres (MCC) classifications

Legend	Colour	Area (km^2) AHC	Area (km^2) MCC
Clear water	Dark blue	15.5	15.2
Turbid water	Light blue	0.2	0.2
Very bright soils	Creamy	36	31
Bright soils	Beige	44	68
Fairly bright soils	Light brown	55	51
Dark soils	Maroon	31	44
Very dark soils	Deep maroon	18.5	17
Sparse vegetation	Bluish-green	45	53
Grasslands	Light green	80	63
Forest and riverine forests	Dark green	98	78
Crops	Yellow	37	40

The difference in the area of pixels between the two classifications is about 88 km^2, which represents 19% of the area studied. This difference is not very high but the two classifications cannot be considered to be giving identical result. It is nevertheless interesting to visually compare the results of these two classifications. It was observed that major units, viz., the Aube valley and secondary valleys, large forests, bright soils of chalky Champagne, dark soils of humid Champagne and water bodies are identified without ambiguity. However, in a more detailed analysis many differences are observed in the boundaries. Hence a theme-by-theme analysis is necessary.

Zones classified as water (and shadow) hardly differ between the two classifications. The cultivated plots are more compact in the classification by mobile centres: no pixels belonging to forest-riverine forests are observed here. The forest group is less extensive in the mobile-centres classification since some of the border pixels are classified in the grasslands group. This in fact represents a modification in the boundary values of digital number between the two groups, values situated at the centre of the three-dimensional histogram which will have to be specifically reclassified. The forest-riverine forests zones become "nibbled" with the grasslands group. On the other hand, grasslands are converted in the MCC to sparse vegetation. In this case also the same boundary effect and the same position of pixels, viz., centre of the three-dimensional histogram, are observed.

Very bright soils, less significant in the classification by mobile centres, are partly integrated in the group of 'bright soils' of the ascendant hierarchic classification. The same effect extends to other bare soils; transfers occur from one group to another, which also takes place for pixels located at the centre of the radiometric histogram.

8.3.4 Conclusion

The method of mobile centres is very fast for computations. Its importance lies in the fact that an approximate idea of the stability of groups is obtained from the number of iterations. It is especially useful when neither field data nor spectral characteristics of objects in the area of investigation are available. However, its thematic interpretation is not easy and demands much time. It serves as a rapid method for reconnaissance study of an unknown region. It should be combined with other methods in order to extend interpretation up to characterisation of the various groups identified.

ne major part (99.8%) of the image consists of 10 classes. However, 4 classes are found to
n 100,000 to 400,000 pixels. They correspond to the central part of the histogram which obviously
nes several objects that cannot be differentiated in the continuum of the histogram. This
strates the limitations of this method based on a single-dimensional approach for deciding the
of classes.

is method, very fast, can be used to roughly divide the image according to various themes, the
eing very general. In the image of Brienne the chalky Champagne region is distinguished from
Champagne and similarly highly chlorophyllian crops, forests, grasslands and three types of
odies are clearly differentiated. Such a method can be useful for a very general analysis.

Multispectral segmentation

ee-dimensional histogram can be directly segmented making use of only the simplified
etric model comprising the objects bare soil, chlorophyllian vegetation, water and shadow (Chap.
7.4).

3–b1 plane of radiometric values, soils are represented by a linear cluster. For the image of
, the linear soil cluster (LSC) is defined by the equation:

$$LSC = 0.56\ b3 + 0.83\ b1 \quad \text{(Fig. 9.3)}$$

the pixels situated on both sides of this straight line and pertaining to the cluster can be
d as a single unit. If more detailed analysis is desired, the cluster thus obtained can be further
nto several parts that will determine soil groups, from very bright to very dark (Fig. 9.4 A to C),
the three groups in the Brienne image (CD 9.2). This depends on the objective identified. In
wing example of Brienne, six groups, viz., 6, 0, 3, 11, 16 and 22, are used (Fig. 9.5).
hundred and forty-four classes of pixels or different types of spectral characteristics exist in
e-dimensional histogram of the IRC image. Of these, the soil cluster comprises 90, which

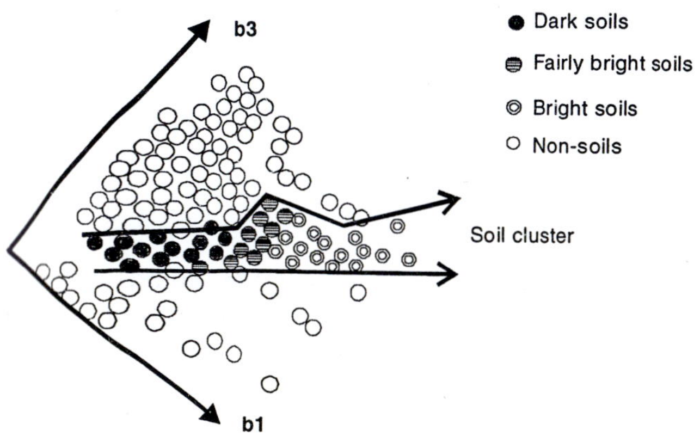

Fig. 9.3: Segmentation of soil cluster in b3–b1 histogram.

9

Supervised Classification

9.1 PARALLELEPIPED CLASSIFICATION

Parallelepiped classification is based on a radiometric model but not on measurement of distance or probability. Every pixel can be represented by its radiometric characteristic on a Cartesian diagram with as many axes as bands. On such a diagram every pixel is situated in an *n*-dimensional hyperspace. Most often we operate in a three-dimensional space to represent the image using three basic colours. However, this classification develops on a two-dimensional space, viz., the monitor screen. Hence the software employed should be able to support three-dimensional information.

If field measured reflectance values are available, they are converted to digital numbers by means of models that use various satellite parameters. These are used as references on a three-dimensional histogram from which parallelepiped classification can be constructed. If field radiometric data are not available, important objects to be interpreted may be defined in the field. Digital numbers of these features can be determined from their geographic location and used as references on the three-dimensional histogram. When no field data are available, the classic radiometric model is employed (Chap. 7, Fig. 7.4).

The classic radiometric model facilitates differentiation between soils, water bodies and chlorophyllian vegetation. Obviously, depending on the regions under investigation, climatic zones, etc., several other features can be added to these three basic objects. Examples are clouds and their shadows, shadows of large topographic elements, bush fire ashes, etc. Towns whose centres often appear as very dark soils are characterised by a mix of various vegetations, bare soils, water, etc. Their identification is based more on structural analysis than textural. It should be remembered that it is the surface of objects nearest to the satellite which is interpreted, viz., surface of bare soils or upper part of canopy in the case of vegetation. In the latter, obviously the tallest vegetation is interpreted. Interpretation of sub-layers of vegetation is possible only when coverage of higher layers is incomplete, as in the case of deciduous forests during fall and winter.

9.1.1 Segmentation of radiometric scatter diagram

Classification amounts to segmentation of the radiometric scatter diagram. Segmentation can be carried out in several ways. In all cases, the radiometric space is divided into volumes by defining cubes or, more often, rectangular parallelepipeds (hence the name of the method). Classes can thus be defined for each band. Their intersection leads to correct segmentation if the various objects we wish to separate do not overlap in the radiometric space (scatter diagram) of the image, something which rarely occurs.

Division of radiometric scatter diagram can be improved by working directly on three-dimensional histogram in which each point represents a specific spectral characteristic in hyperspace. The latter is defined as a parallelepiped that can be isolated, coloured and assigned to the desired class. Thus, parallelepipeds with similar spectral characteristics are combined. It is possible to determine their spatial relationships since they are coloured over a geographic space constituting the image. The classification hence becomes an association of elementary parallelepipeds connected by at least a face, an edge or an apex in three-dimensional space (Fig. 9.1).

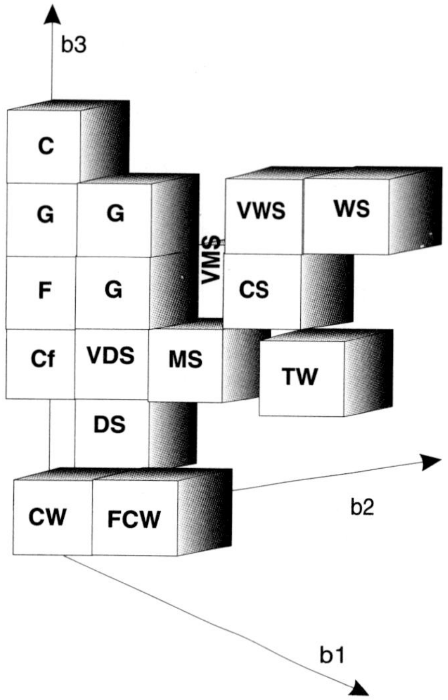

Fig. 9.1: Model of parallelepiped classification.
C: Crops; G: Grasslands; F: Forests; Cf: Conifers; VDS: Vegetation and dark soils; DS: Dark soils; MS: Medium bright soils; VMS: Vegetation and medium bright soils; CS: Bright soils; VWS: Vegetation and white soils; WS: white soils; CW: Clear water; FCW: Fairly clear water; TW: Turbid water.

9.1.2 Single-band segmentation

The plane corresponding to each band can be segmented. However, this is not always easy if histograms are not unimodal nor is it easy to determine the dividing lines for close digital numbers. It is hence necessary to study the histogram of digital numbers of every band (see Chap. 7, sec. 7.1).

■ Construction of classes

A simple division of band b1 of the image of Brienne into three units (dividing boundaries at values 58 and 72) clearly reveals three types of objects, viz., broadleaved forests at non-chlorophyllian stage and water, chlorophyllian vegetation canopies and bare soils (Fig. 9.2 and CD 9.1).

Band b2 of the Brienne image also shows a trimodal histogram. Division can be at values 35 and 62 (Fig. 9.2 and CD 9.1). This image can also be segmented into four parts since an overlapping is observed in high values around 106. The first part corresponds to forests and water bodies, the second to cultivated plots, the third to more or less dark bare soils and the fourth to bare soils.

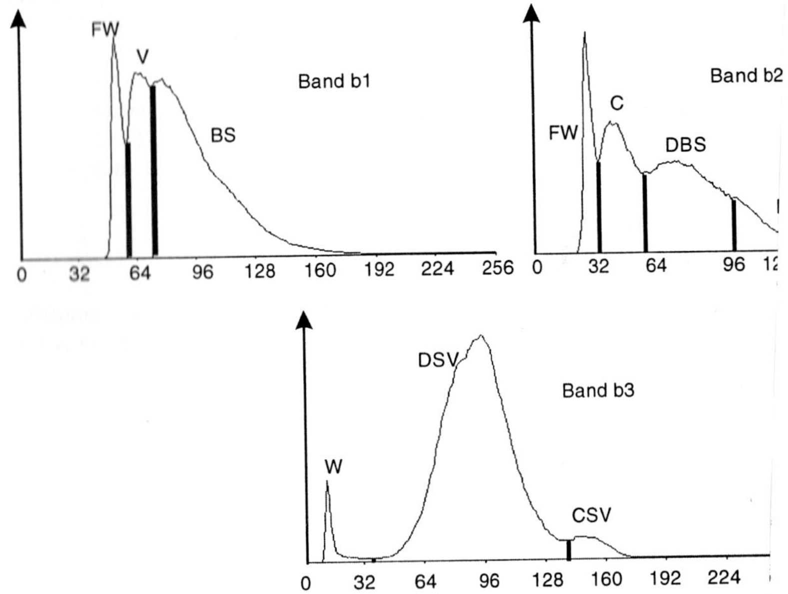

Fig. 9.2: Divisions of b1, b2 and b3 histograms of the Brienn
b1: FW: Forest and water; V: Chlorophyllian vegetation; BS: Bare soils. b2: FW: I
plots; DBS: Dark bare soils; BS: Bare soils. b3: W: Water; DSV: Dark soils and vege
soils and vegetation.

Band b3 of the Brienne image also shows trimodal distribution. Th
second by a very wide gap. Choice of a specific value of digital number a
the two modes is not obvious. Nor can the boundary between the sec
traced (Fig. 9.2 and CD 9.1). Values from 9 to 34 represent water, 136 tc
plots and for rest of the image values between 35 and 135 correspond to
vegetation with different degrees of chlorophyllian activity and covera
identified around value 83. However, interpretation of two classes is no
dark soils and vegetation on the one hand, and of more or less bright sc
the other.

■ Results

When every band is segmented into 3 or 4 classes, an image can be c
3 × 3 × 4 = 36 classes. In fact, only 23 classes are found to contain pi
between classes and that too despite taking fewer classes in each banc
strong correlation between bands 1 and 2 (CD 9.1).

Of the 23 classes, 13 comprise less than 5000 pixels, of which 6
pixels. Classes with very small frequency arise from segmentation of h
limits of the three bands were modified by one value of digital number,
Hence they should not be taken into consideration too formally.

conta
comb
demo
choic
latter
humic
water

9.1.

The t
radion
7, Fig.

■ So

In the
Brienn

Al
colour
dividec
such a
the foll
Tw
the thr

9

Supervised Classification

9.1 PARALLELEPIPED CLASSIFICATION

Parallelepiped classification is based on a radiometric model but not on measurement of distance or probability. Every pixel can be represented by its radiometric characteristic on a Cartesian diagram with as many axes as bands. On such a diagram every pixel is situated in an *n*-dimensional hyperspace. Most often we operate in a three-dimensional space to represent the image using three basic colours. However, this classification develops on a two-dimensional space, viz., the monitor screen. Hence the software employed should be able to support three-dimensional information.

If field measured reflectance values are available, they are converted to digital numbers by means of models that use various satellite parameters. These are used as references on a three-dimensional histogram from which parallelepiped classification can be constructed. If field radiometric data are not available, important objects to be interpreted may be defined in the field. Digital numbers of these features can be determined from their geographic location and used as references on the three-dimensional histogram. When no field data are available, the classic radiometric model is employed (Chap. 7, Fig. 7.4).

The classic radiometric model facilitates differentiation between soils, water bodies and chlorophyllian vegetation. Obviously, depending on the regions under investigation, climatic zones, etc., several other features can be added to these three basic objects. Examples are clouds and their shadows, shadows of large topographic elements, bush fire ashes, etc. Towns whose centres often appear as very dark soils are characterised by a mix of various vegetations, bare soils, water, etc. Their identification is based more on structural analysis than textural. It should be remembered that it is the surface of objects nearest to the satellite which is interpreted, viz., surface of bare soils or upper part of canopy in the case of vegetation. In the latter, obviously the tallest vegetation is interpreted. Interpretation of sub-layers of vegetation is possible only when coverage of higher layers is incomplete, as in the case of deciduous forests during fall and winter.

9.1.1 Segmentation of radiometric scatter diagram

Classification amounts to segmentation of the radiometric scatter diagram. Segmentation can be carried out in several ways. In all cases, the radiometric space is divided into volumes by defining cubes or, more often, rectangular parallelepipeds (hence the name of the method). Classes can thus be defined for each band. Their intersection leads to correct segmentation if the various objects we wish to separate do not overlap in the radiometric space (scatter diagram) of the image, something which rarely occurs.

Division of radiometric scatter diagram can be improved by working directly on three-dimensional histogram in which each point represents a specific spectral characteristic in hyperspace. The latter is defined as a parallelepiped that can be isolated, coloured and assigned to the desired class. Thus, parallelepipeds with similar spectral characteristics are combined. It is possible to determine their spatial relationships since they are coloured over a geographic space constituting the image. The classification hence becomes an association of elementary parallelepipeds connected by at least a face, an edge or an apex in three-dimensional space (Fig. 9.1).

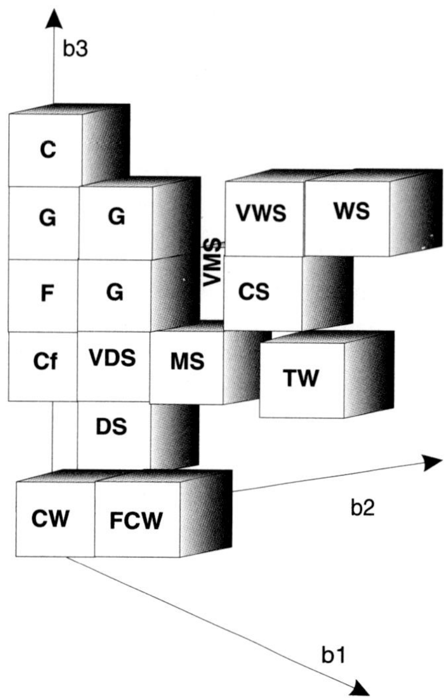

Fig. 9.1: Model of parallelepiped classification.
C: Crops; G: Grasslands; F: Forests; Cf: Conifers; VDS: Vegetation and dark soils; DS: Dark soils; MS: Medium bright soils; VMS: Vegetation and medium bright soils; CS: Bright soils; VWS: Vegetation and white soils; WS: white soils; CW: Clear water; FCW: Fairly clear water; TW: Turbid water.

9.1.2 Single-band segmentation

The plane corresponding to each band can be segmented. However, this is not always easy if histograms are not unimodal nor is it easy to determine the dividing lines for close digital numbers. It is hence necessary to study the histogram of digital numbers of every band (see Chap. 7, sec. 7.1).

■ Construction of classes

A simple division of band b1 of the image of Brienne into three units (dividing boundaries at values 58 and 72) clearly reveals three types of objects, viz., broadleaved forests at non-chlorophyllian stage and water, chlorophyllian vegetation canopies and bare soils (Fig. 9.2 and CD 9.1).

Band b2 of the Brienne image also shows a trimodal histogram. Division can be at values 35 and 62 (Fig. 9.2 and CD 9.1). This image can also be segmented into four parts since an overlapping is observed in high values around 106. The first part corresponds to forests and water bodies, the second to cultivated plots, the third to more or less dark bare soils and the fourth to bare soils.

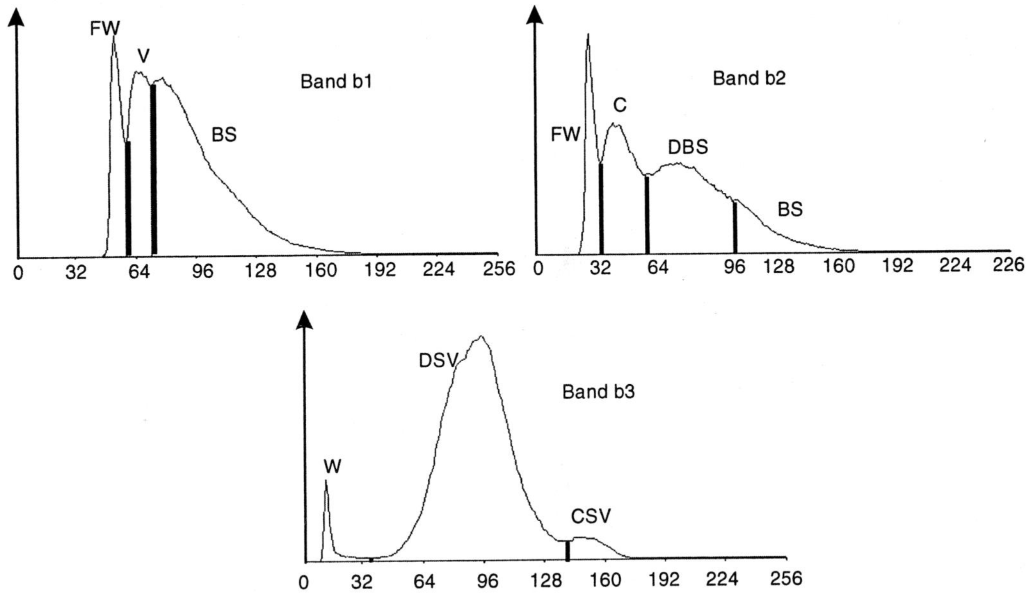

Fig. 9.2: Divisions of b1, b2 and b3 histograms of the Brienne image.
b1: FW: Forest and water; V: Chlorophyllian vegetation; BS: Bare soils. b2: FW: Forest and water; C: Cultivated plots; DBS: Dark bare soils; BS: Bare soils. b3: W: Water; DSV: Dark soils and vegetation; CSV: More or less bright soils and vegetation.

Band b3 of the Brienne image also shows trimodal distribution. The first is separated from the second by a very wide gap. Choice of a specific value of digital number as the dividing point between the two modes is not obvious. Nor can the boundary between the second and the third be readily traced (Fig. 9.2 and CD 9.1). Values from 9 to 34 represent water, 136 to 198 the most chlorophyllian plots and for rest of the image values between 35 and 135 correspond to more or less bright soils and vegetation with different degrees of chlorophyllian activity and coverage. One more class can be identified around value 83. However, interpretation of two classes is not easy since there is a mix of dark soils and vegetation on the one hand, and of more or less bright soils and sparse vegetation on the other.

■ Results

When every band is segmented into 3 or 4 classes, an image can be composed using 3 bands with $3 \times 3 \times 4 = 36$ classes. In fact, only 23 classes are found to contain pixels. Thus redundancy exists between classes and that too despite taking fewer classes in each band. This of course is due to very strong correlation between bands 1 and 2 (CD 9.1).

Of the 23 classes, 13 comprise less than 5000 pixels, of which 6 contain even less than 1000 pixels. Classes with very small frequency arise from segmentation of histograms. If the segmentation limits of the three bands were modified by one value of digital number, the results would change little. Hence they should not be taken into consideration too formally.

The major part (99.8%) of the image consists of 10 classes. However, 4 classes are found to contain 100,000 to 400,000 pixels. They correspond to the central part of the histogram which obviously combines several objects that cannot be differentiated in the continuum of the histogram. This demonstrates the limitations of this method based on a single-dimensional approach for deciding the choice of classes.

This method, very fast, can be used to roughly divide the image according to various themes, the latter being very general. In the image of Brienne the chalky Champagne region is distinguished from humid Champagne and similarly highly chlorophyllian crops, forests, grasslands and three types of water bodies are clearly differentiated. Such a method can be useful for a very general analysis.

9.1.3 Multispectral segmentation

The three-dimensional histogram can be directly segmented making use of only the simplified radiometric model comprising the objects bare soil, chlorophyllian vegetation, water and shadow (Chap. 7, Fig. 7.4).

■ Soil

In the b3–b1 plane of radiometric values, soils are represented by a linear cluster. For the image of Brienne, the linear soil cluster (LSC) is defined by the equation:

$$LSC = 0.56\ b3 + 0.83\ b1 \quad \text{(Fig. 9.3)}$$

All the pixels situated on both sides of this straight line and pertaining to the cluster can be coloured as a single unit. If more detailed analysis is desired, the cluster thus obtained can be further divided into several parts that will determine soil groups, from very bright to very dark (Fig. 9.4 A to C), such as the three groups in the Brienne image (CD 9.2). This depends on the objective identified. In the following example of Brienne, six groups, viz., 6, 0, 3, 11, 16 and 22, are used (Fig. 9.5).

Two hundred and forty-four classes of pixels or different types of spectral characteristics exist in the three-dimensional histogram of the IRC image. Of these, the soil cluster comprises 90, which

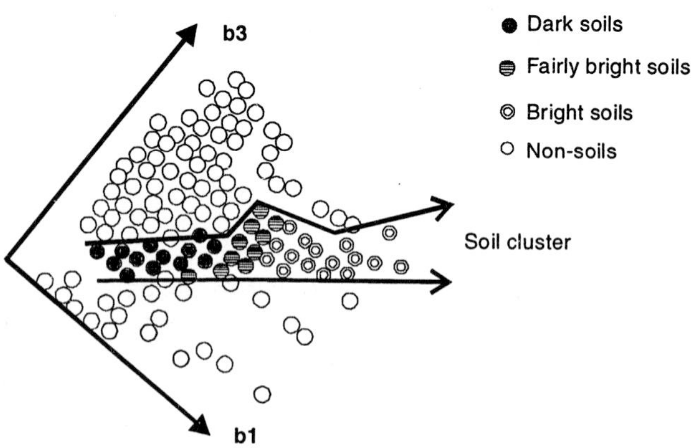

Fig. 9.3: Segmentation of soil cluster in b3–b1 histogram.

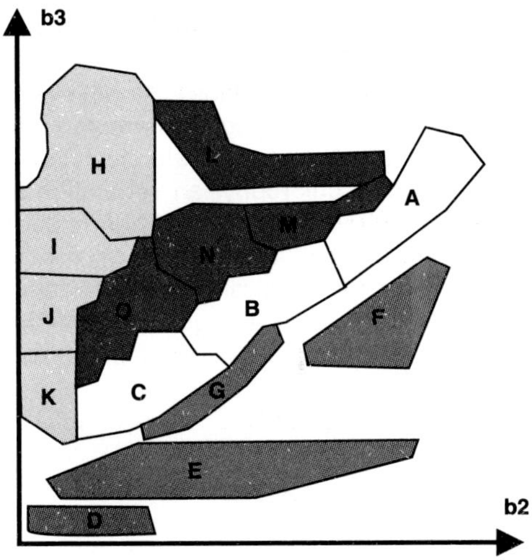

Fig. 9.4: Multispectral segmentation of image of Brienne (representation on axes b2 and b3).
A: Bright soils; B: Soils of medium bright; C: Dark soils; D: Clear water; E: Fairly clear water; F: Turbid water; G: Boundaries of water bodies; H: Crops; I: Grasslands; J: Broadleaved forests; K: Conifers; L: Sparse vegetation; M: Vegetation and bright soils; N: Vegetation and soils of medium bright; O: Vegetation and dark soils.

indicates that soils represent 37% of spectral types. Geographically these 90 classes of pixels correspond to 41.5% of the area under investigation (CD 9.3).

■ Water

In the b2–b3 plane the radiometric values of different types of water always occur below those of the soil cluster (CD 9.3). The digital number value of the near infrared band is primarily related to clarity or depth of water layer. Different groups of water can hence be defined as a function of this value (Fig. 9.4 D to F).

Cloud shadows are very often confused with clear or deep water. A very accurate analysis is needed to differentiate them. In this case, shadows and clear water overlap each other.

Pixels are also observed between soils and water. These are mixed pixels (mixels) in the boundaries of water bodies (Fig. 9.4 G).

■ Vegetation

Radiometric values of chlorophyllian vegetation are situated above the soil cluster towards higher values of the near infrared band (CD 9.3). Two units can be differentiated, one for complete vegetation cover or at least more than 40% and the other for sparse coverage.

The first unit can be further classified according to the magnitude of digital number in the infrared band. The following groups can thus be distinguished:

—Vegetation with high chlorophyllian activity and large biomass characterised by high digital numbers in infrared and low in red (Fig. 9.4 H). In the Brienne example these are crops of beetroot, potato or maize (as the image pertains to the month of September).

—Vegetation consisting of a mix of more or less chlorophyllian plants and low biomass such as grasslands and partly broadleaved trees (Fig. 9.4 I).

—Broadleaved forests comprising highly chlorophyllian tall trees and associated shadows, mixed within a single pixel and giving rise to a mixel. This results in a smaller value in near infrared and a generally low value in band b2 (red) (Fig. 9.4 J).

—Coniferous forests whose canopy structure leads to smaller values in near infrared (Fig. 9.4 K).

The second unit is made up of pixels whose spectral characteristics indicate a mix of more or less chlorophyllian vegetation and bare soil. Several groups are identifiable:

—One group comprising both highly chlorophyllian vegetation and sparse covers. The digital numbers are equally high in the near infrared and the red bands due to the contribution of soils for which the values are greater in band b2 (Fig. 9.4 L).

—A series of groups situated close to the soil cluster, which indicate a mix of spectral characteristics of chlorophyllian vegetation and soils, the latter exerting a strong influence. Several subgroups can be identified depending on whether the vegetation is situated on bright soils or darker ones (Fig. 9.4 M, N and O).

9.1.4 Chorological segmentation

A three-dimensional histogram can be segmented using the radiometric model described above, together with reasoning based on spatial relationships between pixels belonging to a given group, in other words on the basis of chorological laws. More precise classes are defined which make radiometric relationships compatible with spatial relationships.

For tracing boundaries between soil groups that differ in luminance, we can use relationships between radiometric vicinity (on three-dimensional histogram) and geographic neighbourhood. A spatial organisation of soils is discovered, represented by a geographic transition from 'white' soils to very bright soils, bright soils and fairly bright soils. This variation is explained by the fact that the soils are on chalk (of Champagne) and their colour varies inversely with thickness. Thus they vary from LITHOSOLS to RENDOSOLS and CALCISOLS (*Baize and Girard Référentiel pédologique,* 1995). Dark soils correspond to decarbonatisation clays or clayey and moist soils pertaining to the humid Champagne region. Hence there is a correlation between soils and their surface states. These chorological laws of soil can be fairly well determined, based in particular on images in which soils are bare. Such soil organisations have been detected in other regions also, Caux county for example (Burlot, 1995).

Plant canopies in agricultural plots reveal soil heterogeneities, viz., more or less thick soils, occurrence of microtalwegs, etc. Different water bodies have different spectral characteristics which ought to be related to their mode of utilisation. In the 'reservoir-ponds' of the Aube various spectral units parallel to the border are observed on a flank, which are related to depth and nature of the bottom, whereas the water in the central part exhibits a single spectral characteristic. However, differences are observed between various ponds, which may be due to differences in their management and utilisation.

It was decided to retain only three groups for agricultural plots corresponding to different chlorophyllian states. In fact, if they pertained to different crops, the three classes would have been distributed according to plots, but they are often inside a single plot. This hence corresponds to intraplot heterogeneities associated with differences in soils and not due to agricultural practices. Forms of these intraplot heterogeneities, in fact, resemble microtalwegs. This is often confirmed since a given heterogeneity extends across several plots, irrespective of whether they are crops or bare soils.

A total of 27 groups were chosen in the image of Brienne (CD 9.4).

9.1.5 Spectral characteristics of various groups

Results of this classification can be expressed by digital characteristics of each group (Fig. 9.5). However, it should be remembered that this procedure cannot be used rigorously *unless the position of objects*

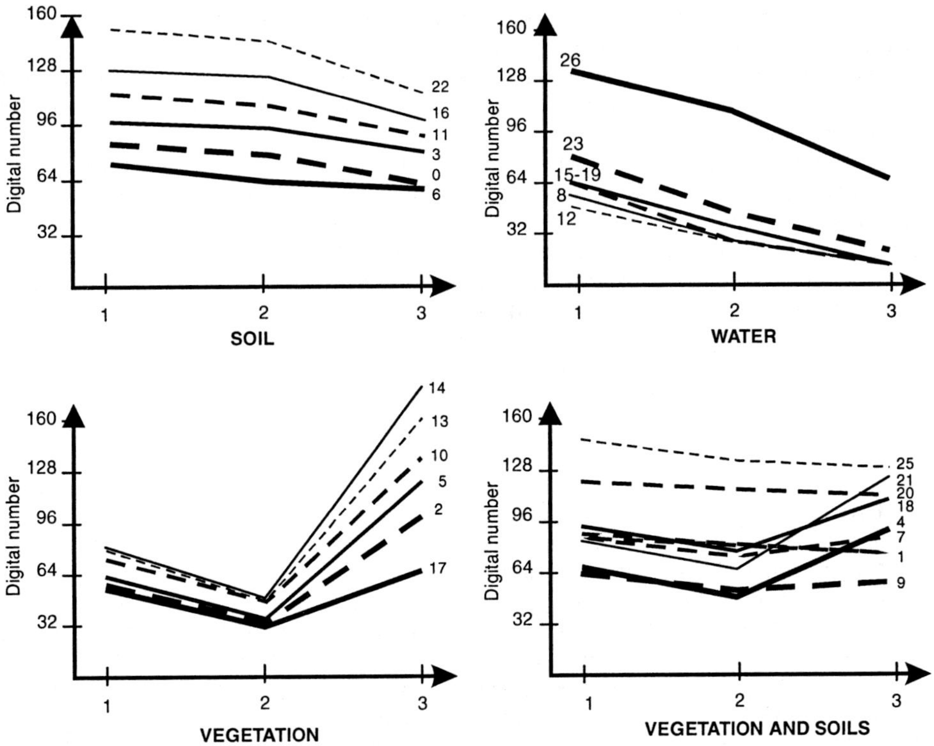

Fig. 9.5: Digital characteristics of 27 units obtained from parallelepiped classification by chorological analysis.

is studied band by band (also see Chap. 8, Figs. 8.1 and 8.2). If position of objects in two- or three-dimensional histograms is studied, relative positions with respect to one another should be used. These objects can be marked by a combination of three values of digital numbers that characterise them *for a given image* but one should be cautious. Thus, it was observed in the image around Brienne that the curves representing the objects in the three bands (Fig. 9.5, Chap. 8, Figs. 8.1 and 8.2) do not directly correspond to those of reflectance (see Chaps. 4, 23 and 25).

In fact, bright soils which have a continuous, convex and ascendant reflectance curve exhibit a convex but descendant curve of digital numbers. Water likewise has no zero values in the near infrared band. The form of digital number curves of groups corresponding to vegetation covers, which is close to that of their spectral reflectance curve, cannot be considered a general case.

Six soil groups are clearly distinguished (Fig. 9.5, soils): white soils (22), very bright soils (16), bright soils (11), medium bright soils (3), dark soils (0) and very dark soils (6). All the digital characteristic curves are convex except for the very dark soils, which are concave.

Of the six types of water (Fig. 9.5, water), very clear (12), clear (8) and fairly clear (15 and 19) waters are similar to one another. Slightly turbid (23) and very turbid (26) waters differ from others. If high precision of classification is not required, only two or three groups can be kept.

All the six groups of chlorophyllian vegetation (Fig. 9.5, vegetation) show small values of digital numbers in band 2, representing absorption of chlorophyll. They are distributed in two groups differentiated by band 1 values. Digital numbers of band 3 indicate a regular decrease from one group to another. The first group comprises highly chlorophyllian cultivated plots (14 and 13) and permanent grasslands or forest clearings (10). The second group is represented by a complex combination of less chlorophyllian permanent grasslands and trees (5), forests with shadows (2) and conifers and plants associated with shadows and water, such as on river banks (17).

The other eight groups consist of vegetation and soil mixes (Fig. 9.5, vegetation and soils).

Three groups, 4, 18 and 21, exhibit characteristics of a plant canopy. Group 21 is relatively the most chlorophyllian and densest since the difference between b3 and b2 is the highest. This type is always associated with cultivated plots either on the borders or throughout the area. It corresponds to heterogeneity in a crop that is less chlorophyllian because in an advanced stage of development or less dense due to boundary effects or soil differences. Type 18 is situated on bright soil and type 4 corresponds to tall trees in forests or heterogeneous grasslands with green and dry plants.

The two types 1 and 20 are intermediate between soils and vegetation: convexity of spectral characteristic of soils (Fig. 9.5, soils) is compensated by concavity of vegetation spectral curves (Fig. 9.5, vegetation), resulting in a straight line. Other types are more or less influenced by plant canopy but types 20 and 25 correspond to bright soils, while types 7, 1 and 9 correspond to increasingly darker soils. Type 7 comprises more chlorophyllian or denser vegetation than type 1.

Type 24 corresponding to cloud shadows is not shown in Fig. 9.5.

9.1.6 Statistics of various groups

If all the 27 groups covered equal areas, there would be an average of 42,666 pixels per group (CD 9.4). This is used as reference for statistical analysis.

Groups with the smallest surface areas, excluding cloud shadows, represent various water bodies and conifers, which constitutes a characteristic feature of the region under investigation.

Next in size are white and very bright soils and dark soils. These groups can evidently be combined with their neighbours and hence classified into three or four units instead of six. This means that bright soils are relatively less numerous than others in the image. However, this is specific to the portion of the image studied: if the portion farther west, in chalky Champagne is analysed, coverage by bright soils would be greater and in the portion farther east in humid Champagne almost completely absent. This shows that for correct interpretation well-defined landscape units should be used (see Chap. 18) if statistical results are to be analysed in terms of areas. However, this also indicates that with fairly simple land use types, landscape units can be readily defined. In the present case, the chalky Champagne region can be characterised by an association of bright soils and the humid Champagne region by darker soils.

The two groups representing agricultural plots not yet covered by chlorophyllian crops in September, occupy less area. Obviously, processing of an image of the same region acquired a little later would show these two as a single group. The influence of date of acquisition on differentiation of groups can be assessed from the LANDSAT TM images of April and May (CD 7.12). This shows the importance of the choice of acquisition date depending on the objective of investigation (see Chap. 16).

Lastly, several groups comprising bare soils with dense or sparse vegetation cover, mature or seedling, mostly have small areal extents. Obviously they vary depending on the climate of subsequent days: for some vegetation will dry and for others develop. A new class could be identified but it would represent a different feature. It is certain that in an image acquired a little later, plots already sown for the next year can be readily separated from plots which will be tilled later and are not yet sown.

9.1.7 Visual interpretation

The preceding interpretation is facilitated by choice of established colours according to the functions identified. Large spatial groupings perceived by colour associations are thus readily identified. If colours are distributed to various groups randomly, spatial organisation is no longer easy to interpret (CD 9.5).

In order to discover or evaluate chorological laws that define spatial organisation of a region, correlation should exist between semantic characterisation of objects and their geographic distribution. Study of reciprocal spatial positions of different groups allows refinement of interpretation.

The number of groups can be increased or decreased according to the objective of classification. The result of 27 classes may be considered too precise for a preliminary investigation of the region. The number of classes may hence be reduced to 5, viz., bright soils, dark soils, vegetation, water and shadows (CD 9.5).

Contrarily, for accurate identification of a theme a group may be taken, isolated by masking and again subjected to parallelepiped classification.

Thus the groups chosen by parallelepiped classification can be precisely defined by means of radiometric differentiation and analysis of distribution and spatial organisation, i.e., by constructing a set of chorological laws.

9.1.8 Quality of parallelepiped classification

Parallelepiped classification is based on radiometric models and eventually chorological laws. Three levels of models were identified.

The simplest is based on segmentation of each of the three bands before combining them. Results are fairly general and may suffice if the aim is to identify major groups. However, this method cannot be employed unless histograms of each band can be segmented with no serious errors. Such is not always the case. If the processing tools available are limited in performance, the single-band method is still useful. Multispectral segmentation is preferable, however.

The model based on multispectral segmentation is more efficient than the preceding but assumes knowledge of the principal spectral characteristics of objects of the region under study. The latter is not always possible, for example when this method is employed just before conducting a detailed field study of the region. The results will hence depend on how effective the model is for the zone under analysis. Care should be taken to choose the groups according to adaptability level of the radiometric model to the conditions of the region. In a poorly known area it is desirable to choose a relatively limited number of groups.

The model based on radiometric and chorological segmentation is evidently the most efficient, but it assumes that spectral behaviour of major objects of the image is known and that certain chorological laws can be applied. At least a general knowledge of the region is necessary in the absence of accurate information. The more certain we are of the chorological laws, the more precise analysis. This certainly is the most suitable method for a thematic specialist having experience in remote sensing or in the region, since he (she) can derive a combined advantage. This method provides the best use of expertise. For the expert, it is the most reliable, fast and least expensive.

It is possible to assess the quality of parallelepiped classification. As in ascendant hierarchic classification (Chap. 8, Fig. 8.5), a modal radiometric value is defined for each class and compared with the values of all pixels belonging to it. Thus pixels classified well or poorly are determined.

9.1.9 Conclusion

Parallelepiped classification ought to be developed based on a correct radiometric model applicable to the region under study. As it is totally supervised, the analyst can build up his (her) interpretation step by step. In this respect, it is very efficient for the expert. It also aids in rapidly defining a classification based on a conventional radiometric model. Lastly, it enables classification of only those groups which correspond to the objective of investigation and grouping of all other components of the image that are not useful for the objective into a single class (for example, coloured black).

9.2 MAXIMUM-LIKELIHOOD CLASSIFICATION

9.2.1 Probabilistic spectral behaviour

Spectral behaviour of an object is neither unique nor specific and changes with time. This results in an uncertainty in discrimination level. In fact, remote sensing data always contain random errors which reduce contrasts between objects. The position of an object cannot be readily marked on an image and locating on the image a region known on the ground is also not easy and may lead to errors (see Chap. 10). Finally, pixel size and positioning (variable depending on the satellite) make it difficult to avoid mixels (see Chap. 13) which, by definition, are poorly discriminated spectrally. Consequently, spectral behaviour of an object varies depending on images and changes with time. It is hence not possible to correlate the n values of digital numbers (n being the number of bands) assigned to a pixel with a definite type of object. In fact, if it is considered that a signature is characteristic of an object and that it does not vary in space and time, an object cannot be said to have a spectral signature at all since digital numbers vary in time and space.

We can, however, consider that a pixel has a 'spectral signature' which is often represented in software programs (including TeraVue) by a set of lines that indicate a digital characteristic. Contrarily, an object which is represented by a pixel population, has no spectral signature. It is dangerous to speak of a spectral signature for corn, for example, since its spectral behaviour varies depending on the region in which it is cultivated and the season.

The probabilistic approach, in fact, replaces the spectral signature of a pixel, described by a simple set of lines, by spectral behaviour described by a probability distribution centred on a mean spectral characteristic.

So it is interesting to study spectral characterisation of objects in terms of probabilities.

9.2.2 Principles of classification

The maximum-likelihood classification is thus based on probabilistic methods. We compute for each pixel its probability to be assigned to a given class rather than to another. The law of affectation thus derived enables minimisation of risk of error by a better utilisation of probabilities pertaining to one class or another. In this method a pixel belongs to a class with a certain probability and not in a binary manner, viz., *'belongs'* or *'does not belong'*, as in the preceding methods of classification.

Consequently, in order to determine these probabilities known zones should be chosen as a priori references and the final result discussed on the basis of appurtenance of a given probability.

■ Bayes decision rule and maximum-likelihood rule

When reference zones are chosen in an image, digital number histograms of pixels contained in each zone can be computed. Each histogram defines a density of conditional probability of a pixel p pertaining to a class Ci. This probability is denoted by $P(p/Ci)$. We can also determine the probability of occurrence of a pixel of a class Cj, indicated by $P(Cj)$. However, this requires that the pixel frequency of class Cj be known before undertaking classification.

If during classification pixel p is associated with class $C1$ when it actually belongs to class $C2$, an affectation error occurs. *The rule of affectation of a pixel p to a class C is known as maximum likelihood when it minimises the mean error of affectation.*

When the error of affectation does not depend on class C, the decision rule is called 'Bayesian' and may be described as follows:

If Ci is the class of appurtenance of pixel p and Cj every other class,

p belongs to class Ci, is equivalent to $P(p/Ci) \cdot P(Ci) > P(p/Cj) \cdot P(Cj)$.

■ Gaussian parametrisation

The conditional probabilities of appurtenance based on the histogram of a reference zone are called empirical probabilities. Their determination is restricted to sampling of classes present in reference zones. Obviously an attempt is made to ensure that the geographic extent of the final classes far exceeds that of reference zones. Hence it is sometimes necessary to determine the parameters of these classes by assuming that every distribution corresponding to the classes under investigation is Gaussian. The probability *P(p/Ci)* in this case is defined by the mean and variance-covariance matrix of pixels of each class. The latter are estimated from the values computed for reference zones.

In remote sensing it is often assumed that every class (defined by reference zones) is a priori equiprobable, i.e.,

$$P(C1) = P(C2) = \ldots = P(Ci).$$

The Gaussian discrimination function *FDi* (*p*), necessitating computation of variance-covariance matrix, can thus be calculated and simplifies the decision rule expressed as:

p belongs to class *Ci*, is equivalent to *FDi* (*p*) > *FDj* (*p*) irrespective of *i* and *j*.

Hence it is necessary to choose reference zones in such a way that they represent the entire object and that the number of pixels of such a zone suffices to correctly estimate the parameters of the Gaussian statistical population which it is assumed to represent.

9.2.3 Rejection threshold

The discrimination function used allots every pixel to one of the classes defined by the reference zones. This may be accepted if the objective of classification is to classify all the pixels even when poorly categorised. However, the importance of this method lies in detecting the validity of classification of every pixel into the class to which it is assigned.

It is hence natural to create a rejection class which receives pixels that cannot be assigned to one of the classes defined by the reference zones. For this, a rejection threshold is defined which corresponds to the admissible lower limit of conditional probability of a pixel *p* pertaining to the class *Cj*. Pixels below this threshold will be assigned to a rejection class.

The rejection threshold can be expressed as percentage. While the threshold can be readily determined in the case of one-dimensional analysis, it is much more difficult for multidimensional situations, the most common in remote sensing. Based on the analytical expression of the Gaussian discrimination function *FDi*(*p*), Monget (1986) showed that a *p*% rejection threshold, fixed for Gaussian densities of conditional probability, is manifested by a threshold of *p*% applied only once on the χ^2-law with *n* degrees of freedom (*n* being the number of bands used for classification). *Hence for 'Bayesian' classification this threshold should also be defined.*

9.2.4 Classification operations

The maximum-likelihood classification requires: 1—a choice of reference zones or nuclei, 2—statistical and radiometric analyses of these nuclei, 3—study of separability table, 4—classification proper, with accurate evaluation of classes by probability study and 5- analysis of performance table after classification is completed.

■ Preparation of nuclei

Nuclei are used to define populations based on which classification can be carried out. A large number of pixels and a homogeneous population are necessary for obtaining the best possible approximation to a Gaussian law. Every nucleus can be defined by several zones separated from one another.

Each nucleus ought to have a frequency of more than 200 pixels, which is not difficult to get in a remote sensing image with millions of pixels, except in some cases where the theme is very precise. It is preferable to choose sample zones in several parts of the image so as to avoid the situation wherein the nucleus represents only one type of region. If the image features to be traced are very small, the software program used should enable their delineation through zooming and for this the zones should be multiplied.

It is also preferable that the population chosen be very homogeneous and defined correctly. The image features retained should be largest possible and in any case the most compact. The zones should also be delineated accurately if the boundaries are sinuous. This is very difficult in the case of clouds and linear features such as roads and watercourses.

As errors are always possible when delineation of image zones is difficult, facility should exist in the software to return to the traces already drawn and hence to erase easily.

Nuclei are defined geographically. This assumes knowledge that such a place exists, established by even minimal field study whereby major objects of special attention can be identified on the ground. However, they may have been forgotten or not properly evaluated due to the perspective effect (see Chap. 18, Fig. 18.1). Definition of nuclei hence depends on the geographic approach. It is useful to take the help of existing thematic maps for a better definition of image zones comprising the nuclei. In fact, in most cases field radiometric measurements are not made.

Another approach is based on combining field observations with radiometric measurements and constructing a local radiometric model that facilitates choice of nuclei, since locations identified in the field and radiometric differentiation are used simultaneously.

In all cases, it is recommended to visually interpret colours observed in the infrared band for delineating image zones of the nuclei. One can always refer to the *interpretation model of IRC images* (Chap. 7, Fig. 7.4).

It is common in a real study to use several types of classification for a given object. Thus ascendant hierarchic classification or classification by mobile centres can be employed to chose the nuclei. It is better to construct more nuclei than necessary and regroup them, rather than taking a small number of heterogeneous nuclei.

Lastly, the visual or thematic evaluation of objects should be followed by their spectral characterisation based at least on their digital numbers in the image under study. This statistical study is mandatory.

■ Statistical and radiometric analysis of nuclei

For correct interpretation of the nuclei selected, which constitute the basis for Bayesian classification, statistical analysis is necessary. Means, standard deviations and variance-covariance matrix of the bands used are essential for computations in classification (see above). Consequently, access to statistical programs during the process of interpretation is imperative.

It is possible to reconstruct the digital characteristic of each nucleus from the mean values of digital numbers of each band through a set of lines, as mentioned in the example of parallelepiped classification (Fig. 9.5) or ascendant hierarchic classification (Chap. 8, Figs. 8.2 and 8.3).

Thus, in the 20 nuclei (characterised by 91 elementary image zones) chosen for the image of Brienne (CD 9.6), the gradual variation from a convex curve to a concave one represents soils with different levels of luminance. In some cases, the standard deviations are not distinct. The standard deviation of nucleus 'white soil' is slightly higher relative to others; this can be corrected by removing the darkest parts of the image zones of this nucleus, which reduces the heterogeneity.

Various types of water exhibit a concave curve of digital characteristic. They are differentiated mainly from digital number values of band 1. The most transparent water has the lowest value while turbid waters containing suspended matter or shallow waters, or waters located on a chalky surface are represented by higher values.

Vegetation is characterised by low values of digital number in band 2. This value is very low for chlorophyllian crops growing in September in this region (beetroot for example), a little less for forests and riverine forests, and becomes higher for ripe crops or for sparse vegetation. The latter two cases represent mixes of soil and vegetation. It may be noted that the spectral characteristics for the nuclei 'forest-riverine forest' and 'dark forest' are very similar. These should be combined and the zone of conifers and shadows defined by a new, more representative nucleus.

■ Separability

A separability table shows the degree of differentiation of pixels used for defining the nuclei. In this table, nuclei are given in rows and the thematic groups defined by various nuclei listed in columns. The table is read column by column. Each column, defined by a number that refers to the name of the nucleus and its pixel frequency, shows the number of pixels of nuclei which have values identical to pixels pertaining to another nucleus. This corresponds to pixels for which histograms of different nuclei overlap.

Thus, for Brienne (Table 9.1), 1,864 pixels constitute the nucleus (no. 3) 'bright soil'. 1,730 pixels are common with the nucleus 'very bright soil', 7 with 'bright soil', 62 with 'fairly bright soil', 1,603 with 'sparse vegetation', 1,682 with 'dry crops', 332 with 'riverine forests and forests' and 159 with 'clouds'. In total, 5,575 pixels have at least a dual appurtenance; as there are only 1,864 pixels, many pixels have multiple identification.

This table hence gives an approximate idea about the accuracy of nuclei selection. If pixels of a nucleus interfere with no other nucleus (in the corresponding column there are only blanks or zeros), its histogram is perfectly distinct from the histograms of other nuclei. Separability is then said to be perfect.

In the image of Brienne, a nucleus such as no. 10 (crop) is well separated from others. In fact, only a very few pixels (151), compared to the number of pixels used for defining it (frequency: 3,593), are common to other nuclei. A similar situation is observed for nucleus 15 (very clear water) which has only 20 pixels common with nucleus 16 (clear water), for a frequency of 3,554 pixels.

Contrarily, a nucleus is less separable if a large number of pixels used to define it are common to other nuclei. Such is the case especially with nuclei 13 (riverine forests and forests), 20 (clouds), 8 (sparse vegetation), 9 (dry crops), 11 (green grasslands) and 12 (dry grasslands).

The separability table shows that in the case of bare soils, many pixels simultaneously belong to different soil types: brighter or less bright. Hence it is necessary to improve the test zones of these different nuclei so that they become better separated.

For the nucleus 'cloud', of 171 pixels 130 are not separable from the nucleus 'white soil', 139 from 'dry crop' and 21 from 'very bright soil'. In total, the 171 pixels can be distributed in 7 nuclei, which represents 301 possible cases.

Many pixels indicate a dual or triple appurtenance. This shows that the nucleus 'clouds' is not satisfactory. Therefore it is not possible to identify clouds by their radiometric characteristics. This is a common case in images (CD 4.2).

It can be seen that total confusion exists between the nuclei 'green grasslands' (no. 11) and 'dry grasslands' (no. 12). The curves defining spectral characteristics from digital numbers are practically parallel. They also show confusion with 'riverine forests and forests' (no. 13) and partly with 'sparse vegetation' (no. 8) and 'dark soil' (no. 5). In the latter case, it is necessary to combine these two nuclei and take them as only one.

One way of deciding which of the two nuclei is to be removed is to mask all features other than these two in the infrared image. It can be seen that the nucleus of green grasslands lies in the centre of the three-dimensional histogram, which is significant since there are few nuclei in the centre. It does not confound with various nuclei pertaining to crops. Contrarily, the nucleus of 'dry grasslands' is much

Table 9.1: Separability between 20 nuclei chosen for maximum-likelihood classification

	Nuclei	1	2	3	4	5	6	7	8	9	10	11	12	13	14	15	16	17	18	19	20	Total
	Pixel frequency	1371	1483	1864	1755	1751	885	848	1667	1478	3593	2691	1419	3703	4433	3554	1490	1471	383	433	171	36443
1	White soil		603	7					2	714											130	1456
2	Very bright soil	49		1730	12	1			377	931	1	2	4	1							21	3129
3	Bright soil	12	1157						52	64				1							5	1291
4	Fairly bright soil		1	62		638			544	27		2	2	2							2	1280
5	Fairly dark soil		1		1306		236	3	393	2		1421	270	133					1			3766
6	Dark soil					923		607	91	1	35	404	1030	79						16		3186
7	Very dark soil					89	855		1	1		6	50	161						109		1272
8	Sparse vegetation	7	1031	1603	1395	768	205	2		202	2	357	1004	60							4	6640
9	Dry crops	81	1200	1682	382	60	24	1	574		3		3	3							139	4152
10	Crops		1				6		2	205		5		44								263
11	Green grasslands		1		10	615	20	1	1034		1		1380	2289	4275				2	11		9639
12	Dry grasslands		1		1	158	44	20	891	2		2640		1527					2			5286
13	Riverine forests and forests		397	332	2	133	276	726	621	9	108	2677	1405		4433			1	2	116		11238
14	Dark forests											8		1244						6		1258
15	Very clear water																1480					1480
16	Clear water															20		935				955
17	Fairly clear water						51	361						1			10		160	23		606
18	Turbid water					1				1		555	75	83				29				744
19	Cloud shadows						52	489				3		89	82			288				1003
20	Clouds	171	368	159	204	326				842	1								4	281		2356
	Number of pixels with multiple appurtenances	320	4761	5575	3312	3712	1769	2210	4582	3001	151	8080	5223	5717	8790	20	1490	1253	171	281	301	60719

more similar to those of soils. Consequently, all image zones that define the nucleus of 'dry grasslands' are simply removed. Pixels belonging to this nucleus are redistributed between the nucleus of 'green grasslands' and various nuclei of soils.

A similar analysis is followed for various nuclei that are poorly classified.

■ Analysis of conditional probabilities of appurtenance

Every pixel has a greater or smaller probability of belonging to a statistical population defined by the pixels of each nucleus. To avoid getting too many poorly classified pixels in the thematic groups of interest, a probability threshold of rejection is defined. Any pixel classified with a probability lower than the threshold is eliminated. It is possible to make the value of threshold variable. The consequences are examined on the three-dimensional histogram and the image.

□ Analysis of three-dimensional histogram giving classification probabilities

The three-dimensional histogram represents the probability of appurtenance of every pixel to a given nucleus. The probabilities of appurtenance are coloured in the decreasing order from white to black (Fig. 9.6) or in TeraVue, from white to black, passing through yellow, orange, red and maroon. Pixels with probabilities lower than the threshold defined by χ^2-parameter are not indicated and are rejected.

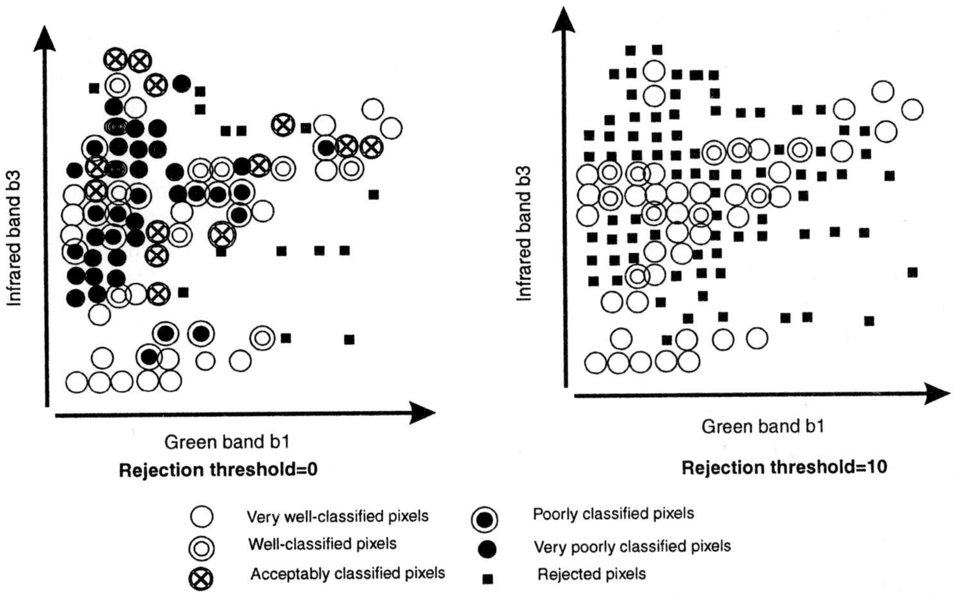

Fig. 9.6: Histograms for bands b1 and b3 giving probabilities of appurtenance of pixels to nuclei, with a rejection threshold of 0 and 10%.

□ Rejects

The rejection threshold can be varied from 0 to 100%. When the threshold is 0, almost all the points are classified and most of them poorly, hence the large number of black or dark pixels in Fig. 9.6. When the threshold is higher (10% in Fig. 9.6), the number of rejected pixels is very large and those classified have a much higher probability.

In practice, the probability level of rejection should not exceed a few per cent.

□ Analysis of probability images

In a probability image the colour of every pixel is dependent on its probability pertaining to one or the other nucleus. The whiter or yellower a pixel, the greater its probability, and hence the better it is classified (CD 9.6). Thus, for a given rejection level, the accuracy of classification of every pixel in the image can be evaluated. This information can also be obtained pixel by pixel. The latter enables identification of pixels assigned to a given nucleus and determination of whether the pixels situated at a given place are well classified.

We can also determine for a given area (a cultivated or a forest zone for example), at what rejection threshold a pixel becomes black, i.e., is rejected. The higher the rejection threshold, the better it is classified.

It can be readily seen that there are regions that are poorly classified or unclassified, using the nuclei defined. It is hence useful to add some new nuclei to represent them. For this, digital spectra of the unclassified zones in the image are investigated. A new nucleus can be defined by choosing it geographically at the place indicated (in black) in the probability image.

□ Example of image of Brienne

'Cloud' is characteristic of a heterogeneous nucleus. During classification with a preliminary set of nuclei it contains 11,790 pixels, which largely exceed the surface area of clouds present in the image. The pixels classified as cloud belong to the nucleus of white soil for the brightest and to the nucleus of dry crops for the darkest. Study of the probability histogram indicates that in the image there are many points which can be allotted to the nucleus 'cloud' (this can also be seen by reading the line 'cloud' of Table 9.1, the separability table). Hence, this nucleus is heterogeneous and overlaps on the neighbouring nuclei. It introduces confusion during allotment of nuclei to various other nuclei. It is better to remove this nucleus in order to reduce confusion. As the clouds observed in the image do not exhibit a specific spectral characteristic it is not possible to specifically identify them.

When all the nuclei and the probability image are analysed systematically with a rejection threshold of 0 (Fig. 9.6), it can be seen that unclassified or poorly classified pixels exist in the centre of the three-dimensional histogram. On the other hand, unclassified zones are distinguished in the image (coloured black—CD 9.6). This indicates that the nuclei retained do not suffice to classify them. If some nuclei are added in the unclassified segments, by choosing the zones that characterise them, classification can certainly be improved.

The nucleus 'dry crops' (no. 9 in Table 9.1) confounds with too many other nuclei. This nucleus is redefined by associating it with the nucleus 'crops'. Hence, a mask was applied isolating these two nuclei in the band of NDVI (CD 9.6). The histogram of the latter band is distinctly found to be trimodal (Fig. 9.7). This nucleus can hence be segmented into three units: the earlier nuclei 'crops' and 'dry crops', as well as a new one 'little chlorophyllian crops'.

■ Quality in nuclei selection

The probability image, the separability table and the spectral characteristics of every class constitute tools for evaluating the quality of classification. They aid in selecting the nuclei and improving them through several iterations. At the end of an iteration several modifications of nuclei may exist corresponding to the following:

1. The chosen nuclei exhibit good quality and are retained. However, modifications of other nuclei may lead to modification of the quality of the first nuclei if the new nuclei created are quite close to the former. The nucleus 'crop' is a typical example.
2. The nuclei are close to one another but spectral characteristics are quite different. They are retained for several reasons, such as thematic. Their quality needs to be re-examined by means of a performance table at the end of classification (see later). This is the case of various levels of bright of bare soils.

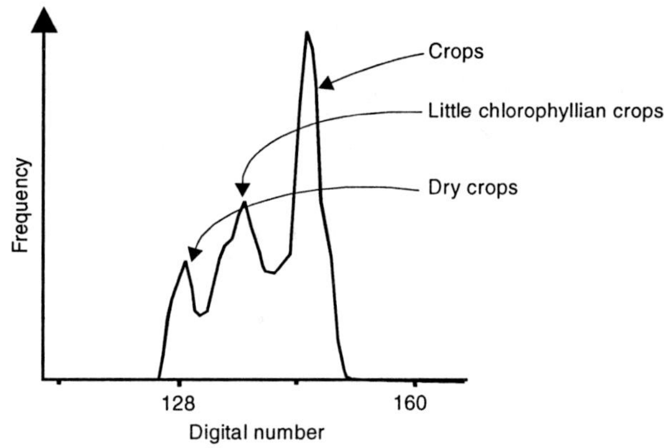

Fig. 9.7: Values of NDVI for the two nuclei 'crops' and 'dry crops', which are regrouped into three classes.

3. The nuclei indicate poor separability and similar spectral characteristics. They may be preserved if necessary but it is prudent to use new image features in defining them to achieve the best possible semantic separation. Such is the case of the nuclei 'dry crops' and 'sparse vegetation'.
4. If two nuclei are very close and their separability negligible, they can be combined into a single nucleus. Such is the case of the nuclei 'dry grasslands' and 'green grasslands'.
5. Some nuclei may be least separable. This can be seen in the line corresponding to incriminated nucleus. It is better to remove them. This is the case of the nucleus 'cloud' for example.
6. It can be seen by observing the probability image that nuclei for which rejected pixels are minimal are absent. 'Vegetation with fair coverage', 'vegetation and dark soils' and 'dense vegetation' are examples.

■ Performance table

Before accepting the final classification, it is important to consider the last quality indicator of this classification, viz., the performance table (Table 9.2).

It may be recalled that a nucleus is *a group of pixels that serves to a priori characterise an object we wish to use for designing a classification procedure and that a thematic group is the population of pixels which at the end of classification, hence a posteriori, is combined into a single group.* After classification, the thematic groups *carry the same name as those of the nuclei used to constitute them during classification.*

If all the pixels of a nucleus are found in the corresponding thematic group, it is considered that the pixels of the nucleus are well chosen and that performance is very good. Contrarily, it often happens that a certain number of pixels chosen to constitute a nucleus are found classified in another thematic group that does not correspond to the nucleus. This indicates that the pixels of the nucleus in question are not correctly chosen. Choice of pixels constituting the nucleus can be modified or nuclei that are too close can be changed.

The performance table is useful for evaluating the homogeneity of every thematic group and hence of nuclei chosen before classification (Table 9.2). Rows in the table represent the nuclei and columns the thematic groups. The distribution of pixels used to characterise the nuclei in various thematic groups can be read from each line. The sum over a line is hence equal to the number of pixels that represent the nucleus.

Classification errors in this table are far fewer than in the separability table. The sum of pixels of nuclei classified into the corresponding thematic groups (first diagonal) is 33,236 or 91.2%. Classification quality is hence acceptable. Each line also gives the percentage of well-classified pixels. Pixels

Table 9.2: Performance table for 20 nuclei of maximum-likelihood classification of the Brienne image

	Nuclei / Thematic groups	1	2	3	4	5	6	7	8	9	10	11	12	13	14	15	16	17	18	19	20	Total	% well classified
1	White soil	1316	26							4											25	1371	96
2	Very bright soil	4	1404	36	1				1	34			1								2	1483	95
3	Bright soil		337	1505	9				2	6											5	1864	81
4	Fairly bright soil			34	1702	10			8	1												1755	97
5	Fairly dark soil				42	1664	39		1			2	3									1751	95
6	Dark soil					34	775	57	1		2	1	13	1						1		885	88
7	Very dark soil					1	33	800					2	1						11		848	94
8	Sparse vegetation		4	19	109	31	1		1432	15			55	1								1667	86
9	Dry crops	2	22	15	7	1			10	1281											140	1478	87
10	Crops						2			3	3587			1								3593	99
11	Green grasslands					3		4	21			2133	377	143	7					3		2691	79
12	Dry grasslands					4		8	25			136	1230	16								1419	87
13	Riverine forests and forests		1			1	8	45	8		15	250	47	2703	571					54		3703	73
14	Dark forests													78	4353					2		4433	98
15	Very clear water															3534	20					3554	99
16	Clear water																1480	10				1490	99
17	Fairly clear water							1										1441	29			1471	98
18	Turbid water											2							381			383	99
19	Cloud shadows							30							2			8		393		433	91
20	Clouds	19	1	1	1		1			26											122	171	71
	Number of pixels of thematic groups	1341	1795	1510	1341	1795	1610	945	1509	1370	3604	2524	1728	2944	4933	3534	1500	1459	410	464	294	36443	91

corresponding to the nuclei of clouds, riverine forests and forests and bright soil are poorly classified. Test zones of these nuclei may hence be improved.

9.2.5 Iterations: a heuristic approach

After an iteration is completed and balance is achieved, it is very often necessary to carry out other iterations.

When a small number of nuclei is taken, separability is generally good but rejections would be many. To overcome this, it is necessary to increase the number of nuclei and avoid ultimately grouping them in a single colour, if several groups have the same function for the objective envisaged.

We can also combine multiple groups, apply a mask and reclassify them. This procedure may be followed in the case of bright soils, clouds and riverine forests and forest zones in the image of Brienne region, since confusion exists between these features.

In the example of Brienne, after three iterations (CD 9.6) for the same image with 22 nuclei, the following results are obtained: some are the same as in the preceding iteration, others have a modified composition, some have been suppressed and some new nuclei have been created. Hence it is imperative to analyse the modifications produced by various iterations in terms of quality of classification.

At the end of the third iteration, the image obtained is coloured so as to better reveal thematic units (CD 9.6). The major units earlier detected from other classifications are again observed. There is no thematic unit with less than 3,600 pixels. The largest class, viz., 'green grasslands' has more than 140,000 pixels. As mentioned earlier, this class was poorly defined since all the pixels having a digital characteristic of chlorophyllian vegetation and not being in a specific group are classified in this group. This group could be redefined if necessary to study a particular theme in detail. However, this would not be simple since this group replaces the concerned boundaries already existing between the 'chlorophyllian' groups. Hence, this group can be isolated by masking and reclassification carried out only on the corresponding pixels. Fifteen different groups were thus obtained (CD 9.7).

For a more detailed analysis of soil bright, which is very useful for estimating susceptibility to erosion, the thematic groups 'fairly bright soils' and 'fairly dark soils' can be reclassified. Each of these comprises more than 100,000 pixels.

Various thematic groups can be readily combined for obtaining a general picture of the region. The combined groups are coloured with a single colour. For example, only seven themes may be retained (CD 9.8): 1—bright bare soils, 2—dark bare soils, 3—in-situ crops, 4—grasslands, 5—forests and woods, 6—water bodies and 7—shadows and very dark soils. The major regions of chalky Champagne with very bright soils and large plots drained by valleys can be readily recognised. Very dark, clayey soils form a transition zone with humid Champagne which comprises grasslands and forests. The Brienne plain consists of bright soils and, further south, lakes and reservoirs.

9.2.6 Classification quality assessment

To analyse the results of maximum-likelihood classification its quality needs to be determined. This is achieved by successively examining the following aspects:

—spectral characteristics of each nucleus,
—separability,
—probability image with rejection level,
—performance table.

Three iterations are carried out, designated ML1, ML2 and ML3, using the maximum-likelihood method, each time improving the nuclei chosen and accuracy of classification (performance table).

The quality of classification was estimated by the same computations as earlier (sum of diagonals) and 94.2% of the pixels defining the nuclei found to be well-classified in their respective thematic groups for ML3 (Table 9.3) compared to 91.2 for ML1 (values used for this computation were taken from a performance table with zero rejection threshold). This is satisfactory inasmuch as the number of thematic groups has increased from 20 to 22.

Table 9.3: Performance evaluation of various classifications.
Values of well-classified pixels for ML1 and ML3 are given in bold; values mentioned in the text in italic.

Classi-fication	Rejection threshold	Well classified pixels	Poorly classified pixels	Rejected pixels	Sum	Well classified pixels (%)	Poorly classified pixels (%)	Rejected pixels (%)
ML1	0	33,236	3,207	0	36,443	**91.2**	8.8	0
ML1	1	32,589	2,970	884	36,443	89.4	8.2	2.4
ML1	3	30,954	2,625	2,864	36,443	84.9	7.2	7.9
ML1	5	29,697	2,505	4,241	36,443	*81.5*	*6.9*	*11.6*
ML1	10	26,522	2,293	7,628	36,443	72.8	6.3	20.9
ML3	0	33,295	2,061	0	35,356	**94.2**	5.8	0
ML3	5	29,990	1,628	3,738	35,356	*84.8*	*4.6*	*10.6*

For the third iteration (ML3), if the rejection threshold is taken as 5 for example, the performance table indicates 10.6% rejection pixels and 84.8% well-classified pixels, whereas for the first (ML1) iteration rejects are more (11.6%) and well-classified are less (81.5%). The percentage of poorly classified pixels decreases from 6.9 to 4.6. Hence, performances of the third iteration are better than the first.

Let us now examine the objective of classification. Are accurate thematic groups preferred even if all pixels are not classified or classification of the entire image preferred even if the thematic groups are not very accurate? Obviously, the more the thematic groups, the closer they are to one another. Consequently, separability is poorer. If it is desired to improve separability, more homogeneous nuclei and hence of smaller frequency should be taken. This makes the method more difficult to apply since it is not easy to correctly adjust the statistical populations using a nucleus having too small a number of pixels.

It is also possible to use the three axes of principal component analysis instead of the raw bands, in order to reduce the dimensionality of the problem.

9.2.7 Conclusion

Maximum-likelihood classification is a mathematically satisfactory method since the pixels are classified using a probability; this approach is extremely desirable in remote sensing. Obviously, the choice of nuclei and judgement of the expert and interpreter influence the classification. It should be noted, however, that the method, in particular its operation in the TeraVue program, provides for monitoring the choice of nuclei during the process by means of statistics and the separability tables. The performance table likewise enables evaluation of the final result. Lastly, the image obtained with various rejection levels gives a numerical value for the quality of results. Classification can be improved by carrying out several iterations. The quality of this classification can thus be determined, which is essential in image processing. However, it is necessary in this case also that a true confusion matrix be established, which allows only comparison of classification results with field data.

10

Image Processing Methodology

Any classification in image processing results in reduction of the quantity of existing information and definition of less number of units known as thematic groups or classes by combining several pixels.

Most often these groups are constrained with reference to their number, thematic significance, statistical coherence, etc.

10.1 OBJECTIVES

In order to evaluate the constraints mentioned above, the objectives need to be defined before processing the images. The objectives determine the characterisation of the field of study, fineness of analysis desired, homogeneity investigated, accuracy required, reliability of the methods, need for diachronic images and the number of groups to be studied. The most common objectives are:

—Preparation of an integrated map and general investigation of the most characteristic units in the field of study. This may correspond to the question: what are the major spatial units observed in the region under investigation?

—Segmentation of the field of investigation contributing to a mission of study. This corresponds to the question: where should we look on the ground for the major thematic units of this region?

—Investigation of a simple segmentation for some specific themes only. This may be represented by the question: where are such objects, for example grasslands or uncultivated lands, located in this region?

—Precise analysis of a single type of object, avoiding others. This may be represented by the question: what are the various types of bare soil identifiable by their hue, in order to determine the types of inputs they receive?

—Diachronic monitoring of an object and its modelling to enable prognosis.

10.2 METHOD

To achieve any of the several objectives the study has to be divided into three phases as in any information processing system:

—status prior to processing: data available and objectives;

—proper processing and choice of methods;

—results of processing: evaluation and presentation.

10.2.1 Input

For a processing technique to be applicable to any investigation, the best possible images should be chosen according to the objectives, environment and theme. An image processing not suited to the objective pursued does not provide correct answers. However, most often images not necessarily optimal for the given objective are analysed for reasons of cost. In such cases it is necessary to take into consideration the accuracy required (which in turn determines the pixel resolution), image periodicity, acquisition date and spectral characteristics of the phenomenon under investigation (which determine choice of most suitable wavelength bands available with the existing sensors) (see Chap. 16).

Basic models appropriate for processing have to be defined. Radiometric or chorological models are mainly used for this purpose.

Radiometric models are established from field measurements in the zone of study or in a zone considered equivalent. If field measurements are not available, a general radiometric model (Chap. 7, Fig. 7.4) can be used. Care has to be taken to adapt the model to the region or theme under investigation.

Chorological models are developed by thematic experts and geographically tested in various sectors of the region or in equivalent regions. If precise chorological models are to be employed, it should be remembered that they are in most cases applicable only for specific zones and cannot be generalised.

10.2.2 Processing

The processing methods to be used are limited by the following technical considerations:

—capacity of the computer to analyse a large number of pixels and speed of execution of classification;

—number of persons and work stations to be employed to obtain the answer, when faster analysis is required (see Chap. 18).

The processing techniques employed are often those given in available softwares. Hence, it is particularly important when purchasing the software to check that it is compatible with the problems to be solved. Costly programs do not necessarily give better results. It is imperative to recognise the 'default' options provided in each and to acquire the correct method pertaining to programmed classification. Thus, principal component analysis rarely gives the same results for the same image with different softwares.

The processing method used should be amenable to evaluation for quality. So it is necessary to be equipped with a statistical tool that provides such evaluation; this is not always easy.

Generally speaking, thematic experts ought to intervene during classification. This intervention depends on various phases of classification and its impact on the results is more or less significant, as has been seen in the preceding chapters. In all cases it is desirable to most explicitly specify the decision rules and phases in which they are applied. This is a condition of repeatability of the method and hence of technology transfer.

It is often necessary to use several processing methods in a single work sequence in order to adapt them to the specific problem posed during analysis of results or during preparation of images for a subsequent processing. As mentioned earlier, a given processing method is incorporated within a more general procedure. Some examples are: return to parallelepiped analysis for better identification of a thematic group, use of a vegetation index to subdivide or to reformulate a group of very similar nuclei, succession of ascendant hierarchic classifications after masking a given theme, etc.

10.2.3 Output

Before utilising the result of a classification, it should be assessed for quality (see Chap. 17). This is done with confusion matrices, usually represented as performance tables. The classification result is compared with external data corresponding to ground observations or reference data acquired from aerial photos or thematic maps (see Chap. 19).

The image processing results ought to be evaluated from the *semantic* (content of each spatial unit retained) as well as *graphic* point of view. It is important to interpret the shape of the map zones obtained for each landscape unit and thematic group. This is carried out by analysis of shape criteria (perimeter and surface area), readily obtained, especially when a geographic information system is available.

Ground reference data should also be acquired according to the objectives envisaged. Collection of such information is often a long process but must be entirely reliable since the type of classification, method and procedure recommended are based on it. The method of validation of results ought to be compatible with the cost and time allotted for the study. It represents one of the main considerations, along with acquisition of data to be processed, in the management of a project.

The results of analysis are conventionally represented as a map, often on a computer monitor which offers many advantages vis-à-vis map presentation on paper. The number of colours in which a map can be displayed is much greater and readily changeable and zoom is always possible. A minimum of 100 pixels is required for representation on a paper map and for defining the smallest map units obtained at the end of classification. This corresponds to a minimal surface area of 4 ha in the case of SPOT-HRV, 9 ha for LANDSAT-TM and even 0.1 ha for resolutions of 2 to 3 m. If we apply the *quarter rule, viz., that any zone should have at least 1/4* cm^2 *on the map*, 100 pixels are needed per 25 mm^2, or 10 pixels per 5 mm. The scale of map representation $E = 1/X$ is hence defined as:

$$X = 2\,p \times 10^3$$

where p is the pixel resolution used for processing, expressed in metres. The following presentation scales are obtained:

Resolution	*Scale*	
10 m	1:20,000	possible utilisation for 1:25,000 maps
20 m	1:40,000	possible utilisation for 1:50,000 maps
30 m	1:60,000	possible utilisation for 1:63,600 maps

10.3 PROCEDURE

Classification procedure varies according to the theme investigated and the concepts of thematic experts. A new view of the Earth can be obtained from remote sensing applying the following rule:

If a unit having a specific and detectable spectral characteristic can be concomitantly defined in space and time, then an entity corresponding to this unit actually exists.

This entity does not necessarily correspond to a theme; it may represent a specific organisation of objects (a pattern), the latter considered very different by various thematic specialists.

It is therefore necessary to identify an object not only by its nature, but also by its spatial organisation and diachronic behaviour. This identification can be done before, during or after processing of images. Thus four steps of image classification can be defined (Fig. 10.1).

10.3.1 Initialisation

Initialisation consists of extracting maximum information from image data by means of statistical, structural, diachronic or chorological analysis. These analyses are *unsupervised* and necessitate no field study. They enable identification of zoning of image features for which a precise legend cannot be associated. While the container can be precisely determined, the content of each zone is often poorly defined.

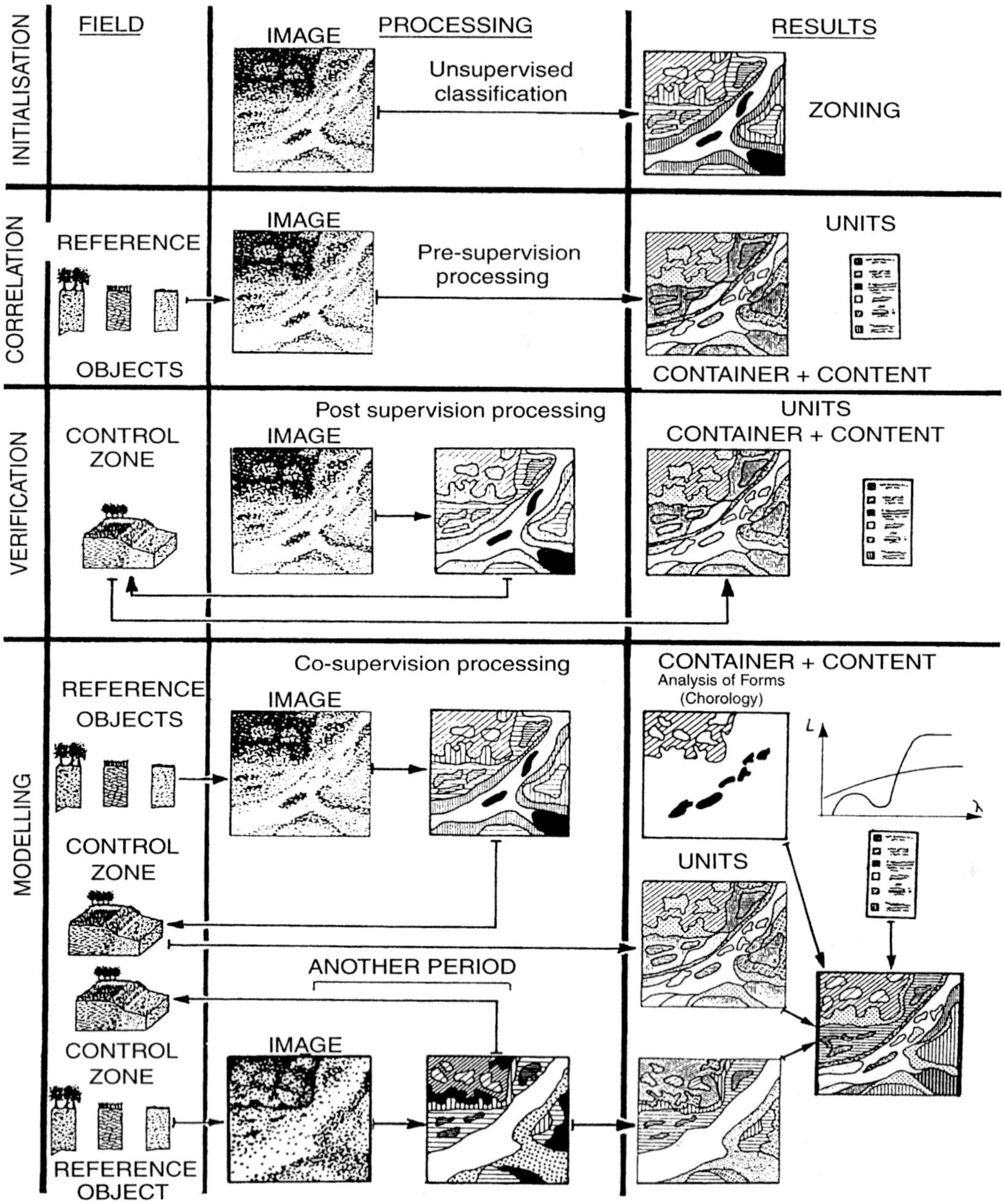

Fig.10.1: Four stages of classification.

10.3.2 Correlation

This stage consists of establishing *references* on the ground, in the laboratory or through investigations at the time of image acquisition, in order to compare the spectral data recorded by the sensor with those collected on the ground. The reference zones facilitate *pre-supervision* processing. For example, a radiometric interpretation model adapted to the region under study can be created. From this stage spatial units are obtained to which we can assign a legend prepared from the reference data acquired prior to processing.

10.3.3 Verification

This stage involves determination by means of ground *control* of the content of the zones delineated by a preliminary *unsupervised* classification. A second *post-supervision* classification is then applied. Maps thus obtained are accompanied by a legend corresponding to units identified and verified on the ground.

10.3.4 Modelling

The modelling stage utilises field references for pre-supervision classification and the units obtained are verified for their content by ground control. *Co-supervision* processing is thus carried out combining ground control and reference control in a to-and-fro movement. Radiometric and chorological models are developed. It is possible to determine all parameters of a *general model,* which enables *generalisations* based on certain rules, and introducing it in a geographic information system. The general model in some cases imposes a diachronic study for incorporating characteristics of objects in it.

10.4 INTERPRETATION OF PROCESSING

The result of image classification is not the end of image analysis. The results need to be further interpreted relative to the initial objectives. This interpretation is generally based on two approaches, viz., radiometric and chorological.

10.4.1 Radiometric approach

Each object ought to be defined by its spectral characteristics and this requires ground reference measurements. If the latter already exist, they can be used but only after verifying that they were acquired in conditions similar to those in which the data will be used. Thus, laboratory measurements and field measurements should be adjusted. Generally, measurements obtained in one region are not directly applicable to another for reasons of topography, climate, latitude, etc.

As the field data most often consist of reflectance measurements, it is necessary to verify that the reference model thus obtained is coherent with the digital numbers of the image processed. This is not always simple.

10.4.2 Chorological approach

Chorology (derived from Greek 'khoros', country or district, and 'logos', logic or science) is the study of relationships existing between the characteristics of semantic (thematic) units identified (intrinsic factors) and their distribution in three-dimensional landscape (extrinsic factors).

Chorological analysis in the case of remote sensing thus comprises geographic-domain investigation of distribution of objects identified in radiometric space. Objects that define the map units obtained from image classification are most often not randomly distributed in geographic space. Some units are located in a particular geographic position, in a particular region or contrarily scattered, but scattering is not necessarily random. Some units show systematic spatial associations: unit A always surrounded by unit B which per se is connected to unit C, and so forth. Frequently the shape of a unit is indicative of its nature, for example digitate indicating a zone of circulation and infundibular representing a zone of accumulation. Other common forms are elongate, compact, segmented, regular, etc.

Shapes and interzone and interunit relationships are readily detectable in images. This is one of the advantages of the latter. Analysis is much easier on computer systems since it is possible to play with various colour contrasts, masks, zooms, etc., which enables evaluation of various relationships.

The data derived from image processing, once interpreted, are used to prepare maps corresponding to the designated objective. However, at present these data often constitute input to another information system.

References

Anonymous. 1994. Guide d'utilisation de TeraVue. Éditions La Boyère. Valbonne, 220 pp.

Baize D, Girard M-C. 1995. Référentiel pédologique. INRA-AFES, 332 pp.

Belluzo G, Girard C-M. 1997. Identification et classification de l'occupation du sol à partir de scènes Thematic mapper: application aux prairies d'une région de champagne humide. Bull. SFPT, 146: 22–32.

Bertin J, Barbut M. 1973. Sémiologie graphique. Mouton, Gauthier-Villars, Paris, 2nd ed., 431 pp.

Burlot F. 1995. De l'interprétation visuelle à l'interprétation automatique des images satellitaires: application aux pédo- et hydropaysages. Mémoire Mastère SILAT, Grignon, 65 pp.

Congalton RG. 1991. A review of assessing the accuracy of classifications of remotely sensed data. Remote Sensing of Environment, 37: 35–46.

Deschamps P-Y, Duhaut P, Rouquet M-C, Tanre D. 1984. Mise en évidence, analyse et correction des effets atmosphériques sur les données multispectrales de Landsat ou de SPOT. Les colloques de l'INRA, **23:** 709–722.

Diday E. 1971. Une nouvelle méthode en classification automatique et reconnaissance des formes: la méthode des nuées dynamiques. Revue de Statistiques Appliquées, 19 (2): 283–300.

Gao BC, Goetza AFH. 1995. Retrieval of equivalent water thickness and information related to biochemical components of vegetation canopies from AVIRIS data. Remote Sensing of Environment, 52: 155–162.

Girard C-M. 1987. Spectral and botanical classification of grasslands: Auxois example. Advances in Space Research, 7: 67–70.

Girard C-M, Girard M-C. 1995. Qualité des méthodes d'interprétation: application à la caractérisation et cartographie d'unités de paysage, qualité et validation des résultats. Bull. SFPT, 137: 62–66.

Girard M-C. 1983. Recherche d'une modélisation en vue d'une représentation spatiale de la couverture pédologique. Application à une région des plateaux jurassiques de Bourgogne. Thèse Doc. ès sciences, Université Paris 7. SOLS, 12: 414 pp.

Girard M-C, Girard C-M. 1989. Télédétection appliquée: Zones tempérées et intertropicales. Masson, Paris, 260 pp.

Jambu M. 1978. Classification automatique pour l'analyse des données. Dunod, Paris.

Monget J-M. 1986. Cours de télédétection, 1. CTAMN, octobre, Sophia-Antipolis.

Monget J-M. 1986. Cours de télédétection. École des Mines de Paris, 165 pp.

Monget J-M, Robertson Y-C. 1992. Two-variable mapping applied to remote sensing data interpretation: a software implementation. In: Remote sensing from research to operation. Proc. 18th Ann. Conf. The Remote Sensing Society, pp. 571–580.

Orth D. 1996. Typologies et caractérisation des prairies permanentes des marais du Cotentin, en vue de leur cartographie par télédétection satellitaire, pour une aide à leur gestion. Thèse INA P-G, 150 pp.

Van Den Driessche R. 1965. La recherche des constellations de groupe à partir des distances généralisées de Mahalanobis. Biométrie, 41 (1): 36–47.

Wilmet J. 1996. Télédétection aérospatiale. Methodologie et applications. SIDES, 300 pp.

11

Structural Processing of Satellite Images

11.1 INTRODUCTION

The most common techniques of satellite image processing, viz., principal component and parallelepiped analyses and ascendant hierarchic, mobile centres and maximum likelihood methods of classification, etc. are all based on *textural analysis* of images. In these methods each pixel is classified individually without taking its neighbourhood into consideration. Methods based on *structural analysis* of images are utilised for three reasons described below.

11.1.1 Boundaries and mixels

The first reason for using structural analysis of images is that the grid formed by pixel positions on the ground is entirely arbitrary since it depends on the satellite trajectory and sensor type. This is one of the reasons why a number of mixels exist whose spectral characteristic is a combination of spectral responses of several objects geographically included in the resolution element (see Chapter 15). This makes delineation of a boundary between two objects difficult even when it is distinct on the ground. Hence, it can be practically assumed that most boundaries identified on images are erroneous. This leads to a significant error in classification. This error will be larger if the map units identified are small or digitate and hence of a longer perimeter. This 'noise' which frequently attains a strip of 1 to 2 pixels along the boundary of the map units should be related to the mapping quality desired. Any superposition of various maps on images by imposing a precision greater than 1, 2 or 3 pixels (or 20, 40 or 60 m for SPOT multispectral images) would be in vain. It is therefore necessary to take the pixel neighbourhood into consideration for this primary reason.

11.1.2 Classification and mapping

The second reason for using structural analysis is as follows: textural methods are efficient as long as a definite spectral characteristic can be defined for an object. In almost all cases, a classification is obtained in which each class comprises several map units that are not compact. The result of such a classification is similar to an impressionist's tableau with a multitude of small patches of various colours, seen from distance, an assemblage of large units. However, no boundaries of these units exist. Fuzzy logic needs to be applied to generate these 'homogeneous units from groups that are heterogeneous by nature' which are called *patterns*. In fact, for these units it is difficult to decide appurtenance of a pixel to a class with absolute certitude.

Moreover, thematic maps exhibit a relatively limited number of zones for each map unit retained and these zones are much more compact than those obtained through textural image processing. These units often correspond to complex groups of objects such as juxtaposition, association, sequence, combination, etc., which comprise several elements. It is therefore difficult to compare a remote–sensing document derived from a textural classification with a thematic (structural) map using the geographic information system, since the nature and number of spatial units—classes in remote sensing and map units in thematic study—are too diverse.

It is hence necessary to find a classification method that enables enhancing compactness of image zones.

11.1.3 Complex map units

The third reason for use of structural analysis arises from the fact that entities characterised by a group of objects related to one another and exhibiting spatial organisation do exist. Such is the case, for example, of a 'valley' which comprises the objects 'water', 'trees', 'grasslands', 'alluvium', etc. In such cases structural image processing has to be employed. However, only very few classification algorithms based on *image structure* exist.

Two methods of structural processing of satellite images based on neighbourhood analysis, viz., OASIS and VOISIN (Didier, 1992; Francoual, 1994; Girard et al., 1980, 1990 and 1991; Gilliot, 1989, 1992 and 1994; Ranaivoson, 1990), are presented below for various applications. Similar methods have also been developed in France, viz., PAPRY (Borne, 1990) and CLAPAS (Robbez-Masson,1994). This is one of the important lines of research at present (Girard, 1995; Girard and Girard,1996; Girard et al., 1997).

11.2 VOISIN

Classification of a pixel should also take into consideration the pixels that surround it and hence its neighbourhood. This neighbourhood may be defined by a certain distance from the pixel under study. It thus consists of investigating *le Voisinage et l'Organisation des Informations Spatialisées par Informatique et Numérisation* (neighbourhood and organisation of spatial data by information technology and digitisation), known as the VOISIN programme.

The objective of VOISIN is to determine the neighbourhood of an image pixel as it gradually increases. Characterisation of a point effectively varies with its surroundings. Let us consider an example: a fisherman (1) is situated in a boat (2), sailing in a lake (3), surrounded by a beach (4), bounded by mountains (5) in a country (6) in Europe (7) on the planet Earth (8) in the solar system (9), galaxy Milky Way (10). These ten levels of neighbourhood represent totally different distances varying from 1 m to 10^{18} km. The fisherman can be defined in ten different manners: a man, a marine, a resident of a country, an European, a Terrestrian, etc. All these definitions are true but important only once, when the perception level taken into consideration is defined (see Chapter 15). The VOISIN model enables us to delineate, through the neighbourhood parameter, boundaries of spatial units whose heterogeneity is defined by a *pattern* identified by a composition vector.

11.2.1 Neighbourhood in window: composition vector

The composition vector is a textural measure giving frequency distribution of various classes in a given neighbourhood. It is described by a local histogram. While this measure does not take the shape into consideration, it is well suited to describe textures of satellite images and facilitates determination of *patterns*. For example, a valley constitutes a repetitive pattern, comprising free-water surfaces, trees (riverine forest), some cultivated plots, grasslands and roads.

The level of neighbourhood is analysed by means of a window as follows (Fig. 11.1). For a square window F_t of size t (t is the number of pixels defining the side of the square), a 'composition vector' $\vec{C_t}$ is defined as the vectorial sum of elementary vectors $\vec{C_i}$. Each of the latter represents a class c_i, balanced by a coefficient p_{ci} equal to the ratio of the number of pixels in class c_i, present in the window under consideration, to the total number of pixels in the window (or t^2).

$$\vec{C_t} = p_{c1}\,\vec{C_1} + p_{c2}\,\vec{C_2} + p_{c3}\,\vec{C_3} + \ldots + p_{ci}\vec{C_i} + \ldots$$

$$p_{ci} = c_i / t^2$$

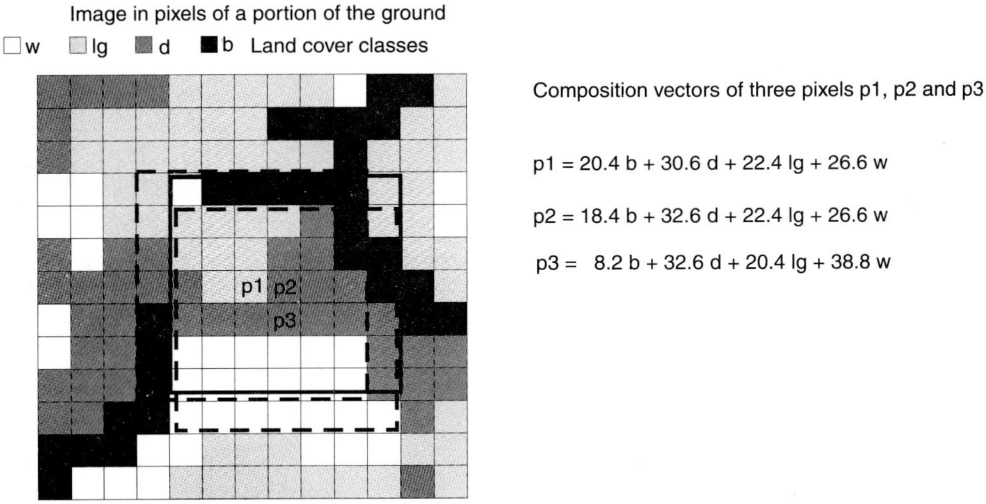

Fig. 11.1: Definition of the neighbourhood by a composition vector.

11.2.2 Method

For neighbourhood analysis, a given point of the image constitutes the central pixel for successive windows. The window size is gradually increased to values of $t = 3, 5, 7, \ldots$ (Fig. 11.2) and difference D between the successive 'composition vectors' $\vec{C_t}$ is computed as: $\vec{C_5} - \vec{C_3}$; $D_7 = \vec{C_7} - \vec{C_5}$; $D_9 = \vec{C_9} - \vec{C_7}$; .... The results are represented on a graph with the successive differences D_5, D_7, D_9, ... on the y-axis and the size of successive windows on the x-axis. These sizes are equivalent to an increasing geographic distance vis-à-vis the central point.

Sizes of successive windows for which differences are close to one another are detected from this curve. Window sizes that have relative minima in D, denoted as t_m, are used for subsequent analysis. For these window sizes, the local histogram remains stable even if the window size changes and permits determination of a *pattern*. It is characterised by the window vector of size t_m. The magnitude of the difference $D_t - D_{t-1}$ is a measure of its heterogeneity (Fig. 11.2). Most often, several local minima appear in the $D = f(t)$ curve, defining multiple patterns. The geographic distance between these various patterns is expressed by the size difference between two windows corresponding to two minima of D. In this way we can

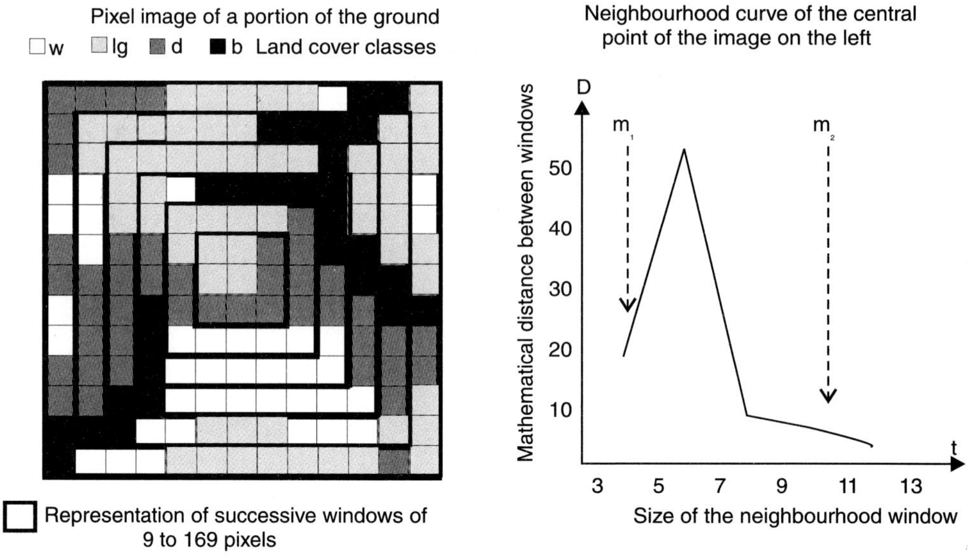

Fig. 11.2: Neighbourhood of a point in an image and representation of its neighbourhood function showing two local minima m_1 and m_2 for windows of 3 and 11 pixels. Minimum m_1 corresponds to a pattern consisting of light grey or dark pixels (lg, d) and extends north-west; the minimum m_2 represents a pattern containing more black (b) and white (w) pixels than the preceding, it extends more southwards.

1) identify patterns,
2) characterise them by their 'composition vector',
3) obtain a measure of their heterogeneity by the value of D,
4) estimate the distance between two patterns that are juxtaposed; this corresponds to the minimal distance on the ground to change from one pattern to another.

11.2.3 Example

At a given point in the space under study, the composition vector for the smallest window F_3 of size 3 × 3 is determined. Composition vectors are then computed for increasing window sizes, viz., 5 × 5, 7 × 7, ..., $(n-2) \times (n-2)$, $n \times n$.

Manhattan distance between each composition vector and that of the next higher size, $|VF_{(n-2)} - VF_{(n)}|$, is computed. We thus obtain the *neighbourhood curve*, which represents change in the neighbourhood when the field of study is enlarged.

When the same object is analysed while increasing the field, Manhattan distance remains zero; this represents a homogeneous pattern. If heterogeneity manifests in the form of appearance of a new object, this distance increases; this increase is more when this object occupies a larger area or several objects appear together. In all cases, increase in distance indicates a change in neighbourhood. If increase in the field of investigation is continued, the Manhattan distance curve passes through a maximum and then decreases to reach a new relative minimum, i.e., a second pattern is observed. Its statistical composition is determined from the composition vector.

Interpretation of *neighbourhood curve* (Fig. 11.3) defined by the relation

'mathematical distance between windows' = f[window size]

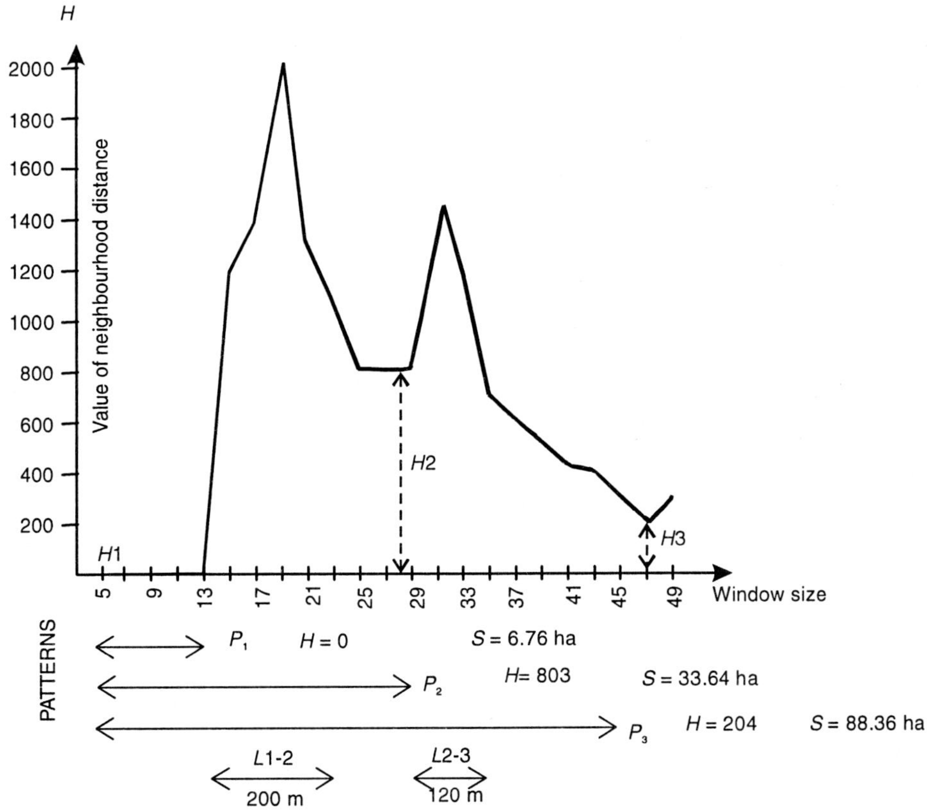

Fig. 11.3: Neighbourhood curve of a classified image. In this curve three patterns *P* can be distinguished, having heterogeneity *H* defined for window sizes of area *S*, the width of the boundary between these patterns of definite homogeneity being *L*.

enables identification of various patterns (*P*), window sizes (*f*) (or area $S = 400 \times f^2$, in m^2 for multispectral SPOT) for which they have stable composition and width of the boundary between two stable patterns (*L*). For each pattern a heterogeneity factor (*H*) can be defined, characterised by a value of mathematical distance between successive windows of a definite pattern.

This model permits characterisation of various patterns situated around a localised point or, by analogy, various planes of a landscape observed on the ground from an observation point. It establishes locations of transitions (boundaries) between patterns in which the mathematical distance changes from one value to another (boundary width), as well as places of maximum heterogeneity. Lastly, it gives a value of internal heterogeneity for each pattern.

VOISIN is a model for heterogeneity characterisation of spatial patterns, leading to a localised geostatistical study using *n* variables. It helps in decision-making for optimal delineation of heterogeneous groups in a given spatial field. It enables correct choice of a representative moving window for a spatial pattern to be used for OASIS classification.

This program also constitutes a tool for making scale changes and enables 'synthesised generalisation' (see Fig. 15.1).

11.3 OASIS

11.3.1 Method

■ Definition

OASIS, *Organsiation et analyse de la structure des informations spatialisées*[1] (organisation and analysis of spatial data structure), is a method of supervised classification based on the fuzzy groups theory, which uses spectral and textural characteristics for heterogeneity analysis.

Classifying all pixels according to neighbourhood parameter is the objective of this method. Each class is characterised by a pattern defined by several classes of objects. This classification is based on nuclei defined by the thematic specialist and assumed to characterise the patterns being searched. The result gives zones that are coherent with the units of a map and much more compact than those given by textural methods. Moreover, high quality classified image is obtained. The OASIS method can be used for raw bands of a satellite image as well as for a single band such as a panchromatic image or a result of classification. It is most efficient in the latter case since the composition vectors correspond to a known assemblage of classes of objects but not to raw spectral characteristics.

A moving window is used to cover the entire image and aids in determining the composition vector for each pixel.

■ Nuclei

Several important patterns, designated *nuclei*, are defined either by investigating their composition derived from the image or by parametrising their composition vector.

It is possible to define the composition of nuclei in satellite images by marking zones corresponding to the patterns which the thematic specialist wishes to detect. The user delineates each nucleus by one or several polygons traced on the computer monitor.

Composition of these zones, determined by the local histogram, is computed and the corresponding nucleus is then defined from its composition vector. It is possible to modify the composition vector or to directly define it in a table.

■ Size of moving window

The thematic specialist defines the window size. Either the window defined by means of VOISIN, which indicates the average size of the patterns, is taken, or it is defined as a function of the object studied. The larger the window, the larger the neighbourhood considered for characterising the pixel and more compact zones obtained. However, if the window is too large, smaller objects will disappear and will be integrated with a more complex reference zone. Thus, rivers are frequently missed with windows of 5 or 7 pixels (CD 11.5 or 11.8). In this case, they are integrated into the reference zones that comprise a few per cent of 'river vector' in their vector composition, as was observed in the CORINE Land Cover study (Chapter 19).

■ Computations

Manhattan distance from each pixel to each nucleus is computed and a pixel is assigned to the nucleus for which the distance is minimum (DIMITRI method, Girard, 1983; Girard and King, 1988). A preliminary classification is obtained which is evaluated by the sum of minimal distances of all pixels

---

[1]This free software can be obtained from the website http://lacan.grignon.inra.fr/resources.

to the nuclei they are assigned to; the smaller this sum, the better the classification. Multiple approximations are made until stability of nuclei is obtained. Computations may be extensive since analysis of millions of pixels is involved depending on the windows (defining neighbourhood) which comprise hundreds of pixels.

■ Iterations

Nuclei of the next iteration are redefined from the statistics of pixels grouped in each nucleus. More nuclei can be added in each iteration and then a new iteration started.

The sum of distances (Σ D) of pixels to nuclei in which they are grouped varies for each iteration and gradually decreases. Several iterations can thus be carried out until the grouping is optimal. The optimum is reached when the sum of distances (Σ D) becomes constant.

The term iteration is employed when only the nuclei change from one classification to another. Contrarily, if the window size is changed or if the initial image is no longer used for making a new classification but an already classified image is used, the term approximation is employed.

■ Results and quality

We thus obtain nuclei which at the end of processing are called *references*. The references define the composition of map units of the final document. Each reference may be defined by a single class or several classes. The former is a homogeneous reference and the latter heterogeneous. The latter corresponds to a pattern whose composition makes the heterogeneity explicit. The pattern reveals all information of the initial classes and hence no information loss occurs. OASIS is not a reduction method and differs completely in this from a smoothing technique.

The result of each iteration is displayed as a classified image (one colour for each class) to which a black and white image of distances of each pixel to the nucleus in which it is grouped is appended (see the example and images in CD11.2 and 11.3). The latter gives a quantitative estimate of the quality of grouping. Its analysis enables identification of well-classified and poorly classified regions. New nuclei more appropriate to poorly classified regions are redefined in such cases.

Compactness of zones obtained varies according to the window size used to define neighbourhood. Form and size of zones vary with window size in a more complex manner (Fig. 11.4) but generally towards greater compactness (expressed by a maximum area for a minimum perimeter). At the end of the process, images with much more compact zones are obtained than in the maximum likelihood classification, for example.

The images can then be readily converted into vector mode for introducing in a geographic information system.

■ Conclusion

OASIS is a non-hierarchic structural classification system based on patterns defined by organisation and neighbourhood of each classified pixel, which creates non-hierarchic units.

Classification by OASIS enables delineation of compact map units from the patterns chosen and model heterogeneity. OASIS, through iteration, determines the best spatial division for a moving window of a given size.

11.3.2 Example

Processing techniques corresponding to this example are given in the CD in index CD 11. Each illustration is indicated by a number 1 to 12, which represents the image band processed.

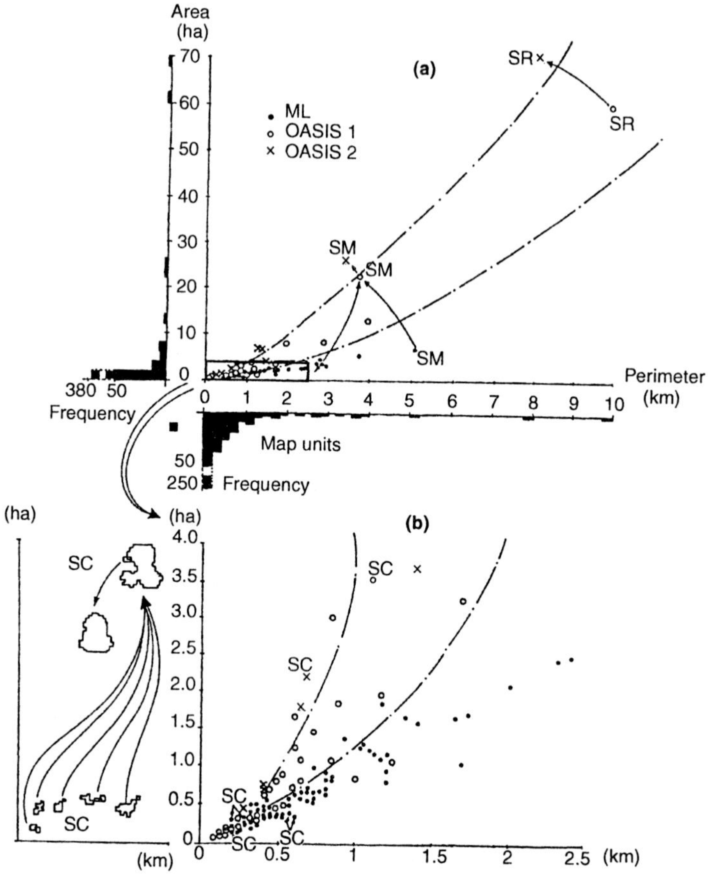

Fig. 11.4: Variation in shapes of map units according to various approximations of OASIS and window size (after Yongchalermchai, 1993).
Points represent shape of units obtained from three types of classification:
ML: Maximum likelihood; OASIS 1: First approximation of OASIS; OASIS 2: Second approximation of OASIS.
SM, SC and SR represent three types of units.
The figure shows their variation during three types of analyses.
a: shows all zones; *b:* represents enlargement of the preceding figure between values 0 to 4 ha and 0 to 2.5 km.

■ Characterisation of nuclei

The image in CD 11.1 was classified by the maximum likelihood method after three iterations (see Chapter 9). Eight nuclei are defined, viz., forest (violet), water (blue), valley (dark cyan), riverine forest (khaki), grasslands (light green), bright soils (light yellow), beige soils (salmon) and dark soils (maroon). The composition of each nucleus is given as a function of 22 classes used for maximum likelihood classification (Table 11.1).

The nucleus 'forest' thus comprises 88% pixels classified as 'hardwood' and 12% pixels classified as 'conifers and shadows' in the maximum likelihood method. A similar situation is observed for other nuclei. However, for better readability of the table, cells are left blank when the value is less than 10%.

Table 11.1: Composition of nuclei from groups of maximum likelihood classification
The values given in the table are percentages read along a row: the sum of a row is equal to 100. Percentages less than 10 are ignored.

	MV3													
Nuclei	1	2	3	4	5	6	11	14	15	16	17	22	23	26
	White soil	Very bright soil	Bright soil	Fairly bright soil	Fairly dark soil	Dark soil	Grass-lands	Hard-wood forest	Very clear water	Clear water	Fairly clear water	Green crops	Conifer sha-dows	Dense vege-tation
Riverine forest							18	18				18	10	15
Grasslands							56							35
Forest								88					12	
Water									35	32	31			
Valleys				14			20						10	
Bright soil	20	38												
Beige soil		15	34	17										
Dark soil				20	28	12								

Consequently, columns corresponding to groups of maximum likelihood classification having values less than 10% are removed.

The nucleus 'water' includes various types of water. 'Forest' mainly consists of deciduous trees and conifers. The compositions of 'valley' and 'riverine forest' nuclei have common groups, viz., 'grasslands' and 'conifers and shadows', which in the case of riverine forest correspond to shadows. However, the nucleus 'valley' also comprises 'fairly bright soils', whereas riverine forest includes 'deciduous' and other chlorophyllian plants. The nucleus 'grasslands' is made up of 'grasslands' and 'dense vegetation'. Various soil nuclei have compositions extending over several groups of maximum likelihood classification, but a continuous variation exists from bright to dark soils.

■ Interpretation of results: first approximation

A first approximate classification is made using OASIS, with a 3 × 3 window of 9 pixels. An image (CD 11.2) is obtained which is less differentiating than that of maximum likelihood classification. Fairly simple groups are obtained: all the pixels assigned to various types of water are classified in the nucleus 'water'. The nucleus 'forest' is slightly less than the sum of the two groups 'forest' and 'conifers and shadows'. A final composition after each approximation is also given, which can be compared with the initial.

A black and white image is also given for each approximation for assessing the quality of classification. This image gives the value of distance separating a pixel from the nucleus to which it is attached. The smaller the distance (represented as blank), the better the quality of classification (CD 11.3). The histogram of distance values extends from 3 to 199 (Fig. 11.5).

The channel that gives classification and the channel that gives its quality can be combined into a single image (CD 11.4). Objects are represented by various colours. Their classification quality is expressed in the same colour of higher intensity when it is good (code 4) and lower intensity when it is poorer (code 0). It can be seen (Table 11.2) that the best-classified patterns, relative to the initial objects, are forest (66 of 93 per thousand well classified), grasslands (140 of 193) and beige soils (73 of 160). This indicates that one object predominates in each pattern (see Table 11.1).

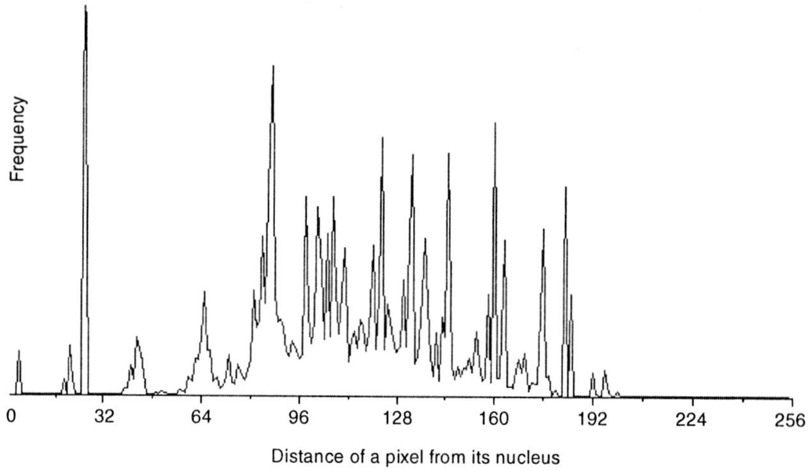

Fig. 11.5: Histogram for first approximation of distances of all pixels classified into their classification nuclei. If a pixel is perfectly classified, its distance is zero.

Table 11.2: Classification quality of 8 patterns of first approximation. Values for 4 quality grades are given per thousand image pixels.

Quality	High (4)	Correct (3)	Poor (2)	Inadequate (1)	Total
Water	0	4	**28**	3	35
Forest	**66**	17	4	6	93
Valley	0	7	0	**10**	17
Riverine forest	1	26	**35**	19	81
Bright soils	0	39	**70**	**65**	174
Beige soils	0	**73**	**85**	2	160
Dark soils	0	35	**144**	65	244
Grasslands	15	**140**	37	1	193
Total	82	341	403	171	997

■ Change of nuclei: second approximation

After preliminary conclusions are drawn, nuclei are redefined in a second approximation either by adjusting them, combining them or taking new ones. In the example considered, the nucleus 'valley' is removed as it is too close to riverine forest. The new composition of nuclei for the second approximation is shown in Table 11.3, wherein rows indicate the composition of nuclei of second approximation from the nuclei of first approximation.

If, for example, a window with a side of 5 pixels, i.e., total 25 pixels, is chosen, the resultant image describes a neighbourhood of 1 ha around each pixel. A second image is obtained (CD 11.5) which consists of more compact units. It is simpler to analyse.

The pixels that were grouped in the unit 'valley' in the first approximation are now found in the units 'riverine forest', 'grasslands' and 'dark soil'. The pixels in 'riverine forest' of the second approximation comprise pixels that belonged to the units 'riverine forest', 'valley', 'grassland' and 'forest' in the first approximation. This is logical in view of the composition of the nucleus 'riverine forest'.

Table 11.3: Composition of nuclei for making second approximation (in rows) from the classes obtained in the first approximation (in columns). Values are expressed as percentage.

	Nuclei 1	Riverine forest	Grass-lands	Forest	Water	Bright soil	Beige soil	Dark soil	Valley	Number of pixels
Nuclei 2	Riverine forest	53	29	12					6	100
	Grasslands	6	92					1	1	3040
	Forest	2		98						1726
	Water				100					2047
	Bright soil					84	14	1		3499
	Beige soil					10	77	13		4939
	Dark soil		2				12	85	1	5053

The image of distances is much clearer, indicating that the quality of classification has improved. The entire image is fairly well classified. It is mainly the boundaries of grasslands, water bodies, forests, etc. which appear as poorly classified (the dark zones). They represent assemblages of more than 7 objects used for classification over a small area (25 pixels or 1 ha). These zones, being heterogeneous, exhibit a wide variation. The histogram of distance values extends from 1 to 186 (Fig. 11.6).

Comparison with the preceding approximation (Fig. 11.5) shows that the range of distances has decreased and hence the classification quality has improved. Comparison of quality of the two approximations (CD 11.6) indicates that poorly classified pixels constitute 3.9% and are localised (in blue) mainly along valleys, which is explained by the removal of the nucleus 'valley' between the two approximations. They are observed in the boundaries between the units 'forest' and 'grasslands'. They are associated with the pixels that have not changed the group (in red: same distance in the two approximations) and represent 5.5% of the image pixels. Improvement in the quality of the image

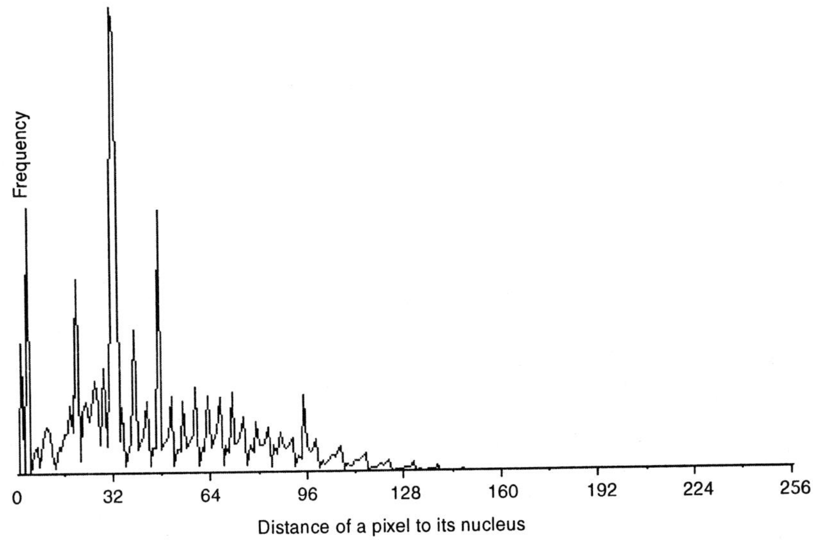

Fig.11.6: Histogram for second approximation of distances of pixels classified to their nuclei. If a pixel is perfectly classified, the distance is zero.

between the two approximations (which corresponds to decrease in distances) is hence greater than 90%.

Comparison of the classification results of the first and second approximations (CD 11.7) shows that 172,973 pixels (15%) have changed group, which is due to the removal of the unit 'valley' and also to modification in window size. There is no change of class in 85% cases.

Lastly, one can compare the characteristics of the composition vectors of the nuclei used to make the second approximation (Table 11.3) with the composition of reference vectors at the end of the second approximation (Table 11.4). Comparison of the two tables shows some small modifications in the composition of vectors. The unit 'riverine forest' consists more of 'forest'—23% instead of 12% initially—and less of 'grasslands'—18% instead of 29%. The reference unit 'beige soil' comprises less 'beige soil'—64% instead of 77%—and more 'bright soil'—20% instead of 10%. The reference unit 'dark soil' is characterised by a composition vector comprising almost all classes, each represented slightly. However, the class 'beige soil', which had a weight of 12% in the nucleus, represents only 3% in the reference unit.

Table 11.4: Characterisation of composition vectors of reference units obtained at the end of the second approximation (row) from the land-cover classes of the first approximation (column). Values expressed in percentage.

	Nuclei 1	Riverine forest	Grasslands	Forest	Water	Bright soil	Beige soil	Dark soil	Valley
Nuclei 2	Riverine forest	55	18	23					4
	Grasslands	12	84				1	2	1
	Forest	4	2	93				1	
	Water				100				
	Bright soil					82	15	3	
	Beige soil					20	64	15	
	Dark soil	2	5	1		1	3	83	5

■ Change of window: third and fourth approximations

For the third approximation, only the window size is changed from 25 pixels (5 by 5) to 49 pixels (7 by 7). The image is little modified (CD 11.8), except that the units are more compact and units of very small surface area have disappeared because absorbed by neighbouring units. Consequently, characteristics of the final units have changed little.

On comparing the classifications as done earlier, it can be seen that the unit 'riverine forest' is still the most heterogeneous, producing maximum changes between the two approximations (CD11.9). This indicates that riverine forest is made up of a mosaic of very diverse units over a small distance, since significant modifications are observed when the window size is changed from 25 to 49 pixels. Other modified pixels pertain to boundaries of the units and hence are boundary mixels (see Chapter 15). The modifications concern 9.3% pixels instead of the 15% earlier.

With respect to quality, the distance histogram varies from 0 to 184 (Fig. 11.7). In fact, most pixels have slightly modified values due to change in window size.

A fourth approximation is made (CD 11.10) by significantly increasing the window size to 225 pixels (15 × 15). Modifications are observed between the nuclei before the approximation and the final reference units (Table 11.5).

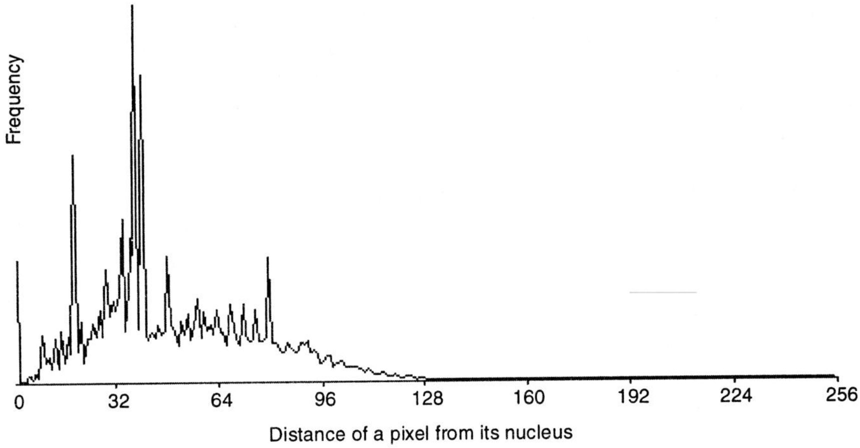

Fig. 11.7: Histogram for third approximation of distances of pixels classified to their nuclei. If a pixel is perfectly classified, the distance is zero.

Table 11.5: Characterisation of composition vectors of reference units obtained at the end of the fourth approximation (row) from land-cover classes of the second approximation (column). Values rounded off to percentages

	Nuclei 2	Riverine forest	Grasslands	Forest	Water	Bright soil	Beige soil	Dark soil
Nuclei 4	Riverine forest	68	12	20				
	Grasslands	22	76				1	1
	Forest	18	15	67				1
	Water				100			
	Bright soil		3			72	21	4
	Beige soil	2	1			11	68	18
	Dark soil	2	6			2	10	80

The quality of results has improved since the distance histogram varies only from 0 to 139 (Fig. 11.8).

Comparison of the results of classification at the end of fourth and third approximations (CD 11.11) shows that the modifications concern 21.4% pixels. This large number indicates that a threshold has been exceeded between a neighbourhood estimated over 49 pixels and a neighbourhood expressed over 225 pixels. Hence the threshold of organisation level has been changed (see Fig. 11.3 and Chapter 15). Evidently, these computations can be made by systematically increasing the window of the neighbourhood and a function of changes defined relative to the size of the neighbourhood. Thus the procedure is similar to that used in VOISIN but applied to the entire image. A similar method is described in Chapter 15.

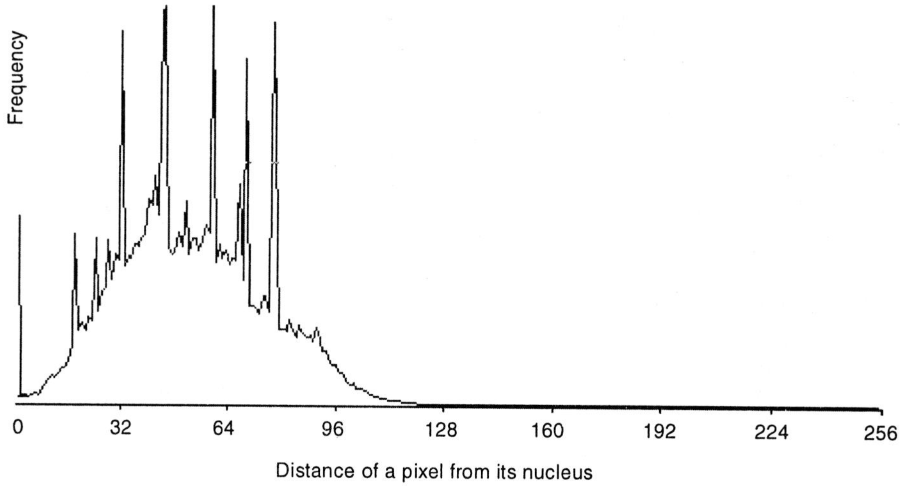

Fig. 11.8: Histogram for fourth approximation of distances of pixels classified to their nuclei. If a pixel is perfectly classified, the distance is zero.

■ Conclusion

In total, modifications have occurred for 32.5% pixels (CD 11.12) between the first approximation with a neighbourhood window of 9 pixels and the fourth approximation with a window of 225 pixels. This indicates that more than two-thirds of the pixels are classified in a stable manner. This shows the advantage of this type of classification when it is desired to delineate relatively compact and composite map units from automatic processing of satellite images. This is a method for making generalisations by synthesis (see Fig. 15.1) and scale changes.

References

Borne F. 1990. Méthodes numériques de reconnaissance de paysages, application à la région du lac Alaotra, Madagascar. Thèse de doctorat, Université de Paris 7, 213 pp.

Didier F. 1992. Analyse structurale des images: étude de l'organisation spatiale et de l'hétérogénéité. Mémoire de Diplôme d'Études Supérieures Spécialisées, École des Hautes Études en Informatique, Université René Descartes, Paris, 29 pp.

Francoual T. 1994. OASIS, notice d'utilisation du logiciel. Laboratoire de Science des sols et Hydrologie de l'INA-PG, 18 pp.

Gilliot J-M. 1989. Analyse thématique d'images par multidensité, DESS EHEI. Université Paris 5, 56 pp.

Gilliot J-M. 1992. OASIS, un système de télédétection sur station IBM RS/6000. Rapport INA-PG, 24 pp.

Gilliot J-M. 1994. Traitement et interprétation d'images satellitaires SPOT, application à l'analyse des voies de communication. Thèse de Doctorat, Université René Descartes, Paris 5, 197 pp.

Girard C-M. 1995. Changements d'échelle et occupation du sol en télédétection. Bull. SFPT, 140: 10–11.

Girard C-M, Girard M-C. 1994. Aide à la cartographie d'unités paysagères par une méthode d'analyse du voisinage des pixels: application en Basse Normandie. Photo-interprétation, 3–4: 145–154.

Girard C-M, Gilliot J-M, Girard M-C, Thorette J. 1997. Comparaison de la cartographie de l'occupation des terres par classification de données de télédétection avec la cartographie CORINE niveau 3: application à une zone au nord-ouest de l'Ile-de-France. Revue internationale de Géomatique, 7 (1): 57–86.

Girard M-C. 1993. Recherche d'une modélisation en vue d'une représentation spatiale de la couverture pédologique. Application à une région des plateaux jurassiques de Bourgogne. Thèse Doc. ès sciences, Université Paris 7. SOLS, 12: 414 pp.

Girard M-C, King D. 1988. Un algorithme interactif pour la classification des horizons de la couverture pédologique: DIMITRI, Science du sol, 26 (4): 101.

Girard M-C, Girard C-M, Rogala J-P. 1980. Automatisation de l'interprétation de l'humidité des sols et interprétation des paysages ruraux. OPIT, 85 pp.

Girard M-C, Mougenot B, Ranaivoson A. 1990. Présentation d'un modèle d'Organisation et Analyse de la Structure des Informations Spatialisées (OASIS), Deuxièmes journées de télédétection: Caracterisation et suivi des milieux terrestres en régions arides et tropicales. ORSTOM, Bondy, pp. 341–350.

Girard M-C, Yongchalermchai C, Girard C-M, 1991. Analyse d'un espace par la prise en compte du voisinage. Gestion de l'espace rural et système d'information géographique. INRA. Florac, 22–24 octobre, pp. 349–359.

Girard M-C, Girard C-M, Bertrand P, Orth D, Gilliot J-M. 1996. Analyse de la structure des paysages ruraux par télédétection. C.R. Acad. Agri. Fr., 82 (4): 11–25.

Ranaivoson A. 1990. Organisation et Analyse de la Structure des Informations Spatiales. DESS, Université Paris 5, INA-PG, Grignon, 36 pp. et annexes.

Robbez-Masson J-M. 1994. Reconnaissance et délimitation de motifs d'organisation spatiale. Application à la cartographie des pédopaysages. Thèse de doctorat de l'École nationale supérieure agronomique de Montpellier, 161 pp.

Yongchalermchai C. 1993. Étude d'objets complexes, sol/plante, à différents niveaux d'organisation: de la parcelle au paysage. Thèse de l'INA-PG, Sols, Grignon, 19: 232 pp.

12

Digital Filtering of Images

Filtering is the action of a filter. From an electronic point of view, a filter is a tool designed to pass or block certain frequency components of an electric signal (Larousse, 1998). Digital image filtering has emerged from the theory of signal processing. It consists of a group of methods that analyse the signal in spatial domain or frequency domain. The field of application of these methods in image analysis is quite vast, involving all stages of processing, viz., preprocessing (noise reduction), detection and extraction of image elements (edge detection), analysis (morphological analysis) and post-processing. Conventionally, two major categories of filters are distinguished depending on whether the operators employed fulfil or not the linearity criteria. Accordingly, they are known as linear and non-linear filters.

12.1 LINEAR FILTERING

12.1.1 Image as a two-dimensional signal

■ Source of digital image: discretisation

A signal (*s*) constitutes the physical basis of information (Kunt, 1981). Mathematically, the signal is represented as a function of one or several variables (*x* and *y* for an image). Depending on the nature of the variables, two types of signals are distinguished: analog signal for continuous variables and discrete or sampled signal for discrete variables. Further, the amplitude of the signal (grey level for an image) can also be continuous or discrete.

An analog signal whose amplitude is discrete is a quantised signal.

A discrete signal whose amplitude is discrete is a digital signal. Let *s* be a digital signal of a variable *k*, designated as $s(k)$. $t(k)$ is an offset or delayed version of $s(k)$ with a lag k_0 such that $t(k) = s(k-k_0)$. A digital signal *s* can be expressed as a balanced sum of offset unit impulses *d* (Fig. 12.1):

$$s(k) = \sum_{l=-\infty}^{l=+\infty} (l)\, d(k-l) \tag{1}$$

A digital image *I* is an example of a two-dimensional digital signal generated from two stages:

—sampling, which is spatial discretisation in *x* and *y* co-ordinates;

—quantisation, which is discretisation of signal amplitude that represents the grey level.

A digital image is hence represented by a discrete and finite zone of discrete amplitude (Fig. 12.3). It constitutes spatial representation of an object of a two-dimensional or three-dimensional scene or of another image (Haralick and Shapiro, 1991).

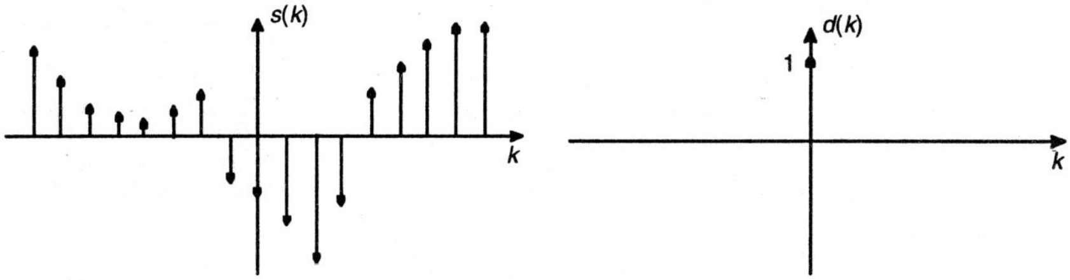

Fig.12.1: A one-dimensional digital signal *s(k)* and a unit impulse *d(k)*.

■ Gridding (tessellation) of a digital image

Division of a continuous space leading to discretisation of an image is done along a grid. A grid is a division of a plane using the same elementary geometric figure known as a cell. Several types of grids or meshes (graphs associated with grids) exist. The most common grids in image processing are square and hexagonal (Fig. 12.2).

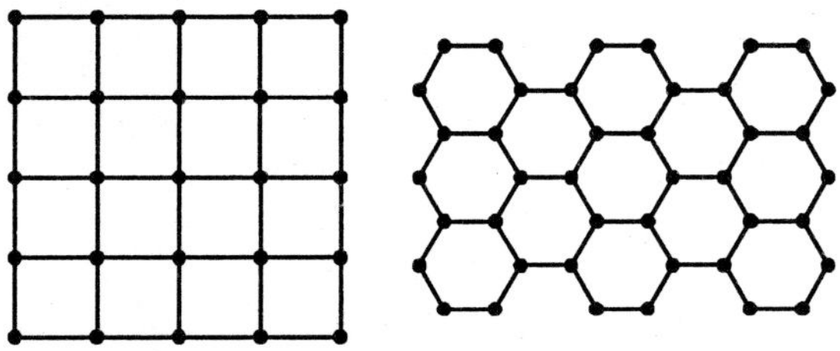

Fig. 12.2: Square and hexagonal grids.

Most systems operate in square meshes. It is possible to simulate a hexagonal grid on a square grid, however, by means of processing. The hexagonal grid offers the advantage of isotropy since, unlike the square grid, the distance between the central pixel and its neighbours is identical in all directions.

■ Digital image format

Let I be a two-dimensional image function:

$$I : N^2 \rightarrow N$$

$$I : (x,y) \rightarrow I(x,y)$$

where $I(x,y)$, an image element with co-ordinates (x,y), is termed a *pixel* (*picture element*). The value of $I(x,y)$ may be a *grey level* or a *colour value* for colour images which generally comprises three components (red, green and blue in the RGB system). The value of a pixel gives information representative of the image in the cell under consideration. Digital representation of an image in a grid format is a matrix in which each cell contains the value of a pixel. This type of representation is known as the raster format (Fig. 12.3). The origin is generally fixed at the top left.

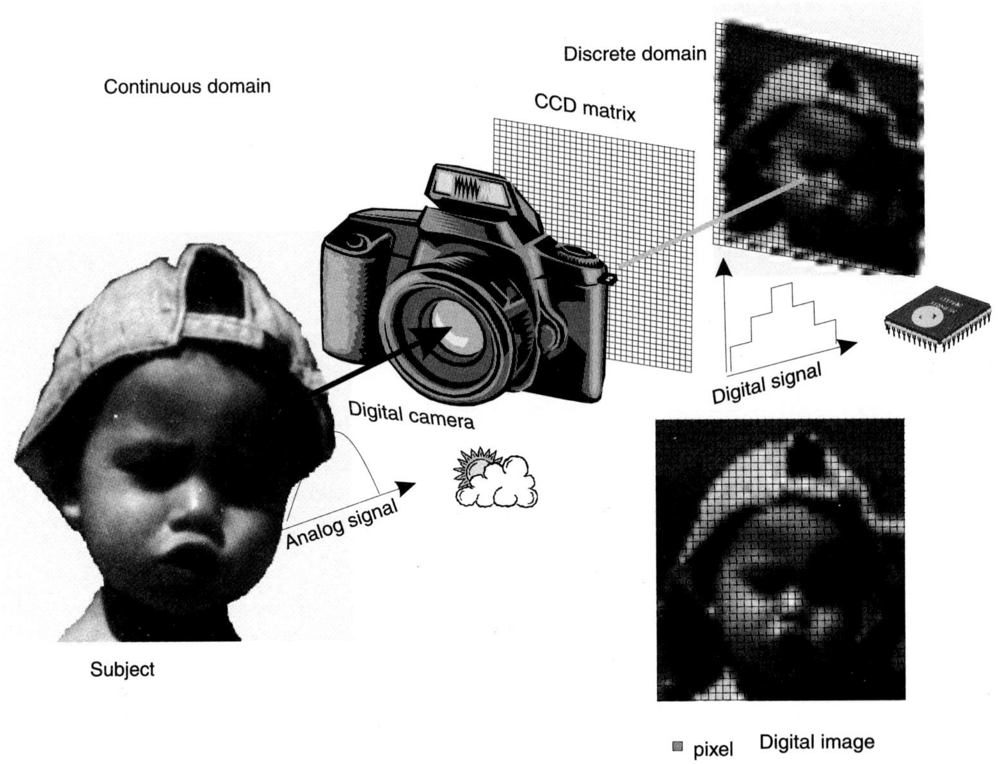

Fig.12.3: Discretisation of information during image digitisation.

■ Coding digital values and quantisation of pixel level

As mentioned earlier, the most common digital representation used for images is the matrix representation. Each image pixel is associated with a cell of the matrix. Like any computer representation of digital values, coding of pixels is based on binary numbers. A bit (binary digit) is the elementary unit of information which can only take the value of 1or 0. For reasons of efficiency, computers process many bits simultaneously. The bits are thus combined into 'machine words' of 8 bits (*octet* or *byte*) of 16, 32 or 64 bits. Coding used in these machine words facilitates representation of various mathematical types of numbers. The discrete mode of representation allows description of only a finite number of values coded with a finite precision (Table 12.1).

An image for which the values have been quantised over various level intervals, coded in an appropriate number of bits, is shown in Fig. 12.4. Coding over an octet is most common for grey level images and provides adequate precision in many cases. Most image-processing systems can manage

Table 12.1: Coded value domains for various bit numbers for different types of numbers

Number type	Number of bits			
	1	8	16	32
Integer with no sign	[0; 1]	[0; 255]	[0; 65,535]	[0; 4×10^9]
Signed integer	X	[– 128; 127]	[– 32,768; 32,767]	[– 2×10^9; 2×10^9]

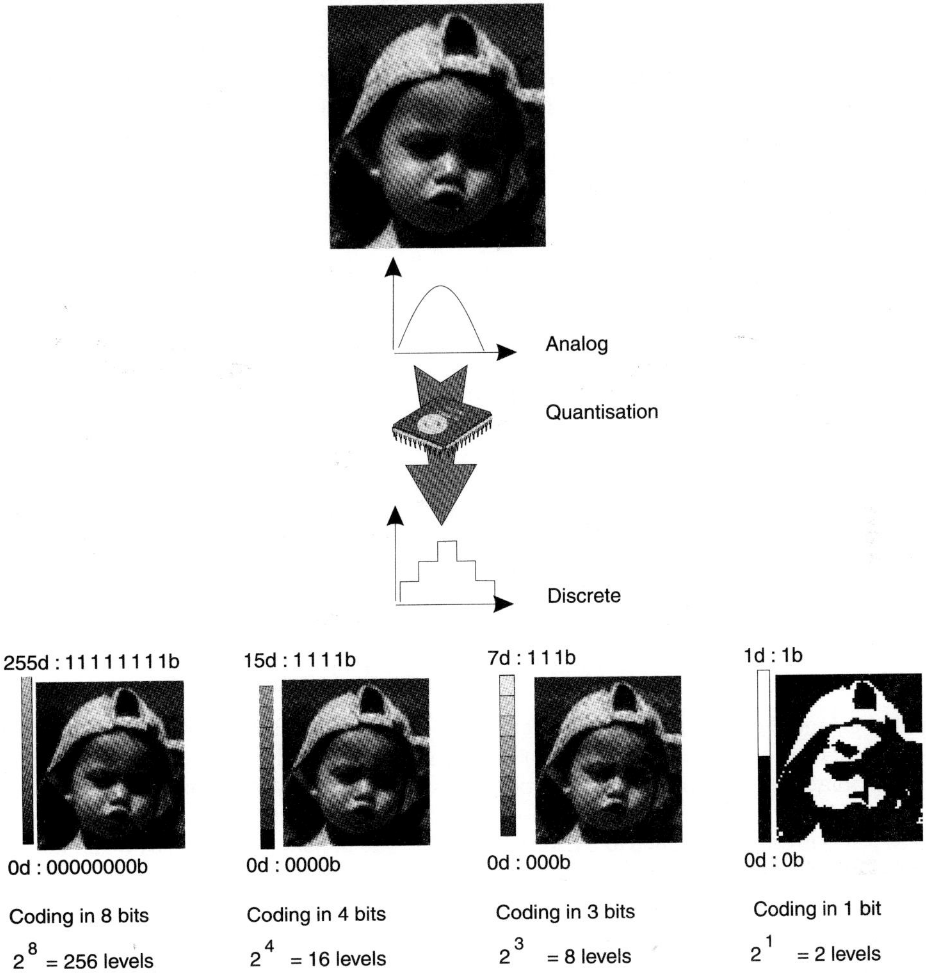

Fig. 12.4: Image quantisation and pixel coding (d: decimal, b: binary).

images in different codes (binary, octet, integer, real) and with different precision levels (1 to 64 bits). It should also be noted that representation of numbers over machine words differs according to the computer architecture (microprocessor), particularly in the order of coding used for the self-same number (high-weight and low-weight octets). Various conventions are employed (Intel/Motorola).

The integer 258, for example, is too large to be coded in a single octet (Table 12.1): 258/255 = 1*255 + 2 (the remainder of integer division is 2). Two octets are hence necessary to code this integer; one is known as high-weight octet (1*255) and the other low-weight octet (2).

2	1	
00000010b	00000001b	in Intel code: high weight octet as primary
1	2	
00000001b	00000010b	in Motorola code: low weight octet as primary

■ Neighbourhood and connectivity of a digital image

As will be seen later, filtration processing is based on computations for each point of the image. These computations use not only the value of the pixel per se, but also the value of adjacent pixels. These are referred to as neighbouring pixels or neighbourhood (Fig. 12.5).

Fig. 12.5: Neighbourhoods V_4 with connectivity 4 and V_8 with connectivity 8 in a neighbourhood of size 3 × 3 (black: the pixel under analysis; grey: neighbourhood pixels).

The definition of neighbourhood depends on the type of grid used (Fig. 12.2) and the metric defined for computations of distance in image space. Two types of metrics are commonly employed in a square grid: distance d_4 and distance d_8, respectively defining the neighbourhoods V_4 with connectivity 4 and V_8 with connectivity 8 (Fig.12.5):

$$d_4\ (A, B) = |x_B - x_A| + |y_B - y_A|$$

$$d_8\ (A, B) = \max\ (|x_B - x_A|, |y_B - y_A|)$$

Spatial extent of the neighbourhood is generally given by the side of the window as a number of pixels. Most often, it consists of an odd number such as 3 × 3, 5 × 5, etc., since centred on the pixel under consideration. The processing time increases significantly with window size. If the window is changed from 3 × 3 to 9 × 9, for example, the side of the window is changed three times, but the total number of pixels is changed from 9 to 81. Increase in the side of neighbourhood by a factor n increases the number of pixels and hence the number of computations by n^2.

12.1.2 Signal-processing systems: convolution product

■ One-dimensional convolution product

Information is extracted from signals by means of signal-processing systems (Fig.12.6). These systems are classified, in the same way as signals, into analog, sampled and digital systems.

Let s be an input signal of system S and t the output signal (Fig. 12.6).

$$t\,(k) = S\,[s\,(k)]$$

Linearity is an important property for signal processing systems. A linear system responds to linearity criteria with two constants, a and b:

$$S\,[a\,s\,(k) + b\,t\,(k)] = a\,S\,[s\,(k)] + b\,S\,[t\,(k)] \qquad (2)$$

Impulse response $g\,(k, l)$ of a linear system is defined as follows.

Let $s\,(k)$ be a digital signal and d a unit impulse (eqn 1).

Fig. 12.6: Signal-processing system.

$$t(k) = S[s(k)] = S[\Sigma\ s(l)\ d(k-l)] = \Sigma\ s(l)\ S[d(k-l)] \tag{3}$$

Substituting $g(k, l) = S[d(k-l)]$,

$$t(k) = \sum_{l=-\infty}^{l=+\infty} s(l)\, g(k,l) \tag{4}$$

This equation shows that response t of a linear system to a signal s can be expressed as a function of impulse response of system $g(k, l)$, where $g(k, l)$ is the response of the system to an elementary impulse d.

Linear systems that are invariant by transfer are especially important for image processing (see sec. 12.1.3).

If $t(k)$ is the response of a linear system S to the signal $s(k)$ and S is invariant by transfer, then $t(k-k_0)$ is the response of S to $s(k-k_0)$, where k_0 is an integer.

For systems that are invariant by transfer, eqn (4) is expressed as a convolution product as follows:

$$t(k) = \sum_{l=-\infty}^{l=+\infty} s(l)\, g(k-l) = s(k) \otimes g(k) = g(k) \otimes s(k) \tag{5}$$

—The impulse response is tilted over the co-ordinate axes

$$g(l) \rightarrow g(-l).$$

—It is shifted by k: $g(-l) \rightarrow g(k-l)$.
—Product $s(l)\ g(k-l)$ is obtained for each sample for all l.
—These products are added to obtain $t(k)$.

■ Two-dimensional convolution product

Image processing is the analysis of two-dimensional digital signals. The definitions given in the preceding section are extendable to two-dimensional signals such as images. The discrete two-dimensional convolution product of finite extent is written as:

$$t(k_1, k_2) = \sum_{l_1=0}^{l_1=k_1-l} \sum_{l_2=0}^{l_2=k_2-l} s(l_1, l_2)\, g(k_1 - l_1, k_2 - l_2)$$

$$t(k_1, k_2) = s(k_1, k_2) \otimes \otimes\, g(k_1, k_2) \tag{6}$$

12.1.3 Spatial filtering

■ Definition of digital filter

Digital filtering applies a linear digital system, invariant by transfer, for modifying the frequency distribution of components of a signal using arithmetic operations of limited precision.

Let I and J be two images and h a convolution mask such that

$$J(x, y) = h(x, y) \otimes I(x, y) \tag{7}$$

Equation (7) represents (with an approximate symmetry relative to the origin) the discrete convolution product introduced by eqn (6).

The operator h is implemented in the form of a matrix. It is referred to as mask, processing window or filter.

■ Normalisation of filter computations

The result of application of the convolution product is not necessarily contained within the interval of values permissible by image coding. Hence it is necessary in some cases to normalise the result. The coefficient of normalisation is generally defined as a function of the weightages of the convolution filter.

■ Neighbourhood analysis by 'moving window'

As mentioned earlier, the discrete convolution product applied to an image during filtering amounts to making for each point a linear combination of grey levels of its neighbourhood. The coefficients of this addition are filter values or weightages. They are stored in the matrix of the processing mask. At each point, a linear combination of the pixel and its neighbours is computed using filter values as coefficients (Fig. 12.7). The mask is 'placed' on the neighbourhood centred on the pixel under analysis and each coefficient is combined to the pixel located below (Fig.12.7).

Each pixel is analysed sequentially, covering the image line by line starting from the origin at the top left up to the last point at the bottom right (Fig. 12.7). The entire operation takes place as though the mask was 'moving' from top to bottom and from left to right passing through each image pixel and hence it is called a moving window for neighbourhood analysis. It should be noted that algorithms dedicated to computers with parallel architecture have been developed for this type of processing in order to reduce computational time. Some linear filters known as separable can be divided into two successive filters, one along rows and the other along columns. This reduces the number of computations to be made.

Unlike point analysis, the result of neighbourhood analysis cannot be progressively stored in the image per se. In fact, the result depends not only on the value of the pixel per se, but also on the values of its neighbours. As the processing of various points in the image is sequential, generally from top to bottom and from left to right, a part of the neighbourhood of a point (position B, Fig. 12.8) will be modified by computations in preceding positions (position A, Fig. 12.8).

To overcome this problem, the result of computation is generally stored in another image. The initial image remains unchanged at any point. These images are often called the source and result images (Fig. 12.7).

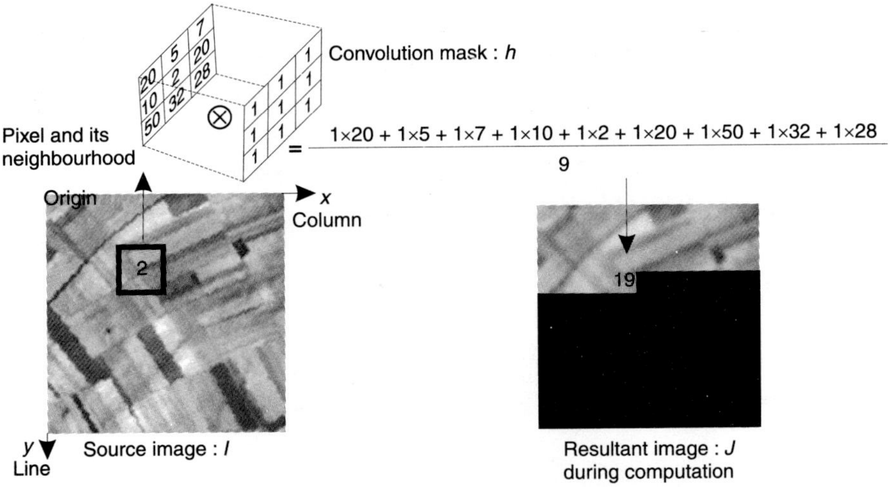

Fig. 12.7: Analysis by 'moving window': example of 'mean' filter.

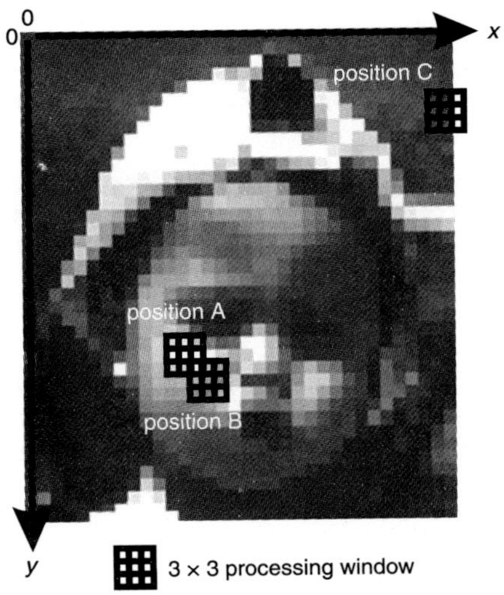

Fig. 12.8: Coverage of neighbourhood zones of pixels.

■ Problem of image borders in neighbourhood analysis

As mentioned earlier, processing by filtering is based on a specific computation of a pixel in its neighbourhood. Digital images are geometrically finite. Definition of neighbourhood for pixels situated at the borders of the image hence poses problems. In fact, no points exist in a part of the neighbourhood (position C, Fig. 12.8). Several classical methods exist to tackle this problem (Fig. 12.9):

—B: analyse only those points for which the neighbourhood is completely defined and replace the unprocessed border by zeros;

—C: consider that the image is surrounded by a boundary *V*/2 (where *V* is the size of the neighbourhood analysed) containing zeros;

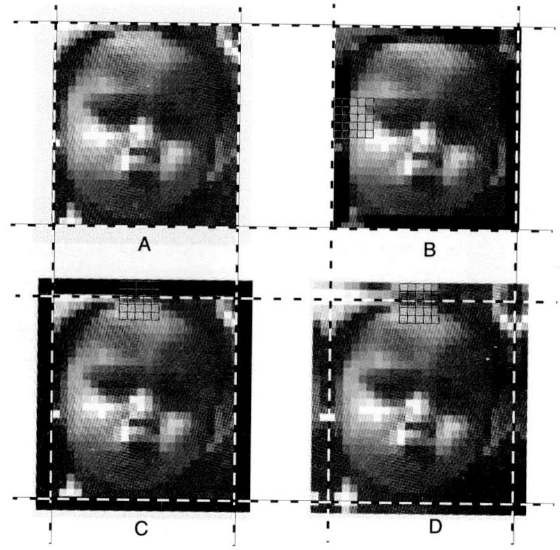

Fig. 12.9: Management of image boundary problem in neighbourhood analysis with a 5 × 5 window. A: Original image; B: analysis frame reduced to 2 pixels (neighbourhood/2); C: Enlargement of image by a 2-pixel frame with 0; D: Enlargement of image by a 2-pixel frame symmetrical to image boundary.

—D: consider that the image is surrounded by a border of size *V*/2, for which the values are the same as in an image symmetrical to its border. A variant consists of taking values of opposite sign to those of the image.

■ Low-pass and high-pass filters

Filters are referred to as low-pass or high-pass according to the range of frequencies they allow to pass through. In the case of filters for image analysis, low-pass filters attenuate level variations in the image and hence smoothen the image, as in an averaging filter used to reduce noise. Contrarily, high-pass filters enhance the variations, as in the case of gradient filters used as edge detectors.

12.2 NON-LINEAR FILTERING

Methods of neighbourhood analysis, which cannot be replaced with a convolution product, are also used in filtering. Such a filtering is called non-linear filtering in contrast to linear filtering. This category includes various kinds of processing.

12.2.1 Order filtering: median

This filtering is based on analysis of pixels of a neighbourhood not with respect to their spatial disposition, but as a function of a statistical measure. This measure is established for a classification in increasing order of grey levels of all pixels of the neighbourhood. The median is an example of an order filter, for which the result is the median value of an ordered distribution of points in the neighbourhood (CD 12.2). For a square window of side N, $M = N \times N$ pixels, the latter are classified in increasing order of grey levels and the position of a point in the distribution forms the order index $i \in [1; M]$. The median is the value with index $(M + 1)/2$.

12.2.2 Morphological filtering

■ Mathematical morphology

The concept of mathematical morphology was developed by G. Matheron and J. Serra in 1967 at the Paris School of Mines (Serra, 1982). The concept of form constitutes the base of analysis in mathematical morphology. It consists of comparing the objects under study with an object of known form, referred to as the *structural element*. The basic operations in mathematical morphology are *group transformations* in all-or-none (Coster and Chermant, 1989). In other words, it is based on union, intersection and inclusion and gives 1 or 0 as the answer.

■ All-or-none group transformations

Mathematical morphology makes use of group relationships (Fig. 12.10), unlike most image-processing techniques based on linear algebraic relationships (Preteux, 1989). These are the conventional group operations that use structural elements.

If A and B are two groups, the conventional group operations would be:

—union: $A \cup B$;

—intersection: $A \cap B$;

—complement $A^C: x \in A^C \Leftrightarrow x \notin A$;

—symmetric difference $A / B = A \cup B - A \cap B$.

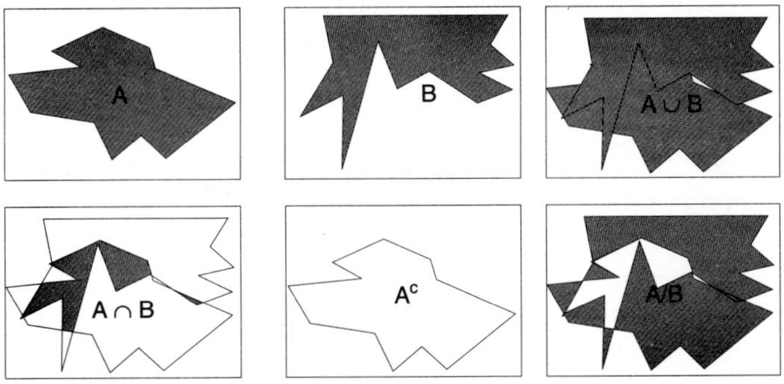

Fig. 12.10: Group transformations.

For an object under study, let us consider A, the group of points in the space containing this object. Let E be the structural element of a known geometry. All-or-none group transformation with a structural element involves 'moving' E onto all the positions in the space under consideration and testing the group relationship between E and A each time. The relationship is either confirmed or not confirmed. It thus constitutes a test of Boolean nature and hence the name 'all-or-none' transformation.

■ Basic operations

Let us assume an image A and a structural element E for these basic operations.

□ Erosion

Erosion is the oldest transformation defined in mathematical morphology. Let the group *A* be defined in space R^2 and *E* a structural element of known geometric figure, say, a circle. In transformation by erosion, the origin of the structural element (here, centre of the circle) is placed on each point in the space and whether *E* is entirely included in *A* or not then checked. The group of points satisfying this criterion forms a new group *B*, called the erosion of *A* by *E* (Fig. 12.11) (CD 12.1).

Erosion of *A* by *E*: $A \odot E$ is group *B* of points $x \in A/[x : E_x \subset A]$.

Fig. 12.11: Basic operations of mathematical morphology.

□ Expansion

By analogy, a transformation by expansion is defined in a dual manner. Let a group *A* be defined in a space R^2 and *E* be a structural element. Transformation by expansion consists of positioning the origin of the structural element at the point under consideration and checking whether *E* has at least one point in common with *A*. The group of points satisfying this criterion forms a new group *B* and is known as expansion of *A* by *E* (Fig. 12.11; CD 12.1).

The expansion of *A* by *E* ($A \oplus E$) is the group *B* of points $x \in A/[x : (A_x \cap E) \neq \varnothing]$.

Transformations by erosion and by expansion are two dual transformations relative to complementation. In fact, if a complementary group of *A* is eroded, A^C by *E*, a result identical to expansion of *A* by *E* is obtained.

$$A^C \odot E = A \oplus E$$

□ Opening

Opening of *A* by *E*: $A \circ E = (A \odot E) \oplus E$ is an erosion followed by an expansion (Fig. 12.11) (CD 12.1).

□ Closure

Closure of *A* by *E*: $A \bullet E = (A \oplus E) \odot E$ is an expansion followed by erosion (Fig. 12.11; CD 12.1).

□ Thinning and thickening

Thinning consists of removing a group of points corresponding to a given configuration of neighbourhood.

If $\{E\}$ is a group of structural elements that gives all configurations of the neighbourhood to be taken into consideration, thinning of the image *A* by $\{E\}$ will be equal to:

$$A \otimes \{E\} = (\ldots(((A \otimes E_1) \otimes E_2) \otimes E_3) \otimes \ldots \otimes E_n)$$

The application of $\{E\}$ is iterated until there is no change in the image. At each iteration, *n* masks are applied for each point if the configuration in the image corresponds to that of a mask the point is removed from the image, while thickening consists of adding points corresponding to a given configuration in the same manner.

□ Skeletisation and trimming

Let A be a group and F its boundary; a point s belongs to skeleton S of A if the Euclidean distance of s to F is reached at least in two distinct points of A. Trimming consists of thinning using masks that aid in removing small irregularities of boundaries, i.e., free short lines at an edge.

■ Mathematical morphology in grey levels

As mentioned earlier, SPOT images are grey level (panchromatic) or colour (multispectral) images which can also be considered as multiple grey level images corresponding to various spectral bands. The methods of morphological analysis described so far are applicable only to binary images and hence only their generalisation to grey level images will be discussed.

The concept of subgraph of a function is introduced to relate to the group definitions of the preceding section.

Let $I(x)$ be a one-dimensional grey level function such that

$$I = x \rightarrow I(x) \in \mathbb{R}^+$$

G_I is the curve of function I, i.e., the group of points $[x, t] \in \mathbb{R}^2$ where $t = I(x)$ (Fig. 12.12). The area under the curve G_I determines a group SG_I called subgraph of function $I(x)$.

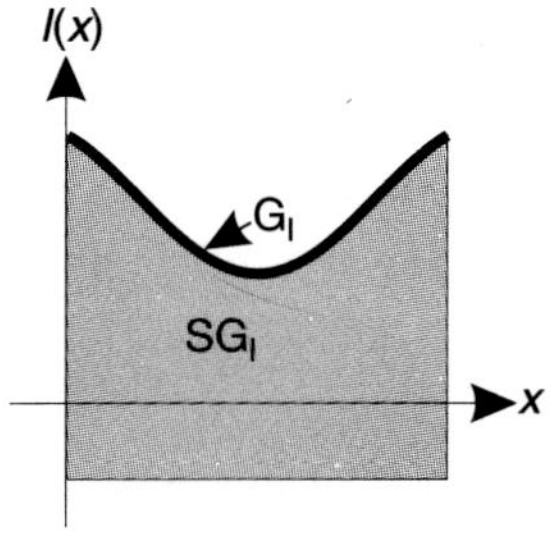

Fig. 12.12: Subgraph SG_I of a function I.

G is defined from SG as: $G = \sup(t: x, t \in SG)$.

As SG_I defines a group, the group definitions described earlier can be used for it. For example, erosion of group SG by a structural element E to give a new group SG_J would be:

$$SG_J = SG_I \odot E$$

A structural element E is associated with each point $P(x, t) \in \mathbb{R}^2$. In a neighbourhood ν defined by E around P, the function $I(x)$ has a minimum and a maximum. For carrying out an erosion by a planar structural element E, each point P is attributed the minimum value of $I(x)$ in E: $I(x) \odot E = \text{minimum}\{I(\nu): \nu \in E\}$. For expansion, the maximum is taken: $I(x) \oplus E = \text{maximum}\{I(\nu): \nu \in E\}$.

Expansion produces a lighter image than the original (CD 12.1) wherein dark details are reduced or disappear. The result of erosion is a darker image in which lighter details disappear (CD 12.1), emphasising the dual nature of erosion and expansion. Opening and closure (CD 12.1) are determined in the same manner as for binary images, combining erosion and expansion (Fig. 12.13).

12.2.3 Homomorphic filtering

Unlike linear systems, non-linear systems are relatively more difficult to analyse and characterise mathematically. It is difficult to analyse by non-linear methods a signal resulting from multiplication or

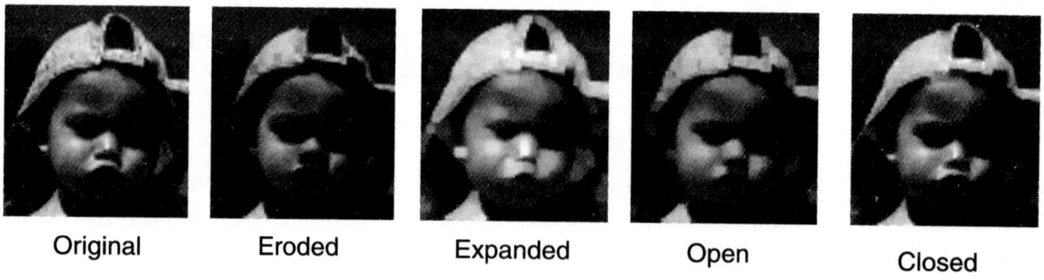

Fig. 12.13: Basic operations in mathematical morphology in grey levels.

convolution of several signals. Moreover, utilisation of linear filters is based on the assumption, crudely sometimes, that the signal is effectively linear. Homomorphic filtering results from generalisation of the theory of linear systems and is based on the principle of generalised superposition (Kunt, 1981). It represents generalisation of linear equations (2), for which homomorphic filtering combines linear and non-linear processing techniques.

12.2.4 Adaptive filtering

Linear filtering has facilitated numerous applications due to simple and powerful mathematical representation. Nonetheless, use of a linear filter is always a compromise, in terms of efficacy, between the action of the filter in the zones to correct the image and its undesirable effect on other zones of the image that need to be preserved. In fact, application of a linear filter is indifferent for all objects of the image. Thus, application of a noise reduction filter, such as an averaging filter, will not only have the desired effect of reducing the impulse noise of the image, but also make the transition zones corresponding to image boundaries hazier (Fig. 12.14).

Adaptive filtering is based on adjusting the structure or weight of the filter according to the situation encountered at each position in the image. The adaptive mean is an example of adaptive filter designed to reduce impulse noise without disturbing the boundaries.

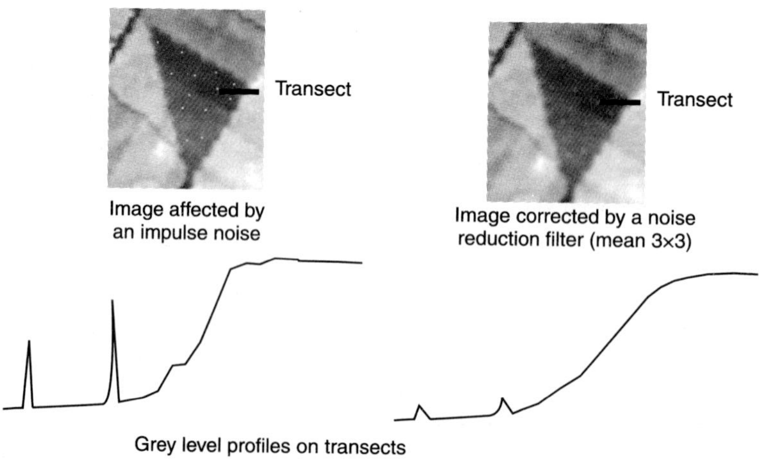

Fig. 12.14: Undesirable effect of noise reduction linear filter at the boundaries.

$$J(x,y) = \frac{\sum_{m=-v/2}^{v/2} \sum_{l=-v/2}^{v/2} k(x+l, y+m) * I(x+l, y+m)}{k(x+l, y+m)} \quad (8)$$

$$k(x+l, y+m) = \begin{cases} 1\_if\_|Ix+l, y+m - Ix, y| \leq preset\_threshold \\ 0 \end{cases}$$

where I is the source image, J the resultant image, v the size of neighbourhood and k a coefficient.

The only difference in eqn (8) compared to a computation of the classical mean is the presence of the coefficient k. The value of coefficient k can be either 1 or 0 and it does not take some pixels of the neighbourhood into consideration for computation of mean. The coefficient is calculated for each position of the neighbourhood and its value directly depends on the gradient between the neighbourhood pixel and the central pixel. If this gradient is greater than a preset threshold, k is fixed at zero to eliminate this neighbouring pixel from computation of the mean. In the boundary zone, only the pixels that are on the same side of the boundary as the central pixel are used for computation of the mean.

12.3 NOISE REDUCTION IN IMAGE PREPROCESSING

12.3.1 Noise in images

Grey level variations in images represent not only the information contained in the scene acquired by the sensor, but may also be due to disturbances in acquisition, for example. These fluctuations, which carry no information, are artefacts associated with acquisition process and constitute noise. Generally, the noise is by nature spatially uncorrelated with the rest of the image and represented by a higher frequency signal than the image (Pratt, 1991). Noise is most often considered a random distribution. A preliminary stage of processing is usually necessary before the analytical phase to remove or at least reduce the noise.

12.3.2 Linear and non-linear filters for noise reduction

Linear and non-linear low-pass filters are used for preprocessing (Fig. 12.15). Various models more or less assume the impulse nature of noises, represented by isolated pixels that differ totally from their neighbours.

Linear and non-linear filtering methods are used for noise reduction in images. As mentioned earlier, use of a linear filter such as the mean filter (CD 12.2) for noise reduction makes boundaries of objects hazier (Fig. 12.14). Non-linear filters enable reduction of this effect. Among non-linear filters, the median filter is least sensitive to impulse noise, while preserving regular transitions between image zones. It has the disadvantage of modifying the geometry of image regions (Fig. 12.15). The median filter (CD 12.2) is very expensive in terms of computation time and the number of computations exponentially increases with the size of the processing window. Fast algorithms such as the pseudo-median filter have been proposed (Pratt et al., 1984). The median filter and its derivatives are very useful for noise reduction. Many other non-linear filters are also employed, among them homomorphic, adaptive and morphological filters (Fig.12.16).

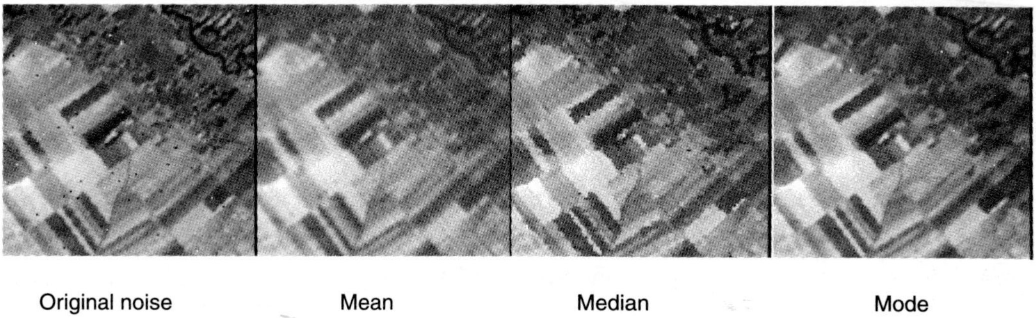

Fig. 12.15: Reduction of salt-and-pepper type impulse noise in a portion of SPOT image by some low-pass filters.

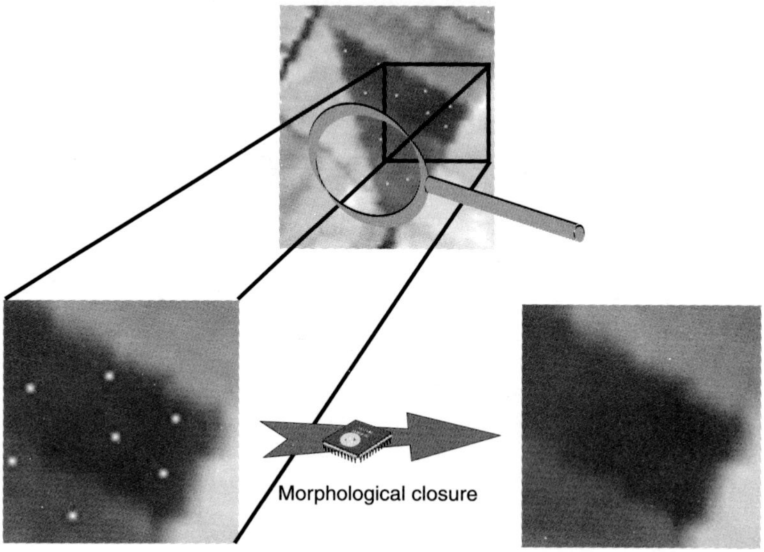

Fig. 12.16: Noise elimination by a morphological filter using a closure operation.

Some linear methods used for noise reduction are based on analysis in the Fourier domain rather than the spatial domain. They are more particularly employed for reduction of periodic noise in images (Pratt, 1991). Such periodic noises may often appear in remote-sensing images acquired with a digital CCD sensor, for example when a defect in the sensor can produce a line effect.

12.4 EDGE DETECTION

An important field of application of filtering in image processing is edge (contour) detection. Image analysis systems proposed by various authors are conventionally structured in three levels: low level for segmentation, intermediate level for constructing approximate geometries and high level for interpretation. In the segmentation stage the image is divided into several parts corresponding to the objects present in the image. Although no universal method of segmentation exists, two approaches can be identified, viz., region and edge methods, depending on whether data inside regions or at the boundaries of regions are used for analysis. An optimal detection of edges is hence necessary for a correct segmentation of the image under study.

12.4.1 Model of an edge

Let *I*(*x*, *y*) be a function describing an image, which gives amplitude variations in space. Discontinuities in amplitude (grey levels or colour components) correspond to edges in an image.

These edges are often important indicators of boundaries of objects in images. However, they may also arise from regions such as shadow zones and noise zones, which contain no objects. An edge can be characterised by its height *H* (Fig. 12.17), its slope α relative to the horizontal, its width *W* and its orientation *O*. If *H* is greater than a certain limit, it can be said that an edge exists. Various types of edges conventionally identified in digital images are shown in Fig. 12.18. If $\alpha = 90^o$, the edge is stepped (Fig. 12.18a).

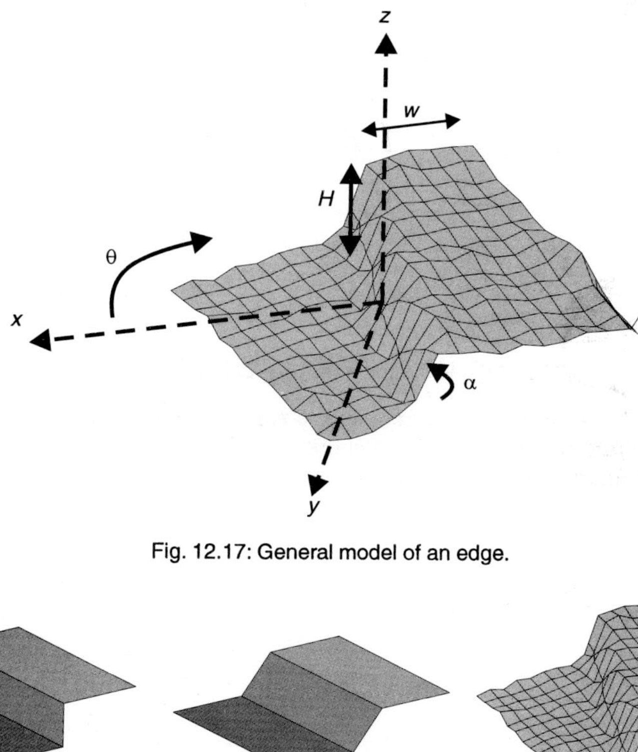

Fig. 12.17: General model of an edge.

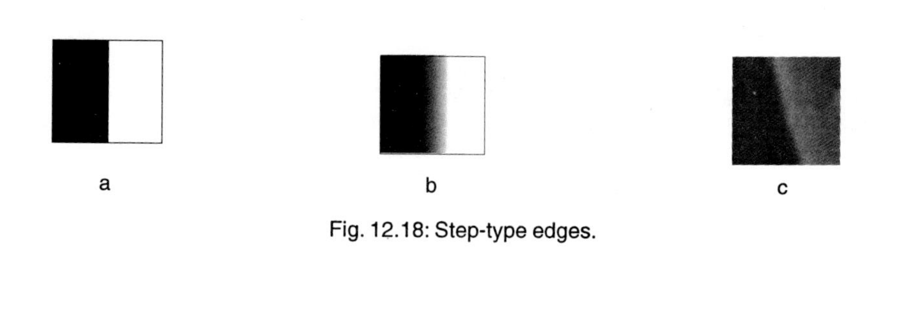

Fig. 12.18: Step-type edges.

Generally, this type of edge is observed only in artificial images. In natural images, discretisation reduces slope and introduces noise; hence stepped edges resemble those in Figs. 12.18 b and c. When $\alpha < 90^o$, the edge is said to be ramp-like (Fig. 12.19). The grey level profile of a line is given in Fig. 12.19 a and b. If the width of a line tends to zero, the edge is roof-like (Fig. 12.19 c).

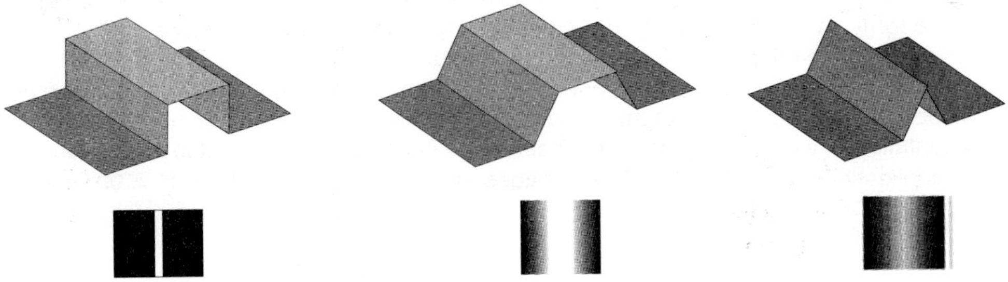

Fig. 12.19: Ramp- and roof-like edges.

Two methods are commonly employed for edge filtering, viz., differentiation operators and comparison with an edge model. In the differential approach, a filter is applied for enhancing the grey level contrast of the image. A second operation, which may be simple thresholding, is used to select the pixels for which the differential is considered sufficient to make part of the edge. Two classes of differentiation operators are identified based on first- and second-order derivatives respectively.

12.4.2 Edge detection by differential filters

As seen earlier, boundaries of objects correspond to zones of steep variations in grey levels of the image. These zones can be detected by studying the derivatives of the function of an image (Fig. 12.20). The gradient technique for first-order differentiation is based on computation of the gradient ∇ along two orthogonal directions and addition of these two gradients. It is given by the equation:

$$\nabla(x,y) = \frac{\partial I(x,y)}{\partial x}\cos\vartheta + \frac{\partial I(x,y)}{\partial y}\sin\vartheta$$

The amplitude of ∇ is given by: $\nabla = \sqrt{(\nabla_x^2 + \nabla_y^2)}$

where ∇_x and ∇_y are the horizontal and vertical gradients respectively. For reasons of efficiency, the amplitude (norm) of the gradient is often obtained from the sum of their absolute values:

$$\nabla = |\nabla_x| + |\nabla_y|$$

The direction of the gradient is given by: $\vartheta = \tan^{-1}(\nabla_x / \nabla_y)$.

Using these definitions in the continuous domain, approximations for discrete data pertaining to digital images are chosen (Rosenfield and Kak, 1982; Pratt, 1991):

$$\nabla_x = I(x, y) - I(x+1, y)$$

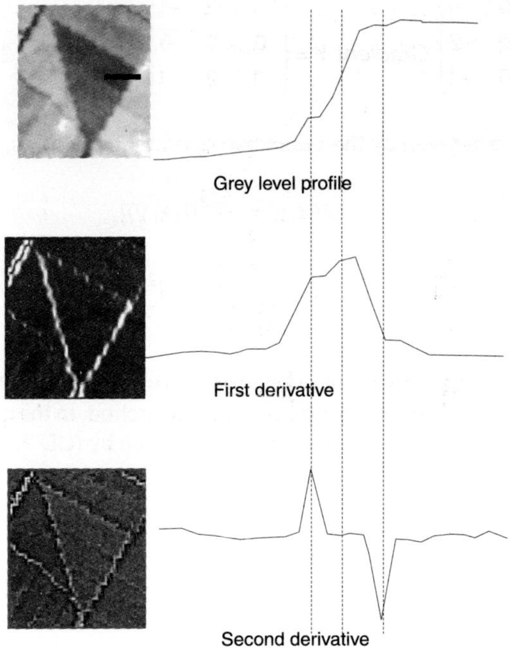

Fig. 12.20: Edges and derivatives.

$$\nabla_y = I(x, y) - I(x, y-1)$$

Several differential operators corresponding to this definition have been proposed:

$$Gradient\ X = \begin{bmatrix} 0 & 0 & 0 \\ 0 & 1 & -1 \\ 0 & 0 & 0 \end{bmatrix} \quad Gradient\ Y = \begin{bmatrix} 0 & -1 & 0 \\ 0 & 1 & 0 \\ 0 & 0 & 0 \end{bmatrix} \quad \text{Difference (CD 12.3)}$$

$$Gradient\ X = \begin{bmatrix} 0 & 0 & 0 \\ 1 & 0 & -1 \\ 0 & 0 & 0 \end{bmatrix} \quad Gradient\ Y = \begin{bmatrix} 0 & -1 & 0 \\ 0 & 0 & 0 \\ 0 & 1 & 0 \end{bmatrix} \quad \text{Difference separated by a pixel (CD 12.3)}$$

$$Gradient\ X = \begin{bmatrix} 0 & 0 & -1 \\ 0 & 1 & 0 \\ 0 & 0 & 0 \end{bmatrix} \quad Gradient\ Y = \begin{bmatrix} -1 & 0 & 0 \\ 0 & 1 & 0 \\ 0 & 0 & 0 \end{bmatrix} \quad \text{Roberts (CD 12.3)}$$

$$Gradient\ X = \begin{bmatrix} 1 & 0 & -1 \\ 1 & 0 & -1 \\ 1 & 0 & -1 \end{bmatrix} \quad Gradient\ Y = \begin{bmatrix} -1 & -1 & -1 \\ 0 & 0 & 0 \\ 1 & 1 & 1 \end{bmatrix} \quad \text{Prewitt (CD 12.3)}$$

$$Gradient\ X = \begin{bmatrix} 1 & 0 & -1 \\ 2 & 0 & -2 \\ 1 & 0 & -1 \end{bmatrix} \quad Gradient\ Y = \begin{bmatrix} -1 & -2 & -1 \\ 0 & 0 & 0 \\ 1 & 2 & 1 \end{bmatrix} \quad \text{Sobel (CD 12.3)}$$

The second derivative is given by the Laplacian operator defined as:

$$J(x,y) = -\nabla^2[I(x,y)]$$

$$\nabla I^2 = \frac{\partial I^2}{\partial x^2} + \frac{\partial I^2}{\partial y^2}$$

The Laplacian J is zero if I is constant or changes linearly with amplitude. J changes sign at the inflection point of I (Fig. 12.20). Hence change of sign is searched. In the discrete domain, depending on the neighbourhood considered, the Laplacian filter is given by (CD 12.3):

$$Laplacian\_V4 = \begin{bmatrix} 0 & -1 & 0 \\ -1 & 4 & -1 \\ 0 & -1 & 0 \end{bmatrix} \qquad Laplacian\_V8 = \begin{bmatrix} -1 & -1 & -1 \\ -1 & 8 & -1 \\ -1 & -1 & -1 \end{bmatrix}$$

12.4.3 Edge detection by other filters

■ Optimal filter

The Canny filter, for example, uses an analytical method to obtain differentiation operators (Canny, 1986; Pratt, 1991). It consists of obtaining an optimal differential filter for a given type of edge. The Canny filter is based on a stepped one-dimensional edge model, of height h_e, with a Gaussian-type additive white noise n, having a standard deviation σ_n: $I(x) = A * h(x) + n(x)$. This technique can be used for two-dimensional images by applying it in two orthogonal directions. If h is the impulse response of the filter and I the image of the edge, the response of the filter on the edge is given by the convolution integral:

$$H = \int_{-L}^{+L} I(x)h(x0 - x)\,dx$$

The image is convoluted with h and the edges are detected at the maximum of $I(x) \otimes h$. h is chosen so as to satisfy certain criteria for correct detection, correct location and unique response.

For a correct detection, the signal/noise ratio is maximised so as to minimise the probability of detection of false edges and the probability of non-detection of true edges.

$$\frac{\text{Signal}}{\text{noise}} = \frac{h_e}{\sigma_n} S(h) \qquad \text{where} \qquad S(h) = \frac{\int_{-L}^{0} h(x)\,dx}{\int_{-L}^{L} [h(x)]^2\,dx}$$

For a correct location, the distance between the points detected and the centre of the true edge ought to be minimal.

$$Location = \frac{h_e}{\sigma_n} L(h), \text{ where } L(h) = \frac{h'(0)}{\int_{-L}^{L} [(h'(x)]^2 dx}, \ h'(x)h \text{ being the derivative of } h(x).$$

To obtain a unique response, there should not be multiple responses for a single edge. If x_m is the distance between the peaks of gradient in the presence of only noise, it is fixed at a certain proportion of the width L, i.e., $x_m = k^*L$. It hence consists of maximising the product $S(h)L(h)$ under the condition $x_m = k^*L$. No analytical solution exists for this. For large values of x_m, the Canny function can be approximated by the derivative of the Gaussian operator.

■ Conditional filtering

The mean conditional filter is applied only if the mean of the potential grey levels of a line differs at least by 5% from the grey level mean of the pixels outside the line. This technique is derived from visual perception systems that differentiate the lines compared relative to the base. The median conditional filter is derived from the preceding filter. The mean values are replaced by the median measures. The results are better but the computation time longer.

■ Morphological filters

□ Morphological gradient

Morphological gradient is defined as the difference between the expanded and eroded versions of an image: $G = [I \oplus E] - [I \odot E]$. Like the gradient defined in linear filtering, the morphological gradient can be used for detection of edges in images.

□ Top-hat form: Detection of highways in SPOT images

The IGN is in the process of developing geographic databases (BDCarto, BDTopo, BDRoutière) for the territories of France. The amount of information required for preparing such digital maps is high, viz., several hundred billion octets. Under these conditions, problems of initial collection of data and updating these databases are of great significance. In this context, remote-sensing images constitute an important source of information. One of the important elements of these geographic databases is the network of communication lines, viz., roads, highways, river channels and railways. The images furnished by SPOT and especially SPOT 5 have a sufficient resolution enabling detection of such networks. Many methods based on edge detection or, more specifically, detection of lines can be used for this purpose (Gilliot, 1994). In particular, a mathematical morphological approach based on the top-hat shaped operator is useful.

The top-hat form is defined as the difference between the initial image and its opening by E: $J = I - [(I \odot E) \oplus E]$, followed by thresholding (Fig. 12.21). During opening by structural element E, grey level peaks whose width is less than that of E are eliminated. Valleys are detected in a similar manner using the difference with closed image: $J = I - [(I \oplus E) \odot E]$. Top-hat transformation enables selection of grey level peaks not only according to their width, but also according to their height, because of thresholding:

$$J(x) = \{x : I(x) - [\{I(x) \odot E) \oplus E] > \text{Threshold}\}$$

The term top-hat shape is used since the transformation preserves only the zones that enter into a cylinder whose diameter corresponds to the size of the structural element and the height corresponds to the threshold.

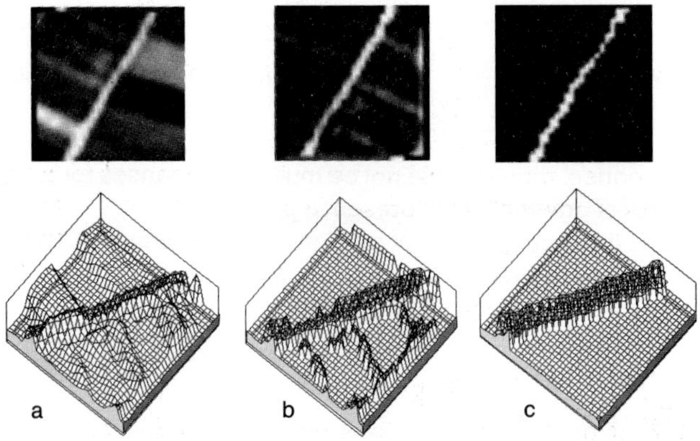

Fig. 12.21: Portion of highway network in SPOT image (a) and 3D view of difference between (a) and its opening in (b), and result after thresholding in (c).

A morphological skeletisation is applied on the binary image resulting from the top-hat form filtering and aids in obtaining edges of unitary size. A trimming operation eliminates small segments with a free edge.

12.4.4 Closure of edges

Edge detection by filtering most often results in discontinuous boundaries due to noise and imperfection of the filtration operators. In fact, application of a differential operator provides in the result the intensity of a potential edge. The choice of a proper threshold value will transform the amplitude of a gradient into a Boolean result, representing the presence or absence of an edge. This stage of thresholding (Fig. 12.22), depending on the threshold chosen, furnishes a more or less continuous or more or less noisy edge.

A stage of so-called edge closure is hence essential for reconstructing the initial boundary (Gonzalez and Woods, 1992). Various methods based on graphic exploration techniques are used in this stage.

12.4.5 Other edge detection methods

■ Parametric methods based on an edge model

These methods make use of a priori knowledge about boundaries of simple geometric form that can be represented by a simple mathematical function. Hough proposed a method for detection of lines, which is known as the Hough transformation (Ballard and Brown, 1982; Maitre, 1985; Wang, 1990). The problem of detecting collinear points in the image (Cartesian [*y*, *x*] space) is replaced by a problem of detecting 'clusters' of points in a parametric [ρ, θ] space (Fig.12.23). The relationship between the two spaces is given by the equation:

$$\rho = x \cos \theta + y \sin \theta$$

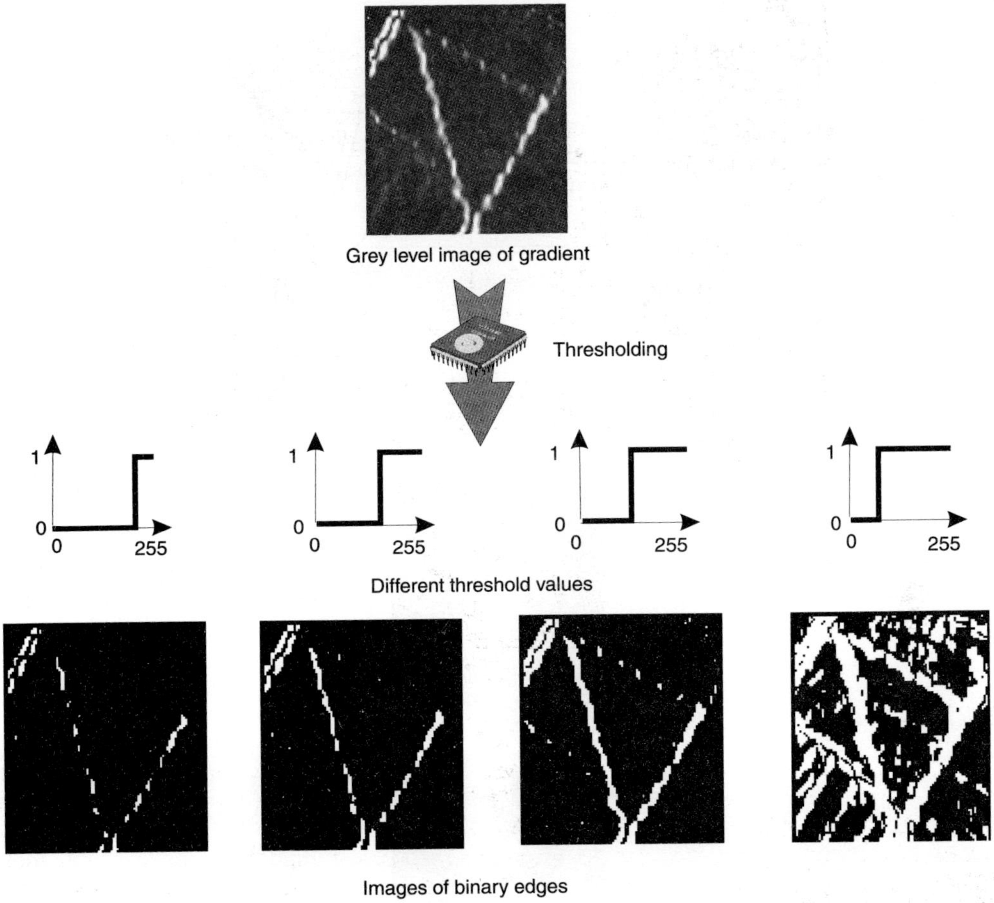

Fig. 12.22: Thresholding the amplitude of a gradient with various values of threshold for obtaining an image of boundaries.

An infinite number of straight lines can be drawn through each point (x_i, y_i), all of which satisfy the equation $\rho = x_i \cos\theta + y_i \sin\theta$. The space defined by ρ and θ is often known as parametric space (Fig. 12.23). A straight line in the Cartesian space corresponds to a point in the parametric space. A point in the Cartesian space corresponds to a curve in the parametric space (Fig. 12.24). This curve represents all possible straight lines that can pass through this point. In this way all possible straight lines or candidates at each point are determined and this stage can be compared to a vote. Search for local maxima in the battery of lines is used to determine those straight lines which have had the highest votes and which hence are present in the image. Thus roads present in the SPOT image are reconstructed. This reconstruction of a road network calls for a knowledge-based system (Gilliot et al., 1993).

■ Markovian methods

Stochastic methods have mainly been used for textural classification of images. The Markovian fields constitute a family of stochastic models whose definition is essentially based on the concept of

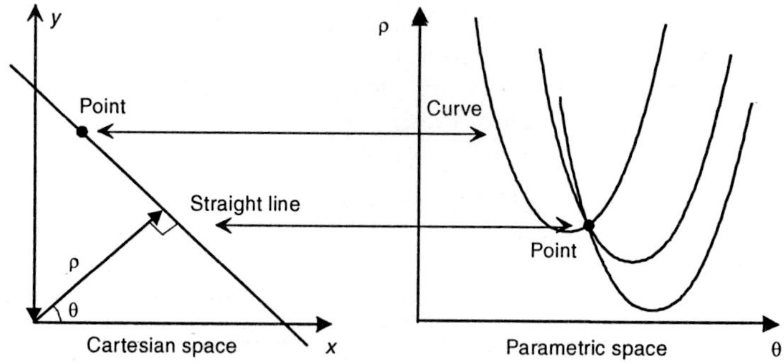

Fig. 12.23: Hough transformation.

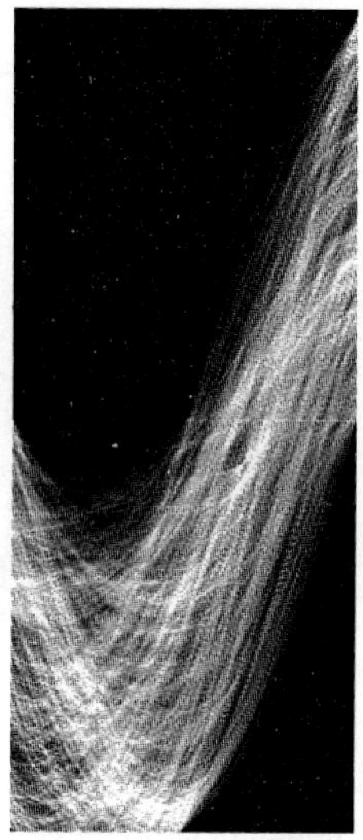

Fig. 12.24: Parametric space or battery of lines.

neighbourhood which, as seen earlier, is fundamental for detection of boundaries and, more generally, in analysis of images. Methods based on Markovian fields have thus been developed for edge detection (Cocquerez and Phillip, 1995).

■ Methods of active edges or snakes

This method of image segmentation was proposed by Kass in 1988 (Kass et al., 1988). The boundary under investigation is created from a curve evolving under constraints for gradually reconstructing its exact form. The initial curve is started in the proximity of the boundary and then deformed according to an iterative process, which is stopped depending on a convergence criterion generally related to the stability of the curve during iterations. The active edge is modelled by a curve C:

$$C = \{p(s,t);\ s \in [a,b], s \in [0,T]\}$$

where p is a point, s its curvilinear abscissa, a and b the extremities of the curve and t the iteration time.

An energy function $E(C)$ is computed at each iteration for the curve C. We search for the curve C which would minimise this energy so as to provide a good approximation to the boundary in the image. For this, E is expressed as a function of the curve, the image and interactions between the two in the vicinity of the curve (Cocquerez and Phillip, 1995):

$$E(C) = E_{\text{internal}}(C) + E_{\text{external}}(C) + E_{\text{image}}(C)$$

Various categories of *snake* are identified depending on the constraints applied on the curve (fixed or free extremities, closed curve). In addition to continuous formulations, discrete formulations of snakes also exist which can be readily implemented. The boundary is described by a suite of points that represent apexes of a polygon approaching it. Conventional methods of edge extraction consist of two phases: detection, which provides often an incomplete edge, followed by closure of edges. The method of active edges, making use of general information, aids in directly obtaining the closed and regular edges with a good robustness vis-à-vis anomalies in the image (Cloppet-Oliva, 1996).

■ Neural network methods

Methods known as neuromimetic, based on mechanisms of biological vision, attempt to implement systems of artificial vision in the form of neural networks capable of edge detection.

Drawing from the well-known visual illusions which make edges that are not drawn appear (Fig. 12.25), Grossberg (Grossberg and Mingolla, 1985) suggested the existence of virtual edges in the mechanism of biological vision. This consists of a low-level, rapid and attentive mechanism that allows strengthening of edges of objects in the vision process. Artificial neural networks operating on similar mechanisms have been proposed. Various possible boundaries depending on various orientations are first detected using directional gradients and constitute the state of entry of neurons. From this initial state, continuation of boundaries is obtained through a mechanism of activation and inhibition between neighbouring neurons. Rules used for continuation of edges are simple geometric laws detected in biological vision systems (in the direction of a boundary, perpendicular to an edge). Several assumptions are made and a comparison mechanism is used to identify the most probable boundary.

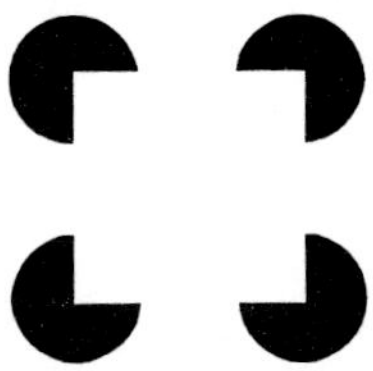

Fig. 12.25: Kamizsa square: a visual illusion showing virtual boundaries of a square that is not drawn.

12.4.6 Evaluation of edge detection operators

As can be seen throughout section 12.4, several edge detection operators exist. It may be necessary to evaluate their relative performance for a given problem. For this purpose, the results of various operators are compared with reference data. The reference may be taken from a map or another processing or may be the result of a manual segmentation of image. An example pertaining to detection of linear elements in SPOT images for automatic identification of communication lines (Gilliot, 1994) is presented below. A manual segmentation of the main communication lines was done such that it can be used as a reference. Let C be the group of points of the edges comprising the communication lines and C its complementary group of points where no edge exists (Fig. 12.26).

Let D be the detection of an edge by an operator.

$p\,(D/C)$ is the conditional probability of detection of an edge, knowing that this edge effectively exists.

$p\,(D/\overline{C})$ is the conditional probability of detection of an edge knowing that in fact no edge exists. It represents the probability of false detection. The various probabilities can hence be defined as:

P_E, the probability that an operator makes an exact detection;

P_F, the probability that an operator makes a false detection;

$$P_E = \int_s^{\infty} p(D/C)\partial s; \qquad P_F = \int_s^{\infty} p(D/\overline{C})\partial s$$

where s is the threshold value.

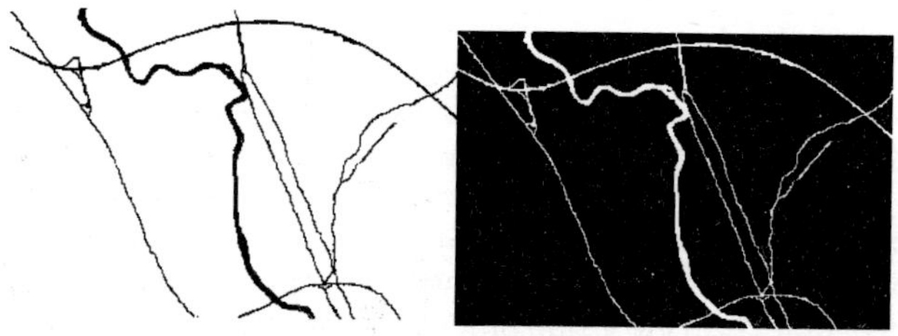

Fig. 12.26: C and $\overline{C}$, groups of points of edges and non-edges respectively.

$p\,(D/C)$ and $p\,(D/\overline{C})$ are determined for each of these images and for different threshold values. Then $p\,(D/C)$ is traced as a function of $p\,(D/\overline{C})$ for various operators (Fig. 12.27).

The topmost curve in the graph, which corresponds to top-hat shape, is the one that possesses the highest probability P_E in the left zone of the graph where P_F is small, and confirms the visual analysis of the significance of top-hat shape as a detector of lines. In fact, the right-hand side of the graph corresponds to high probability of false detection and hence is not interesting.

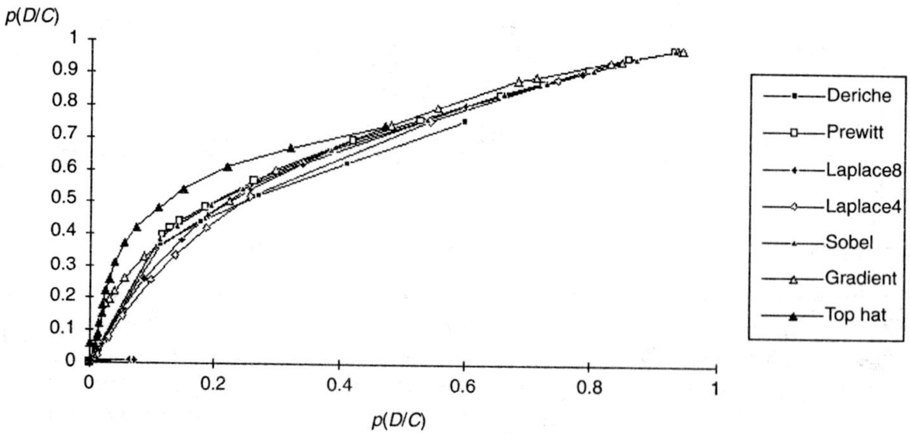

Fig. 12.27: Relationship between $p(D/C)$ and $p(D/\overline{C})$ for various edge detection operators.

12.5 CONCLUSION

Digital filtering, which is very useful for preprocessing of images, also finds important application for segmentation of remote-sensing data using boundaries. The methods based on linear or non-linear mathematical tools call for global or local approaches (adaptive methods). The most efficient methods simultaneously combine general as well as local criteria. As mentioned earlier, in 'natural' images such as remote sensing, the result of filtering rarely corresponds to all boundaries of image objects. The large variability of the signal of these data, the noise inherent in the acquisition process and regulation of operators (thresholds) can explain the imperfect results. Most often, the segmentation method employed uses an operator in combination with others in sequence of processing methods. Less sequential approaches such as dynamic programming, neural networks, *snakes*, etc. have also come into vogue recently.

References

Ballard DH, Brown CM. 1982. Computer Vision. Prentice-Hall, 515 pp.

Canny J. 1986. A computational approach to edge detection, IEEE Pattern Analysis and Machine Intelligence. PAMI, 8 (6): 679–698.

Cloppet-Oliva F. 1996. Analyse d'images de cultures cellulaires obtenues par microscopie optique: Application à des images de neuroblastomes de souris. Thèse de l'Université René Descartes, Paris, 5, 264 pp.

Cocquerez J-P, Philipp S. 1995. Analyse d'images: filtrage et segmentation. Masson, 457 pp.

Coster M, Chermant J-L, 1989. Précis d'analyse d'images. Presses du CNRS, 560 pp.

Gilliot J-M. 1994. Traitement et interprétation d'images satellitaires SPOT: Application à l'extraction des voies de communication. Thèse de l'Université René Descartes, Paris, 5, 207 pp.

Gilliot J-M, Stamon G, Le Men H. 1993. A knowledge-based system in image processing for communication networks study in aerial images, a tool for cartography automation. IEEE Systems: Man and Cybernetic. SMC-93, Le Touquet, pp. 77–82.

Gonzalez R, Woods R. 1992. Digital Image Processing. Addison Wesley Publ., 716 pp.

Grossberg S, Mingolla E. 1985. Neural dynamics of form perception: boundary completion, illusory figures and neon color spreading. Psychological Review, 92 (2): 173–211.

Haralick RM, Shapiro LG. 1991. Glossary of computer vision terms. *Pattern Recognition*, 24 (1): 69–93.

Kass M, Within D, Terzopoulos D. 1988. Snakes: active contour model. Int. J. Computer Vision, 1: 312–331.

Kunt M. 1981. Traitement numérique des signaux. Dunod, 402 pp.

Larousse. 1998. Le petit Larousse illustré.

Maitre H. 1985. Un panaroma de la transformée de Hough. Traitement du signal, 2 (4): 305–317.

Pratt W. 1991. Digital Image Processing. Wiley-Interscience, 2nd ed., 698 pp.

Pratt W, Cooper T, Kabir I. 1984. Pseudomedian filter. Proc. SPIE Conf., Los Angeles, CA January.

Préteux F. 1989. Éléments de morphologie mathématique: Approche fonctionnelle pour images numériques. Cours de DESS.

Rosenfeld, Kak. 1982. Digital Picture Processing. Academic Press, London-NY.

Serra J. 1982. Image Analysis and Mathematical Morphology. Academic Press, London-NY.

Wang F. 1990. Improving remote sensing image analysis through fuzzy information representation. Photogrammetric Engineering and Remote Sensing, 56 (8): 1163–1169.

13

Geometric Transformation of Remote-sensing Images

Geometric deformations develop between the scene observed by the satellite sensors and the raw image received at the receiving stations. These deformations need to be corrected in order to identify the geographic reality on the ground. Remote-sensing data are often used for analysing the natural media. They are hence interpreted and processed with the objective of producing maps that provide spatial information regarding the problem at hand. It is useful to make this interpretation amenable for geometric superposition on other maps in order to compare or utilise them conjunctively. For this purpose, a geometric correction is applied either before interpretation so that data from various bands can be used, such as in the case of diachronic analysis, or after interpretation for making statistical comparisons or to integrate them in a geographic information system.

13.1 METHODS OF GEOMETRIC CORRECTION

13.1.1 Causes of geometric deformation

One type of deformation arises from errors generated during image acquisition. These errors can be due to platform instability (rolling, pitch) leading to a drift relative to the theoretical orbit of the satellite (see CD: SPOT system) or due to aberrations of the sensor resulting in non-linear distortions of the image (somersault effect). Knowledge of sensor operation and use of precise flight path parameters of the satellite facilitate corrections for these errors in part. These corrections are generally carried out by image preprocessing centres.

Another type of geometric deformation is a direct consequence of the curvature and topography of the Earth. The image, which by nature is two-dimensional, is only a planar projection of a three-dimensional view of the zone of observation. In practice, these complex deformations cannot be replaced by simple geometric transformations.

13.1.2 Parametric and interpolation methods

Two main categories of methods used for correction of geometric deformations of remote-sensing images can be distinguished (Teissier and Stamon, 1994). These are: parametric methods and interpolation methods. The former are based on modelling the phenomena that produce deformations. The latter are not concerned with mechanisms of origin of distortions but only assume that these deformations can be determined by means of a polynomial approximation. Parametric methods necessitate knowledge of the operative mechanisms and their parameters. They do not directly use the image and are specific to each sensor-platform pair. These methods are complex to employ and hence much simpler interpolation methods are mostly used. They are more general than parametric methods, requiring no information about the sensor-platform system, and essentially based on the geometry of the image. Only these methods are described below. Intermediate methods, essentially polynomial but which also partially take the scene parameters into consideration, are also used.

13.1.3 Global or local modelling of interpolation methods

Depending on the mathematical model used, interpolation methods are of two types: methods that correct for global deformations and those that correct for local deformations of the image. In the first type of methods, a unique mathematical model is determined and uniformly applied to the entire image. In the second, several sub-models are applied locally to various regions of the image. Polynomial approximation methods are most commonly employed but provide corrections only for global distortions. Transformations that can be replaced by these simple geometric operations or a linear combination of these operations are called linear transformations of the first degree or order 1 (formulated by a first-degree polynomial). More complex transformations, non-linear, are modelled by second-degree or higher-order polynomials.

The method based on spatial modelling using finite elements is the simplest of the local methods. It makes use of a regular division of the image space into cells (tessellations) and generally involves a linear interpolation inside triangles. More recent methods, such as the multiquadric (MQ), have proven more efficient for processing stronger local distortions. They are based, in fact, on a global model that integrates local distortions.

Satellite images are generally characterised by geometric distortions that progressively vary in the image. Such distortions can be analysed by a global model. That is why the polynomial models most often give satisfactory results. Aerial photographs are subjected to additional distortions such as vibrations, speed variations and topographic effects. In such cases local methods have to be employed.

13.1.4 Direct and inverse transformations

Digital images are defined in a discrete space. The geometric transformations generally employed are based on Euclidean geometry and are defined in a Cartesian space using real co-ordinates. This involves an application of $N^2 \Rightarrow R^2$. The co-ordinates of the transformed image can hence be real in some cases. They ought to be converted into integers for assignment to a digital image. This conversion, generally to the nearest integer, may lead through rounding off to a situation wherein some addresses in the final image are not covered. The final image will hence comprise gaps (Pratt, 1991). To overcome this problem, inverse transformation is often employed. For each address in the final image, the nearest address in the initial image is computed by inversion of the transformation equation. Coverage of all pixels in the final image is thus ensured.

13.1.5 Interpolation in geometric transformation

Creation of a geometrically corrected image by an interpolation method implies two types of interpolations, viz., geometric and radiometric. Geometric interpolation is used to model geometric transformations and enables computation of co-ordinates for each pixel in the rectified image. For each position in the rectified image, the value to be assigned to the pixel has to be determined and this is done by the second interpolation, viz., radiometric.

13.2 INTERPOLATION METHODS

13.2.1 Radiometric interpolation

Utilisation of inverse transformation enables analysis of all the pixels in the transformed image, but the address obtained in the source image is no longer necessarily integer. It is hence necessary to determine the method by which the value of the pixel in the transformed image is interpolated from the nearest pixels in the source image.

■ Interpolation to the nearest neighbour

From the address computed by the inverse transform, the value of the pixel can be determined by taking the value of the nearest address in the image source. This is the simplest interpolation method, known as nearest neighbour interpolation (Fig. 13.1). Although this method does not provide a smooth image, it is fast and offers the advantage of preserving the initial value of the pixel (Fig. 13.4). This property is often appreciated by thematic specialists who attempt to model the characteristics of objects from their radiometric properties.

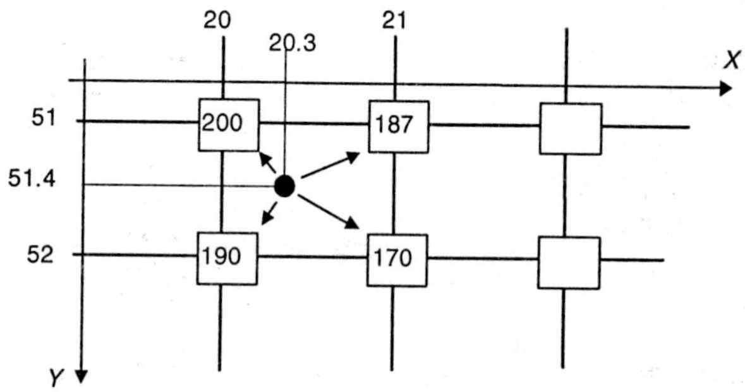

Fig. 13.1: Interpolation by nearest neighbour method. For the computed address (20.3, 51.4), the nearest address in the source image (20, 51) is determined. The value of this pixel (200) is used in the resultant image.

■ Bilinear interpolation

In this method, the value of the pixel is computed as the weighted average of the values of the four nearest pixels. The nearest pixel has the highest weightage for computation of the value. It thus consists of a bilinear interpolation which can be divided into two successive linear interpolations in two orthogonal directions (X, Y) in the image. Let us consider the example of Fig. 13.1 and shift the top left pixel of co-ordinates (20, 51) to (0, 0) (Fig. 13.2). The analysis is carried out in two stages. Four values are interpolated, two at a time, along X-direction in the plane VOX and then the two intermediate values between them along Y-direction in the VOY plane. Thus in the VOX plane, two straight lines represented by the equation $V = a \times X + b$ can be defined as follows:

$200 = a1 \times 0 + b1 \quad \Rightarrow b1 = 200$

$187 = a1 \times 1 + b1 \quad \Rightarrow a1 = 187 - 200 = -13$

$Vi_1 = -13 \times 0.3 + 200 \quad \Rightarrow Vi_1 = 196.1$

$190 = a2 \times 0 + b2 \quad \Rightarrow b2 = 190$

$170 = a2 \times 1 + b2 \quad \Rightarrow a2 = 170 - 190 = -20$

$Vi_2 = -20 \times 0.3 + 190 \quad \Rightarrow Vi_2 = 184$

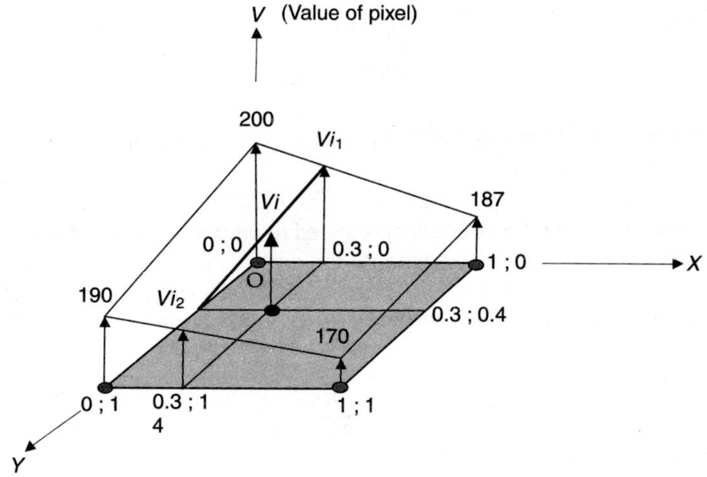

Fig. 13.2: Bilinear interpolation.

The linear equation relating Vi_1 and Vi_2 in the plane *VOY* is then considered using the equation $V = a \times Y + b$.

$Vi_1 = 196.1 = a3 \times 0 + b3 \quad \Rightarrow b3 = 196.1$

$Vi_2 = 184 = a3 \times 1 + b3 \quad \Rightarrow a3 = 184 - 196.1 = -12.1$

$Vi = -12.1 \times 0.4 + 196.1 \quad \Rightarrow Vi = 191.26$

The value 191 is assigned to the resultant pixel. This method is slower than the nearest neighbour method and the radiometric values are not preserved. However, the 'staircase' effect of boundaries in the preceding method is reduced (*anti-aliasing* effect) and a better visual display with more continuous transitions is obtained (Fig. 13.4).

■ Bicubic interpolation and convolution method

The preceding method, defined with a 2 × 2 neighbourhood, can be extended to larger neighbourhood sizes. Nonetheless, due to complexity of algorithms and computational speed, generally the neighbourhood size is limited to less than or equal to 4 × 4. For larger sizes, the assumption of linearity is not valid and in such cases cubic interpolation functions, such as cubic B-spline, are applied. Convolution methods are useful for some geometric analyses (see Chapter 12). Especially in the case of enlargement operations, the convolution method is an efficient means of making interpolations (Fig. 13.4).

13.2.2 Geometric interpolation

■ Global models: polynomial methods

□ First-order geometric transformations

Translation

Translation of an image relative to origin is done by the equations:

$$\begin{cases} x_t = x + \Delta_x \\ y_t = y + \Delta_y \end{cases} \quad (1)$$

where (x_t, y_t) is the point shifted from the source point (x, y) by Δ_x in x and Δ_y in y.

Symmetry
If symmetry with respect to any axis is defined, operations relative to a vertical axis or a horizontal axis are most often employed. In these specific cases, as the axis coincides with a row or column of pixels, the symmetry simply becomes an exchange of values between two addresses of the image (Fig. 13.3). For a symmetry of horizontal axis, defined by $y = S$, the relation is:

$$\begin{cases} x_s = x \\ y_s = 2S - y \end{cases} \quad (2)$$

Fig. 13.3: Symmetry relative to a horizontal or vertical axis.

Homothety
Scale changes of the image can be obtained by means of multiplication factors:

$$\begin{cases} x_h = a_x x \\ y_h = a_y y \end{cases} \quad (3)$$

If the multiplication factor is greater than one, an enlargement of the image is obtained (Fig. 13.4). If it is less than one, the image is reduced. As emphasised earlier, interpolation will be necessary for determining the values of the initial image.

In the case of enlargement by an integral factor, the convolution method can be used.

For enlargement by a factor of two, for example, the image is at first transferred into an image twofold larger in X and Y co-ordinates, wherein each pixel is separated from its neighbours in row and in column by zeros (Fig. 13.5). The image is then filtered with a convolution mask to make interpolation for replacing the zeros. For an interpolation to the nearest neighbour, the following convolution mask can be used:

1	1
1	1

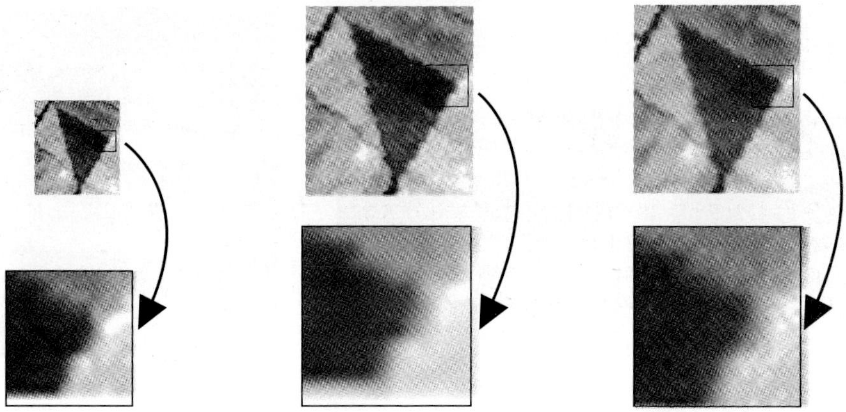

Source image

2 × enlarged image; interpolation to the nearest neighbour

2 × enlarged image; bilinear interpolation

Fig. 13.4: Image enlargement over an agricultural plot extracted from SPOT image of Brienne.

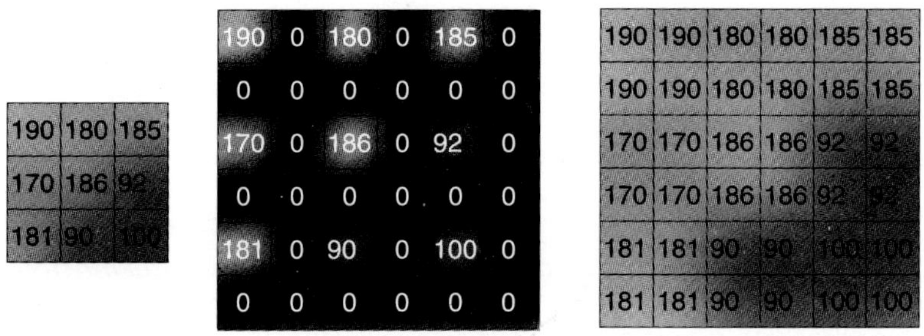

Original image

Enlarged image

Image after convolution

Fig. 13.5: Enlargement by convolution with interpolation to the nearest neighbour.

Based on the same principle, other convolution masks are used for higher-order interpolations.

Rotation

Equations for rotation with respect to origin are defined as:

$$\begin{cases} x_r = x\cos\theta - y\sin\theta \\ y_r = x\sin\theta + y\cos\theta \end{cases} \qquad (4)$$

A rotation relative to any arbitrary point is readily resolved by combining the rotation with a translation. The size of the destination image ought to be larger so as to contain the entire zone after rotation (Fig. 13.6).

For example, for a rotation of 45°, the size of the image should be increased by a factor of the square root of two. For some particular angles (90°, 180°), rotation is simply replaced by a symmetry and corresponds to a simple exchange between two addresses.

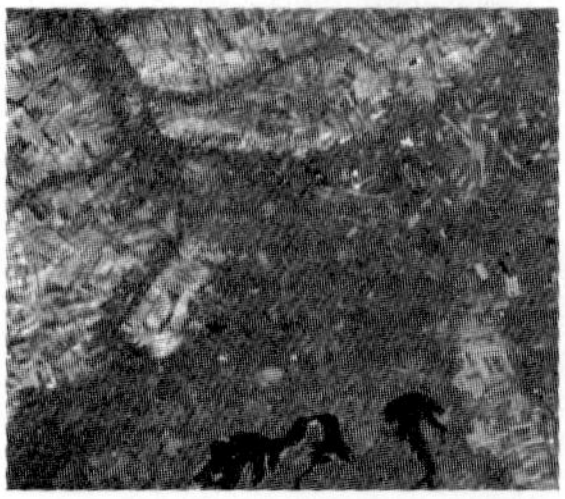

Original image

Rotation by 20°
in same image

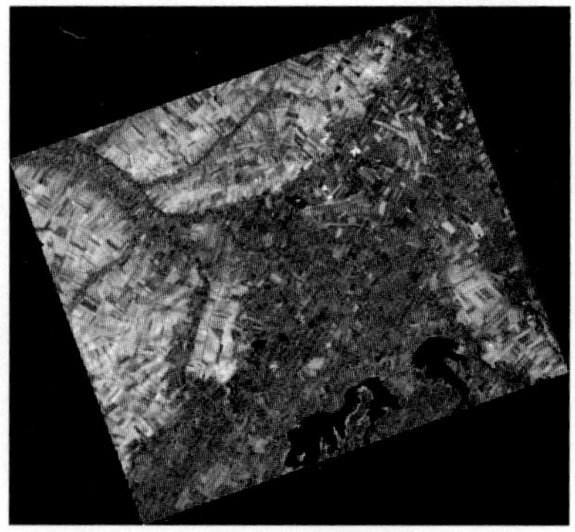

Rotation by 20°
in enlarged image

Fig. 13.6: Rotation of images.

Composition of first-order transformations

The preceding geometric operations, viz., translation, rotation and homothetic conversion can be combined for defining more complex transformations. Complex transformations can be defined in the form of a linear algebraic combination of elementary operators. It should be noted that these operations are not commutative and hence the order in which various operators are applied is important. For example, a transformation T1, comprising a translation followed by a rotation (Fig. 13.7), will not always give the same result (depending on the parameters of operations) as a transformation T2 consisting of the same rotation followed by the same translation.

First-degree polynomial modelling of any linear geometric transformation

While the operations are not commutative, the parameters can however be modified to obtain the same result. In the preceding example (Fig. 13.7), use of another translation (TR2) allows defining a

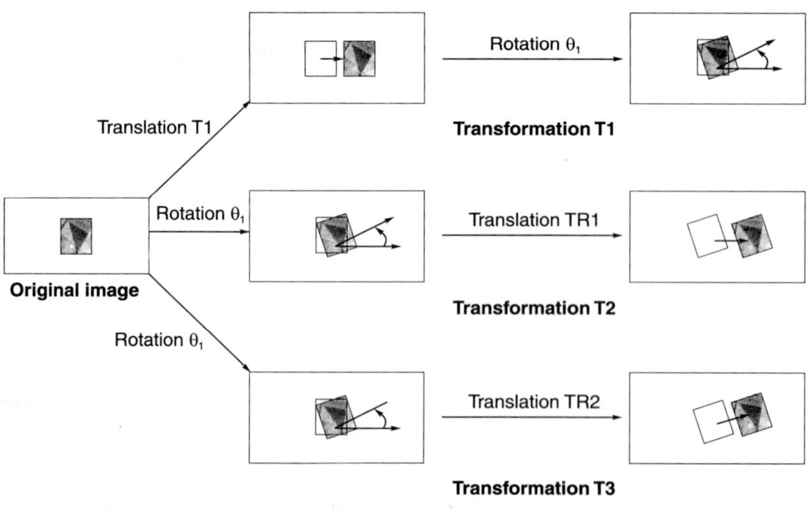

Fig. 13.7: Non-commutative nature of compositions of geometric transformations: example of composition of a translation and 20° rotation with respect to the centre of the image.

transformation T3 equivalent to T1. Any linear transformation can be replaced by a succession of three basic geometric operations such as translation → homothety → rotation. Thus by linear combination, a transformation matrix can be established, using which any linear geometric transformation can be determined. Substituting $x = x_h$ in eqn (4) and by successive substitutions in eqns (3) and (1), we get:

$$\Rightarrow \begin{cases} X = x_h \cos\theta - y_h \sin\theta \\ Y = x_h \sin\theta + y_h \cos\theta \end{cases} \qquad \Rightarrow \begin{cases} X = (a_x x_t)\cos\theta - (a_y y_t)\sin\theta \\ Y = (a_x x_t)\sin\theta + (a_y y_t)\cos\theta \end{cases}$$

$$\Rightarrow \begin{cases} X = a_x(x + \Delta_x)\cos\theta - a_y(y + \Delta_y)\sin\theta \\ Y = a_x(x + \Delta_x)\sin\theta + a_y(y + \Delta_y)\cos\theta \end{cases}$$

$$\Rightarrow \begin{cases} X = (a_x \cos\theta)x - (a_y \sin\theta)y + (a_x \cos\theta\Delta_x - a_y \sin\theta\Delta_y) \\ Y = (a_x \cos\theta)x + (a_y \cos\theta)y + (a_x \sin\theta\Delta_x + a_y \cos\theta\Delta_y) \end{cases}$$

$$\Rightarrow \begin{cases} X = C_1 x + C_2 y + C_3 \\ Y = C_4 x + C_5 y + C_6 \end{cases} \qquad (5)$$

Any linear transformation can thus be established by means of this single equation defined by six constants (3 in X and 3 in Y).

3D-visualisation of remote-sensing images

A remote-sensing image is a view of a three-dimensional scene. If altitude of the scene at every point is known, an image representing perspective view of the ground can be created. This view can be especially useful to replace the image in its topographic context.

Altimeter data is provided in digital format by a digital elevation model (DEM). The image needs to be geographically calibrated over the DEM. Geometric transformation that provides this pseudo-3D view consists of computation of projection on a plane. The DEM defines a surface given by the equation $Z = f(X, Y)$ (Fig. 13.8a), where Z is the altitude at every point (X, Y). A rectangular network is thus defined. By connecting the points in space with one another by straight-line segments, facets are determined (Fig. 13.8b). An approximation of the real surface is obtained by a combination of facets (Fig. 13.8c). For two-dimensional representation of this surface, it is projected onto a plane. Various

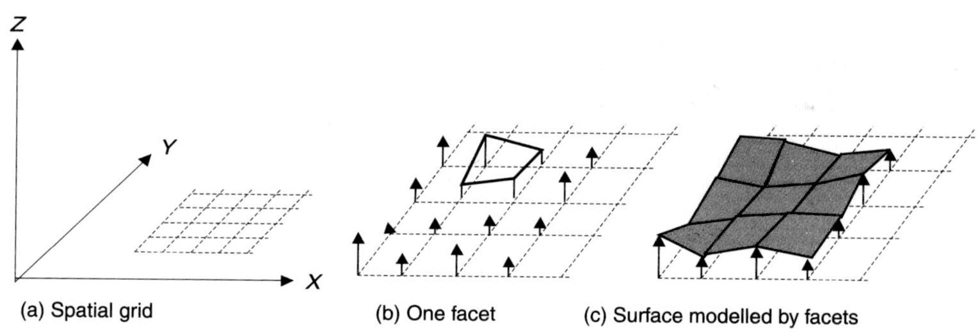

Fig. 13.8: Example of modelling a surface in space by facets.

projections can be used for this representation and cavaliers perspective is one of the simplest (Rousselet, 1985). Projection of a point (x, y, z) onto a point (u, v) on the plane of projection (Fig. 13.9) is given by the equation:

$$\begin{cases} u = x + ky\cos\alpha \\ v = z + ky\sin\alpha \end{cases} \tag{6}$$

A projection plane has to be chosen for this purpose. Any segment contained in this plane is represented as the true magnitude. Any segment perpendicular to the plane of projection is represented by a segment making an angle α with the horizontal. α is the angle of flight. The length of segments is balanced by a multiplication coefficient lying between 0 and 1. This is known as the coefficient of reduction of perspective. An additional multiplication coefficient may be used for varying the scale and this is known as the scale factor. Intensity at each point of projection may represent the altitude or a combination of altitude with an image (Fig. 13.10). The image is then said to be draped or 'mapped' over the DEM.

This intensity can be regulated at will according to the desired output effects, such as shadows as a function of the sun's position. Managing the problem of concealed parts is an important point for obtaining a realistic output. As a matter of fact, some points of the scene are hidden by other points and hence the method of image creation should be able to manage masking of concealed points.

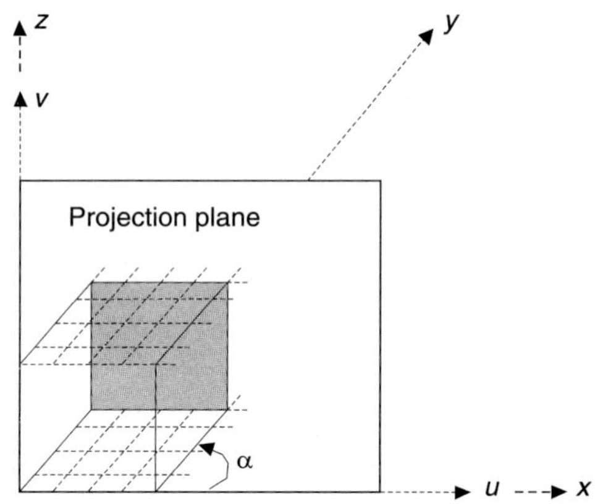

Fig. 13.9: Projection from a cavaliers perspective of a DEM volume.

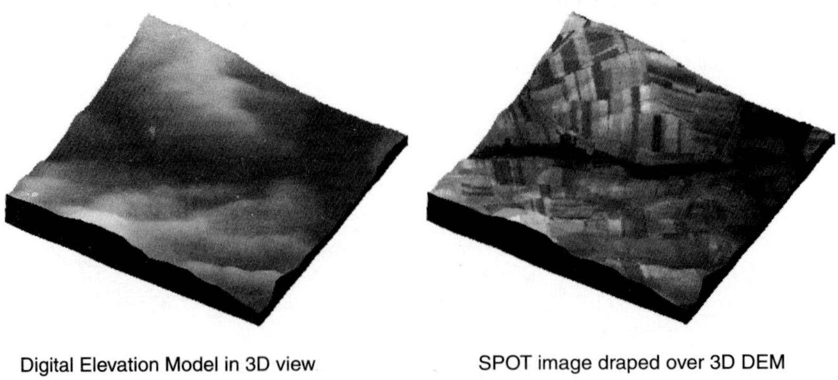

Fig. 13.10: Pseudo-3D visualisation of a SPOT image.

A simple software known as 'painter's algorithm' is sometimes used for this purpose. Masking of concealed points is entirely managed by the order of display of points. It consists of displaying the points according to their distance from the observer, with the farthest points displayed first. When a nearer point is drawn, if it is higher in *z* than the points behind, it masks them by display over them.

Three-dimensional animations can be created using this model. A trajectory in space is defined and an image corresponding to the projection computed at each point. The system thus carries out a sequence of animations, image by image. Computational time for each image may be relatively long depending on the image size and desired output effects. However, such animations can be achieved almost in real time on the latest computers equipped with 3D graphics and large live memory. This enables interactive exploration of the zone under study in 3D space.

□ Second-order geometric transformations

Complex non-linear geometric deformations

Most often, distortions in remote-sensing images cannot be restored by simple, i.e., linear, geometric transformations. Complex variations in the platform's trajectory and sensor response lead to non-linear distortions of the image (Fig. 13.11).

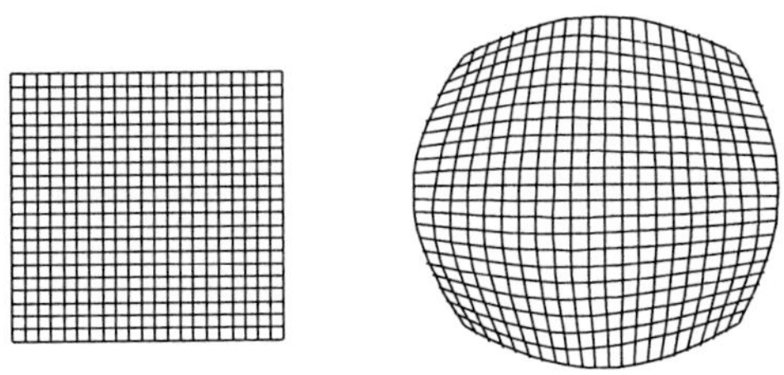

Fig. 13.11: Distortion (somersault effect) due to a sensor aberration.

Conditions of image acquisition and nature of terrain give rise to complex image distortions. Depending on the platform altitude and angle of view (see Chapter 2), deformations can be large or small. Images obtained by satellites, orbiting at high altitude with a small field of view, provide a better geometric quality than aerial photographs (Jensen, 1986). When one image is compared with another image or a map, it must necessarily be geometrically deformed for superposing on the latter. An acquisition of a scene is in effect a projection on a plane. The large number of parameters involved in data acquisition and the geometry of the terrain observed (Earth's curvature and topography) make the projection complex (non-linear deformation) and highly variable depending on the image. It is difficult to mathematically reconstruct it from acquisition parameters. This geometric transformation can be approximately estimated by a polynomial modelling by means of a second- or higher-order polynomial.

Second-degree polynomial modelling

A second-order polynomial is used for modelling such complex transformations:

$$\begin{cases} X = C_1x^2 + C_2y^2 + C_3xy + C_4x + C_5y + C_6 \\ Y = C_7x^2 + C_8y^2 + C_9xy + C_{10}x + C_{11}y + C_{12} \end{cases} \qquad (7)$$

As mentioned earlier, in modelling first-order deformations, the coefficients of the polynomial can be directly related to the elementary geometric distortions involved (rotation, translation, homothety). This is not so in the case of second-order modelling and no physical meaning can be given to each coefficient. Transformation is not continuous and the polynomial equation provides an approximate estimate. The coefficients are so computed as to minimise the mean quadratic error between a set of image points and a set of geographically similar points used as references. These reference points are generally known as Ground Control Points or GCP. They are obtained from a spatial reference dataset, such as a map or another image, which represents the undistorted image. An equation of an order higher than two can also be used but inversion of the relation becomes more complex.

□ Conclusion on polynomial methods

Depending on the number of points used for solving the polynomial equation, the method is referred to as either interpolation or approximation. If the number of control points is strictly equal to the number of unknowns in the polynomial expression, the system of equations is exactly determined and termed interpolation. In this case, the position of the control points computed by the model is the same as that in the reference image. There is no residual error since the model is exactly determined on the control points. If the number of control points is greater than the number of unknowns, the polynomial coefficients are approximated using various points. The model no longer corresponds exactly to the control points and a residual error appears at every point relative to the reference position. This deviation is an indicator of the quality of the model.

A polynomial approximation of an order higher than one may give rise to undesirable image distortions (such as stretching, compression). To overcome these problems associated with extrapolation, the control points should be chosen all around the zone to be corrected. For a n-degree polynomial, the number of control points required to solve the system of equations is $(n+1)(n+2)/2$. The minimum number of control points hence is 3, 6 and 10 for first, second and third degree polynomials. For example, in the case of a first-degree polynomial for which 6 constants are to be determined, three control points allow formulation of 6 equations (3 in X and 3 in Y), sufficient for solving the system.

■ Local models

Polynomial approximation provides a global model and hence corrects deformations applicable only to the entire image. In the case of images with large local distortions, a rectification technique based on use of an image grid ensures better results.

□ Finite element modelling

In image processing the finite element method is mostly used as linear interpolation in a triangular grid. The grid divides the image into elementary regions (finite elements) which correlate the source image with the reference image. A local distortion model is applied region by region. Division of image (gridding) can be done automatically, using Delaunay triangulation (Fig. 13.12) for example (Worboys, 1995). A grid covers the surface under analysis by an arrangement of non-overlapping polygons. Grids can be distinguished as regular (repetition of the same elementary polygon) and irregular.

Delaunay triangulation and the Voronoi diagram (Thiessen polygons) are the most commonly used tessellations. A triangular grid can be created using control points. Triangles are so constructed

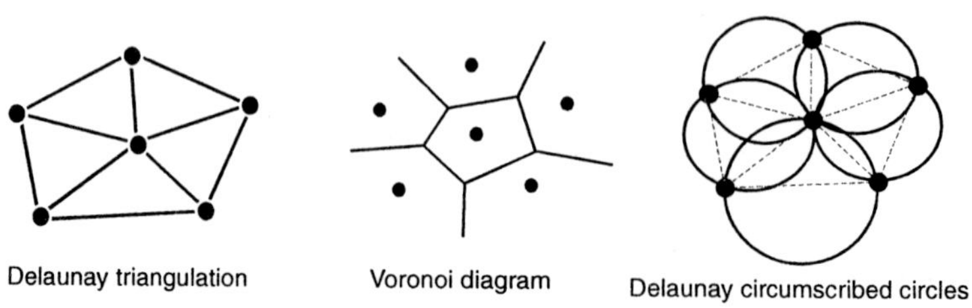

Fig. 13.12: Delaunay triangulation.

that the circle circumscribes a triangle and contains nodes of only this triangle. The triangles should be as regular as possible. Polynomial coefficients are determined for each triangle and rectification is applied to the image points, triangle by triangle. Triangulation can also be fixed to correspond to the known zones of deformation. Higher-order polynomial approximations are also used. This method is well suited for locally linear distortions.

□ Multiquadric functions

Multiquadric functions were introduced by Hardy (1971) and are used for image rectification. The hypothesis assumes that a mathematical surface and more generally any surface can be approximated to a given precision by a sum of mathematically defined elementary surfaces, in particular those of quadratic form (Hardy, 1977). These methods are intermediate between global and local methods. As a matter of fact, they use a global model that takes local variations into consideration, by integrating a term which is computed between the point to be relocated and the control points (Fogel, 1996):

$$\begin{cases} X = \sum_{i=1}^{N} a_i \sqrt{(x-x_i)^2 + (y-y_i)^2 + R^2} + P_a(x,y) \\ Y = \sum_{i=1}^{N} b_i \sqrt{(x-x_i)^2 + (y-y_i)^2 + R^2} + P_b(x,y) \end{cases} \quad (8)$$

where (x, y) is the point under consideration, N the number of control points, (x_i, y_i) control points, a_i and b_i constants specific for each control point and P_a and P_b polynomial terms which are not always used.

It may be noted that $(x-x_i)^2 + (y-y_i)^2$ in the main term of eqn (8) represents the distance between the point under study and the control points. The optimal value of the term R^2 depends on the type of distortion. It may also depend on the distance between the control points or their scatter, for example.

$R^2 = 0.6 \min \sqrt{(x_i-x_j)^2 + (y_i-y_j)^2} \; \forall i, j \in [1 \ldots N]$ for example, depends on the minimum distance between the control points. R^2 can also be replaced by a set of R_i^2, one for each control point.

These methods are found to be very efficient and Hardy later (1990) cited nearly a hundred references regarding their use for rectification of remote-sensing images.

13.2.3 Conclusion

Polynomial methods based on global modelling of deformations are well suited for satellite images exhibiting mainly global distortions. Contrarily, in the case of aerial photographs, local distortions are often more severe and the orthophotographic approach becomes necessary. Methods using division of the image by gridding (tessellation) are good for local linear distortions. This is the type of deformation generally observed in scanned images, comprising distortions due to folding.

13.3 IMAGE RECTIFICATION

13.3.1 General problem

Application of the methods described in the preceding section for remote-sensing images corresponds to the same general problem, viz., placing an image in spatial conformity with other spatial data.

These methods include relocating, rectification, placing in conformity, georeferencing (relative to a map), warping and stretching.

13.3.2 Systems and data for rectification

■ Systems of geometric correction

Various processing systems are used for applying geometric corrections.

□ Image enhancement systems
Most image enhancement systems (such as Photoshop©, Image'in©) include simple geometric transformation functions. Two modes of making these transformations are currently in vogue: one assigns digital values to the transformation parameters (ex: angle of rotation) and the other interactively carries them out using tools that enable moving, turning and enlarging the image on the computer monitor. A visualisation mode called the 'transparency' mode is also quite common. It allows viewing another image 'below' the image to be deformed. The image can also be visually adjusted over a reference image (by linear deformation).

□ Image processing systems
Image processing systems such as Visilog© (Noesis, 1997), Khoros© (Khoros, 1997) contain an additional module of deformation (warping) by control points which can be used for rectification of remote-sensing data.

□ Remote-sensing systems and geographic information systems
Special processing programs for remote-sensing data, such as Erdas© (Erdas, 1997), ER/Mapper© (ER/Mapper, 1997), TeraVue© (Teravue, 1997), generally have a versatile module that allows manipulation of raw images, interactive choice of control points, working with cartographic co-ordinate systems, displaying cartographic data, assembling images (mosaic), viewing the image in three dimensions using DEM, etc. Some geographic information systems such as ArcInfo©, ArcView© (Esri, 1997), Mapinfo© (Mapinfo, 1997), Geoconcept© (Alsoft, 1997), Idrisi© (Idrisi, 1997) also employ more or less versatile sophisticated modules. While they all enable use of polynomial models, some also propose models based on local triangulation. However, programs comprising rectification software based on multiquadric functions are rare, except for some specific algorithms (MQReg, 1997). Most versatile systems are equipped with an orthophotographic module.

■ Data used for rectification

□ Level of correction
Images supplied by organisations that distribute remote-sensing data are already often subjected to preprocessing. SPOT images, for example, may have been preprocessed to correct for acquisition errors or to directly provide a georeferenced image, conforming to a conventional cartographic projection system; such an image is known as a spatial map. We can also obtain geographically rectified and grouped images that form a mosaic covering a large area, an entire district for example. In such cases, radiometric corrections are often necessary to reduce response variations between images comprising the mosaic. According to the level of processing, various products can be identified (see Chapter 2).

Aerial photographs, because of their wide field of view, are much more deformed than satellite images. Moreover, given the lower altitude of their acquisition deformations related to topographic variations cannot be ignored. Combined use of cartographic reference and altimetric reference to produce a geographically conformable image is known as orthophotography.

Data available for processing may be of different qualities depending on the remote-sensing product. The cost of these various products, of course, is not equivalent. According to the number of images and frequency of processing, one can acquire rectified images that can be used directly or set up one's own sequence of rectification. In the latter case, reference data for image rectification are mandatory.

□ Reference data used for corrections

The same principles of rectification are used irrespective of the nature of reference data, its projection, co-ordinate system or physical support. Most often, the reference system is another remote-sensing image or a map.

• Another image

In order to compare two images or to combine them in a single processing, the two images can be rectified relative to a reference map or simply one image rectified with reference to the other. In the latter case, no reference map is needed. The images in such cases cannot be superposed on a cartographic base and are not corrected for geometric distortions. The rectified image will inherit the distortions of the reference image. These comparisons or combinations of images can be used, for example, in diachronic analyses or for combining data from various sources.

• Reference map

Whenever possible, it is more useful to rectify images with respect to a reference map. This effectively enables superimposition of the result over other maps, more accurate spatial control on the ground and its direct integration in a geographic database in the form of GIS, where it can be compared with other data. Co-ordinates of the control points are determined in a cartographic projection system, usually the Lambert system in France. Depending on the method of determination of these co-ordinates from a map, two cases are identified, viz., using a paper map or a digital one.

Cartographic co-ordinates from a paper map

When digital maps are not available, the co-ordinates can be determined directly from a paper map. For this, the map should have reference points in its projection system. Maps often contain a square grid representing the co-ordinate system. The grid may be drawn completely or more often, to avoid overloading the map, only crosses at grid intersections or marks for boundaries are indicated. While co-ordinates defined in a projection system are almost invariably represented in large-scale maps, such is generally not the case in small-scale maps in which geographic co-ordinates (longitude and latitude) are marked instead. The distance of a control point from the nearest grid points is measured with a scale as accurately as possible. It should be noted that the grid lines are not parallel to the borders of the map and hence the measurements should be imperatively made parallel to the grid lines. Knowing the scale of the map, a simple three-point rule is used to determine the co-ordinates in the projection system of the control point. The precision of the co-ordinates obtained is limited in this type of measurement. At best, a precision of 0.5 mm in reading the scale can be expected, which is equivalent to 25 m on a 1:50,000 IGN map for example. Several errors accumulate in this measurement: errors in reading the scale, in determining the position of a point (such as centre of a road) and in referring to the grid.

These measurement errors can be reduced using a digitising table. At first the table is calibrated, in general, on the grid and then the co-ordinates of the control points are directly determined using the mouse. Advantages of the digitising table, compared to the preceding method, are: minimisation of reading errors given the very precise pointing of the mouse, elimination of errors in referring to the grid and availability of an adjustment model for correcting deformations due to the paper.

Map co-ordinates derived from digital data

Most publishers nowadays supply digital versions of their maps in raster (image) or vector form. Raster data are scanned maps. Because of their nature, they can be used directly in all image correction systems. The National Geographic Institute (*Institut géographique national, IGN*) supplies a range of scanned maps. Scan 50 (IGN, 1997) or 1:50,000 maps, for example, are scanned at 254 dpi (dots per inch) giving a resolution of 5 m reportedly for a geographic precision of 10 m. When only a paper map is available, it can be scanned but several problems need to be taken into consideration. For large format maps, an A0 scanner may be necessary. Deformations of the paper map, in particular its folds, may lead to a deviation of several mm in width of a map with a theoretical grid. These deformations can be corrected with a rectification model using the theoretical grid as reference. A finite element model is well suited in such cases. It should also be noted that digitisation of a map, IGN for example, is not freely permitted and is subject to copyright. Vector data are most often used in cartographic systems and in GIS. Some rectification programs are also capable of processing these data. The 'BDcarto' and 'BDtopo' of IGN (IGN, 1997) are the most widely used vector products in France for these corrections. Digital Chart of the World (DCW) is the world vectorial base map and can be used for small scales.

Utilisation of digital reference data offers some advantages. A high precision of map reading is obtained thanks to software zoom. Images can be effectively visualised nearly to a pixel level and pointing error almost corresponds to data resolution. As in the digitising table, reference errors are eliminated since the software 'measures' the position of the pixels. The possible distortions of reference data can be corrected when a paper map is digitised. The advantage of a cartographic base lies in verification of rectification by a posteriori superposing of the base on the rectified image.

• Altimetric reference (DEM) for correction of parallax errors due to topography

Image distortions due to topography are proportional to the relief and angle of view (Fig. 13.13).

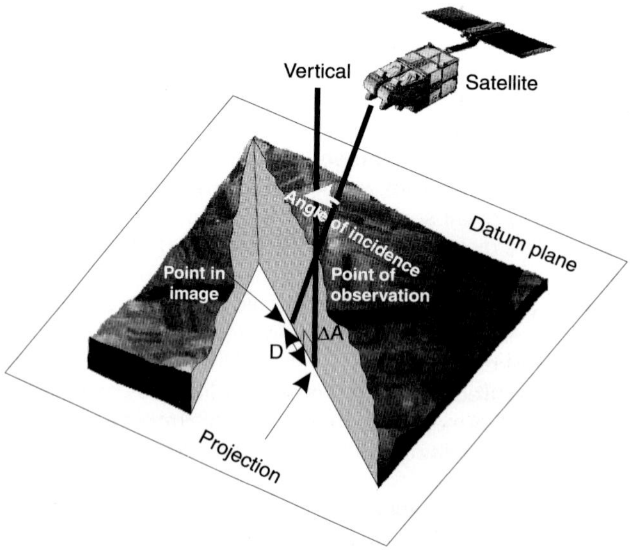

Fig. 13.13: Parallax error due to relief and angle of vision.

For a point viewed vertically, the position in the image coincides perfectly with the vertical projection and no deformation due to relief occurs. For a point viewed at an inclined field of view, the parallax effect (see Chapter 14) gives rise to an offset between the position in the image and its vertical projection. Let us consider a datum plane (at minimal altitude, for example) on which the relief of the zone is projected. The size of the pixel and its position on the projection vary with altitude relative to the datum plane and with slope. For an altitude difference from the datum plane ΔA and an angle of incidence of the satellite α, the offset between the observed position of the point and its vertical projection is given by $D = \Delta A \tan \alpha$. For an angle of 20° and altitude difference of 150 m, D is equal to 55 m or nearly 3 pixels in the SPOT image. For zones of high relief variations, these parallax errors should be corrected if a georeferenced image is required. For this, altitude, angle of incidence and local slope should be known. A digital elevation model is used to make these corrections.

13.3.3 Formulation of rectification model

■ Choice of control points

The control points should belong to objects visible both in the image to be rectified and the reference map. Some parts of a map represent elements that are not present in the physical medium (administrative boundaries for example) and not observed in remote-sensing data. A large difference in resolution may also give rise to the fact that features present in the reference do not appear in the image (or vice versa) or appear with a variable form, making a precise location difficult. A difference in date of acquisition of the reference and the image may also have the same effects, since image features may appear, disappear or be modified by human activity (roads, deforestation) or natural phenomena (such as floods). Cloud cover may mask some features. Generally, objects that can be identified with no ambiguity and located precisely are chosen as tie points. Objects whose form and spatial organisation aid in locating a reference point at the level of an image pixel are used. Temporal stability of morphological characteristics of the objects is essential for comparing data not necessarily acquired on the same date. Constructions and management elements are often preferred over natural objects, since they are more regular in geometry and stable. Buildings and especially road networks satisfy these criteria. Borders of agricultural parcels or forests are avoided since they are not stable. Objects used for calibration are at the limit of image resolution and may exhibit contrast variations, making them difficult to precisely locate pixel by pixel. Hence, alignments and intersections are used as reference for locating a precise point in the image. A set of aligned points, rather than a single point, is relied upon. Intersections of road networks are therefore most commonly used reference points. If the reference document is another image, a map is useful to mark the reference points in the images. In some zones, only a few or no control points may be precisely located in the image. This is particularly the case in areas of least human activity and in methods using reference forms (group of pixels) rather than existing reference points (Hubault, 1994).

■ Number of control points

In the case of first-order rectification, the polynomial equation has three constants to be determined (3 in X and 3 in Y). Three control points are therefore needed to formulate 3 equations to determine these 3 unknowns. For a second-order rectification, six points are also necessary. This is the minimum number of points mathematically required for solving the polynomial approximation. More number of points than the minimum are recommended so as to reduce possible errors related to quality of reference data and location of points. A dozen points are generally used for a second-order correction of a slightly deformed satellite image (Fig. 13.14). On the other hand, taking too many control points for a polynomial model is not useful. In fact, since it is a global model, while the number of points is sufficient

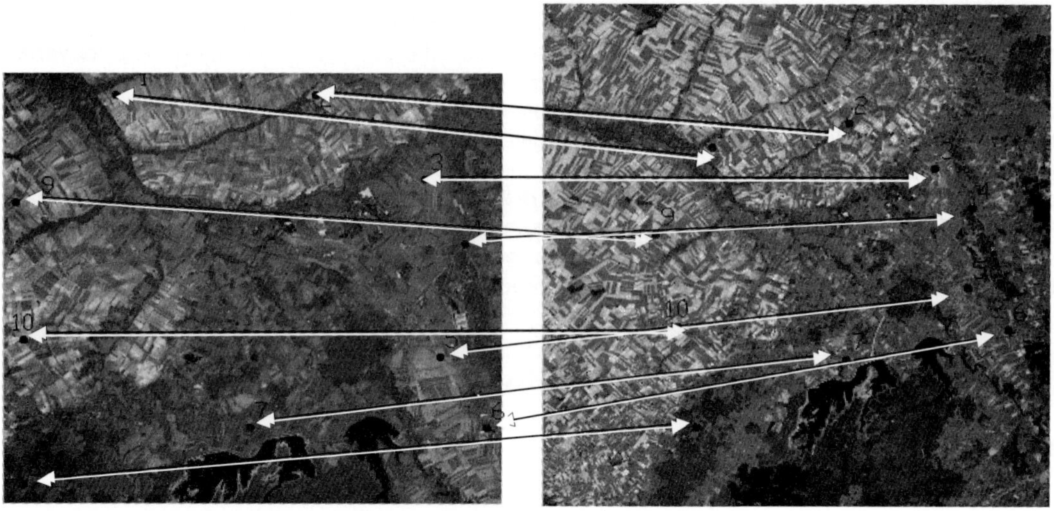

Reference : georeferenced SPOT image (CD-ROM) LANDSAT TM image to be rectified (CD-ROM)

Fig. 13.14: Second-degree rectification of a TM image over a georeferenced SPOT image: location of ten control points.

for regular distribution, rectification cannot be improved locally by adding points in the poorly rectified zones. In such a case, the entire model will be altered and other parts of the image not calibrated correctly. In the case of a locally too irregular image, a local deformation model can be used with more number of points, covering various distortion zones. If topographic variations of a zone are not negligible in the scale of the image, use of an orthophotographic method with DEM provides high-quality rectification with no need for too many points.

■ Distribution of control points in the image

In higher-order polynomial approximations the extrapolation problem may give rise to undesirable image distortions depending on the spatial distribution of control points. The image may then be globally stretched or compressed in several directions. A regular distribution of points on the boundaries of the zone to be corrected will minimise such distortions (Fig. 13.15a). Concentration of points in one or several zones of the image should be avoided (Fig. 13.15b).

High-accuracy restitution cannot be obtained for the entire region when the zone under analysis significantly differs in altitude and orthophotographic and DEM softwares are not available. Mean altitude of zones of importance is determined and control points in the image are distributed according to this altitude. Accuracy of restitution is optimal at this altitude and gradually decreases with distance from the mean altitude.

■ Identification of control points

After defining the type, number and general position of control points in the zone under study, their co-ordinates in the image and the reference have to be determined. Rectification programs most often comprise an interface that aids in interactively identifying the points. If such facility is not available, it is necessary to determine these co-ordinates by means of a simple program of visualisation of images and transfer them to the rectification system, generally through an ASCII file. In integral systems, the

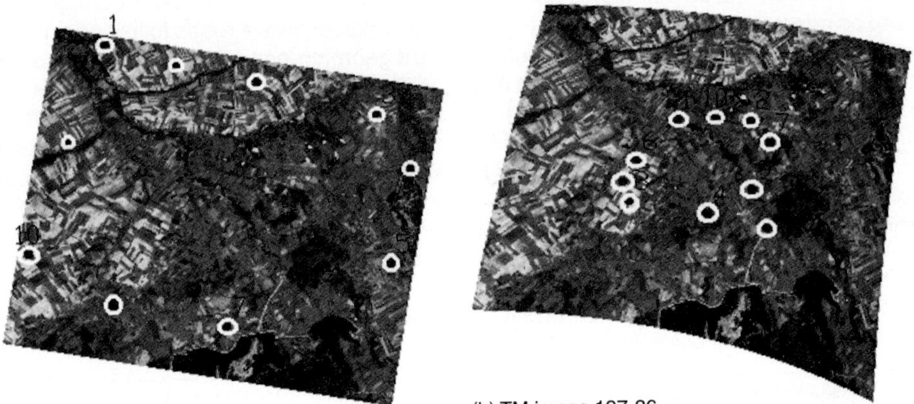

(a) TM image 197-26 (b) TM image 197-26

Fig. 13.15: Distribution of control points in the zone to be corrected: The same TM 197-26 image after rectification over the same reference image (georeferenced SPOT) with the same number of control points, distributed around the image in case (a) and at the centre of the image in case (b).

image and the reference are displayed simultaneously. Any part of the zone can be zoomed or moved. Some software programs enable simultaneous view of several reference images or several views with different level of zoom. All the control points are displayed in these views as and when they are identified. In some cases, residual error in X and Y at each tie point is displayed as a small vector. These data facilitate correction of the model during identification. It is necessary to determine the correct level of zoom for identification of points to an adequate precision but not necessarily to the pixel level, because of local contrast variations (see the section on 'Choice of control points'). Several types of error may arise during identification of control points: errors of pointing, location and interpretation.

Pointing errors are less common if a sufficient zoom level is used. Errors of pointing with the mouse depend on the level of zoom.

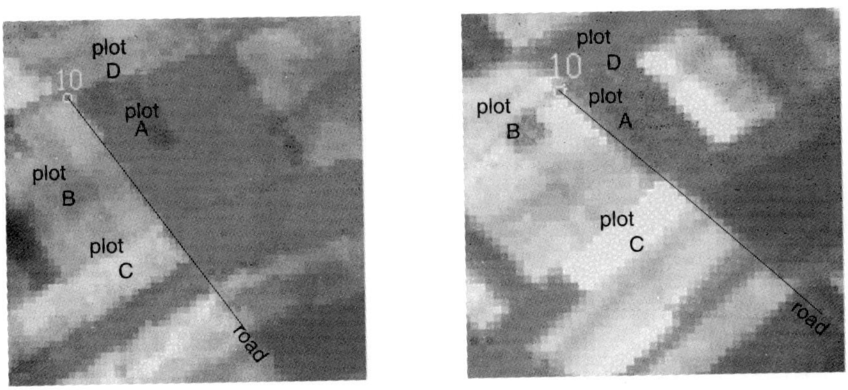

(a) SPOT reference image (b) TM image to be corrected

Road used for locating point Number 10
shown in the two images

Fig. 13.16: Error of location of tie point No. 10.

Errors of location of a tie point arise due to incorrect interpretation of the pixel configuration of the object used as reference. In the example of Fig. 13.16, an intersection of roads following an agricultural parcel was chosen as the control point. The reference is a georeferenced SPOT image with a resolution of 20 m and the image to be calibrated is a LANDSAT TM image with a resolution of 30 m.

The first rectification model (Table 13.1a) has shown a larger error in *X* for point No. 10. It can be noted, using a map, that point No. 10 has been incorrectly located in the TM image. In fact, the resolution of the TM image does not allow a clear view of this small road. It is located based on the boundaries of the parcel. However, a dark band can be noticed in this image, to the left of the road in the area of plot A. This band has caused confusion with plot boundary where the road really exists. A correction of point No. 10 has minimised the residual error of this point (Table 13.1b).

Table 13.1: Variation of rectification model after correction of an erroneous point

(a) Model with erroneous point 10

Points	Residual in *X*	Residual in *Y*
1	0.25	–0.046
2	–0.65	–0.391
3	–0.16	0.323
4	0.66	0.132
5	0.67	0.083
6	–0.82	0.207
7	0.20	0.289
8	0.58	0.230
9	0.75	0.593
10	–1.50	–0.840

(b) Model with corrected point 10

Points	Residual in *X*	Residual in *Y*
1	0.39	–0.059
2	–0.29	–0.424
3	–0.29	0.336
4	0.39	0.157
5	0.48	–0.065
6	–0.39	–0.248
7	–0.09	0.320
8	0.20	0.263
9	–0.14	0.677
10	–0.24	–0.957

It may be noted that the correction of error in *X* of point 10 reduced the same error for most other points of the model. In fact, in the case of polynomial methods, only one global model exists, computed for all the points, and hence an error for one point leads to an error in the entire model. Errors of location may also be generated by an incorrect interpretation associated with the scale of objects in the data. Difference in resolution of the images and effects of cartographic generalisation of maps (thickness of representation of some objects not to scale) make it necessary to use, in the case of linear objects, the axis of the objects, but not their boundaries for fixing control points. Only the image points present on the axis of the object can be considered geographically similar to the reference points.

Identification errors arise when the object identified in the image does not correspond to that chosen in the reference data. For example, in a zone of high urbanisation road intersections are 'mistaken'.

■ Final resolution of rectification

Resolution of the rectified image produced generally corresponds to that of the reference image. Some systems enable determination of this resolution. This facility is particularly useful when multisource data, which most often are of various resolutions, are integrated. While most processing programs require that images to be combined should be of the same resolution, some are capable of dynamically combining images of different resolutions.

■ Verification and improvement of rectification accuracy

One way of verifying the accuracy of rectification by an interpolation method is an a priori verification of the accuracy of the approximation model. This is estimated from the residual errors at each point. A posteriori verification can be made by superimposing the rectified image over its reference image. If residual errors are too large, the model should be corrected. Various approaches can be followed for this: verifying the points commencing with those having the highest residual errors; adding points if they are too small in number or if their distribution is not regular in the zone (see section on distribution of control points); removing or replacing the most ambiguous points; checking the distribution of points in the zone according to altitude and verifying that altitude differences are acceptable for a polynomial model. If this is not so, DEM should be used along with an orthophotographic program.

13.3.4 Conclusion

Image rectification is an important stage for using remote-sensing data. Choice of reference data and methods used depend on the scale of study. The nature of remote-sensing data directly determines the mode of rectification to be employed. Resolution of the image, acquisition parameters (altitude, angle) and relief of the zone of study are also equally important in choosing a model. Numerous studies have been carried out on development of more efficient models. Local or intermediate models, such as the multiquadric, have enabled improvement of spatial accuracy of rectification. Rectification can be applied for images of different resolutions derived from the same sensor or from various sensors. Multiscale models based on multiresolution analysis of images using wavelet transformation have been developed (Djamdji, 1993). The mosaic consists of rectifying several images together so as to cover an extensive field. It facilitates studies at district or regional levels. Integration of rectified images in a GIS often constitutes the final stage of rectification. It opens many possibilities of analysis by facilitating combination of remote-sensing data with other geographic databases.

References

Alsoft. 1997. http://www.alsoft.fr.

Djamdji JP, Bijaoui A, Maniere R. 1993. Geometrical registration of images—the multiresolution approach. Photogrammetric Engineering and Remote Sensing, 59: 645–653.

Erdas. 1997. http://www.erdas.com.

ER/Mapper. 1997. http://www.ermapper.com.

Esri. 1997. http://www.esri.com.

Fogel DN. 1996. Image registration using multiquadric functions, the finite element method, bivariate mapping polynomials and the thin plate spline. National Center for Geographic Information and Analysis, University of California USA. Internal report n° 96-01, pp. 1–44.

Hardy RL. 1971. Multiquadric equations of topography and other irregular surfaces. J. Geophysical Research, 76 (8): 1905–1915.

Hardy RL. 1977. Least square prediction. Photogrammetric Engineering and Remote Sensing, 43 (4): 475–492.

Hardy RL. 1990. Theory and applications of the multiquadric-biharmonic method. Computers and Mathematical Applications, 19: 163–208.

Hubault J. 1994. Corrections géométriques utilisant des formes d'appui d'image à image ou d'image à cartes; un critère d' automatisation. Séminaire SFPT-RSS, Qualité de l'interprétation des images de télédétection pour la cartographie. Grignon, pp. 9–11.

Idrisi. 1997. http://www.idrisi. com.

IGN. 1997. La gamme de produits geomarketing IGN.

Jensen JR. 1986. Introductory Digital Image Processing, a Remote-sensing Perspective. Prentice-Hall, 379 pp.

Khoros. 1997. http://www.khoros.unm.edu.

Mapinfo. 1997. http:www.//mapinfo.com.
MQReg. 1997. Multiquadric rectification, http://pollux.geog.ucsb.edu/~fogel/mqreg.html.
Noésis. 1997. http://www.noesisvision.com.
Pratt WK. 1991. Digital Image Processing, Wiley-Interscience, 698 pp.
Rousselet M. 1985. Graphisme 3D, Éditions techniques et scientifiques françaises, 223 pp.
Teissier L, Stamon G. 1994. Geometric ortho-rectification of flash radar images. First IEEE International Conference on Image Processing, Austin, Texas, USA
Teravue. 1997. http://lacan.grignon.inra.fr
Worboys MF. 1995. GIS Computing Perspective. Taylor & Francis, 376 pp.

14

Fundamentals of Aerial Photography

It is necessary to review the knowledge base of aerial photography for two reasons. Firstly, visual interpretation has regained its place in the gamut of methods of interpretation of remote sensing data such as satellite images, satellite photographs, aerial photographs, etc. Secondly, satellite remote sensing including development of satellite missions, image interpretation, relief measurements, etc., has emerged on the basis of data, concepts and methods concerning aerial photography: acquisition of aerial photos, their quality, organisation of flight paths, photogrammetric measurements and interpretation of aerial photos. Initially, in fact, satellite remote sensing combined photo-interpretation and photogrammetry by associating physicists and aeronautical engineers.

Fundamentals of photo-acquisition, stereoscopy and photogrammetry are presented below.

14.1 PHOTO-ACQUISITION

Acquisition of aerial photos during a denominated mission flight is dependent on three factors:

1) Factors related to the data to be generated: focal length of the objective of the receiver, format, coverage and scale.
2) Factors concerned with the region under study: area to be studied, maximum altitude.
3) Factors pertaining to platform: speed of aircraft, stability in flight, etc.

A good aerial photograph should satisfy three demands. Firstly, it should furnish maximum information, which necessitates correct choice of optics of the camera used, combination of film and filters and conditions of acquisition. Secondly, the photographs should cover the entire region under study. Thirdly, it should facilitate accurate measurements of length and area, which necessitate metric cameras and precise geometric analysis of the entire sensor-platform system.

14.1.1 Photographic sensors

Aerial photographs can be taken with an amateur photo-camera in an aircraft but the photos will not be of excellent quality for several reasons.

For a colour emulsion, the photos appear with dominant blue colours and a yellow filter has to be used to avoid this mist which is due to atmospheric disturbances (Chapter 1). Further, there can be reflection of the porthole glass and a better angle has to be chosen to avoid this reflection or the camera has to be directly inserted in the wall of the aircraft cabin. In this case, the photograph will not be vertical. Hence, the aircraft has to be turned so that the camera is vertical when clicked manually, which, of course, would be an acrobatic feat. Alternatively, the camera should be kept in a vertical position under the aircraft wing (but provided with a long clicking mechanism) or directly in the floor of the cabin.

Basics of emulsions and films are mentioned in Chapter 3.

■ Metric cameras

Special cameras, known as metric cameras, were developed for acquiring aerial photos. They represent cameras for which any geometric and chromatic distortion can be detected and measured. These are employed for obtaining systematic photographic views, which can be subsequently processed for relief restitution.

A navigational telescope was attached to these metric cameras so that the navigator of the aircraft could verify the flight path. Today GPS (Global Positioning Systems) are used for this purpose. All data about the flight, viz., its speed, position, etc., are recorded in an electronic control unit.

Aerial photographic cameras are characterised by film formats ranging from 35 mm (24 × 36 mm photo) to 70 mm (55 × 55 mm photo) and for metric cameras (Fig. 14.1) 240 mm (230 × 230 mm photo). Thickness of the film on the base should be uniform to avoid distortions in measurements to be made subsequently on the photographs. This is ensured by a depression in the focal plane punched with numerous small holes. These cameras are equipped with an automatic trigger which should be regulated according to flight plan: altitude, coverage, flight height and speed. Film magazines should contain very long film, 120 m for example (or 500 snaps), to avoid frequent changing. An automatic system of film movement should be regulated as a function of speed of aircraft and the 'base', i.e., distance covered between two successive photographs. Exposure times are very short, 1/10 to 1/1000 s. Roll due to aircraft motion during photo acquisition should thus be avoided to preclude photo blurring. For an exposure of 1/100 s, displacement is of the order of 20 cm for an aerial photograph and 80 m for a satellite image. Compensating mechanisms are presently used. Correction for roll is obtained by a displacement of focal plane in the direction of flight during the exposure time with a velocity v:

$$v = V \times f / H$$

where V is the flight speed (for example, $V = 720$ km h^{-1} or 200 m s^{-1}), f the focal length of the metric camera ($f = 152.4$ mm) and H the flight height above the ground ($H = 6096$ m).

For the example given, $v = 5$ mm s^{-1}, and the focal plane should be displaced by 5/100 mm during the exposure time of 1/100 s.

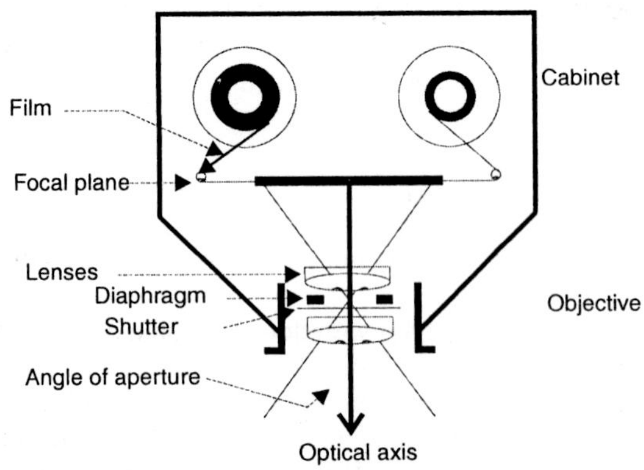

Fig. 14.1: Schematic diagram of a metric camera.

Aerial cameras are equipped with an objective (Fig. 14.1) comprising a shutter, a diaphragm, filters and an objective which most often consists of an assembly of several lenses. The objective is defined by its focal length which varies from 90 to 300 mm. Focal lengths of 125 and 152.4 mm are common in cameras used by the National Geographic Institute (IGN). A camera is also defined by its field of view α which varies from less than 75° to more than 120°.

The optical axis of the objective is perpendicular to the plane of the film at the centre. Consequently, aerial photos are conical projections whose field of view is dependent on the angle of aperture α and flight altitude H and whose scale E is dependent on the focal length f and flight altitude. Focusing is evidently at infinity and hence the distance between the film and the optical centre of the objective is equal to the focal distance.

■ Types of cameras

Various types of cameras are used for several specific cases.

Large format cameras were developed for satellites such as the Gemini and Apollo missions. They had a focal length of 305 mm and took 230 × 460 mm photos. The European Space Agency had a program with a metric camera mounted on SPACELAB during a flight of space shuttle. The camera constructed by Germany had a focal length of 305 mm and 230 × 230 mm size photos (Schuhr et al., 1984). The aim of this program was to study the possibilities of interpretation of spatial data for preparation of maps. Preparation of current maps based on satellite data shows that the result is positive (SFPT special number 99, 1985).

Multispectral cameras have also been developed. They consist of four objectives equipped with various filters and take photos on black and white films. The four photos taken at the same time are viewed simultaneously with four filters to obtain colour composites. Thus composites with 'real colour' and 'false colour' or infrared colour are prepared. Later, several missions were launched in which four Hasselblad cameras were combined, each with a different filter, since these cameras were less expensive.

Panoramic cameras are also available which, equipped with a rotating objective, photograph a long strip perpendicular to the flight path. The disadvantage of these photos is that the scale changes continuously. However, the Apollo missions photographed the surface of the Moon using these cameras. A camera of this type used by NASA has a focal length of 610 mm, aperture angle of 120° and film length of 2000 m.

Several other cameras are marketed but not described here. Only digital cameras, which are undergoing continuous development for usage by amateurs as well as professionals, need to be mentioned. Digitisation of data is evidently an important advantage since nowadays data is processed using computers. It may be noted that aerial photos can also be digitised by scanning. Modern scanners provide high-resolution data, comparable to the geometric resolution of aerial photos in black and white or colour.

Criteria for comparison between aerial photos and digital images (Table 14.1) have been developed in recent years since digital images ought to provide significant improvements.

Table 14.1: Comparison between photographs and digital images

Operation	Photographs	Digital images
Data acquisition	Film	CCD digital sensors
Storage	Film or paper	Magnetic tapes, CDs, hard disks
Interpretation	Visual	Computer processing
Transmission	Post, Fax	Radio, networks
Visualisation	Projection, paper maps	Video projection, television, computer monitors
Reproduction	Photo printing	Electronic printing, ink jet, laser

14.1.2 Types of aerial photos

Several types of aerial photos can be distinguished depending on the position of optical axis with respect to the vertical and motion of the aircraft. The main aim in aerial photography has been to take photos as fast as possible in the best meteorological and geometric conditions and with as wide a field of view as possible. This objective has remained almost the same for satellite images, aside from increasing revisiting capability.

■ Vertical photos

The optical axis of the camera is vertical (Fig. 14.2). When the ground is horizontal, the relationships existing between the ground (point *Y*) and the photo (point *y*) are homothetic. In vertical photographs, the central point of the photo (*p*) coincides with the nadir (*P*), point at the vertical of the optical centre of the camera. The optical axis passes through the nadir for these photos. This position is used to acquire satellite images. The homothetic ratio *H*/*f*, where *H* is camera height above ground and *f* the focal length of the camera, is equal to the inverse of scale. The photo is a conical projection.

Knowing the aperture angle of the camera and the flight height, the field of view of the photograph and area from which the information is gathered by the sensor are derived.

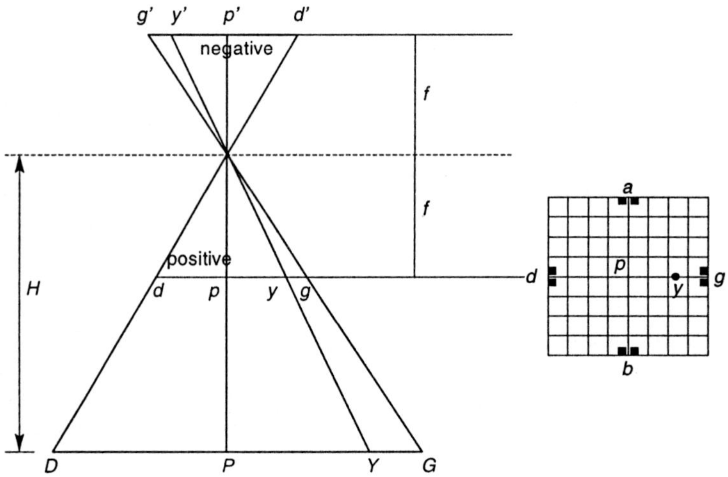

Fig. 14.2: Schematic diagram of taking a vertical photograph.

■ Oblique photos

When the optical axis of the camera forms an angle of more than 5° with the vertical, oblique photographs are obtained. The nadir and central point of the photograph are not coincident.

A square outline on the ground appears as a trapezium in the photo. These photographs can be distinguished as panoramic oblique and low oblique (Fig. 14.3). In panoramic photos the nadir is no longer photographed and the horizon is visible in the photo.

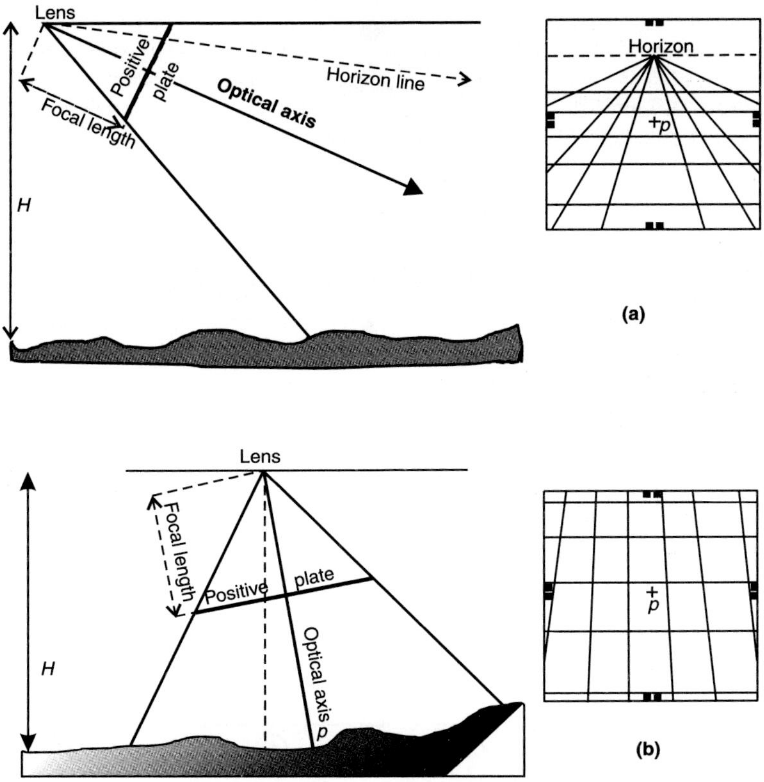

Fig. 14.3: Panoramic (a) and low oblique (b) photographs.

Two oblique (at 60°) panoramic photographs together with a vertical photograph constitute a trimetrogon system (Fig. 14.4). They enable rapid reconnaissance mapping. They were employed prior to satellite images to obtain aerial photographs of Nordic regions where one can fly only for short periods. For example, such a mission was launched to photograph Spitzberg. Thus strips separated by 25 to 30 km could be photographed at a flight height of 8000 m.

The horizon does not appear in low oblique photographs. Two photographs can be taken to the left and right of the camera or one in front and the other in the rear (Fig. 14.5). In the latter case a convergent view is obtained. Depending on the angle of inclination of the camera, 100% coverage can be achieved. With the advent of super-wide angle cameras interest in low oblique cameras has decreased. However, this concept has been applied in one of the SPOT-5 projects with the objective of obtaining a stereoscopic pair from a single pass of the satellite in nearly 10 s.

■ Emulsions and formats

Emulsions were described in Chapter 3. Conventionally emulsions are classified as panchromatic and infrared (black and white), colour and colour infrared (CIR). Aerial photographs are acquired in all these four emulsions in IGN.

The most common format used at present in France is 23 cm × 23 cm, with 1:30,000 scale. Earlier formats of 18 cm × 18 cm with 1:25,000 scale and still older ones of 13 cm × 18 cm are also available.

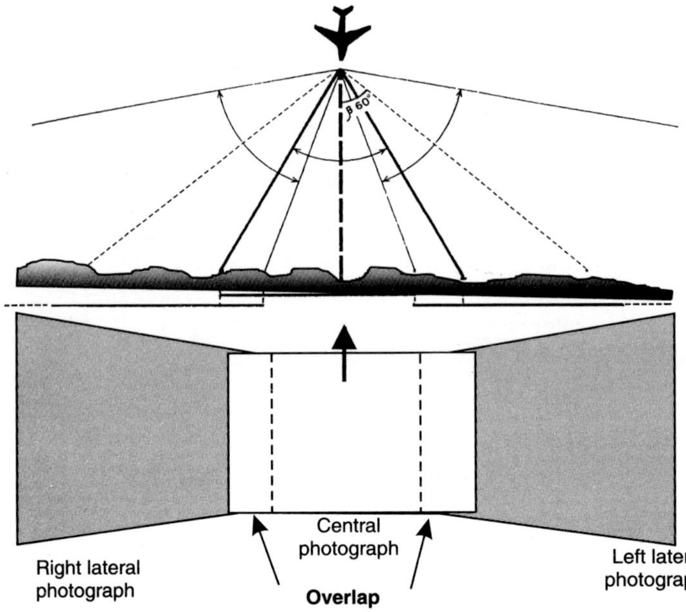

Fig. 14.4: Trimetrogon photography.

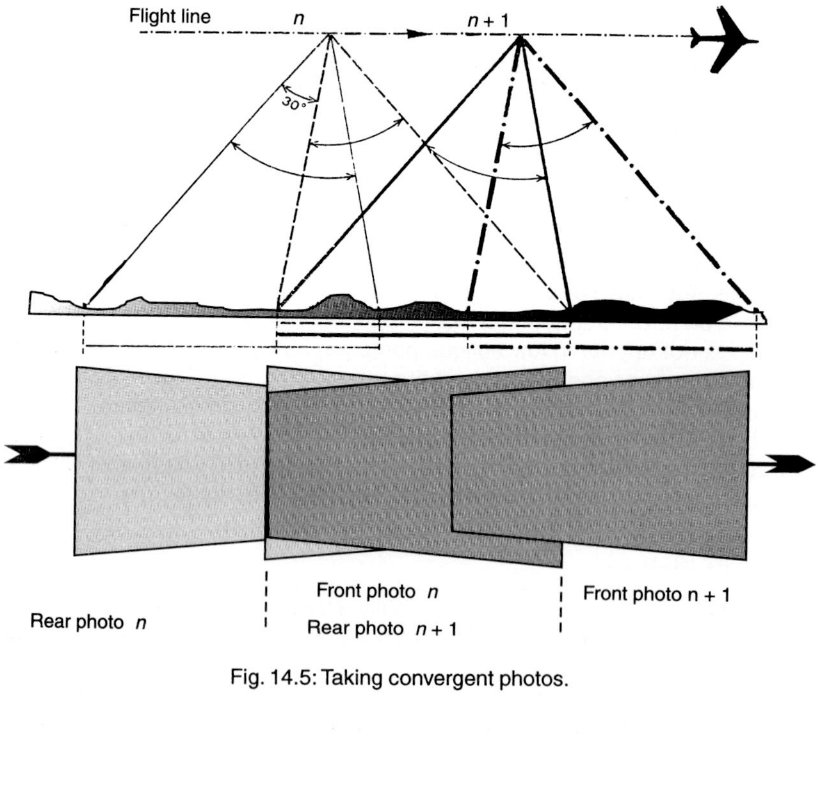

Fig. 14.5: Taking convergent photos.

It may be noted that aerial photographs sold by IGN can be enlarged by about 25 to 40 times. Enlargements are carried out by IGN up to scales of about 1:1000.

14.1.3 Regular missions

The Hurel-Dubois aeroplane, with a flight height of 6000 m, was specially designed for aerial photography and proved very stable. In France, regular missions of aerial photography were undertaken for a long time on propeller bombers B 17 ('flying fortress') with a flight height of 8000 m. The Mystere 20 aircraft with an operational altitude of 12,500 m and range of 4000 km was also used. Today various aeroplanes are used for aerial photography. A pilot, navigator and photographer man an aircraft for a photographic mission. They are equipped with a navigational telescope for viewing the ground below the aircraft, two cameras, an intervalometer for fixing correct spacing between photographs, a GPS, as well as an apparatus for determining the aircraft attitude and height. Three hatch doors are therefore required in the cabin floor, which sometimes pose problems in designing the aircraft profile. This results in higher costs, especially when the mission employs other sensors such as metric cameras, some of which do not accept glass or quartz structures as in the case of middle or thermal infrared scanners for example. Aeroplanes used for remote sensing cannot be employed for other missions.

When roll, pitch or tilt occur in flight, the track of the aeroplane and hence of the camera is modified, which changes the area covered on the ground by photography (Fig. 14.6). Thus a spherical surface gives the deviation of the axis of the metric camera with respect to the vertical for each photograph. Sometimes, stabilisation is achieved using gyroscopes and servomotors, which ensure verticality of the metric camera in spite of aircraft oscillations.

Aerial photographic surveys have several disadvantages:

—choice of date of photography is not certain since flight plans depend on meteorological conditions;

—area covered is relatively small for a given mission;

—cost is relatively high.

The advantage of aerial photography still evident to date is its very high geometric resolution, which probably won't be enjoyed much longer.

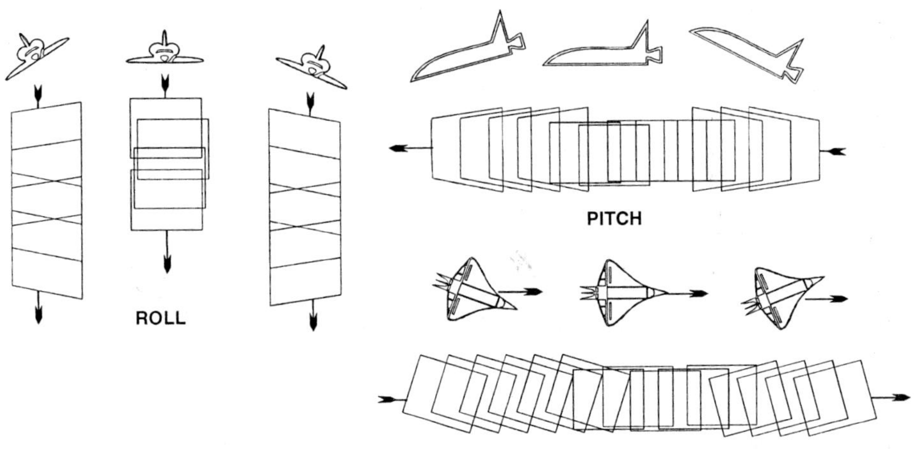

Fig. 14.6: Roll, pitch and tilt of aeroplane: modifications in ground coverage by aerial photography.

One foresees that with miniaturisation of electronic components of sensors, the possibility of operating sensors outside the cabin and use of GPS that give very precise position of the aircraft, application of aerial remote sensing will witness new developments.

14.1.4 Organisation and overlap of aerial photos

The National Geographic Institute (IGN) conducts regular aerial photographic surveys in areas corresponding to 1:50,000 topographic maps of France or districts (to the requirements of National Forest Inventory, IFN, in particular). The survey consists of a series of photographs taken successively, if not the same day, at the shortest possible time. The successive flight lines are parallel to one another (Fig. 14.7), thus providing a strip of photographs.

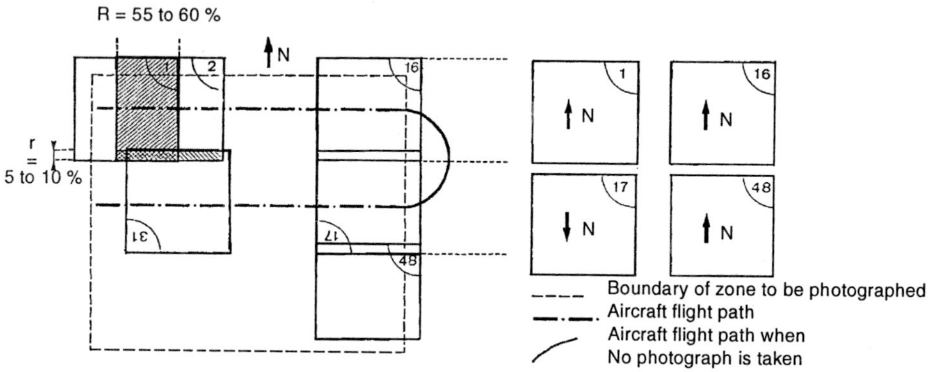

Fig. 14.7: Positions of aerial photos in a survey: overlap.

The photographs should have an overlap for obtaining a stereoscopic view. This successive overlap (*endlap*) (*R*) is 55 to 60% and may go up to 80 to 90%. Photographs of each strip overlap a few grid cells of the adjacent strip. The lateral overlap (*sidelap*) (*r*) varies from 10 to 20%. In most cases the strips are taken from west to east. In this direction north is to the left of the aircraft whereas, when it takes a half-turn to photograph the adjacent strip farther south, south is to its left. As some acquisition parameters are indicated on the borders of a photograph, they represent north and south alternatively. Thus, when aerial photographs of IGN are viewed, numbers of photographs are alternatively observed in the direction north then south.

In the case of high mountains, as in the Alps, surveys are carried out from north to south. For surveys launched for a specific objective, flight directions can be arbitrary.

An overlap of 50% ensures stereoscopic view; below this value, no ground object is seen at two different angles and hence a perspective view cannot be obtained. Contrarily, if the overlap is 60% or more, ground marker points, which are situated in three photographs, can be identified. Consequently, the resolution is improved. If a constant flight height is taken relative to the sea level and if the photographed zone shows high relief, it is possible to observe portions of the ground in three or two photographs. If they exist and are observed in only one photograph, a perspective view cannot be obtained and some parts may not be photographed at all (Fig. 14.8). For this reason, another set of photographs of the strip is taken at a different altitude.

The minimum clearance between the flight line and the maximum altitude Z_m required for a minimum overlap of 50% can be deduced. Z_0 corresponds to the altitude below which the overlap at any point would be at least 60% (Fig. 14.9).

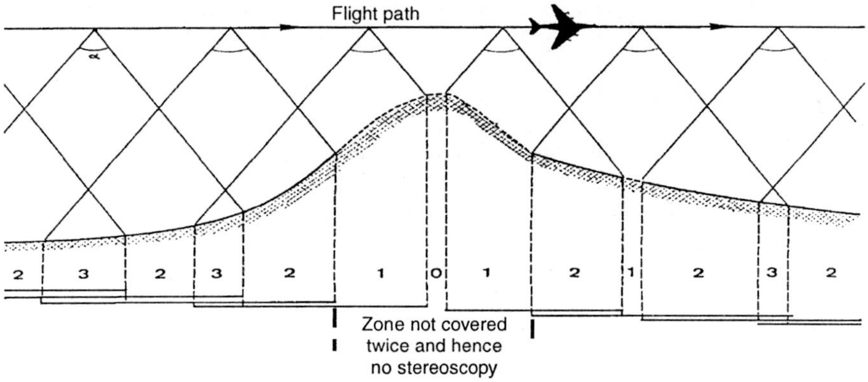

Fig. 14.8: Various overlaps of a zone of aerial photography.
α—aperture angle of camera; numbers 3, 2, 1 and 0—number of photos covering a zone designated by a horizontal line.

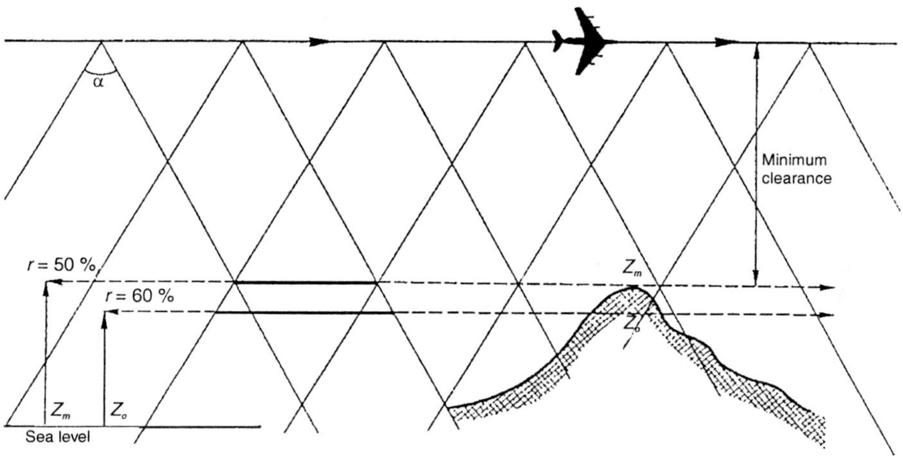

Fig. 14.9: Relationship between altitude and overlap.

The distance separating two successive photographs (Fig. 14.10) on the same flight line, known as air base B, depends on the aperture angle α of the camera, the overlap R and the flight height above ground H.

The format F of the photograph is computed as:

$$F = 2H \tan \alpha / 2$$

Since $B = F(1 - R)$, we get:

$$B = 2H(1 - R) \tan \alpha / 2$$

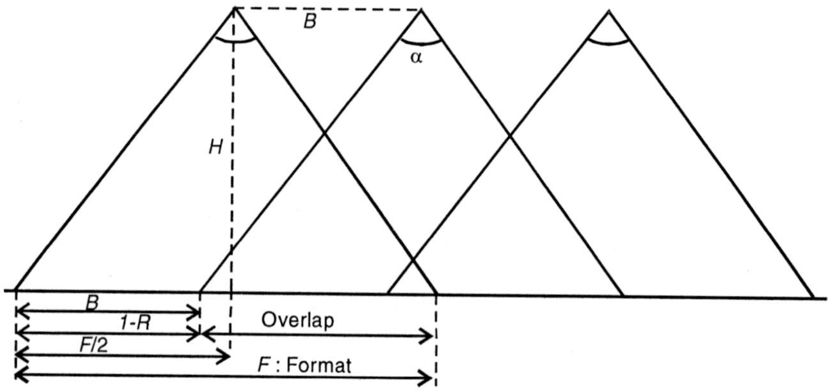

Fig. 14.10: Computation of air base *B*.

The base-height ratio *B/H* is defined as the 'characteristic' that depends only on the aperture angle, which is constant for a given sensor. This ratio is used to determine the overlap. For aerial coverage of France, the base-height ratio is close to 0.58. In the case of SPOT satellite, when images are taken at a day's interval, this ratio is 0.5 and for images taken with an inclination of 27°, it is 1.19 (see Chapter 2 and CD: SPOT system).

The distance between two flight lines is computed in the same way as for *B*, but by taking the sidelap *r*, the values of *H* and α remaining the same.

14.1.5 Flight plan (example)

A flight plan is determined from various parameters mentioned earlier regarding the types of photographs and onboard cameras, and specific characteristics of the platform, here aircraft. Among others, its height and speed should also be taken into consideration. All these parameters can be computed. As an example, information obtained from IGN for a photographic mission available for public sale is given below:

1988 Bourg-Saint-Andéol-Nyons.
Date: 2 June 1988 from 7 h 21 to 7 h 59.
Photographs 1 to 99.
Focal camera: 15 UAg. 1041. Focal: 151.83.
Altitude above ground: 4590 m, above sea level: 5090 m.
Scale 1:30,000
Aircraft: Mystere 20.
For this survey, with one photograph out of two, a 60% overlap is possible.

14.2 STEREOSCOPY

14.2.1 General

Sharpness of visual discrimination ε is defined as the minimum angular distance between two points to be seen as separate. It depends on the illumination, shape, colour and contrast of the object and acuity of the observer's vision. On average, $\varepsilon = 1'$, which corresponds to a distance of 5 μm on the retina.

In monocular vision, the eye estimates distances from modifications in convergence of the lens. Distances are also interpreted with reference to normal objects to which the memory assigns a value.

Estimation of perspective is also based on the position of shadows. An illusion of 3D relief is obtained when the shadow is situated at the bottom and on the right, which indicates that the light source is on top and to the left. If the observed photograph is rotated by 180°, the object and shadow are interchanged and we get the impression that the perspective is reversed. This is observed in images when no other reference objects exist, which our brain identifies as reverse to the impression of relief. Consequently, cloud masses or images of planets give this type of impression. Conventionally a geographic map is drawn with the north showing upwards and the south downwards. In such a reference frame, the sun is never at the top left in the northern hemisphere. Hence, shadows never occur at the bottom right, i.e., south-east. This is one of the reasons for the difficulty in observation of aerial photographs and more so in the case of satellite images acquired over relief, when they are compared with shadowed topographic maps. In fact, shadows in topographic maps are situated in the south-east, whereas no shadow exists in this exposure.

In binocular vision, distance is estimated by convergence of optical axes, i.e., parallax, of the eyes, and the relief is estimated by parallax. This perception of relief is a complex phenomenon but can be represented by a simplified model.

If an observer looks at a point *A* (Fig. 14.11), his left eye *G* and his right eye *D* accommodate a distance d_a. The lines of vision of the two eyes converge at *A* at an angle α. If the observer now looks at point *C*, the distance of accommodation would be d_c and the angle of convergence γ. The segment *AC* will be viewed at an angle β. A relationship exists between d_a and α on the one hand and between d_c and γ on the other.

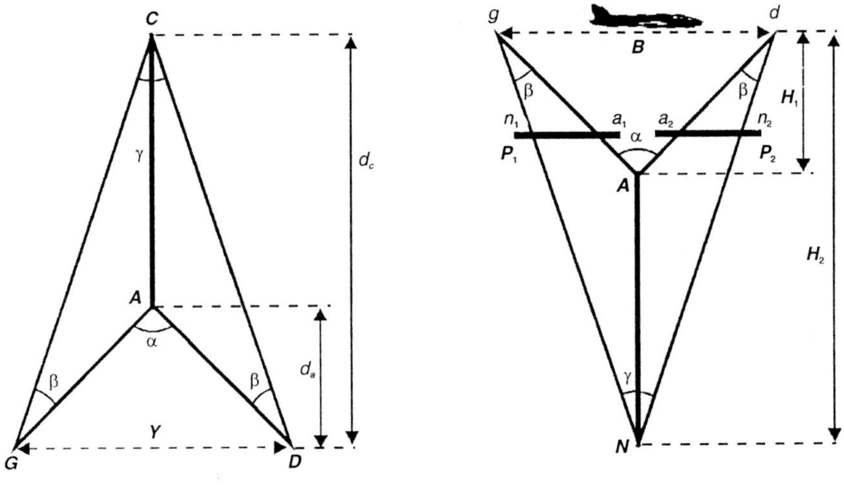

Fig. 14.11: Perception of relief.

When aerial photos separated by a distance *B* (base) are taken, the image of point *A* at height H_1 is viewed at a_1 and the image of point *N* at height H_2 is viewed at n_1 in photograph P_1. Segment *NA* is seen under angle β and its image in photograph P_1 is segment $n_1\,a_1$. Similarly, in photograph P_2, segment *NA* is seen under angle β and its image is segment $a_2\,n_2$. Distances n_1a_1 and a_2n_2 are parallaxes which can be measured and from which distance *NA* can be derived. This provides an estimate of topography. Thus two photographs (or images) in which the same point is seen under two different angles of view, suffice for obtaining the perspective view.

14.2.2 Stereoscopic vision

When two photographs overlap the same region, in which objects *A, B* and *C* are situated at the same altitude and object *D* at a different altitude, the four objects will be observed in a different sequence in the two photographs: *a, b, d, c* in the left photograph and *a, d, b, c* in the right (Fig. 14.12). In the same photograph, segments *ab* and *bc* are equal since they are at the same altitude, but segments *ad* and *dc* are not.

When a stereo pair is viewed, one notices that the vertical scale is greater than the horizontal scale, resulting in exaggeration of relief. This results in a false estimation of slopes, heights, etc. On the other hand, a stereoscopic pair enables detection of small differences in slope or breaks in slope. Hence it is useful for visual interpretation.

A nomogram (Fig. 14.13) connecting three parameters, viz., true slope (α), apparent slope (β) and vertical exaggeration of relief (e_v), is used to determine one parameter from the other two.

The apparent slope (β) is measured from the aerial photograph. We can fix a pin at the base of the slope and, under the stereoscope, tilt it until an impression is obtained that it is in contact with the relief model, and finally measure this slope with a protractor. The true slope can be measured in a topographic map or in the field with a clinometer. On the nomogram, the point of intersection of the lines of apparent slope (moving along the perpendicular to the abscissa) and true slope (moving along the inclined line) is located, and its position on the ordinate gives the vertical exaggeration of relief. For common stereoscopic plates, the relief exaggeration is about 2.5. With such a value, an apparent slope of 55° corresponds to a true slope of 30° and an apparent slope of 20° to a true slope of 8°.

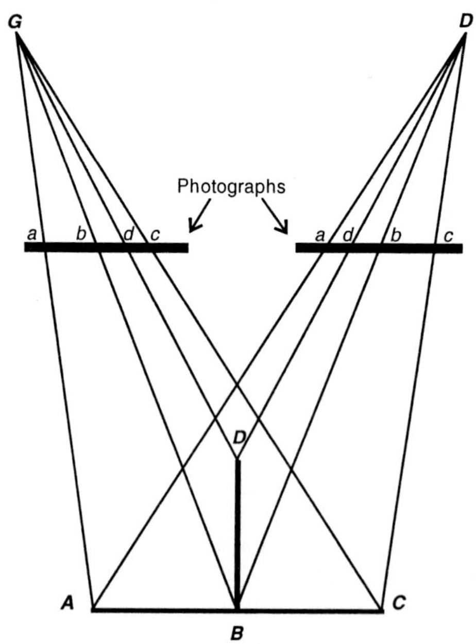

Fig. 14.12: Perception of relief from two aerial photographs.

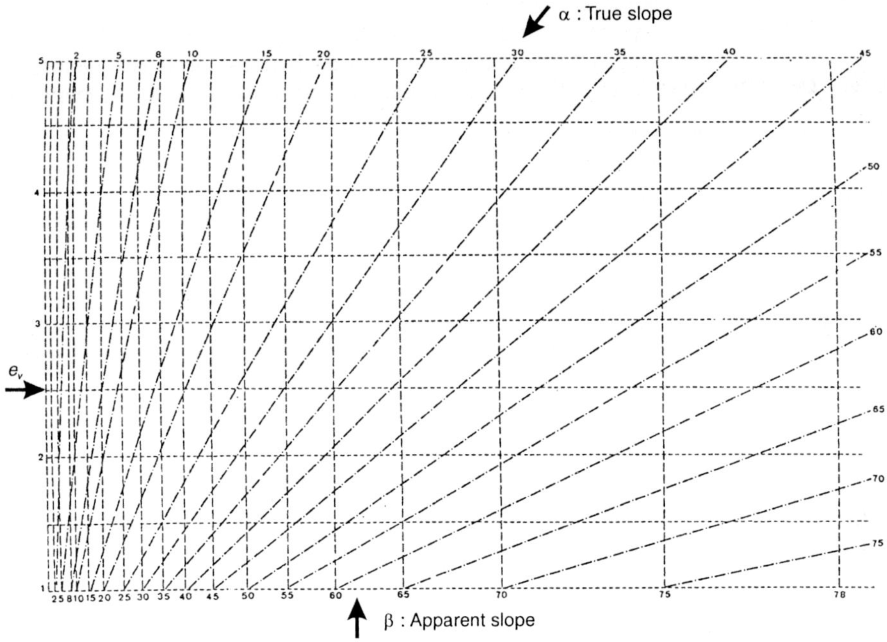

Fig. 14.13: Nomogram of true slope (α), apparent slope (β) and vertical exaggeration of relief (e_V).

In practice, this value of relief exaggeration corresponds fairly well to the impression in the eye and can be recommended for drawing sections or three-dimensional diagrams to represent geomorphology and landscapes.

Relief exaggeration is due to six factors, three dependent on the sensor-platform system and the other three on the stereoscope-observer system.

—Air base *B*: If this increases, relief exaggeration increases. Thus, in the case of satellite images, largest base and hence highest vertical exaggeration is obtained by taking SPOT images with –27° and +27°. This is necessary for morphological interpretation of these images.

—Focal distance *f*: If this increases, vertical relief exaggeration increases. This is an important factor at present, since photographs and images of increasingly longer focal distances are being used: the latter have changed from 70 mm to 125 mm, then to 152 mm and even 300 mm.

—Flight height *H*: If this increases, vertical exaggeration of relief decreases. This is the case of satellite images and photographs.

—Interocular distance *y* is connected with the observer. If *y* decreases (eyes closer), vertical exaggeration of relief increases.

—Eye-to-photograph distance *d*: If this increases, as in some stereoscopes, vertical exaggeration of relief increases.

—Overlap of photographs *R*: Overlap *R* is inversely proportional to the distance between the principal points (centre of each photograph from the stereo pair): it is the separation *s* of the photographs over the support on which they are observed (Fig. 14.14). If *R* decreases, or *s* increases, vertical exaggeration of relief increases. Some stereoscopes have magnification which enhances the virtual overlap. The latter decreases the vertical exaggeration of relief.

The relationship between these factors can be summarised as:

$$e_v = BHd / yRH$$

which can be represented in various forms:

—Depending on the two systems, stereoscope and flight, $e_v = (Bf/H)\,(d/yR)$;

—or $e_v = (f/H)\,(Bd/yR)$; this is the product of scale and a hyperstereoscopy factor;

—or using the base-height ratio (B/H), focal length of the camera f and d of the stereoscope, $e_v = (B/H)\,(fd/yR)$.

14.2.3 Stereoscopes

A stereoscope aids the observer to fix one eye on each photograph of the pair and see them as only one photograph. Further, the distance d between the eye and the photograph is maintained constant.

A stereoscope (Fig. 14.14) consists of a double optical system (lenses, mirrors, prisms, etc.) mounted on a rigid frame supported on legs. In this way, distance d is fixed. The optical system is such that the virtual image is cast at infinity and consequently stereoscopic vision is obtained without eye-

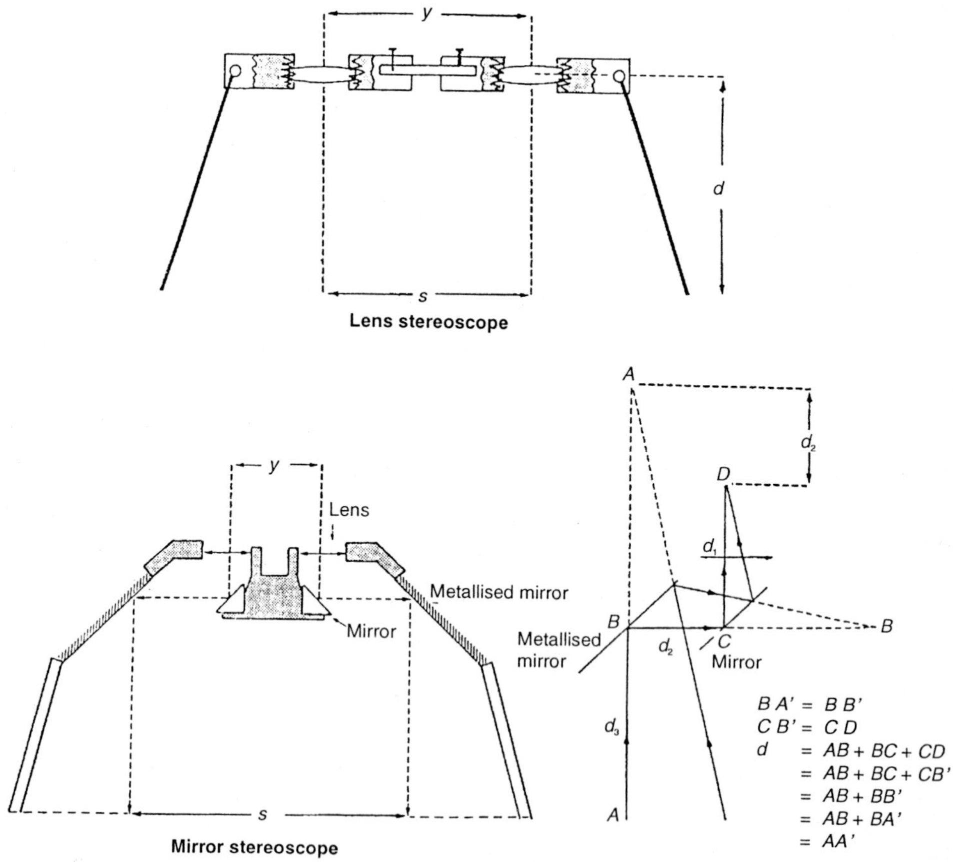

Fig. 14.14: Types of stereoscopes.

strain. Moreover, in most stereoscopes the interocular distance *y* is variable and can be adjusted to suit the observer.

A simple lens stereoscope is made up of two achromatic convex lenses. The focal length is equal to *d* corresponding to the height of the stereoscope above the plane on which the stereo pair is placed. The only inconvenience of this stereoscope is that it does not facilitate vision of the entire area of the photographic pair of format 23 cm × 23 cm.

A mirror stereoscope comprises two metallised mirrors, two prisms, two lenses and two eyepieces. It enables viewing the entire area of the photographic pair of format 23 cm × 23 cm.

In some stereoscopes the optical part is fixed on an arm and the photographic pairs are arranged on two different planes. They facilitate analyses of several stereo pairs consecutively without changing the arrangement.

Many models exist in which the scale of the photograph can be changed and hence stereoscopy of photos of different scales can be obtained and/or interpretation can be made on a topographic map, thus integrating a camera lucida.

These systems, more and more bulky, have to face stiff competition from digital systems, which are highly developed for satellite images and furthermore because photographs can be digitised.

14.2.4 Anaglyphs and vertographs

Some other methods are also used for obtaining stereoscopic view.

Most common is an anaglyph, a figure comprising two traces of the same object, one in blue-green (seen under a particular angle) and the other in red (seen under a different angle). Green and red glasses are used to get a 3D view. The left eye sees through a red transparent film and hence perceives, as black, only those lines that are drawn in green-blue and not those in red. The right eye sees through a green-blue transparent film and perceives, as black, only the red lines. Thus the same object is seen as black under different angles. This suffices to produce a sensation of three dimensions. If the glasses are reversed, a pseudo-stereoscopic view is obtained which gives the impression of a reversed relief: valleys correspond to high points and peaks to basins.

A vertograph (polarised platen viewer) is based on polarisation of light. The observer wears polarising eyeglasses. One glass allows only vertically polarised light and the other only horizontally polarised light. Thus, each eye perceives only one image. The two images are projected onto a screen, one with vertical polarisation and the other with horizontal polarisation. Once again, each eye perceives a different image, since viewed at different angles and the observer obtains perspective view by synthesising them. The principle of the vertograph is very useful at present for stereo restitution.

14.3 PHOTOGRAMMETRY

Photogrammetry includes all aspects concerned with obtaining quantitative information from photographs, viz., measurement of distances, altitudes, etc. It provides geometric descriptions of locations and extensions of phenomena interpreted from photographs. Photogrammetry is an accurate and complex science described in many books; hence we present below only the basics needed for understanding how results of interpretation of aerospatial photographs and images can be linked to a precise geometric space.

Photogrammetry is employed for numerous applications such as restoration of historic monuments and ancient musical instruments, analysis and surveillance of some structures (dams, ship propellers, etc.) as well as in medicine. Its most common use, however, remains production of topographic maps.

14.3.1 Orientation of photographs

A vertical photograph is usually oriented in the same way as a map, i.e., when it is on a horizontal plane—which is most common—north is in the farthest position from the observer and when it is in a vertical plane, north upwards. For quick appraisal, it is considered that in the Northern Hemisphere shadows are more or less directed northwards. Otherwise, points identifiable in the photo and a topographic map should be taken for identifying directions. Segments are drawn to connect various points and the photograph arranged in such a way that the directions of the corresponding segments are parallel. It will be noticed that the directions of the segments do not correspond exactly when high relief is present.

14.3.2 Use of a stereo pair

A stereo pair consists of two photographs having a certain percentage of overlap. The two photographs 1 and 2 should be placed in the order in which they were taken during the mission; otherwise a pseudo-stereoscopy is obtained. The centres *A* and *B* of each photograph or their principal points are marked (Fig. 14.15). On each photograph, the images *a* and *b* corresponding to the principal points are marked and the straight lines *Ab* and *Ba* are drawn. The two photographs are adjusted in such a manner that these two straight lines are coincident. To achieve good stereoscopic vision the distance *Aa* and *Bb* should be equal to the interocular distance *y*.

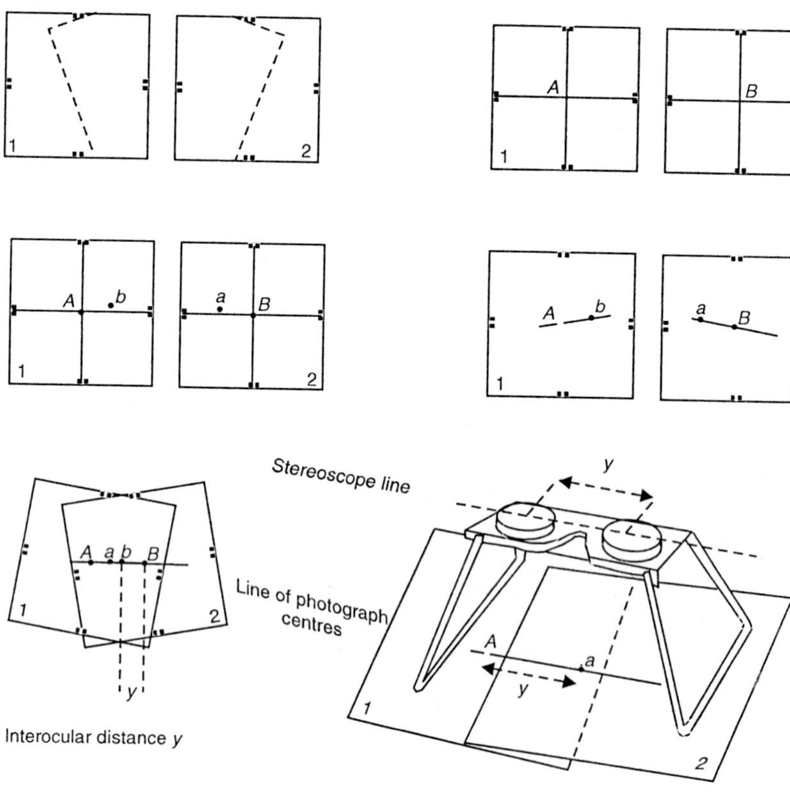

Fig. 14.15: Use of a stereo pair.

14.3.3 Scale of a photograph

■ Measurement of scale

The scale of a map is defined as the ratio of a distance measured on the map to the same distance on the ground. The scale of a photograph can be defined in the same way. Let *N* be a reference plane of altitude *n* above sea level at a distance *H* from the objective of the camera (Fig. 14.16). For example, point *G* and points *U, V, P* and *D*, projected respectively from the points (on the ground) *O* (photograph centre), *R* (river), *A* (plateau) and *C* (hilltop) are located on the reference plane. On the photograph, these points are located at *g, o, v, p* and *d*. Scale of the photograph *E* is hence given by:

$$E = og/UG \quad \text{or} \quad E = f/H$$

Hence, at a given altitude, such as for example above a contour line, the scale of a photograph is constant if the axis of the camera is strictly vertical.

The focal length of the camera and height of the platform hence determine the scale. In the example of the Bourg-Saint-Andéol-Nyons regular mission (sec. 14.1.5), the scale of photographs on the reference plane was:

$$E = 151.83/4590 \quad \text{or} \quad E = 1{:}30{,}231$$

This precision of computation assumes there was no relief (not at all the case for this mission) and that the aircraft flew at exactly the same height.

■ Scale variations

In practice, variations occur in the scale of aerial photographs due to defects in the objective of the camera, deviations in the optical axis relative to the vertical and topographic fluctuations.

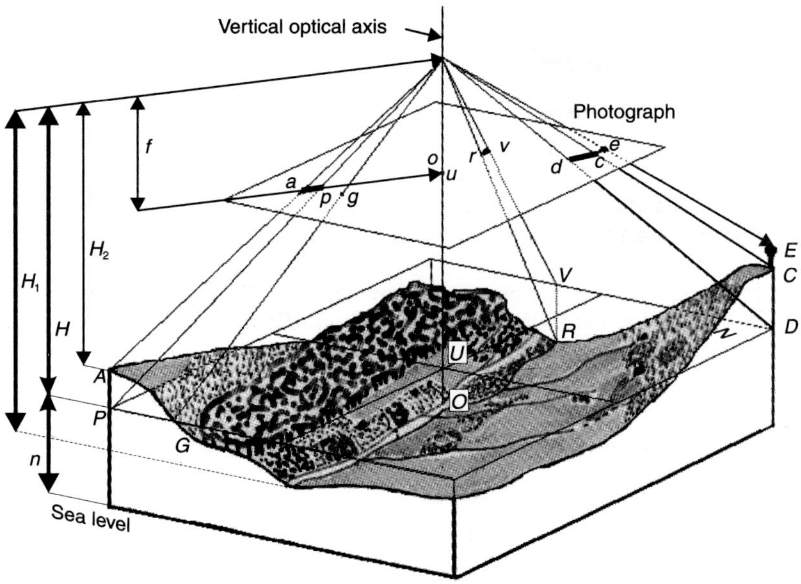

Fig. 14.16: Geometry of a vertical photograph.

Scale variations due to defects in the objective are relatively very small. Evidently, camera objectives are never perfect since small distortions always exist. However, these are totally negligible for specialised aerial photographs. Nevertheless, wide-angle optics of commercial cameras should be carefully checked.

Scale variations due to fluctuations in inclination are more common. In fact, these variations are essentially due to modifications in the attitude of the aircraft (or satellite): common causes are roll, pitch and tilt (see Fig. 14.6). For example, an inclination variation *i* in the angle of an 18 cm × 18 cm photograph, taken with a focal length of 125 mm, causes a displacement *d* that results in a relative error *er* in the distance measured on a half-diagonal (Table 14.2).

Table 14.2: Scale variations due to inclination variations (in degrees)

Inclination *i*	Displacement *d*	Relative error *er*
1G	1.2 mm	1%
3G	3.6 mm	3%
5G	5.8 mm	4.8%

Large variations in scale are produced by relief variations.

For the plane passing through *O* and *R*, situated at the same altitude (Fig. 14.16), the scale of the photograph is given by:

$$E1 = or/\,OR \qquad \text{Hence, } or = E1 \cdot OR \quad \text{and} \quad E1 = f/H_1.$$

For points *U* and *V*, which are equivalent to *O* and *R* but at a greater altitude and hence closer to the aircraft ($H < H_1$), the scale would be:

$E = ov/UV$, and as $OR = UV$, we get $E = ov/OR$, i.e., $ov = E \cdot OR$ and $E = f/H$.

Consequently,

$$or/ov = E1/E, \qquad \text{i.e., } or/ov = H/H_1.$$

Since *or* is smaller than *ov* (Fig. 14.16), scale *E*1 for *OR* is smaller than scale *E* for *UV*. Thus the points *A*, *C* and *E* on the photograph will be at a larger scale than those of the reference plane *N*.

As a result, the scale of a photograph can be determined only when the photographed zone is entirely at the same altitude. In fact, scale can be measured only on a reference plane assumed to represent, at best, the mean altitude of the region under study. Contrarily, scale can be measured at a point as well as on a contour line.

The concept of scale can be generalised as follows (Fig. 14.17). If the height between the objective of the camera and the ground is measured starting from the objective, an inverse relationship is obtained between scale and this height.

If sensor-object height decreases, scale E increases.

In a single photograph, height decreases when passing from a plain to a hill and to a mountain; hence, scale increases (Fig. 14.17, right).

For a given terrain, with the same camera (same aperture angle), height decreases according to whether the platform is a satellite (800 km), a space shuttle or an aircraft (8 km). Hence for this terrain, *scale increases when passing from satellite photograph to aerial photograph* (Fig. 14.17, left).

Similarly, the *effective field of view is proportional to this height*: for an aircraft, the effective field of view is smaller than for a shuttle or a satellite.

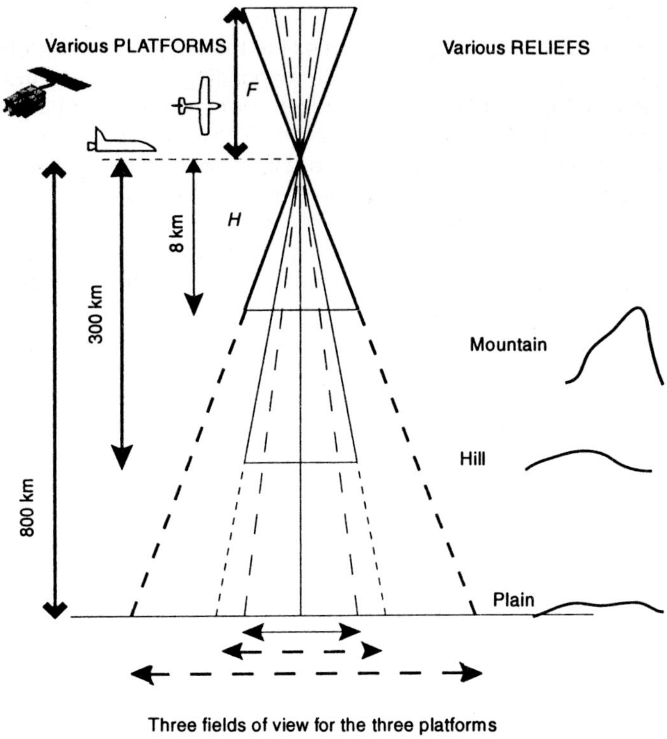

Fig. 14.17: Relationship between scale, height of camera and field of view (in this diagram, focal point of camera is fixed and altitudes varied).

14.3.4 Aerial photo mosaics

If the photographs are assumed to be really vertical and after identifying the principal points of all the photographs, they can be positioned with respect to one another on a topographic map. Thus, an *assembly* of aerial photos is obtained. This facilitates a quick recovery of photographs. Such assembled charts can be consulted for regular missions of IGN.

A mosaic corresponds to an assembly of aerial photographs prepared for obtaining a view of the entire area of an aerial survey.

■ Non-controlled mosaic

A non-controlled mosaic is prepared by matching the edges of aerial photos relative to one another. Obviously, only one-half of each photograph is required. The other half is preserved for stereoscopic observation. A satisfactory result is obtained if we attempt to determine only a small field of view that necessitates only a small number of photos. In fact, due to scale distortions, we cannot have a map from these mosaics. Correct matching is obtained over a single strip of coverage since overlap is sufficiently high, but this is rarely so between strips. If the relief is too high, the result is rarely acceptable; however, an impression of continuity is obtained and this facilitates marking reference points. Stereoscopic study of such a mosaic can be carried out using the other part of the photographs.

■ Semi-controlled mosaics

To prepare semi-controlled mosaics, first a mean scale of all the photographs is determined. Secondly, a kilometric square grid is drawn on a large sheet, corresponding to the area studied, with the mean scale of the mosaic. Using a topographic map, co-ordinates of the photo-centres are determined and marked on the grid. Lines of centres are drawn on the grid. The central points on the grid are made to coincide with the centres of the photographs and, by rotation, the lines of centres on the sheet and the photographs are matched. Quite often some parts are not in contact between two successive photos. This is secondary since the main aim is to maintain angles and distances correct as far as possible. Photographs can be divided into groups in such a way that the best possible matching is achieved between them; such a division can be made, for example, at changes in altitude. Stereoscopy can be obtained with these mosaics if two sets of photographs are available.

■ Controlled mosaic and rectification

Controlled mosaics are prepared based on the same principle as given above but using rectified photographs.

Rectification refers to reconstruction of a photograph for which the optical axis of the camera is strictly vertical and the scale is uniform irrespective of altitude of the terrain. Photographic rectification is obtained through projection by distorting the photo in such a manner that images of four marker points of the photo coincide with the corresponding four points on a topographic map. An assembly of these rectified aerial photographs is also known as photomap or orthophoto. These mosaics provide more information on land cover and land use than topographic maps.

14.3.5 Radial distortions

As a photograph is a conical projection, distortions during projection are of the radial type. The water tower *CE* on the ground in Fig. 14.16 projects as *ce* in the photograph. If it were a planar projection, i.e., an orthogonal projection, points *C* and *E* ought to coincide in the projection, but in the photograph *ce* is a segment and not a point. The nearest point from the photo centre *c* has the lowest altitude. Flattening of the water tower *CE* in the photographic plane has the effect of shifting the lowermost points towards the centre of projection and elevated points towards the circumference.

Points *P*, below *A*, and *D*, below *C*, evidently appear coincident on a map. However, on an aerial photograph, each corresponds to a straight-line segment, *pa* and *dc*. These segments are formed since the highest points are pushed outwards and the lowest points towards the photo centre, conforming to conical projection.

The higher the points, the farther they are from the centre.

For two objects of the same height, the lengths of the two corresponding segments on the photograph depend on their position in the photograph. A segment in the centre of the photograph will be smaller than on the border. Measurement of segment length does not directly give the height of an object.

The farther from the centre, the larger the distortion.

Thus, for a focal length of 125 mm, a scale of 1:25,000 and at 10 cm from the photo centre, an altitude difference *dh* gives rise to a radial distortion *dr* for a true scale *E* (Table 14.3).

A consequence on flight plan is that in the case of high relief difference, one is forced to take repeated photos by changing the reference altitude, in order to cover the entire zone by stereoscopy (case of insufficient coverage) and to obtain photographs of approximately equal scales.

As mentioned earlier, scales vary and the same length on the photo does not represent the same distance on the ground. So it is evident that two slopes which appear similar are not necessarily equal on the ground. Therefore, one has to be cautious in evaluating slopes in a photograph. Nonetheless,

Table 14.3: Radial distortion of a photograph due to altitude difference

dh	*dr*	*E*
10 m	0.3 mm	1:25,000
100 m	3.2 mm	1:24,200
500 m	19 mm	1:21,000

the best interpretation of morphology and relief of a region is achieved from stereoscopic vision and a three-dimensional virtual model.

Stereoscopic vision is obtained by observing two successive photographs. As the aircraft moves between the two photos, the optical centre of the camera is also displaced. Thus, objects seen in the first photo are not at the same position in the second photo and their radial distortions are not represented by the same segments. Based on these various considerations, three-dimensional relief can be reconstructed.

None of this applies to sun-synchronous satellites such as SPOT.

14.3.6 Parallax and altitude determination

In two successive photos 1 and 2 separated by a distance equal to the air base B, the centres N_1 and N_2 of the two project as n_1 and n_2. Point A projects on the two photos as a_1 and a_2 (Fig. 14.18). The relative displacement of point A between the two photos is called parallax p. To represent this parallax, a pseudo-image of point A is drawn at the focal distance of the camera. The difference in angle of view of point A in the two photos determines the value of parallax at the focal distance f in the pseudo-image:

$$p = a_1\,n_1 - a_2\,n_2$$

We get the relation $f/H = p/B$. As B and f are constants for a given stereo pair, i.e., $Bf = K$,

$$H = K/p$$

Altitude and parallax are inversely proportional.

If there is no difference in altitude for a stereo pair, H is constant and hence p is constant. Any two homologous points will be separated by the value of p. Contrarily, if an altitude difference exists between point B and point A, their heights H will differ and hence the parallaxes also differ. Altitudes can be computed by measuring parallaxes in the two images. For this, the distance between the images a and a' of point A is measured in each photo of the stereo pair. A micrometric screw gauge is used to measure distance to a tenth of a millimetre.

Thus, parallaxes of three points A, B and C in a stereo pair were measured and the following values obtained:

$$aa' = 7.73 \text{ cm} \qquad bb' = 7.95 \text{ cm} \qquad cc' = 8.17 \text{ cm}$$

A is at an unknown altitude, B at 500 m and C at 600 m. Consequently, the difference in parallax between B and C is:

$$cc' - bb' = 8.17 - 7.95 = 0.22 \text{ cm}$$

for an altitude difference of 600 – 500 = 100 m.

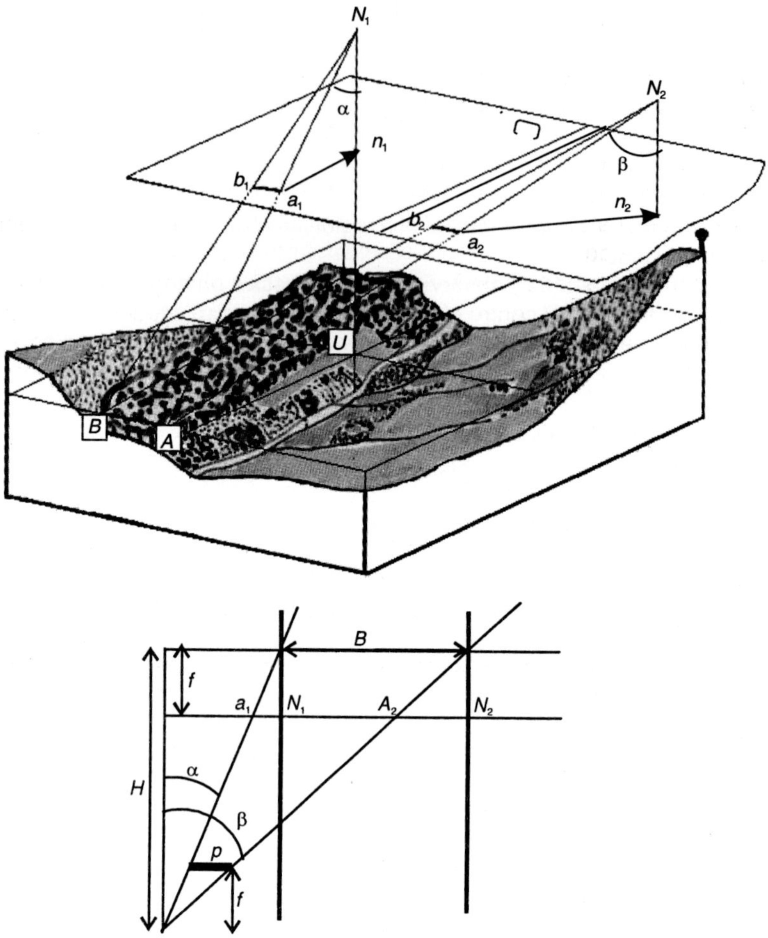

Fig. 14.18: Parallax.

Thus, a parallax difference of 0.22 cm corresponds to an altitude difference of 100 m. As $bb' - aa' = 7.95 - 7.73 = 0.22$ cm, it can be inferred that the altitude of A is 400 m.

14.3.7 Stereo restitution—Orthophotos

Restitution consists of conversion of a conical projection to an orthogonal projection. Aerial photographs are thus converted into maps. Stereo-restitution instruments facilitate this conversion of photos into maps. At present, most topographic maps are refined using aerial photographs or, more recently, satellite images.

Stereo-restitution equipment comprises three components, viz., a projection system, an observation system and a plotting system.

The underlying principle is to reconstruct an elevation model (or plastic image) exactly similar to the object by creating a projection system based on flight parameters and ground reference points for which *x, y* and *z* co-ordinates are accurately known. The procedure includes a relative orientation of the photos, determination of scale and absolute orientation. Preparatory operations, which hitherto took a very long time, are now done using microcomputers that provide much more accurate computations than earlier.

A hyperstereoscopic viewing system enables the operator to fuse, over an elevation model, two cells into a single one for a given altitude. While modifying the *x* and *y* co-ordinates of a cell, the operator can follow a contour line, since if he (she) moves away from it the two cells get dissociated. The method of anaglyphs or the method of vertographs is presently used.

Lastly, a tracing system permits drawing the details to be mapped (planimetry, contour lines, etc.) using either a pantograph or a computer by storing in memory the co-ordinates of points situated on a contour line. Nowadays, the contour line is directly drawn on the three-dimensional image.

Digital stereo-restitution systems also exist, which can be directly used for analysis of digital images, such as those of SPOT, by means of digital correlation of images. A digital elevation model (DEM) is obtained in this way. A DEM consists of a series of points for which geographic position and altitude are determined.

Orthophotos are obtained automatically using a differential rectification process that corrects for scale variations of aerial photographs. Some specific elements of topographic maps are often drawn on orthophotos. This product is developing rapidly since it can be used in geographic information systems. Orthophotos generated from aerial photographs or satellite images can thus be superposed over one another. Recent data-processing software includes systems that enable this transformation.

14.3.8 Historical background of aerial photography

Study of the Earth's surface from an elevated view dates back to the Bronze Age. The first interpretations of such views were discovered at Mont Bego and Val Camonica in the Alps. Shepherds of this period carved a landscape in the rock that overhangs a valley with fields, farm enclosures (planted or empty), cattle and habitations. It is noteworthy that they depicted rectified shapes of fields, viz., squares representing fields that were distorted by perspective.

In the Middle East, land survey records existed, since a perspective plan of the village Telloh (4000 B.C.), inscribed on baked clay bricks, was found.

In Chaldeens, land surveys must have been carried out minutely for fiscal reasons.

Up to the 16th and 17th centuries, maps with imaginative drawings could be distinguished from very accurate land survey records in which figures of various objects such as trees, bushes, fences, etc., are depicted, giving a synoptic view similar to that of aerial photos.

In 1812, Nicéphore Niepce invented a procedure for obtaining negatives and positives. In 1839, Louis Daguerre used them to make direct positive proofs from the black camera, the 'daguerreotype' could not be printed in multiple copies: photography existed. In 1840, Francois Arago, Director of the Paris Observatory, advocated the use of photography for topographic surveying. In 1841, William H. F. Talbot patented a system for producing a negative that enables printing of multiple positives.

The first known aerial photograph was taken in 1858 from a balloon at a height of 80 m above Petit-Clamart by Félix Tournachon, known as Nadar. He demanded patent for invention 'of a new system of aerostatic photography'.

On 13 October 1860, James Wallace Black and Samuel A. King took a balloon photograph from 365 m above Boston (United States), which is the earliest existing aerial photograph to date. In 1862, Col. Francois A. Laussedat prepared plans for Paris fort using ground and aerial photo-topographic sketches. During the War of Secession in America between 1861 and 1865, photographs were taken from balloons.

The term photogrammetry came into existence in 1876.

After 1878, when collodion plates were replaced by gelatine-silver bromide emulsions, many aerial photos were taken. The earliest known aerial views of Paris were taken by Tallandier and Ducom on 19 June 1885 during the rise of a free balloon. The first aerial photograph taken from a kite is credited to E.D. Archibald, an English meteorologist (around 1882).

In 1886, in North America, aerial photos were used for topographic mapping. In 1890, A. Batut of Paris had published a book on aerial photography. Between 1854 and 1898, the Vallot brothers mapped Mount Blanc using aerial photos.

In 1897, Scheimpflug conducted aerial surveys along parallel traverses at a constant altitude with an airship. Photos taken at a regular interval covered more than half of the zones and two adjacent strips partially covered their boundaries.

Thiele, a Russian surveyor, used aerial photos for topographic purposes in Trans-Caucasus from 1898 to 1908.

On 28 May 1906, G.R. Lawrence photographed San Francisco immediately after the earthquake and the great fire with a panoramic camera.

The first photomap, dated 1908, was prepared by the Italian scientist Tardivo for archaeological investigations.

On 24 April 1909, the Wright brothers took the first motion photograph from a plane over Centocelli (Italy).

Aerial photographic technique developed during the First World War of 1914–1918. Development slowed down after the war and again continued with the advent of equipment and methods of photogrammetry, which enabled preparation of aerial maps covering large areas. An aerial survey was carried out from 1919 to 1923 for resettlement of 1500 communities. The United States Department of Agriculture (USA) launched a programme of aerial coverage in 1937.

In the Proceedings of the French Academy of Agriculture of 1919, Captain Bouché outlined the possible applications of aerial photography for agriculture: 'reconstruction of land records, details of land division, particulars of crops in a farm, inventory of trees in orchards and other plantations...Study of form and use of land according to geologic formation, surface relief, investigation of slopes, etc'. All these are visible in aerial photographs taken at altitudes between 500 and 1000 m.

In 1942, Kodak company produced black and white infrared emulsion and later colour emulsion. More than fifteen years later, false colour emulsion, now known as colour infrared (CIR), was developed.

The Army Geographic Service (SGA) up to 1940 and later National Geographic Institute ensured aerial coverage of all of France and for a number of years, overseas regions. At present, the entire territory is photographed on average every five years or less. Aerial photographs are sold at IGN.

14.4 CONCLUSION

Aerial photography has long been employed for topographic and thematic investigations for which a large number of surface maps have been prepared using aerial photographs. They are used as guide maps and essentially for delineating boundaries between map units.

The first interpretations of colour emulsions were made using aerial photographs, which constituted a step forward in the direction of development of modern remote sensing. Aerial photographs have become complementary to satellite images. The latter evidently have the advantage of a much wider field of view, while the small angle of vision of some sensors gives images close to an orthogonal projection. However, photographs even today have the advantage of better resolution than images, but perhaps not for long.

For several years satellite photographs have been acquired which can be processed and interpreted as aerial photographs, particularly with respect to stereoscopy.

Use of the term 'aerospatial remote sensing' clearly indicates the complementary nature of these sources of information.

References

Société de photogrammétrie et télédétection. 1985. Bull. SFPT, 99, Saint-Mandé, 69 pp.

Ducher G. 1985. Intérêt, exploitation et avenir des prises de vues photogrammétriques de Spacelab. Bull. SFPT, 99: 13–14.

Girard C-M, Girard M-C. 1985. Interprétation du paysage à petite échelle à partir de clichés de la chambre métrique spacelab. Approche botanique et Pédologique, Bull. SFPT, 99: 41–51.

Girard M-C, Girard C-M. 1970. Cours de photo-interprétation, Polycopié. INA-PG, Grignon, 208 pp.

Konecny G. 1985. La mission photogrammétrique Spacelab-1. Bull. SFPT. 99: 5–12.

Schuhr W, Engel H, Konecny G, Lohmann P, Schuring A, Wu J. 1984. Investigations of metric camera data quality. Inter. Archives of Photogram. Remote Sens., 25 (part A1): 64–69.

D

QUALITY ASSESSMENT

15

Scale Changes

15.1 INTRODUCTION

The term 'scale' is often used ambiguously. When a manager envisages processing a 'large-scale' problem, he implies a vast spatial field. From the cartographic point of view, such vast areas are represented on a 'small-scale' map, from 1:250,000 to 1:1,000,000 for example. Thus the term has opposite meanings. We prefer to use the term scale only in its cartographic sense, i.e., as the ratio between a distance on the map and the corresponding distance on the ground. The main categories of scales used in analog (paper) maps are given in Table 15.1.

Moreover, the concept of scale hardly has any meaning in processing of remote-sensing data since these data correspond to pixels semantically represented by digital numbers. For about a decade the question of 'scale change' has been the subject of several investigations and discussions consolidated by theses and scientific articles as well as in seminars and study reports (GSTS-CNRS, CEMAGREF, International Journal of Remote Sensing, SFPT, etc.).

Table 15.1: Major categories of map scales

Scale category	Dimensions
Local	1:10,000 or higher
Regional	1:10,000 to 1:50,000
National	1:50,000 to 1:250,000
Continental	1:250,000 to 1:1,000,000
Global	1:1,000,000 or lower

15.2 SCALE AND ORGANISATIONAL LEVEL OF MEDIA

The objects of the Earth's surface exhibit many levels of organisation, from molecule to biosphere. A level of organisation corresponds to the set of relationships and functions between objects that manifest a geometric, semantic and functional coherence, defining a new entity or new pattern. A level of perception corresponds to the tools, methods and estimates employed for detecting a given organisational level. A visual observation using a microscope (perception level) enables study of a vegetative tissue (organisational level). However, perceiving the organisation of a forest canopy requires an aerial view. The levels of spatial organisation of vegetation as well as the order of magnitude of its objects and means of perception used are shown in Table 15.2.

Organisational levels are related to the components of ecosystems, which are of two types, viz., abiotic and biotic. Abiotic components are climate, geologic substratum, hydrological regime, soil,

Table 15.2: Levels of spatial organisation of the plant world.
Each level includes properties of the lower level besides having new properties resulting from the structure of association of lower-level components

Organisational level(*), Nature of objects	Dimensions (order of magnitude)	Examples	Means of perception	
Macromolecule	10^{-9} m	DNA		
Organite	10^{-6} m	Chromosomes, leucoplast	Electron microscopy	
Cell	10^{-6} to 10^{-4} m	Cell	Microscopy	
Tissue	10^{-4} to 10^{-2} m	Parenchyma		
Organ	10^{-3} to 1 m	Leaf	Laboratory	
Individual organism	10^{-6} to 50 m	A daisy	Photography	Spectrometry
Subpopulation	10^{-3} to 10 m	A spot of daisies in a grassland	Field observation	
Natural population	10^{-3} to 10^{2} m	Daisy population in a group of grass-lands, not necessarily limitrophs	Photography	Radiometry
Plant community or synusia	10^{-1} to 10^{3} m	Grasslands on rich soil	Airborne and satellite	
Vegetation formation or biome	10 to 10^{4} m	Grasslands and pastures	observations	
Vegetation landscape	10^{3} to 10^{4} m	Agricultural plains		
Ecological region	10^{4} to 10^{5} m	Euro-Siberian region	Aerial photography	
Biogeographic region	10^{6} to 5×10^{6} m	Holarctic zone	Multispectral data	
Biosphere	4×10^{7} m			

(*) From lower to higher

etc., and biotic components are flora, fauna and man. The latter can be used as a basis for mapping. Generally, climate, geomorphology and substratum belong to small-scale features, while fauna, vegetation and soil are of large scale.

This hierarchic division of objects can be continuous or discontinuous depending on the organisational level; some changes in level correspond to thresholds or stages, as shown in Fig. 15.1. Cartographic representation of various levels of perception is done differently depending on whether a perception level is between two thresholds or at the level of a single threshold.

In the case of organisational levels for which changes occur continuously and for small ratios of reduction, changes in perception levels can be obtained without changing the number of units by moving from particular to general (note that the reverse approach of general to particular is prohibited). To change the scale of a map from 1:25,000 to 1:100,000 without changing the number of map units, we can use a mechanical method such as photographic reduction, change of zoom factor, subsampling, smoothing, etc.

Contrarily, when changes in organisational level correspond to a threshold, the number of units needs to be decreased. For this, we can proceed either by regrouping the units according to methods

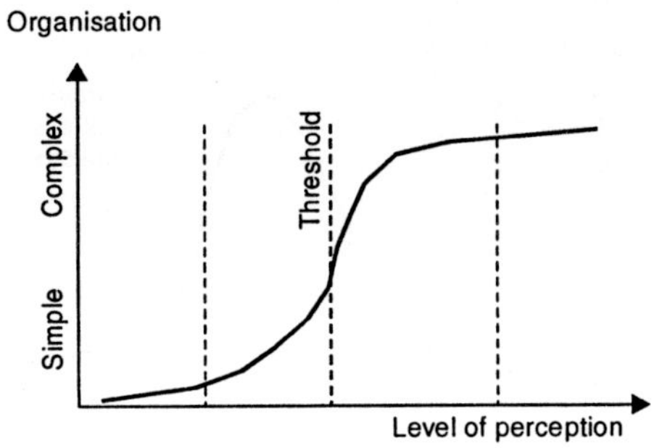

Number of units	No change	Decrease Simplification of criteria	Decrease Establishing new concepts
Transformation	Mechanical reduction	Logical generalisation	Synthesised generalisation

Fig. 15.1: Organisational and perception levels and scale changes (after Guillobez and Bertrand, 1995).

of aggregation of classes, provided organisation of the latter is hierarchic (see Chap. 8), or by decreasing or simplifying the attributes of each map unit and regrouping. Thus a logical generalisation is obtained. If the thresholds are high or many, a new interpretation has to be made based on new concepts, as in the case of changing from a thematic map to a landscape interpretation. This method necessitates an in-depth analysis, knowledge of concerned processes and description of chorological laws. Thus a synthesised generalisation is achieved (see Chap. 11).

It is therefore necessary to be cautious in using the term 'scale', which is ambiguous except when reserved for the final restitution stage of a spatial dataset. It is hence advisable to be concerned with the *quantity of information which is a function of spatial field, resolution and level of analysis* and with *levels of organisation, means of perception and precision.* These levels should be coherent with spatial resolution, level of analysis (number of attributes used for each pixel: in remote sensing, spectral bands analysed), the field (extent of study area), size and nature of objects under study. Adequacy between levels of organisation and perception is essential: cellular structure is not studied with binoculars but rather with a microscope!

15.3 SPECIFIC ASPECTS OF REMOTE SENSING

15.3.1 Resolution

In remote sensing, objects can be characterised by two thresholds of resolution (Puech et al., 1995), as illustrated in Fig. 15.2.

The *minimum resolution* characterises a given object by a given radiometric value. This threshold is reached when the ratio between the resolution r (pixel size, minimum size of observation) and the size of the object T is about 2/3, $r = 2/3\ T$. Below this resolution, the constituents of the object are differentiated but not the entire object.

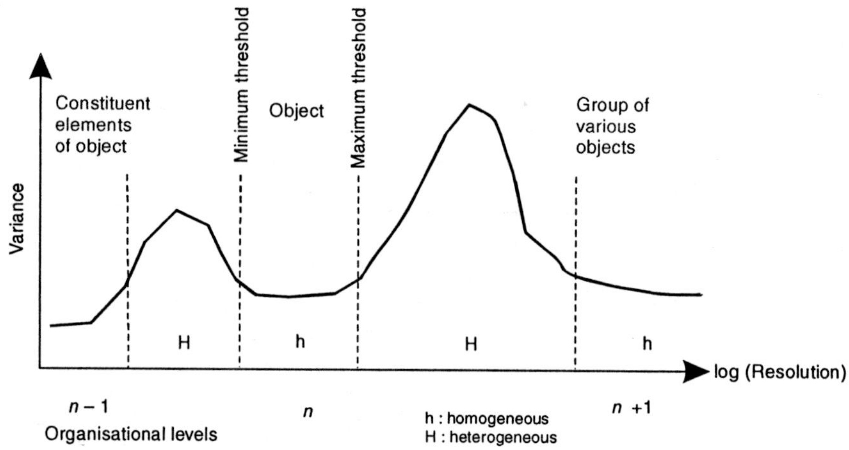

Fig. 15.2: Theoretical model relating variance of digital numbers with resolution (after Puech et al., 1995).

Thus a tree can be identified by a given resolution but, if the latter increases, the branches, leaves and shadows produced by tree tops are perceived and not the tree as an object.

The *maximum resolution* is reached when the pixel is larger than the object observed and does not permit characterisation of the latter by a precise digital value. A mixel is obtained in this case.

Thus, the maximum resolution of a tree is reached when the pixel size is such that there is a mixing between the tree and the openings or skirts.

This model explains how forest covers that appear homogeneous in LANDSAT MSS images (resolution 4424 m^2) become heterogeneous in LANDSAT TM images (resolution 900 m^2) or SPOT images (resolution 400 m^2). This type of model was shown with the software VOISIN for determining the most adequate window size (or over sampling resolution) for identification of a complex object (Chap. 11).

15.3.2 Adequacy between level of observation and level of organisation: Image segmentation

When a remote-sensing image is to be classified, it is desirable to ensure the best possible adequacy between pixel dimensions and sizes of objects to be studied. In other words, remote-sensing data most suited for a given investigation should be chosen (see Chap. 16). As a matter of fact, the basic elements to be analysed, viz., the pixels are situated between two levels of organisation rather than corresponding to one of them. Sometimes the data available with the required repetition rate are at an inadequate level of observation and it becomes necessary to go below the pixel resolution for obtaining a correct classification. Moreover, mapping, one of the most common applications of remote sensing, assumes certain conceptualisation of data as well as a minimum size of units facilitating their representation graphically. Normally, the pixel dimensions in remote sensing do not correspond to the dimensions of map units in the final maps and many mixels exist in the borders. Lastly, it may also be desirable to formulate scientific objectives of data corresponding to other levels of organisation for simulating the data before their acquisition by new sensors and preparation for use, to evaluate their possible applications.

Several solutions exist for restoring the levels of satellite observation with levels of organisation: either a pixel is resolved into its constituent elements, or pixels are combined so that they correspond to a higher level of organisation, or isolated pixels are not classified but their neighbourhood is taken into consideration.

The objective of segmentation is to divide an image into zones corresponding to the objects present on the ground, these zones serving as a base for subsequent processing. Masking (Chap. 7) is an example of segmentation. In a TM or SPOT scene covering a montane region it may be desired to segment the image into various groups according to exposure of slopes. These groups represent an observation level intermediate between the level perceived by the pixels and that of the study, which corresponds to a regional level.

The conceptual model of landscapes implied in image segmentation assumes that they constitute a mosaic of homogeneous objects, irregular in shape, of dimensions greater than a pixel. The segmentation procedure consists of minimising the intragroup distance of digital numbers of the segmented zones. It is assumed that the objects constituting a landscape have a low variance and a single value of internal variance. However, verification of the latter is rare since various land-cover categories may have very different degrees of internal variance.

This is one reason for using the maximum-likelihood method for segmentation. A joint hierarchic model can be employed for taking into consideration various levels of organisation present in a landscape. The joint hierarchic model combines on the one hand, the assumption that a mosaic of discrete objects exists, as in a homogeneous-constituents model, and on the other, the assumption that an explicit hierarchy exists that defines relationships between these objects. The hierarchic levels correspond to the organisational levels of various processes existing on the ground and to the chorological laws determined during the study.

For example, in an agricultural region, the first level may correspond to variations within a parcel (heterogeneities in the canopy) due to effects of competition between individuals and microclimatic and edaphic conditions on the latter. The attributes correspond, for example, to differences in vigour or development of aerial phytomass.

The next level will be that of parcels: plots occupied by various crops. In this case, the canopy occupying a plot is considered a unit in which intraplot heterogeneities are smaller than interplot heterogeneities. The causative factors would be agricultural calendars, phenotype characteristics, rotations and farming systems, etc. The attributes correspond to crops occupying each plot.

The next higher level corresponds to organisation of plots in space according to chorological laws based on major types of soils, geomorphology, regional climate, socioeconomic conditions, etc. The attributes would be, for example, combinations of various land covers that define landscape units.

Some examples illustrating various approaches to achieving better correspondence between organisational and observation levels and the requirements of accuracy are presented below.

15.4 RESOLUTION OF PIXEL INTO CONSTITUENTS OR DESCENDANT APPROACH

The presence of mixels in remote-sensing data gives rise to errors in pixel-by-pixel classification. In fact, mixels are characterised by digital numbers corresponding to none of the themes to be classified (mapped) or have the same values as those of pure pixels belonging to other themes. This situation may hence lead on the one hand, to semantic errors in assigning pixels to one or the other class using a distance computation for example, and on the other, to geographic errors in determining boundaries between two classes. While this boundary can be identified using filters (Laplacian or another), mathematical morphological operators (Chap. 12) or interpolation methods, the thickness of the boundary line may not be less than 1 pixel and it is not directly known in which of the two intermediate zones it is to be drawn.

Consequences of these errors in mapping can be significant, particularly when mapping is used to estimate the area of a given land cover or to monitor the diachronic variations in areal extent. To minimise these errors, it is necessary to identify boundaries of negligible thickness compared to the size of map units and the latter should be drawn as precisely as possible vis-à-vis the zones to be delineated. This assumes that mixels can be resolved into their constituents. Two types of mixels can be defined (Fig. 15.3):

—those corresponding to a boundary between two units: *boundary mixels*;

—those corresponding to a mix of constituents pertaining to two or more different units: *mixed mixels*.

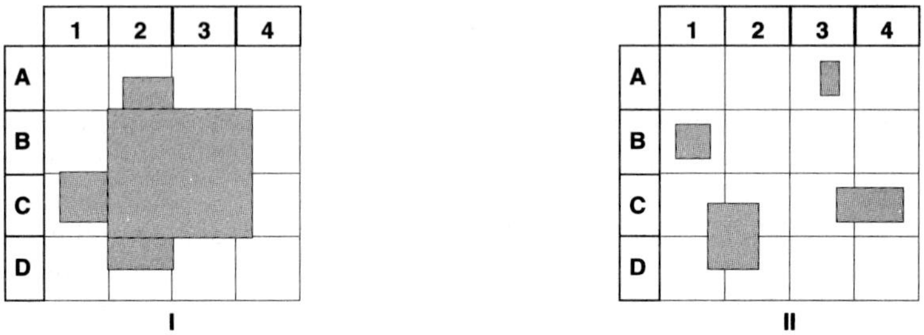

Boundary mixels	A2, B4, C1, C4, D2
White pixels	A1, A3, A4, B1, D1, D3, D4
Grey pixels	B2, B3, C2, C3

Boundary mixels	A3,B1, C1, C2, C3, C4, D1, D2
White pixels	A1, A2, A4, B2, B3, B4, D3, D4
Grey pixels	

Fig. 15.3: Types of mixels.
Figure shows two thematic units, grey and white, extending over 16 pixels marked in rows (A to D) and columns (1 to 4). I: Boundary mixels, II: Mixed mixels.

The digital value of a pixel corresponds to a double integration of geometric and radiometric parameters of radiance of various units present in it. It is hence considered that in the radiometric space corresponding to the spectral bands under analysis, any digital number of a pixel lying between the values pertaining to two pure map units represents a combination of these two units in various proportions. If the geographic dimensions of the units under consideration are greater than the geometric resolution (pixel dimension), as in the case of Fig. 15.3 (I), there is high probability that the mixels are boundary mixels. Contrarily, if the dimensions of the units are less than those of a pixel, they represent mixels comprising a mix of units (Fig. 15.3 II).

Boundary mixels can be identified by the following method proposed by Blamont and Gregoire (1994). It is based on the concept of local heterogeneity defined by the standard deviation computed from the digital values in a 3 pixels × 3 pixels window. This value is assigned to the central pixel of the window. Computation is done for the band in which variations in digital numbers of units are high, for example the infrared band (XS3).

Clusters of digital values of units are determined either from field reflectance measurements, or from zones identified in the image as uniquely belonging to one of the units under consideration. Convex envelops and major axes of these clusters are drawn (Fig. 15.4). The dividing line is represented in this example as corresponding to a composition equal to 50% of each of the two pure units. In

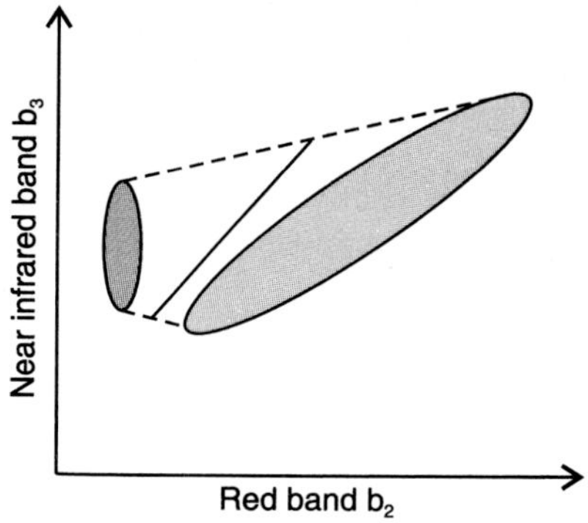

Fig. 15.4: Clusters of digital values of two pure units and the dividing line.

similarity between separation of two units in radiometric space and their geographic distribution, it is estimated that the smaller the distance of a pixel from this dividing line in radiometric space, the higher the probability of a mixel being a boundary mixel.

These pixels are then resolved according to the proportions of various units. Since the digital value of a pixel is the weighted average of radiance values of the various units it comprises, the pixel can be resolved into its constituents. For a pixel Ω of radiance L, several possible in-situ combinations (compatible mixtures) of homogeneous units Ω_k exist, giving the value L. Some examples of compatible combinations are given in Fig. 15.5.

Thus, only the value of L and of digital values of pure units need be known to determine surface areas corresponding to each. The existing pure units have first to be characterised. For this, several methods belonging to two major groups can be employed.

The first group consists of methods in which clusters of digital numbers are represented by points in $\Re^n$ cluster of k units. In this case the signal measured by the sensor is considered a simple sum of signals reflected by each component k of radiance $L(k)$. This assumes that radiance values $L(k)$ are known; supervised classification or tables of spectral characteristics can be used for this purpose. Various constituents in the pixel Ω are regrouped at the most into $n + 1$ classes (n being the number of bands). If $k = n + 1$, a unique solution is obtained; contrarily, if $k > n + 1$, infinite solutions are possible. Various statistical methods (such as analysis of textures and variations in radiance around each pixel or multitemporal regression) need to be employed to complete the information missing in such cases. Such methods assume that variability in radiance values is sufficient. Quantisation errors can be as high as 75% depending on the types of media and number of bands available.

The second group consists of methods that represent clusters of digital values of pure units by extended domains and boundaries by thresholds in the $\Re^n$ cluster of k units. These thresholds must first be defined, for which we can use mean and standard deviation, minimum and maximum values in each band, threshold indices, distances between classes, etc. The percentage of overlap of a homogeneous unit k in a pixel Ω can be computed by a measure of similarity or by calculation of distance.

The disadvantage of these methods lies in assigning 0 or 100% of the pixel to a unit, due to which heterogeneity of the composition of pixels is not taken into consideration. Only the methods based on fuzzy groups or neural networks avoid the disadvantages of binary classification of linear threshold models. These methods involve study of all possible proportions (that can be defined by a lower limit and an upper limit) of various units contributing to the value L, based on knowledge of digital value of pure units. Rather than assigning a single composition to a pixel, the degree of similarity (intensity) of its appurtenance to various units is evaluated.

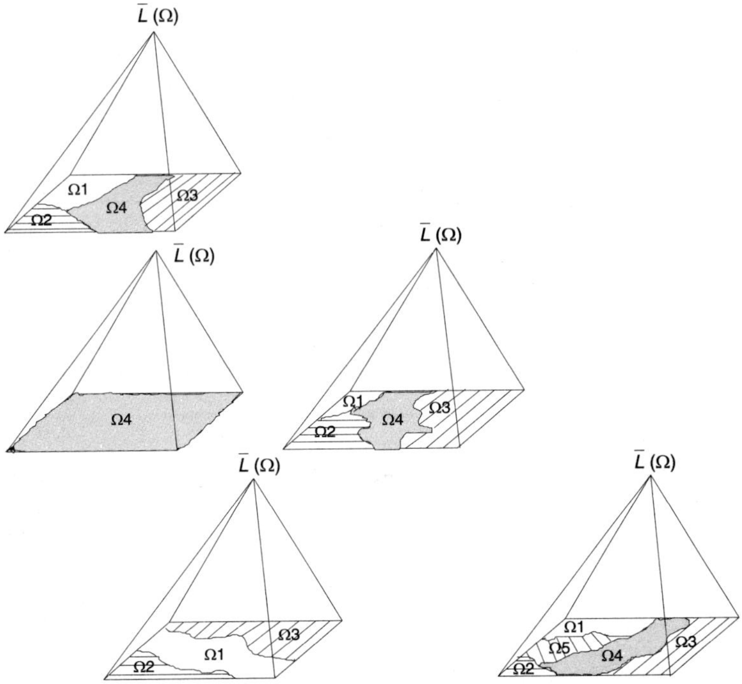

Fig. 15.5: Examples of compatible combinations for a pixel Ω of radiance *L* (after Grégoire, 1995).

Methods using fuzzy groups have the disadvantage of lengthy and time-consuming computations whereas neural network methods, once tested, are much faster. In neural networks, the difficult task is to choose a balance between various datasets used for defining subpixel compositions. The latter is done by successive approximations which minimise differences between the subpixel compositions obtained and the true subpixel compositions of a dataset.

The methods of quantisation of proportions of various units consider that these pure units are radiometrically different, i.e., each unit corresponds to a distinct group of digital values. This assumption may not always be valid when the units are defined based on field data. For example, let us consider a single agronomic unit, viz., 'permanent grasslands', for a scene acquired on day *j*; it corresponds to at least two groups with different digital values: grass plot yet to be harvested and that just harvested.

Moreover, to extract clusters of units or their proportions inside mixels from a matrix of digital numbers of pixels of *n* bands, it is necessary to have a sufficient quantity of pixels covering the entire range of possible proportions or a large number of spectral bands in which each unit is characterised by a specific cluster of digital numbers. In practice, it is most unlikely to observe in a scene several pixels with different proportions of units having constant digital values while the number of spectral bands is generally limited. One solution is to use diachronic data of the same region with a greater number of spectral bands; however, this does not solve the general problem of distinction of groups of different digital values for characterising a land cover unit.

15.5 AGGREGATION OF PIXELS OR ASCENDANT APPROACH

This method consists of generating a new dataset, corresponding to a more global level of organisation, from a given dataset (such as ground measurements, remote-sensing data of high geometric resolution, etc.).

An objective of this approach could be to obtain digital values of pure units. Fischer (1994) used as basic data field reflectance measurements over various crops taken at the rate of one measurement for every 7 days during a crop season. Individual temporal profiles of vegetation index (NDVI) computed from these data showed that they were all of the same form irrespective of the type of crops. This enabled formulation of a model of temporal variation of NDVI for homogeneous canopies. Using the temporal profile models of winter crops on the one hand, and spring crops on the other, he computed regional temporal profiles of NDVI such as would be acquired by a satellite of low geometric resolution like NOAA/AVHRR or SPOT/VEGETATION. Such data can be subsequently integrated in a model of yield estimation for example.

Another application related to change of organisational level could be to determine the spatial composition of spectral units corresponding to a higher level from an image corresponding to a lower level of organisation, classified pixel by pixel, supervised or not. A higher level unit would be heterogeneous, comprising several spectral classes derived from the preliminary classification, some with, others without thematic significance. One or several classes of the first classification could be dominant in the composition of the unit. Knowing this spectral composition of units, the entire image is analysed using a moving window whose size is chosen according to the degree of heterogeneity of the unit to be mapped (see VOISIN model, Chap. 11). Too large a window leads to excessive generalisation, while too small does not adequately take into consideration the spectral classes for characterising a unit.

It may also be desirable to simulate data to a coarser observation level from data acquired at higher resolution (MSS, TM or SPOT). These data are degraded to obtain an observation level (here, a pixel) of dimensions similar to those of a satellite with lower geometric resolution, such as NOAA/AVHRR. Degradation methods are varied. One may start with a subsampling followed by computation of mean or use a spatial filter followed by a resampling by cubic convolution. Various examples were compared by Justice et al. (1989) for 5 × 3 pixel blocks. These included choice of the value of central pixel, mean of the first 4 values in a row, mean of 15 values of pixels, mean of 4 non-adjacent pixels chosen according to a constant pattern, median of 15 values, etc. The method of using the mean has the disadvantage that pixels of very low or very high luminance are eliminated and, on the other hand, values that did not exist in the original data are generated.

A new orientation in obtaining classification results, more precise or usable in more applications, involves making an ordinal classification (dense, moderate or scarce forest coverage) in addition to cardinal (forests, crops, villages, etc.) or using non-dominant characteristics to be assigned to classes or second-choice classes (IDRISI software, for example).

In the first case, pixels of an image of low geometric resolution are compared with corresponding pixels obtained from degradation of an image of high geometric resolution which, in addition, is used for classification of non-degraded pixels. Thus, information on subpixel composition and, for example, on the proportion of a specific class inside a pixel is obtained.

In the second case, non-dominant characteristics (or second-choice classes) that can be assigned to pixels or polygons are identified. For this, we can use models of combination of radiometric values permitting determination of proportions of the various initial classes (see above) as well as neural networks or a linguistic type of notation during verification of land cover on the ground. In all cases precision of non-dominant characteristics (second-choice classes) is dependent on the precision of the initial classification and hence efforts of improvement ought primarily to be concentrated on the latter.

In all these examples, the question of exact location of pixels and their correlation between data of various sources is not resolved. The same applies to test methods and validation of results. Some authors propose ground verification not over pixels, which are difficult to locate in situ, but over polygons. In such cases, one should be cautious in evaluating the precision of interpretation results and, in particular, in computing the Kappa estimator (see Chap. 19).

References

Blamont D, Grégoire C. 1994. Données télédétectées et précision de la cartographie des limites intra-pixellaires. Bulletin SFPT 137: 98–102.

Fischer A. 1994. A simple model for the temporal variations of NDVI at regional scale over agricultural countries. Validation with ground radiometric measurements. Int. J. Remote Sensing, 15: 1421–1446.

Girard C-M. 1995. Changements d'échelle et occupation du sol en télédétection. Bull. SFPT, 140: 10–11.

Gregoire Himmler C. 1995. Étude de l'hétérogénéité sub-pixellaire des milieux naturels observés par radiométrie multispectrale: application à la modélisation de l'état et de l'activité de la végétation. Thèse Dr. Université Louis Pasteur, Strabourg 1, 145 pp. + ann.

Guillobez S, Bertrand R. 1995. Cartographie et changement d'échelle, le point de vue du naturaliste, propositions d'applications en cartographie informatique. Bull. SFPT, 140: 8–9.

Justice CO, Markham BL, Townshend JRG, Kennard RL. 1989. Spatial degradation of satellite data. Int. J. Remote Sensing, 10: 1539–1561.

Puech C, Doumerc F, Lieutaud A. 1995. Identification des objets selon l'échelle: apport des outils de SIG. Bull. SFPT, 140: 18–19.

Woodcock CE, Strahler AH. 1987. The factor of scale. Remote Sensing of Environment, 21: 311–322.

16

Criteria of Choice for the User

The important questions faced by the user of remote-sensing data are summarised in this chapter. The various chapters of this book in which these questions are discussed in detail are referenced.

16.1 WHAT DATA TO CHOOSE?

Development of operational applications of remote sensing demands results that satisfy certain quality criteria. But before reaching this stage and even before commencing a study, it is imperative to ask the following four questions and obtain clear answers.

16.1.1 What is the question posed and the problem to be solved?

Some proposals do not clearly indicate the nature of the problem, which has implications on the nature and pertinence of the methods used for solving the problem. For example, is the objective of study to map flooded zones or to detect zones likely to be flooded every year?

16.1.2 Are remote-sensing data apt to answer the problem?

Some proposals specifically mention use of remote-sensing data although it is not fully justified; *a contrario*, this is not because these data are not foreseen that they may not be useful. It must be clearly understood that remote-sensing data cannot replace the field work but most often enhance the value of the latter by making it more effective (aiding in classifying a region into homogeneous zones, marking zones in which a phenomenon may be generated, selection of control and validation points, spatialisation and generalisation of limited data, etc.).

16.1.3 What are the data acquisition conditions most likely to provide an answer?

The characteristics of existing data (platform, sensor, spectral resolution, geometric resolution, date of acquisition of scenes, etc.) and technical (acquisition, processing) and thematic (knowledge of phenomena or objects to be detected and mapped) information must be identified. This assumes a good understanding of the problem to be investigated, in addition to acquaintance with remote sensing.

16.1.4 Are remote-sensing data responding to the conditions defined and problem posed available?

To ensure availability of the required data, organisations concerned with acquisition of aerial and satellite data need to be consulted. If their response is NO, a special acquisition may be requisitioned if possible in view of time and financial constraints; if the answer is YES, the most suitable data must be chosen.

Answers to these questions are facilitated by the Centre of the Earth Observation (CEO) established by the European Union to assist individuals and organisations in the use of remote-sensing data and to provide information concerning types of data available, costs, dissemination policy, presentation of service societies, etc. The same applies to developing contacts through the Internet (see principal addresses in the annexe): for example, consultation of image catalogues of SPOT and DALI, now called SIRIUS (see the text on SPOT satellite in CD).

16.2 CRITERIA FOR CHOICE OF REMOTE-SENSING DATA

16.2.1 Choice of spectral and geometric resolutions

Choice of spectral and geometric resolutions must be made according to the type and phenomena to be detected (see Chap. 15). For this it should be remembered that a definite relationship exists between organisational level and observational level of a phenomenon. For example, attempts to detect phenomena associated with pigment modifications of vegetative tissues (chlorosis, effects of diseases or parasite attacks, etc.) in spectral bands other than the visible will be ineffective unless a relationship exists between such modifications and other characteristics of tissues, such as water content. In the latter case, reflective middle infrared or thermal infrared data may be useful. However, it should be noted that if the relationships between parameters detected by remote sensing and phenomena studied are more complex, the results will be less precise.

On the other hand, it should also be remembered that organisational levels are characterised by thresholds. Choice of data of higher resolution is not always the best solution for identifying a phenomenon. For example, a sparse forest canopy, which appears homogeneous in LANDSAT MSS data, will look very heterogeneous in SPOT data, making its mapping difficult. In the case of MSS, the pixels exhibit the combined effect of radiance of treetops and lower storey vegetation, while in SPOT images the pixels represent either treetops or lower storey vegetation.

If there is a choice between several datasets, it is advisable to take images with the desired geometric and spectral resolution but with a large field of view in accordance with the phenomena to be studied. This avoids or reduces mosaicking and permits replacing the phenomenon in its regional context. This possibility of analysing a phenomenon in a vast geographic field is often advantageous and validates the data for a wider range of applications. For example, wetlands over a large region can be mapped (see Chap. 21) more readily with LANDSAT TM images (scene covering 185 × 185 km) than with SPOT data (60 × 60 km scene), while SPOT images or aerial photos can be used for more detailed studies of small territories.

16.2.2 Choice of date of acquisition

Date of acquisition must be selected so as to facilitate ready detection either because the phenomena are directly observable or because those not directly observable can be interpreted using simple relationships. In order to study various soil phenomena it is preferable to choose, if possible, seasons in which soils are bare so as to take advantage of differences in their radiance values, permitting direct interpretation of their surface states. Contrarily, a period in which differences in vegetative growth manifest is useful for analysing physicochemical characteristics of soils such as earlier flowering in well-drained soils rapidly warmed up in spring, difficult sprouting of young plants, aqueous stress of crops in shallow soils, chlorosis over carbonate outcrops, etc. If variations in water regime of soils are of interest, a dry period after rains should be selected. If crop inventory is to be prepared, dates corresponding to various physiological states and phenological stages should be used, which necessitates good knowledge of modes of growth and development of species and agricultural calendar. Quality of results will be better in the following situations:

—*with 2 or 3 dates rather than one*, especially when studying phenomena exhibiting large variability such as crop identification, differentiation of diverse categories of permanent grasslands, processes of floods or drying of soils, or when evaluating temporal variation of objects on the terrestrial surface (magnitude of deforestation or development of towns, etc.);

—*when acquisition is made in key periods for a theme*, such as flowering of rapeseed or sunflower plants, ripening of cereals, facilitating their differentiation from other crops. Differences in dry state of bare soils can be detected after a rainy season; differences in thermal exchanges between various land covers become visible after a period of frost or snowfall when it is melting. As much as a thick layer of snow hinders observation of soil surface, that much a thin layer provides useful information on variations in thermal inertia or microclimatic conditions.

When diachronic data are employed it is preferable to use, if possible, data from a single satellite and (in the case of SPOT) acquired at not too different inclinations, so that geometric and radiometric differences (and hence corrections) are minimal. Radiometric correction is imperative if classifications are undertaken with bands corresponding to various dates. In this case, data can be calibrated on targets considered invariant (same reflectance throughout the year) such as water bodies, concrete, pure conifer forest, etc. *Caution: radiometric and some geometric corrections (to ensure superposition) lead to modifications in radiometric values and influence thematic classifications undertaken subsequently.*

Moreover, natural radiometric changes (physiological states and phenological stages of vegetation) between data of different dates must be taken into consideration in classification of land cover modifications, even when corrections related to data acquisition conditions such as sun's height and atmospheric transparency variations have been applied previously. In fact, seasonal variations of the same vegetation canopy may be represented by larger radiance variations than those due to replacement of a vegetation community by another. Use of scenes acquired on the same day cannot completely eliminate errors due to interannual climatic variations (drier or cooler years). Utilisation of the mean of radiance values obtained from a large number of scenes acquired on different dates within a year enables elimination of this seasonal effect and the influence of date of acquisition on detection of changes, but costs of study increase considerably. We feel that it is preferable to compare the results of classifications undertaken for each date separately rather than obtain a classification from bands corresponding to different acquisitions.

For some applications it may be interesting to combine satellite data from various sources for enhancing the geometric and spectral accuracies (resampling of LANDSAT TM images with SPOT panchromatic images, combination of ERS-1 and LANDSAT TM or SPOT data). However, the radiometric values of pixels are modified and the resultant digital values cannot be interpreted easily.

16.2.3 Choice of bands

Choice of bands is possible only if several types of data are available for the zone under study:

—to study *phenomena concerning the terrestrial surface* the visible and infrared (near and reflective middle) bands can be recommended;

—for *subsurface phenomena*, spectral bands characterised by penetration (microwave frequencies) or covering a certain volume (thermal infrared) should be chosen.

When multiple spectral bands are available for a given scene, bands with *high dynamic range for the phenomena to be detected* must generally be used. For this, it is necessary to analyse histograms of values for the training zones identified. It is advisable to retain only those training zones that correlate least with one another to obtain maximum information.

Correlation coefficients between bands are not given once for all. They vary according to the date of acquisition. For example, correlation matrices for a SPOT scene of the western Paris Basin (KJ 39/250) of 11 March 1995, for a 900 × 1000-pixel segment of the same image and for a scene acquired on 26 July 1995 over the same zone are shown in Table 16.1. A stronger correlation between band XS3 on the one hand, and bands XS2 and XS1 on the other, is observed for March when there is much bare soil and non-chlorophyllian vegetation, than for the image segment and for the scene of June when relatively more chlorophyllian vegetation is present.

Lastly, presence of a mask changes the correlation matrix of a scene or a segment relative to that of unmasked images but the influence of masking is more or less significant depending on the number of pixels involved (Table 16.2). In fact, for the segment of the preceding SPOT scene masking of pixels for which the digital value in XS3 is greater than 101, corresponding to chlorophyllian vegetation (3% pixels in the image segment), modifies the correlation between XS3 and bands XS2 and XS1. On the other hand, masking of pixels with values in XS3 lower than 14, corresponding to water bodies (0.1% pixels in the image segment), does not change the correlation matrix.

It is hence advisable to compute the correlation matrix between bands before undertaking classification, especially in the case of programs in which the order of indication of bands influences the result of classification.

Caution: A classification will not necessarily be better when the maximum possible number of bands is used. In fact, adding one or two more bands, if they are well correlated with the bands already taken, has risk of bringing more noise than signal. Classification in this case is likely to be less satisfactory than with a limited number of uncorrelated or less-correlated bands.

Table 16.1: Correlation matrix between bands XS3, XS2 and XS1 for two SPOT scenes and an image segment derived from one of them

	Scene A (11-03-95)			Scene A segment			Scene B (26-06-95)		
Bands	XS3	XS2	XS1	XS3	XS2	XS1	XS3	XS2	XS1
XS3	1			1			1		
XS2	0.49	1		0.23	1		0.20	1	
XS1	0.69	0.95	1	0.30	0.97	1	0.44	0.94	1

Table 16.2: Influence of a digital mask as a function of size

	Scene A image segment			Masked image segment XS3 > 101			Masked image segment XS3 < 14		
Bands	XS3	XS2	XS1	XS3	XS2	XS1	XS3	XS2	XS1
XS3	1			1			1		
XS2	0.23	1		0.31	1		0.23	1	
XS1	0.30	0.97	1	0.36	0.97	1	0.30	0.97	1

For example, addition of TM7 to bands TM5, 4, 3 and 2 for classification by the maximum-likelihood method under the Gaussian assumption of training zones corresponding to 10 grassland units, results in:

—an increase in number of well-classified pixels for 3 units only,
—an identical number of well-classified pixels for 3 other units,
—less number of well-classified pixels for 4 other units.

16.3 CHOICE OF TYPE OF PROCESSING

16.3.1 Visual or computer interpretation of data

The first choice to be made about processing is between visual or computer interpretation. In many cases visual interpretation is faster and less expensive, especially when units are identifiable by shape or textural environment. Contrarily, this type of processing ought to be carried out most rigorously so as to be most objective and unbiased. This assumes use of standardised description keys (see Chaps. 5 and 18) and, if possible, verification of results by various persons. For example, in the collaborative program of the European Community on environmental information, the 3-level CORINE Land Cover mapping was carried out by visual interpretation of IRC satellite images, based on a 44-item nomenclature for which the content of units is described precisely (see Chap. 19). This visual interpretation was, of course, accompanied by additional field verification, imperative for identification of some units involving definition of land-use concepts.

Choice of visual interpretation instead of computer interpretation also depends on the quantum of data to be processed, the precision expected for the final map and the time and financial resources available. It is sometimes preferable to be equipped with a good photographic laboratory rather than information systems necessitating air-conditioning, regular power supply and maintenance, which may not always be available in all countries.

16.3.2 Supervision of digital classification methods

■ Choice of main methods

When units of homogeneous composition are to be classified, digital methods of classification pixel by pixel are chosen (see Chaps. 7, 8 and 9). If the units are complex, it is advisable to use methods based on analysis of pixels neighbourhood, such as CLAPAS, OASIS, PAPRI, VOISIN, etc. (see Chap. 11), or a plot-wise classification.

■ Selection of useful part of image, level of precision and mask

A classification may not be concerned with the entire image. Preliminarily, the image segment corresponding to the zone of study must be selected. If the result of classification has to be superposed on a map or another classification result, geometric restitution (see Chap. 13) must be carried out *after and not before* the classification, since usually restitution modifies the digital values of pixels, leading to errors in interpretation.

The desired level of precision should also be defined. In a number of cases, for example classification of regional units, it is not strictly necessary to classify the data with full resolution; a sampling of every 2 or 3 pixels is better. Likewise, the number of groups for classification depends on the subject of study. It is not necessary to identify groups outside the theme. It is hence essential to clearly specify before undertaking classification the objectives pursued.

Before commencing a classification on a given theme, it is recommended to mask the known areas not pertaining to the theme of interest. This avoids subsequent confusions and errors in classification and permits full exploitation of the discriminative power of the method for the single theme under consideration. Several masking procedures (see Chap. 7) can be employed according to the characteristics of the objects to be masked and the remote-sensing data available. If a mask obtained from exogenous data such as maps has to be applied, the mask must be geometrically rectified vis-à-vis images and not vice-versa. Otherwise, the digital values of images may be modified and ultimately give rise to large errors in the results of classification.

■ Adequacy between statistical characteristics of data and classification method

In digital classifications, attention should be paid to the characteristics of the spectral bands and population of the data to be processed. In fact, any statistical method applied beyond its domain of utility leads to errors: Gaussian assumption for a non-Gaussian population, uncorrelated bands for correlated bands, etc. The data may have statistical parameters corresponding to none of the statistical hypotheses of the classification methods. It is recommended in such cases to undertake other possible processing techniques (e.g., generation of new uncorrelated bands) or, in their absence, apply a processing method whose structure is closest to that of remote-sensing data, keeping in mind that this may be a source of errors.

■ Selection and verification of training zones

Training zones and nuclei necessary for application of the supervised classification (see Chap. 9) ought to be chosen in sufficient sizes (at least 30 and preferably more than 100 pixels) for statistically characterising the units they represent. This is not always feasible in view of relative dimensions of objects vis-à-vis those of pixels. For this reason, it is often necessary to use several polygons (or zones) for characterising a nucleus or a training zone. An attempt is made to at least isolate the *most homogeneous* pixels inside the boundary of a nucleus, eliminating boundary pixels.

Subsequently, during analysis of the confusion matrix of a supervised classification, the percentage (or number) of well-classified pixels in the category which each is considered to represent is determined for each training zone. It is necessary to repeat the classification by modifying the number and characteristics of training zones if the latter are not well-classified to 70%; otherwise, during the final classification there would be many unclassified points and large errors of affectation (see Chap. 17). In this case, it is necessary to:

1) verify that no error is made in identification of the zone or its boundaries (training zones ought to have as far as possible homogeneous digital values or, for structural classification, similar composition vectors);
2) remove small-size training zones that are not or, are poorly classified (less than 70% pixels in the predicted class), and repeat the classification.

■ Validation of classification

At the end of classification, the results need to be validated. Validation must be done on a set of data other than that used for classification (see Chaps. 10 and 17). This can be done in two ways (see Chap. 10, Fig. 10.1):

—Conducting only one field survey for acquiring ground data prior to processing of remote-sensing data, preserving part of the data for validation. This assumes that objectives of study are clearly defined and all classes identified at the start.

—Carrying out a new field survey after data processing and collecting information for validating the results of classification. This approach is more expensive than the first but offers the advantage of verifying new classes identified during processing for which no reference data were acquired previously.

16.4 CONCLUSION

The discussion presented above shows the importance of recognising the advantages and limitations before using remote-sensing data. It also shows that it is essential to organise various phases before and after proper processing. To achieve this pragmatic approach the personnel responsible for the study must have training as a project leader (Master SILAT).

17

Quality of Interpretation

17.1 INTRODUCTION

Remote sensing has emerged from an experimental phase to an operational phase for a number of applications. So it is now imperative to furnish an estimate of quality of interpretation on infographic and cartographic documents derived from it. This aspect has been discussed in many articles, especially in works by Congalton.

The definitions presented below are taken from the project report 'Accuracy of databases of the National Geographic Institute' and a paper by David and Fasquel (1997).

—*Precision (p)*: 'The degree of agreement between a measured quantity or an estimate and the expectation (mathematical) of this quantity or estimate' (David and Fasquel, 1997). It indicates fluctuations of a series of measurements around their mathematical expectation and is denoted by the standard deviation of the mean of measurements. It is represented in Fig. 17.1 by a double arrow, which gives an estimate of the 'diameter' of the group of measurement points. If this arrow is long, precision is low; if it is short, precision is high. It can be estimated by the *intragroup distance.*

—*Bias (b)* measures the deviation between expectation of a series of measurements and the nominal value, i.e., between the centre of gravity of the group and the expected value; viz., reference value or nominal value. In Fig. 17.1 it is represented by the distance between the centre (x) of the target (nominal value) and the centre (y) of the group of measured points. Bias is high if the distance between x and y is large and low if it is small. It can be estimated by the *intergroup distance vis-à-vis the reference value.*

—*Exactitude (e)*: 'The degree of agreement between a measurement or an estimate of a quantity and its nominal value' (IGN, 1997). It combines the two concepts of bias and precision as:

$$p^2 + b^2 = e^2$$

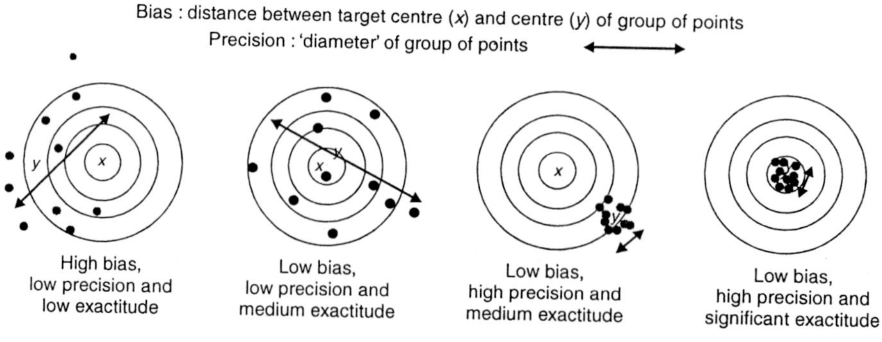

Fig. 17.1: Schematic diagram showing precision, bias and exactitude; black dots represent measurements and centre of target (point x) represents the nominal value (after David and Fasquel, 1997).

It is often denoted by the mean square error. The nominal value by definition serves as reference value. To be exact, a value must be precise and unbiased.

Irrespective of *data quality* (for which the organisation involved in acquisition and dissemination of data is responsible) and *adequacy of data* with the problem to be solved (see Chap. 16) (for which the person carrying out the study is responsible), an interpretation is characterised by geometric accuracy and semantic accuracy.

17.2 GEOMETRIC ACCURACY

The geometric accuracy of an interpretation is expressed by *geometric precision* defined as the estimate of fluctuations of differences between the nominal positions (positions in the nominal ground) and the positions obtained in the data (IGN, 1997). Geometric precision is divided into two components, one related to position and the other to shape.

17.2.1 Precision of position

Positioning precision is resolved into precision of *punctuate position* and precision of *linear position.* The former is represented by the following indicators:

—mean of errors in *x, y* and *z* co-ordinates, which gives an estimate of bias;

—regular grid of regionalised bias circumscribing the data set;

—exactitude estimated by *root mean square (RMS) error*;

The root mean square error is given for *x, y, z, xy* (planar) and *xyz*, accompanied by the corresponding precision estimates. Computational formulae for a sampling of *n* points of nominal ground (on a given date) and homologous points in the dataset are as follows:

$$RMS_x = \sqrt{\frac{1}{n}\sum_{i=1}^{i=n} (X_{i,NG} - X_{i,DS})^2}$$

$$RMS_y = \sqrt{\frac{1}{n}\sum_{i=1}^{i=n} (Y_{i,NG} - Y_{i,DS})^2}$$

$$RMS_z = \sqrt{\frac{1}{n}\sum_{i=1}^{i=n} (Z_{i,NG} - Z_{i,DS})^2}$$

$$RMS_{xy} = \sqrt{\frac{1}{n}\sum_{i=1}^{i=n} (X_{i,NG} - X_{i,DS})^2 + (Y_{i,NG} - Y_{i,DS})^2}$$

where $X_{i,\ NG}$, $Y_{i,\ NG}$ and $Z_{i,\ NG}$ are co-ordinates of a sampling point of nominal ground, and $X_{i};\ _{DS}$, $Y_{i};\ _{DS}$ and $Z_{i};\ _{DS}$ are co-ordinates of their homologues in the dataset.

The corresponding precision estimates are given by:

$$\sigma_{(RMS_x)} = \frac{RMS_x}{\sqrt{2n}}; \quad \sigma_{(RMS_y)} = \frac{RMS_y}{\sqrt{2n}}; \quad \sigma_{(RMS_z)} = \frac{RMS_z}{\sqrt{2n}}; \quad \sigma_{(RMS_{x,y})} = \frac{RMS_{x,y}}{\sqrt{2n}}$$

—sample size (n) for which each of the preceding computations is made;

—rejection threshold.

As the method of punctuate verification is not applicable to linear objects, *precision of linear position* needs to be designated by special indices:

—degree of agreement between the trace of the dataset and that of reference (after rectification of boundaries),

—estimate of planar *RMS* (in x, y).

17.2.2 Precision of shape

Precision of shape refers to the geometric elements (distances, areas, curvatures, volumes, etc.) constructed from known co-ordinates, for which the precision of planar and/or altitude positioning is also estimated. A description of estimated parameters, intensity, unit, sampling size and rejection level should be furnished.

17.2.3 Reliability

Reliability describes the statistical possibility of detection of random or systematic errors resulting from a geometric correction. We can evaluate the *uncertainty* by y pixels of the complete boundary line of a class or of a segment of the line for x pixels in width. It depends on the compositional heterogeneity of pixels and hence partly on the pixel size and partly on the degree of diversity of land-use types inside a pixel. The methods of resolution of pixels into their constituents (see Chap. 15) seek to reduce this uncertainty.

It is advisable to give for an image the reference points of ground control, their number, type of transformation applied, standard deviation, mean square error and, if necessary, the method of resampling.

17.3 SEMANTIC ACCURACY

17.3.1 Definition

Semantic accuracy refers to the attributes of classes. Four parameters can be distinguished to characterise errors concerning a theme.

■ Nature of classes

Classes can be differentiated as qualitative in nature corresponding, for example, to various land covers (forests, crops, grasslands, towns, water bodies, etc.) and of ranked qualitative nature (very dense forest, dense forest, sparse forest). Errors concern affectation of a point in a classification to a category other than that to which it belongs on the ground or in a reference map. In the first case (*grasslands* instead of *forest*), the nature of confusion or error differs from that of the second (*very dense forest* in place of *dense forest*), since it concerns very different objects. It will be seen later that magnitude of affectation errors depends not only on the nature of classes but also on the ultimate use of classification.

■ Frequency of affectation errors

This parameter refers to the number of affectation errors in a class or group of classification. How to estimate this frequency from a error matrix will be described later.

■ Importance of affectation errors

Some affectation errors are generated from a greater mistake and may result in more serious consequences than others. For example, mistake between a *coniferous forest* and a *water surface* (due to low radiance levels in the near infrared band in both cases) will be greater than that between a *pure coniferous forest* and a *mixed deciduous and coniferous forest*. The latter example illustrates the problem of defining class themes (establishment of typology), presented in the next section (17.3.2).

■ Source of errors

Affectation errors may be related to characteristics of data (undiscriminating spectral bands for the theme studied, pixel dimensions inapt for the class themes, etc.), independent of the intrinsic quality of data, to acquisition conditions or to classification algorithms used (see Chaps. 7, 8 and 9).

■ Precision

Semantic precision is differentiated into user's precision and precision of attributes.

Semantic precision is 'the degree of conformity of values of units of data with those of their homologues in nominal ground' (National Committee on Geographic Information, 1993). It represents the degree of agreement between a class assigned to a pixel in the classification and its true class known from ground or other reference sources. For example, classifying a *permanent pasture* in the theme *grasslands* will be less accurate than its classification in *permanent grasslands* but more accurate than classification as *agricultural land*.

User's precision corresponds to estimation of accuracy of classification for a given usage. In the preceding example, in an inventory of grazing areas, user's precision will be low when it is classified as an *agricultural land*.

Precision of attributes indicates the degree of agreement between various class themes derived from a series of identifications of a single entity. This is designated as *percentage affectation error* in the error matrix.

17.3.2 Establishment of typology

This phase should commence even before undertaking classification. In fact, classes to be used should respond to the objective identified or the terms of contract of study. The main difficulty arises when the themes to be mapped do not correspond to identifiable physical or spectral characteristics. For example, unique spectral discrimination between permanent meadow and herbaceous and mowed tracks of an aerodrome is not possible, since objects here are to be differentiated not in terms of constituents but their use.

For pixel-by-pixel classification, any typology must be exhaustive and the classes mutually exclusive: a point should be classified in *one and only one class*. This poses a problem for boundary pixels which may either be left unclassified or regrouped in classes in which they are incorrectly classified. This condition does not apply to classifications based on fuzzy logic for which the points can pertain to two or more classes and in which second-choice affectations can be used for delineating map boundaries.

A hierarchic typology between classes is desirable because it enables regrouping of classes identified on the ground but corresponding to a higher precision level than that required for the study.

Thus classifications corresponding to various precision levels can be obtained and associations between classes established.

It must be ensured that the precision required for a study corresponds to the precision level of the remote-sensing data used or the precision achievable on the ground using existing observation methods and/or measurements.

17.3.3 Establishment of error matrix

An error matrix is established between the data produced from classification and the reference data (*independent of those used for preparation of the document to be tested*). A number of questions arise immediately.

■ Collection of reference data

Reference data can be acquired either from ground truth measurements or aerial photographs or other remote-sensing data different from those used for classification; even maps obtained by other methods may be used for this purpose. Collection of reference data may be easy as in identification of various land-cover types or complex as in measurement of soil surface roughness, chlorophyllian biomass, foliar indices, tree heights, density of plant covers, etc. This assumes in all cases that the reference data are *precise*. However, it is not always easy on the ground to measure classes of density of tall trees in forests.

Moreover, it is sometimes difficult to locate reference data for estimating accuracy. An example of zoning of landscape units into various organisational levels in Normandy can be cited. No map document corresponding to this problem existed earlier. Neither ground data nor aerial photographs were useful for investigating landscape units. No significant statistical data existed in the form of maps. Only semantic accuracy but not cartographic could be estimated using studies of rural geography on the one hand and rural economy on the other.

■ Number and size of samples

While thousands of reference points can be readily obtained in satellite data, it is very time consuming and expensive to verify these points on the ground. The number of points to be sampled depends on the sampling procedure, desired precision of estimation, number of pixels present in a category and estimated precision of classification. Hence the smallest number of ground verification points that enable statistical analysis should be chosen. In other words, an average number that is statistically correct and practically feasible should be selected.

Points that are geographically representative of the data analysed are chosen as reference. A random stratified sampling can be considered but a random choice of reference pixels leads to high costs for acquisition of information on the verification pixels (field trips, inaccessibility, etc.). Moreover, precise location of points of verification vis-à-vis image pixels is often difficult.

To obtain a confusion matrix, a sufficient number of points should be used in each category, totally ignoring how the confusion matrix will be distributed among various classes. This necessitates a number much higher than that indicated by equations based on binomial distribution or normal approximation of binomial distribution. These methods are not appropriate for estimating the number of control points since they compute the number of samples necessary for assessment of global precision or, at best, assessment of global precision for only one class. The number of control points is hence to be determined based on experience and practice. Congalton (1991) recommended 50 points. Our personal experience suggests that *at least* 30 correctly classified control points per category (along the diagonal) are necessary, i.e., taking poorly classified points into consideration, in all about 40 points per class. These numbers, of course, correspond to *independent* samples. If such is not the case, the number of control points must be increased. More than 250 reference pixels are necessary for estimating the mean precision of a class at about 5%. If the number of reference samples is less, the confidence interval of the precision obtained increases considerably. This means the number of points is sometimes limited simply because the unit is least represented in the zone of study and it is not possible to find a sufficient number of points.

It is not compulsory to have the same number of control points in each category, provided a minimum number of samples is available for each. It is recommended that a larger number of samples be taken for important classes in the study and fewer for those of secondary interest. A small number of samples can be taken for classes exhibiting a low variability or lower risk of confusion, as shown by analysis of two- or three-dimensional histograms (see Chaps. 4, 7, 8 and 9).

On the other hand, the size of control sites ought to be compatible with the resolution (pixel) of remote-sensing data, which varies from a few decimetres or metres to several kilometres (see Chap. 2). This is in contrast to existing methods of measurement or assessment commonly used in field, often point-by-point measurements. These need to be modified (or redesigned) according to the remote-sensing data used. For example, an areal sampling (distributed randomly, systematically or along strips) is preferable to a point-by-point sampling, and measurements in extended areas rather than at points.

While CORINE Land Cover type (see Chap. 19) categories can be readily described, measurements of Leaf Area Index over areas comparable to those of a SPOT pixel are very difficult. In the latter case, other methods such as optical densitometer measurements should be employed! Among the solutions adopted are analysis of optical density of layers sensitive to red or near-infrared in aerial photographs acquired at low altitudes or that of a traverse with a SPOT Cimel simulation radiometer using a ULM. In many cases, small size ground sampling must be compensated by a large number of repetitions. For example, the Cimel radiometer of SPOT simulation measures, at a height of 1.5 m, the reflectance of a circular area of about 7 dm^2. Spectral characterisation of a plot of more than a hectare requires measurement of at least 100 different points taken randomly.

As a general rule, it is better to make an assessment of classification precision with a sufficient number of reference pixels. In the absence of sufficient points, a less precise assessment is better than no assessment at all!

■ Sampling mode

The mode of sampling is an important part of selection of control points, since the confusion matrix must be representative of the entire image. The choice is between random and stratified sampling. Stratified sampling consists of preliminary zoning of the region according to a theme (geomorphological units, separation of marine region from terrestrial, etc.), followed by random or systematic sampling inside these zones. Both have advantages and disadvantages.

—*Random sampling* is amenable to use of conventional statistical methods. However, it is not always easy to employ since it may lead to choice of inaccessible points in montane regions or zones with little communication network. Moreover, there may be risk of straddling points in two units or of a preferential sampling in large size units of low thematic interest and an underestimation of small size units of high thematic importance. These units can be taken into consideration only by significantly increasing the number of points, which involves more time and additional expenditure. Lastly, this type of sampling can be conducted only after completion of classification, in other words, often too late for locating control objects in the field (harvested crops, various drainage conditions, etc.).

—*Stratified sampling* has the advantage of distributing samples in various classes of thematic interest. It offers the same disadvantages as the preceding method if points inside strata are sampled randomly. This disadvantage can be avoided by systematic sampling. Contrarily, there could be risk of samples being no longer independent or of introducing bias in some stratified tests (see below, use of Kappa statistic in analysing a error matrix).

The most efficient solution consists of combining random, systematic and stratified samplings. However, caution is needed in the statistical methods employed since they may be beyond the region of validity. Lastly, common methods of assessment are based on comparison of pixels with control points, and not image zones with zones on the ground.

17.3.4 Discussion of error matrix

■ Elements of error matrix

An error matrix comprises the following components:

—*in columns*, data derived from classification,

—*in rows*, data of reference classes (ground truth or other data).

Overall accuracy denotes the number of correctly classified individuals (or pixels) (correlation between groups and classes) divided by the total number of individuals (or pixels).

Error of exclusion, omission errors computed along columns, corresponds to the distribution of pixels of a group, derived from interpretation of remote-sensing data, in the various classes of reference data.

User's accuracy is the percentage of individuals (or pixels) in a group derived from classification, correctly classified vis-à-vis the reference data. It is equal (in %) to '100% - error of omission'.

Error of inclusion, commission errors computed along rows, is the distribution of a reference class among various groups derived from interpretation of remote-sensing data. It differs from error of omission of error matrices established for vector data (David and Fasquel, 1997). In fact, in the latter the omission error denotes absence of data, whereas for an error matrix of a thematic classification, the commission error concerns existent data whose identification is erroneous.

Producer's accuracy corresponds to the percentage of individuals (or pixels) of a reference class correctly classified by remote sensing; it is read along the rows. It is equal (in %) to '100% - error of commission'.

An error matrix controlling a classification should always be given for each class, along with risks of confusion with any other class.

■ Precision estimator

The precision estimator, known as the Kappa coefficient, varies from 0 to 1; it accounts for errors in rows and columns (Congalton, 1999). It enables global assessment as well as at each class level and accounts for two types of errors, viz., omission (deficits) and commission (excesses) (Table 17.1).

$$Kappa = \frac{\left| N\sum_{i=1}^{r} X_{ii} - \sum_{i=1}^{r} (X_{i+}X_{+i}) \right|}{N^2 - \sum_{i=1}^{r} (X_{i+}X_{+i})}$$

where r is the number of rows and N the total number of observations.

It should be remembered that this formula applies to *random sampling* and *comparison of pixels* but not of polygons of control points, i.e., only to independent samples in classification. In the case of a systematic or non-aligned systematic sampling, the resultant bias for Kappa is small (it may hence be used) whereas contrarily the variance estimator is biased. In the case of stratified sampling, another estimator needs to be defined. Stehman (1996) proposed a KS estimator for stratified sampling, provided the stratification pertains to regions covering several classes and not to classes.

Table 17.1: Notation of an error matrix for computation of Kappa statistic

		Classification			Total lines
		Unit 1	Unit i	Unit n	
Reference	**Unit 1**	x_{11}	x_{1i}	X_{1n}	N_1
	Unit i	x_{i+}	x_{ii}	X_{+i}	N_i
	Unit n	x_{n1}	X_{ni}	x_{nn}	N_n
Total columns		M_1	M_i	M_n	N

The Kappa estimator expresses the proportional reduction in error obtained by a classification compared to error obtained by a completely random classification. A value of 0.75 indicates that the classification employed eliminates 75% of the errors obtained by a procedure working completely at random.

The Kappa estimator applies to evaluation of accuracy for classifications pertaining to cardinal units. For ordinal units, it is preferable to use other estimators such as χ^2-test or F-test, which measure the degree of agreement between the true distribution and theoretical distribution of classes.

■ Example of error matrix

This example is taken from a Ph.D. thesis on Lower Normandy region. The classification by maximum-likelihood method in a LANDSAT TM image pertains to mapping of various physiognomic units of permanent grasslands identified on the ground from the height and chlorophyllian state of the canopy. Reference points were chosen after delineation of marsh zones. It thus represents random sampling after stratification.

Figures given in Table 17.2 correspond to the number of pixels, unless indicated otherwise.

Overall accuracy: 14,021 / 16,353 = 0.857

The Kappa coefficient is computed here as follows:

N = 16,353

N^2 = 267,420,609

Table 17.2: Example of error matrix

		Classification						
	Physiognomy	Tall chlorophyllian	Tall slightly chlorophyllian	Heterogeneous	Short chlorophyllian	TOTAL	Producer's precision (%)	Error of omission (deficits) (%)
GROUND TRUTH	Tall chlorophyllian	2 109	3	25	4	2 141	98.5	1.5
	Tall slightly chlorophyllian	8	4 912	376	0	5296	92.7	7.3
	Heterogeneous	874	955	2 709	61	4 317	58.9	41.1
	Short chlorophyllian	5	10	11	4 291	4 599	99.4	0.6
	TOTAL	2 996	5 880	3121	4 356	16 353		
	User's precision (%)	70.4	83.5	86.8	98.5			
	Error of commission (excesses) (%)	29.6	16.5	13.2	1.5			

$$X_{ij} = 2109 + 4912 + 2709 + 4291$$
$$= 14{,}021$$

$$\sum_{i=1}^{r} (X_{i+}X_{+i}) = \{1 \times (3 + 25 + 4)\} + \{8 \times (376 + 0)\} + \{(874 + 955) \times 61\} + \{(5 + 10 + 11) \times 1\}$$
$$= 114{,}635$$

Kappa is equal to:

$$K = [(16{,}353 \times 14{,}021) - 114{,}635] / [267{,}420{,}609 - 114{,}635]$$
$$= [212{,}932{,}413 - 114{,}635] / [267{,}420{,}609 - 114{,}635]$$
$$= 229{,}170{,}778 / 267{,}305{,}974$$
$$= 0.857$$

The Kappa coefficient and overall accuracy are equal.

A good accord is generally seen between the map obtained from classification of satellite data and ground truth observations. However, some classes contain more errors than do others; this is particularly so for group 'heterogeneous', which has accumulation of errors of commission (13.2) and errors of omission (41.1)[1]. The errors of commission consist of pixels corresponding on the ground mainly to the class 'tall and slightly chlorophyllian' and affected to the class 'heterogeneous'. On the other hand, the errors of omission consist mainly of pixels corresponding on the ground to the class 'heterogeneous' and affected by classification to the class 'tall and slighty chlorophyllian'. These confusions and the low producer's precision for the class 'heterogeneous' might necessitate revision of definition and modification of test-zones of reference or elimination of some. In fact, a better adjustment of digital number values in thematic classes may produce better precision.

A confusion matrix controlling a classification should always be given for each class, along with risks of confusion with any other class.

Contingency tables (error matrix) for a complete real case study will be given in an example of CORINE Land Cover (see Chap. 19).

17.3.5 Limitations of conventional methods of accuracy assessment

The conventional methods of classification accuracy assessment have the following defects:

—It is assumed that each point (of classification, such as a control point) can be unambiguously assigned to only one theme which is subsequently compared with the theme obtained from another method. However, it is not always easy to affect a pixel to a given class or to associate a point on the ground to a specific class.

—Information regarding the significance of errors is limited to distribution of incorrect attribute of a class but does not indicate their degree of severity.

—Users of maps derived from classifications require more detailed information on thematic aspects than that provided by conventional methods of error evaluation. For example, they need a detailed analysis of confusion, class by class, or estimates of severity of errors committed for a specific application.

That is why other methods of classification quality assessment are presently being developed. These developments concern the nature, frequency, origin and significance of semantic errors and

[1] The different values of these two errors, not symmetric, should be noted.

make use of fuzzy logic and simulation studies. An example of qualitative evaluation of semantic accuracy for a classification is given in Chap. 19 concerning CORINE Land Cover.

17.4 CONCLUSION

To carry out a study corresponding to operational applications of remote sensing, it is imperative to know the quality of the data to be used and to provide a quality assessment of the product data (thematic interpretations, maps, etc.). This necessitates a preliminary study of the problem posed and a feasibility analysis of how remote-sensing data can solve it.

Ready-made solutions should not be expected from remote sensing.

No method can provide a complete analysis of a problem and only a combination of several approaches will be most constructive. Hence, remote sensing is not a means for *eliminating* ground studies but for making them more effective and validating them for spatial extension. An analysis of quality/cost ratios, computed exhaustively, enables choice of the most appropriate methods of study. Establishment of an organisation such as the Earth Observation Centre (EOC) situated in the research centre of the European Community (Ispra, Italy), which has among others improvement of seller/buyer relations and rendering advice to users as a mission, ought to facilitate this choice. In fact, EOC does not supply remote-sensing data but provides information on where to procure them. This free service concerns not only remote-sensing data but also any ground information necessary for interpretation, validation or calibration of spatial data. Internet communication with EWSE (*European Wide Service Exchange*) enables location of research organisations, societies or companies capable of answering your needs. EWSE is a temporary arrangement that will be replaced by an operational system, INFEO (*Information on Earth Observation*). The EOC ought to become in the long term an information tool for persons and organisations interested in earth observation and some of its objectives should become operational in the near future.

It must be remembered that unsuitable remote-sensing data, inappropriate classification methods and inadequate ground (or reference) information *cannot give good results.*

References

Cnig. 1993. Qualité des données géographiques échangées, Groupe de travail "qualité des données géographiques échangées", September, 22 pp.

Congalton RG, Green K. 1999. Assessing the accuracy of remotely sensed data: principles and practices. Lewis Publ., Boca Raton—London—New York—Washington DC, 137 pp.

David B, Fasquel P. 1997. Qualité d'une base de données géographiques: concepts et terminologie, Bull. d'info. de l'IGN, 67: 1–51.

Gopal S, Woodcock C. 1994. Theory and methods for accuracy assessment of thematic maps using fuzzy sets. Photogrammetric Engineering and Remote Sensing. 60: 181–188.

Stehman SV. 1996. Estimating the Kappa coefficient and its variance under stratified random sampling. Photogrammetric Engineering and Remote Sensing, 62: 401–407.

E

APPLICATIONS

18

Agrolandscapes

It is possible to interpret and locate major geographic units in satellite images with a synoptic view. Visual interpretation of these images is similar to structural processing, few developed and not common in remote-sensing softwares. The role of interpreter thus assumes great importance since he can integrate and define units whose content varies in given proportions and with a given spatial organisation, viz., patterns.

The methods of integrated analysis of satellite images and small-scale photographs acquired from aeroplanes, stratospheric balloons or spaceships are based on the concept of 'landscape' adapted to these interpretations.

18.1 LANDSCAPE INTERPRETATION

18.1.1 Concept of landscape

Several definitions exist for the term 'landscape' covering various realities and related to the sensitivities and modes of perception from space (Rougerie and Beroutchachvili, 1991; Lizet and de Ravignan, 1987). Landscape investigators, and especially those of the School of Landscape (Donadieu, 1993), emphasise on the importance of perspective view which organises the landscape elements according to various planes in the space seen by the observer. However, it is possible to analyse landscapes in a vertical view. In this case, the notion of flight point and relative position of landscape elements is lost: trees do not conceal the landscape but the lower strata and soil; topographic variations no longer limit the view.

In this book, we have followed a rural approach with the aim of management. In fact, need was felt for developing spatial models for division of a region into various units to which information aiding their management may be attached. At present, many divisions of the territory exist: small agricultural zones, forest regions, climatic zones, administrative boundaries, etc. It is not clearly known how these boundaries were drawn. Consequently, it is now difficult to know whether thirty years later and after many variations, they are still valid. Chorological laws prevailed in their constitution are not known and hence these various segmentations of the territory cannot be spatially analysed.

18.1.2 Landscape: synthetic descriptor in remote sensing

The concept of landscape (Henin, 1994) generally refers to 'an extent of land presented by nature to an observer', independent of its meaning in painting. In the definitions of the Larousse and Petit Robert dictionaries, emphasis is on the terms: nature, point of observation and extent of terrain.

With regard to the concept of landscape, landscape specialists emphasise aspects of sensitive perception, giving priority to perspective view. However, it is difficult to define a landscape object in the former case.

Deffontaines (1985) defined landscape as follows: 'portion of region viewed by an observer wherein a combination of facts and interactions are inscribed, which is perceived at a given moment only as the global result'.

We ourselves consider landscape an '*extent of land observed in the vertical direction, in which the main subject is the natural environment*'. Landscape analysis is objective, but perspective view is absent since the observer is situated above the object observed.

Landscape, as an integral descriptor of environment, is '*the entirety of components of the environment whose spatial organisation is studied (in a geographic sense): type of units, dissemination of these units over the region studied, spatial distribution, neighbourhood associations and hierarchy between them*'.

This spatial structuring is defined according to a theme such as agronomy, hydrology, pedology (see Chap. 23), geology or viticulture, etc. The theme enables choice of the most useful landscape variables for determining spatial units corresponding to them.

Thus, hydrolandscapes are defined as:

'*A combination of landscape elements: natural vegetation, timber zones, agricultural plots situated in a geomorphological unit associated with natural water flow (elementary watersheds, drainage patterns, etc.), controlled by topography (slope and exposure) and surface states of soil, whose spatial organisation permits definition of an entire (or part of) a watershed or a drainage system, i.e., hydrological and spatial structure*'.

Subgroups can be defined as hydrolandscape units and hydrolandscape elements.

18.1.3 Agrolandscape

Rural landscape is the resultant of a combination of factors of environment and human actions with regard to it. In a given spatial field, environmental factors observed at a given resolution are organised spatially and their distribution constitutes a source of information. For a given object, the main environmental components can be defined which, once modelled, give landscape units. The latter can also be analysed by seasonal or decennial diachronic iterations.

The concept of agrolandscape is based on the assumption that a spatial, but not random, organisation of environmental factors exists. Spatial organisation of a rural environment is related to its utilisation by man, and the human actions are imprinted in the geographic space. On the other hand, the environmental factors constitute the base of spatial organisation. Consequently, human actions, adapted over the course of time to environmental factors, can be identified at least partially in the agrolandscape by means of spatial analysis.

In studies concerned mainly with rural environment, agrolandscapes are defined as '*a combination of plots (agricultural or forest) and landscape elements, viz., natural vegetation, timber zones, topography (slope and exposure), drainage system, surface states of soil, whose spatial organisation enables definition of (or part of) a land: social and spatial structure*'.

Subgroups can be defined as agrolandscape units and agrolandscape elements. In each case we deal with *patterns: 'spatially organised groups of various landscape elements'*. Landscape units are units characterised by specific patterns (described by a composition vector, see Chap. 11). These are not homogeneous, but heterogeneous zones that can be described only by a combination of criteria of multiple origin based on environmental factors.

18.2 LANDSCAPE ANALYSIS

Interpretation of satellite images for preparation of landscape maps must be done keeping in view the purpose for which it will be used. In fact, such a map should be readily readable by anyone who consults it. So it is desirable to represent components of map in such a way that reader can readily identify major units which he (she) already knows or can locate on the ground.

18.2.1 Different points of vision

The point of vision of an observer (pedestrian, for example) is very different from that of a satellite (Fig. 18.1) for three reasons:

—difference in angle of observation,
—difference in field of view and hence in reference zones,
—difference in resolution.

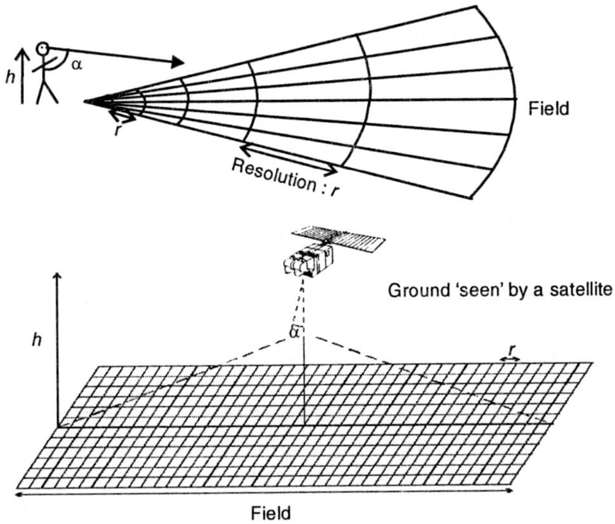

Fig. 18.1: Landscape seen by an observer and a satellite.

■ Angle of observation

An observer views a landscape with a highly inclined angle. On a plane, he (she) sees the landscape around him (her) over a distance of 1.5 to 15 km with a height of 1.5 m or at a very large angle α, of about 90° (arc cos 0.001 = 89.9°) when he (she) observes the horizon. On a mountain, the angle may decrease (arc tan 0.3 = 72°) since, for a given distance, the height may reach 500 m or more due to topographic variations. However, in general the angle of view is large. Contrarily, vision from an aeroplane or satellite is close to vertical. Rarely does an obstacle obstruct the field of view in this case, whereas a screen of trees or more or less high relief suffices to conceal part of landscape from the observer on the ground. Consequently, the point of vision differs between an observer and a satellite. On the ground, friends are recognised from their appearance; from a satellite, they can be recognised by 'implantation of their hair on the skull'!

■ Field of view

It is estimated that in ideal cases an observer can see up to about 20 km. This is due to opacity of the atmosphere which absorbs solar radiation. Nevertheless, there may be special situations when the sky is particularly clear or when contrast between objects observed is very high. Such is the case, for example, of snow-covered mountains exhibiting specular reflectance significantly differing from that of mountains with no snow.

The field of view in satellite images generally covers several tens or even thousands of km^2. Reference areas in such a case are more numerous and the landscape structure is more readily seen than from an observation point on the ground.

■ Resolution

Resolution varies according to the distance of an object from an observer. While the resolution at the foot of the observer is a few centimetres, it increases when the object is removed to a distance. An object appearing as square at the foot becomes a trapezium when it is far (Fig. 18.1). Similar effect is also observed when wide-angle photographs are taken with a camera (see Chap. 14, Fig. 14.3 a).

In the case of high-resolution satellite images (such as SPOT HRV or LANDSAT TM) the resolution is almost the same at any point of the field. A square in the image is slightly distorted into a lozenge in a map, due to the rotation of the Earth and the type of projection used (see Chaps. 2 and 13); however, variations are very small. The same is not true with low-resolution satellites and microwave systems (see Chap. 2).

18.2.2 Principal components of landscape

In satellite images, at the resolution of a pixel, various objects are acquired with their true dimensions. This is not the case with common maps in which objects we wish to emphasise are distinctly enlarged and, hence, are not in the same scale as other features on the map. Thus, on road maps, roads usually have a width of 1 mm which may correspond to 200 m, for example. Considerable distortions also occur in ground photographs due to the angle of view (see Chap. 14).

Hence, roads are not clearly detected in satellite images. But often a highway can be discerned not only because it is wide enough but also because superimposed to other components of landscape.

The main landscape component in satellite images is the land-cover unit and its use by man (see Chap. 17). Strong correlation often exists between land cover and major geomorphological units (mountains, valleys, etc.) which constitute the principal landscape components perceived by an observer on the ground. Hence, satellite images have to be interpreted to depict the following features on the final maps:

—forests,
—valleys and drainage patterns,
—relief zones,
—lots,
—bare soils,
—towns.

■ Forests

In temperate zones, forests are quite readily recognised in a satellite image when they are dense enough and the date of acquisition is carefully taken into consideration. Appearance of forests when deciduous species lose their leaves differs significantly from that during foliation period. Nevertheless, they are readily recognised although their spectral characteristics differ radically.

What can be defined as forest and what appears clearly in images is a group of trees or bushes whose organisation in layers and in density contributes to 100% coverage and reveals certain heterogeneity due mainly to shadows of tall tree canopies on adjacent ones.

It is also necessary to differentiate between forest, grove, tree alignments or riverine forests zones. This is inferred from dimension and form of blocks and their location.

In landscape interpretation, a minimum size is used to demarcate forest blocks. For the European CORINE Land Cover mapping, the minimum size used was 25 ha and 100 m in width for linear objects. In most precise cases, it is considered that at least 25 pixels are needed for characterising an object. This corresponds to a block of 1 ha with a pixel of 20-m resolution or an alignment of trees of the order of 160 m (8 pixels in length and 3 in width).

In satellite images, one can particularly differentiate organisation of forest blocks with respect to one another, their respective dimensions and distance to be covered to go from one block to another, as well as obstacles encountered in the path. Hence, possibilities of movement of various animal species can be readily determined; this leads to ecological applications of landscape analysis.

■ Valleys and drainage systems

Major rivers or streams can be detected by their reflectance if they are wide. The composition of pixels in these zones is characteristic of water. Further, a particular geometric disposition of these bodies is needed for the square grid imposed by the satellite to intercept the watercourses in such a way that connected pixels are observed, enabling delineation of the latter.

If the river is very narrow (less than 10, 20 or 30 m), for each pixel there will be a combination of spectral characteristics of water and surrounding objects such as roads, wooded strips, field, earthen banks, etc. Water will not be directly detected, since only boundary mixels occur along the rivers. This is very common for a number of watercourses. While this presents major difficulty in computer processing, streams can be more readily traced by visual interpretation since the human brain and eye can interpolate between the points where there is no water and the points of mixels. The interpreter has the capacity to take a decision of affecting a mixel to the theme 'watercourse' through study of its neighbourhood and according to pre-existing and memorised models. Thus, tracing meanders is very easy by visual interpretation but much more difficult by software.

Tracing valleys is a still more complex operation but can be readily achieved by visual interpretation if care is taken to clearly define the 'valley' landscape. Delineation is based on the fact that the width of the zone is much smaller than the length. Moreover, the relief in such zones is low and, when it exists, its traces can be observed in the land cover. Thus valleys often comprise grasslands, sand pits, riverine forests, poplar plantations, water zones associated with water table, etc. These neighbouring objects and their distribution characterise a valley. Roads are most often outside of water, similar to railway tracks, or these communication lines are bordered by excavations and fillings, which can be interpreted from their shadows or from large brightness of mineral products of filling. Agricultural lots also often show some relationship with valleys: in many major valleys, plots are perpendicular to the watercourse so that all plots are assured access to the river (irrigation, drinking water for animals, pisciculture, etc.). In major river-beds, plot boundaries of grasslands take the form of meanders and are circular (see Chap. 5, Fig. 5.4). All these features enable ready visual demarcation of valleys.

When there are terraces, it is not always easy to identify them since the relief is flat and not seen on images. Nevertheless, they are differentiated because edge of a terrace runs across a wooded border or because plots of various terraces are not arranged identically. Often, the dimensions of plots are not equal and their size may be related to the width of the terrace or differences lie in diverse orientations of plots.

Interpretation of smaller valleys is more difficult. As in an aerial photograph, adequate detail is not available in stereoscopic views for identify whether the bottom of a valley is flat or curved or whether alluvial fan exists or not. Asymmetric valleys cannot generally be delineated, except through size of plots. Lots are commonly used to interpret drainage patterns. However, shadows of slopes can also be useful. To use this effect, images acquired at low solar angle must be selected, i.e., winter images in Europe, and images taken with a large incident angle for SPOT.

Talwegs are readily interpreted under forest cover in satellite images, if the image is taken in winter, or over bare soils. In fact, soil cover is thicker in the bottom of talwegs due to accumulation of colluvium. Consequently, water reserve is larger than in slopes and soil surface is more humid or more organic, all these contributing to a darker image of the base of talweg. Talwegs with a density at least equal to that observed on a 1:100,000 or 1:50,000 topographic map can be easily delineated. Sometimes talwegs are detected based on differences in vegetation within a plot during anthesis, from its level of growth or ripening. In fact, in these phenological stages plants need to rapidly mobilise various nutrients which they can more likely obtain in colluvium at the base of talweg than on slopes. Depending on the case, vegetation flowers earlier or develops leaves rapidly and hence plant cover on soil increases; or it may die early or become dry. These important modifications in vegetation tissues are readily identified in near- and middle-infrared bands. Quite often talwegs comprise, at the base, plots in the form of strips or at least rectangles, whereas slopes are occupied by forest. Contrast occurs between forest and crops and enables demarcation of talweg boundaries.

These constitute indirect interpretations derived from well-known agronomic models, based on spectral and chorological laws.

■ Relief zones

Relief can be primarily analysed by stereoscopy. In fact, hypsometric maps in countries in which they did not hitherto exist are now prepared with the help of satellite images. Precision is of the order of 5 to 10 m in altitude and area. It can be improved with civil satellites which have a resolution of 5 m for SPOT and about 1 m for others. Precision of aerial photography seems likely to be attained shortly in satellite images as well.

Relief can also be determined from shadows, especially in winter images. However, it should be noted that shadows in images are northwards whereas in topographic maps they are south-east. In fact, we have the impression of perspective view when the sun and hence the source of illumination is at height on the left (north-west). Consequently, in satellite images, the first view is often interpreted by our brain with an inverse relief. It is hence necessary to accustom oneself to see relief in correct perspective. In winter, shadows and zones exposed to the south are better illuminated and so relief is perceived fairly correctly. However, as the sun is lower than in summer, it is common to find a slope entirely in shadow. Contrarily, these disadvantages are eliminated in summer images to a great extent.

Relief is also reinforced by the diversity of vegetation developing at various altitudes or on slopes of different magnitudes. In fact, ecobioclimatic conditions vary according to the altitude: vegetation does not develop in the same way and plant communities are not identical. On slopes and depending on the degree of exposure, vegetation does not receive the same quantity of light and hence plant communities also differ from each other. Thus it is possible to make correct estimation of altitudes and slopes by studying vegetation.

Relief can also be inferred from land-cover maps. In fact, relationships exist between altitudes and spatial distribution of plant communities: mineral zones comprising rocks, boulders and pebbles and rupestrian species, mountain pastures, coniferous forests and deciduous forests. According to the depth of the phreatic water table, it is possible to differentiate various types of terraces and silviculture zones. Slopes are also detected by the shapes of plots, which in some landscapes approximately follow the elevation contours or, contrarily, are perpendicular to them. Lastly, it is quite common to observe forest belts on steep slopes and on top of slopes just in front of plateaus and such zones are easy to identify. They give an indication of plateau boundaries and steep slopes.

■ Agricultural lots

In satellite images, boundaries of lots can be differentiated if sufficient contrast exist between two land-cover types. For example, two plots of cereals cannot be distinguished if they do not differ in some characteristics such as phenological stage. Such may be the case since they are not of the same species or the same cultivar or not sown at the same time or did not develop at the same rate due to differences in soil and hence did not have the same appearance on the day of image acquisition.

Thus, in images as in aerial photos, only cultivated plots can be detected but not basic plots. Breaks can be identified. Contrarily, variations in lots over years can be identified using diachronic images. Forest divisions can also be detected in some cases with images taken on satisfactory dates.

Lots are clearly differentiated in satellite images if plot sizes are greater than 2 mm on the image interpreted. Thus, in a 1:100,000 SPOT image, plot sizes ought to be more than 200 m by 200 m, i.e., 4 ha. In this case, plots less than 4 ha are qualified as small. However, using images with a resolution of 5 m or less it becomes possible to interpret vineyard plots, often very small in size.

It is useful to select classes for plots according to existing plots in the field under study. Thus these classes may be chosen such that the intermediate size is the most frequent (mode). Bertrand (1994) used the following values for plot sizes for analysing the entire region of Yonne district (CD 18.1):

—small plots: less than 4 ha,

—medium plots: 4 to 60 ha,

—large plots: more than 60 ha.

Likewise, most visible forms only are used for interpretation. Strip-form, rectangular and polygonal shapes were chosen for the study of Yonne.

Plots are not necessarily correctly identified on the ground due to the inclined view of the observer. Thus it is almost impossible to identify in the field circular plots or other dispositions of historic plots extending over several kilometres.

■ Towns

Towns are useful in locating the landscape relative to various maps since they are almost invariably indicated in the latter.

They are represented in images by house roofs, often dark, and shadows produced by tall buildings. Roads are characterised by variable luminance values. Rivers or channels and 'green belts' comprising more or less chlorophyllian vegetation according to season are often observed in towns. Almost all types of objects are observed in urban zones: hence they cannot be interpreted using a reflectance model. An urban agglomeration most often comprises all themes used in classification.

However, in infrared colour images, urban zones appear in blue to cyan shades. Various urban zones can be distinguished by the degree of heterogeneity of shades. Often the town centre is indicated by dark shades since roads are narrower and hence shadows larger. Large commercial or industrial areas are usually very bright. Lastly, parks and gardens are readily identified as red, provided they are sufficiently large in size.

A town is a common feature in which mixed mixels (see Chap. 15) occur in large number since most of the pixels in it are represented by various objects. Towns can be better identified using structural analysis, easy for the eye. However, if magnification is not adequate, roads cannot be readily detected with naked eye unless they are large avenues. Contrarily, in some cases, the last axis of principal component analysis or mathematical filtering enables detection of boundaries of units and hence identification of various communication lines (see Chap. 12 and CD 12.3 and 7.13).

Villages are easy to identify in a contact printing of satellite image. They are differentiated by their heterogeneity, by hues close to those of bare soils and by the fact that communication lines often converge towards them.

Hence a satellite image can be taken as a base map, especially in regions for which no maps are present and when there are no resources to procure information layers of the cartographic databank of IGN.

18.2.3 Descriptive format and statistics

Once the principal components are set up, specific aspects of study (such as agricultural, pedological, hydrological, ecological, etc.) need to be highlighted and a legend prepared for each objective. A complete legend comprising all variables, along with their details, must be established so as to obtain a description format of landscape units (Table 18.1).

Each map zone is systematically described by filling the format. Various statistical analyses can then be made and corresponding output maps obtained. The GIS tool would be very useful for exploiting all the resources of landscape analysis.

Zone T167 of agrolandscape unit 1 in Table 18.1 can be described as follows (see Table 18.2). It comprises no rivers, no water bodies, no forests or trees, no habitations. Between 5 and 35% very bright soils cover slopes and plateaus in this zone. More than two-thirds of the zone is occupied by

Table 18.1: Example of description format for agrolandscape units interpreted visually in satellite image (after Bertrand, 1994)

Map zone	Landscape unit No.		1	9	17	Etc.
	Zone No.		T167	S169	1442	
Water	Rivers		0	1	0	
	Lakes, ponds		0	0	0	
Habitat type			0	2	3	
	Hedges		0	0	0	
Forest/Trees	Thickets		0	1	2	
	Groves		0	2	1	
	Blocks		0	0	0	
Soil colour	Very bright		2	0	0	
	Medium bright and dark		0	0	2	
	Large valley		0	0	0	
Morphology	Slope, hillock		3	2	2	
	Talweg		0	4	0	
	Plateau		3	0	4	
	Mixed forest		0	0	2	
	Deciduous		0	1	1	
	Coniferous		0	0	0	
Land cover	Crops		2	3	4	
	Villages, built-up zones		0	1	1	
	Grasslands, heaths		0	2	2	
	Vineyards and orchards		4	0	0	
	Riverine forests, swamps		0	3	0	
		Small	5	2	2	
Agricultural plots	Size	Medium	0	2	4	
		Large	0	0	0	
		Polygonal	2	3	3	
	Form	Quadrilateral	4	0	3	
		Strip-form	0	0	0	
Non-agricultural plots			0	3	2	

vineyards or orchards and the rest by crops. All the plots are small with more than two-thirds quadrilateral and the rest polygonal.

18.3 EXAMPLE: AGROLANDSCAPE OF YONNE DISTRICT

The integrated or landscape method of analysis is useful when a large area is to be studied with various objectives. Such is the case of district-wise and regional databanks now being prepared for administrative management of these regions.

The example presented below pertains to a study conducted in Yonne district (CD 18.1). It is based on the agrolandscape concept. The following aspects are discussed:

—method of description,
—interpretation procedure: delineation and description of map zones,
—assessment of visual interpretation
—classification of map zones: agrolandscape units,
—verification.

18.3.1 Method of description

The variables selected and codified concern three principal aspects of landscape: land cover, geomorphology and plots.

These three principal components of landscape, which indicate the physical characteristics of natural environment and impact of human activity, are well suited for defining agrolandscape units. It has been possible to infer these principal components from four SPOT images.

By means of stereoscopic vision, possible in the boundaries of two adjacent images, major geomorphological features, viz., talwegs, plateaus, slopes, valleys, hillocks, crests and peaks of a region can be readily identified and isolated and their spatial organisation understood. Four morphological units were identified: large valley, slope, talweg and plateau. The concept of slope intensity was not taken into consideration because too difficult to correctly estimate without a digital elevation model.

For zones not covered by stereoscopy, geomorphology was determined making use of shadows and spatial analysis: tree alignments along watercourses, wooded borders of plateaus, etc. Interpretation of 1:100,000 topographic maps of the IGN was used for additional information.

Size and shape of agricultural parcels and non-agricultural zones have been described by visual interpretation. Size-wise classes of agricultural plots (small, medium and large) were set up according to the norm for parcels in Yonne district. The mean quantisised parcel corresponds to the most common plot size in agricultural zones (zones of major crops). Size limits were so fixed that small plots are not rare in these zones and large ones rather marginal.

Land-cover types were identified by textural and structural analyses of images. The following themes were distinguished:

—deciduous and coniferous forests,
—mixed forests,
—crops,
—grassy areas (mowed meadows, grasslands, fallow lands, etc.),
—vineyards and orchards,
—built-up or excavated (quarry) zones,
—riverine forests and swamp zones.

On some dates distinction between plant covers may be difficult: vineyards and crops in summer, crops and woods in spring, winter crops and grasslands in winter, etc.

These three series of variables were codified depending on their coverage in the map zones according to the 0 to 5 scale indicated in Chapter 5. Other variables were determined and combined with these principal components for refining and strengthening the description of map zones.

Among structural elements of a landscape, water occupies first place in exorheic regions. In fact, watercourses represent the main flow in natural environments; their presence characterises a number of landscapes.

Trees also constitute a distinct element of landscape. Structure of tree populations may take varied forms (hedges, groves, thickets, strips, blocks) and result from interaction between man and natural environment.

Soil is also a determinative factor of landscape. Its effect manifests through vegetation, agricultural practices and drainage patterns.

Description of the state of soil surface, and especially its colour, by remote sensing (see Chap. 23) can be envisaged (Yongchalermchai, 1993). However, in this study description of soils is not dealt with since the four images used were acquired in various seasons and soils were not viewed under identical conditions (abundance of bare soil, degree of coverage by crops).

As far as colour is concerned, soils are classified only in two categories, viz., medium and dark; very bright soils which are detectable in any season.

The drawings in Table 18.2 are visual representations of definitions of objects. The distinction between a large valley and a talweg corresponds to the real situation of the district: the criterion of width greater than 1 km facilitates characterisation of the valleys of Yonne river and its principal tributaries (Cure, Armançon, Serein).

18.3.2 Interpretation procedure: delineation and description of map zones

Interpretation was carried out by two interpreters using photographic contacts of images and was restored directly on transparent films. It was hence necessary to ensure pertinence of boundaries, reliability of interpretation and a single level of analysis.

■ Pertinence (local contrast)

Demarcation of map zones is based on local contrasts of principal variables of the description system. *Contrast* (Girard, 1983) is defined as *the ratio of semantic (mathematical) distance to geographic distance.* It is hence evident that these boundaries are more or less distinct.

Interpretation extended to description of areal units: tracing the boundaries is therefore imperative, even in situations of very large variations in the variables. However, if the interpretation is to be used in a flow modelling or a detailed spatial analysis, it is necessary to describe these boundaries by their distinctness, form and contrast.

Boundaries of map zones correspond mainly to variations in land-cover use and to transitions from one geomorphological form to another (often manifested by distinct break in slope).

Some boundaries, which define outlines of 'composite' map zones having a more complex organisation, are validated by differences observed in secondary variables such as change of size or shape of parcels, peripheral occurrence of bright soils, interruption in a hedge system, etc.

■ Reliability of interpretation

Reliability must be ensured in any interpretation program. The attitude of the interpreter towards satellite images should remain similar in time and space. This demands a priori definition of precise rules of interpretation and elaboration of a common model to serve as a reference to interpreter throughout the interpretation stage. This condition necessitates use of a general and reproducible method of interpretation.

Table 18.2: Definition and coding of variables used for description of agrolandscape units

WATER	
River: permanent watercourse	0 = absence 1 = presence assumed or low discharge 2 = major watercourse
Lake or pond: water body of more than 200 m^2	0 = absence 1 = punctuate presence 2 = dominant presence
TYPE OF HABITAT	
Visible structure of built-up area	0 = absence 1 = scattered 2 = combined 3 = mixed (scattered and combined)
FORESTS/TREES	
Hedge: tree alignment Wooded grove: isolated wooded plots of compact form and area less than 25 ha Forest strip: wooded plots of highly elongated form and area less than 100 ha Forest block: more or less compact woods of more than 25 ha	0 = absence 1 = sparse presence 2 = dominant presence 3 = dense presence (only for hedges)

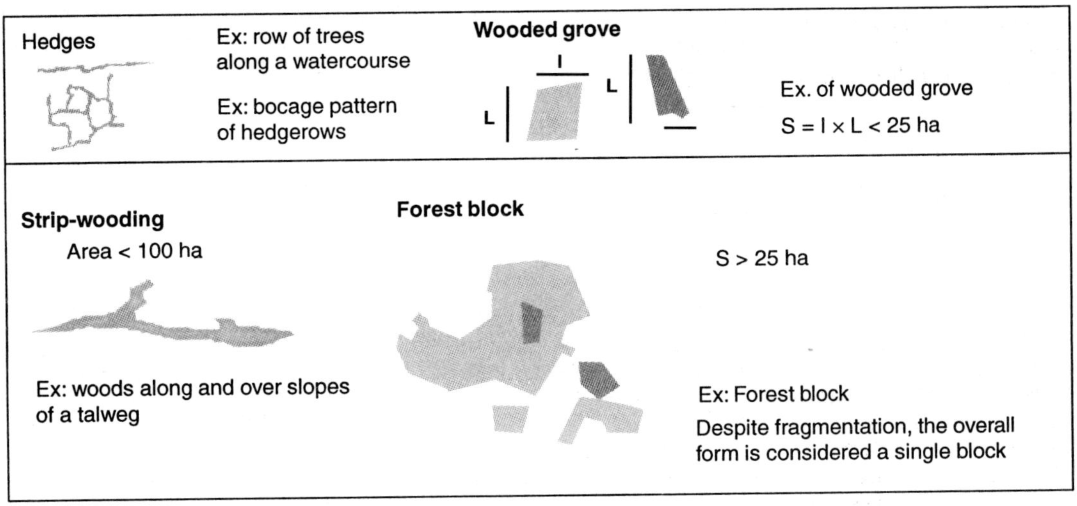

COLOUR OF SOILS	
Very bright: chalky soils Medium and dark: non-chalky soils	0 = absence 1 = diffuse presence 2 = dominant presence

MORPHOLOGY

Large valley: flat-bottomed valley of width greater than 1 km	0 = absence 1 = 0 – 5%
Valley edge: hillock, regular slope and edge of talweg of width greater than 500 m	2 = 5 – 35% 3 = 35 – 65%
Talweg: bottom of talweg at least 1 km in width	4 = 65 – 95%
Plateau: relatively flat and horizontal zone, bounded by distinct breaks in slopes	5 = > 95% (in % area)

Large valley

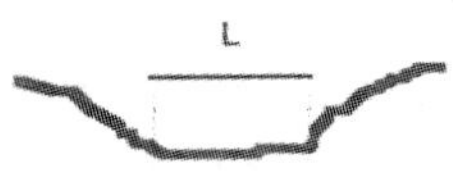

Ex: profile across a large valley L > 1 km

Talweg, Slope, Plateau

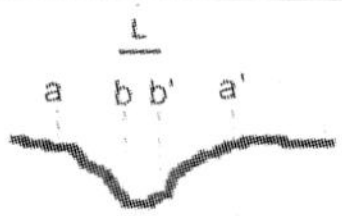

a, a', b, b' represent breaks in slopes (projection of lines of break)
b and b' can be combined (V-shaped form)
L = bb' < 1 km

talweg **slope** **plateau**

1 If ab + ab' > 500 m, talweg and valley edge can be differentiated

if bb' = 0 (talweg without flat bottom, V-shaped), area of talweg is estimated by 'extending' it onto valley edges such that a belt of 200 m in width be classified as talweg

2 If ab + ab' <50 m /everything' (ab and a 'b') is classified as talweg

LAND COVER

Mixed forest: mixed population of deciduous and coniferous trees	0 = absence
Deciduous: deciduous population	1 = 0 – 5%
Conifer: coniferous population	2 = 5 – 35%
Crops: cultivated agricultural area (other than vineyards, orchards and grasslands)	3 = 35 – 65%
Village, built-up zone: anthropic area, constructed, urban or excavated (quarry) zones	4 = 65 – 95%
Grassland, heath: grassy area	5 = > 95%
Vineyard, orchard: grape plantations or fruit trees	(in % area)
Riverine forests: moist zone with tree stands (poplar, alder)	

PLOT

➢ According to **size**:		
Small:	area < 4 ha	0 = absence
Medium:	4 ha < area < 60 ha	1 = 0 – 5%
Large:	area > 60 ha	2 = 5 – 35%
➢ According to **shape**:		3 = 35 – 65%
Strip-form:	rectangular plot with L > 7 × l	4 = 65 – 95%
Quadrilateral:	plot with 4 sides parallel 2 by 2	5 = > 95%
Polygonal:	plot with 3 or 5 sides	(in % of total area)
➢ According to **nature**:		
Agricultural:	crops, grasslands, vineyards	
Non-agricultural:	forest or built-up zone	

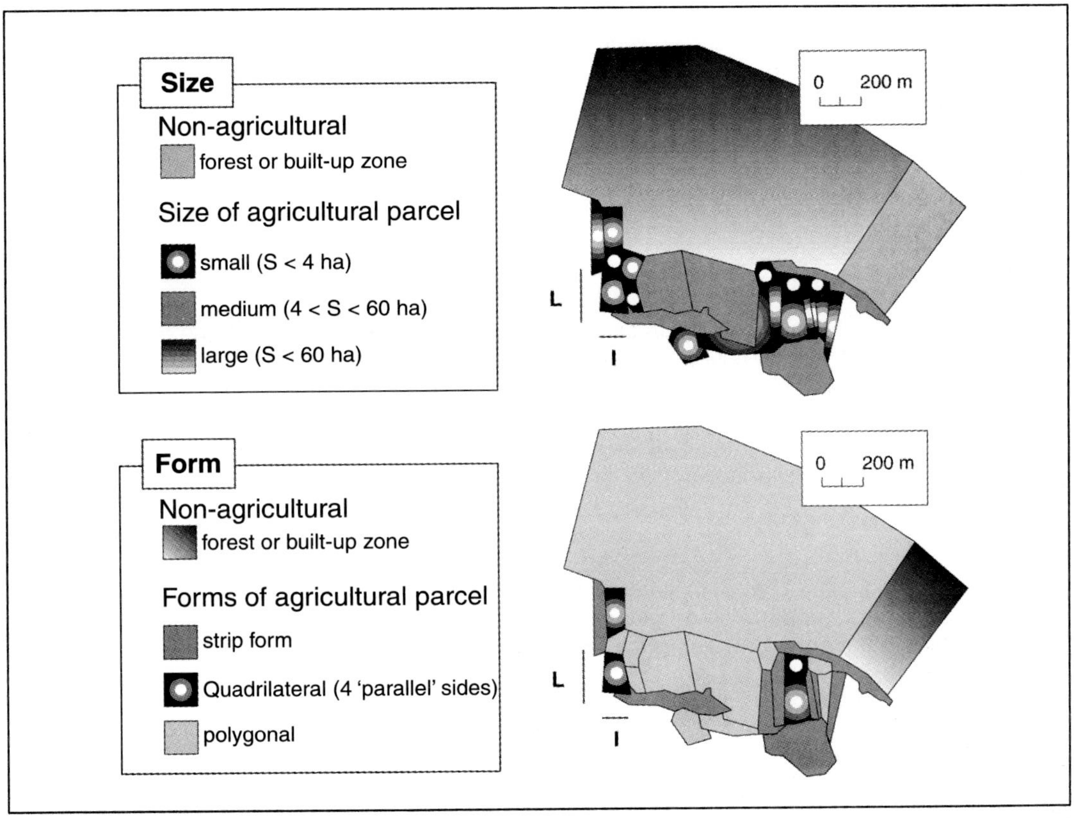

The interpreters proceed with an experimental (or validation) phase of the method by parallel interpretations over the same test zones and comparison of the results. Discussions following this study enable them to standardise their procedure and calibrate the variables on the characteristics of the district under investigation.

As each interpreter operates on the commonly formulated and approved model, it can be ensured that interpretation is carried out identically as it progresses and over all the zones of the district covered by four satellite images.

■ Resolution and level of analysis

Interpretation was made on the 1:100,000 maps but the desired restitution product was a 1:250,000 (1 pixel = 0.08 mm) agrolandscape map of Yonne district. In this example, the *one-fourth rule* was used according to which *it is assumed that at less than 2.5 mm × 2.5 mm (one-fourth of a cm^2) a map zone is not readily legible on a map.* It may hence be inferred that there is no need to acquire data at higher resolution.

It follows that the resolution required for delineating agrolandscape units corresponds to more than 39 ha (6.25 hm × 6.25 hm) on the ground, i.e., 31 × 31 pixels.

This value indicates only an order magnitude since the decision to draw a map zone is also dependent on its shape and its contrast in the landscape. Thus, for elongated forms, minimum distance between pseudo-parallel boundaries will be 2 mm, or 25 pixels in the image.

This concept of spatial resolution is very important since it imposes a particular level of analysis for agrolandscape. In fact, an object, even of high contrast (such as a deciduous grove), with an area less than 39 ha, does not by itself constitute an agrolandscape unit but is an integral part of the agrolandscape unit that surrounds it. On the other hand, it represents a strong landscape element which, together with others, defines a spatial organisation specific to a map unit of agrolandscape.

18.3.3 Assessment of visual interpretation

■ Importance of pattern

Direct visual interpretation of images on paper maps enables observation of a large geographic field in its entirety. Each object, identified by certain characteristics and patterned in a similar manner, is instantaneously relocated in its spatial environment. The interpreter can infer textural form of objects, their relative size, density, connectivity and spatial distribution. Detection of patterns in the image is related to the model of representation of the references. It is hence imperative to be explicit in defining and coding the description variables used for agrolandscape.

■ Duration

For the entire district (about 7120 km^2), the duration of complete visual interpretation, including choice of description criteria, preparation of grid, experimentation in test zones, tracing and description of map zones was of the order of 6 weeks for an interpreter.

■ Use of topographic maps and aerial photos

Totally differing views of images acquired on different dates may be the cause of certain difficulties in interpretation of land cover and plot distribution data. Choice of date of image is hence very important (see Chap. 17). In the absence of images acquired on a key date, it may be necessary to employ other sources of information such as aerial photos. However, cost vis-à-vis accuracy required must be computed. They should be used only when detailed geometric information of the area is desired.

Photographs may be used for validating the results obtained from interpretation of satellite images. Shapes of plots and evidently, morphology of a region are better differentiated if care is taken to purchase a stereoscopic coverage (whose cost is twice higher). However, land cover data is not necessarily more readily detectable. The significance of satellite images lies in obtaining data several times in a year.

The 1:50,000 topographic map is often adequate for estimation of morphology required for description of agrolandscape if the output map is of 1:250,000.

■ Semantic accuracy

Each map zone was described by 28 variables, the modalities of each having been coded by a strict and rigorous system with a simple and unambiguous procedure. While the precision of estimation of area (coded in 6 categories from 0 to 5) may appear low, it is sufficiently strong for reducing errors and disparity between various interpreters. It must be noted that the semantic accuracy thus obtained is largely adequate considering the description of each agrolandscape unit by a combination of 28 variables.

18.3.4 Classification of map zones: agrolandscape units

Principal component analysis was carried out with 1111 map zones, each characterised by 28 variables. Axis 1 of principal component analysis differentiates between:

—non-agricultural plots, deciduous and coniferous blocks on the one hand,

—and polygonal plots, medium and dark soils, crops and medium size plots on the other.

The first structural element of a cluster of points is the contrast between agricultural and non-agricultural units (essentially forests and secondarily anthropic zones). Thus 311 non-agricultural units and 800 agricultural units were identified.

Subsequently, the minimum sorting distance method was employed (Girard, 1983). The first iteration was based on the choice of initial reference zones, viz., the nuclei. Detailed visual interpretation enabled step by step representation of structure and characteristics of the landscape of the district. Information of the region under study acquired during this stage was evaluated for choice of nuclei.

After several iterations (13 in this specific case), classification was stabilised and 28 agrolandscape units were obtained (CD 18.1). Histogram analysis of each variable enabled description of the semantic content of each agrolandscape unit detected by this classification.

Besides the semantic significance of these units (homogeneity and originality), their spatial distribution is a second criterion of pertinence. In most cases, the chorological laws manifest distinctly. This is very important in view of analysis of small regions (a higher level of analysis) characterised by a specific spatial organisation of landscape units (CD 18.1).

18.3.5 Verification

Verification of the results of the entire study on agrolandscapes of Yonne district was carried out on 5% of map units. The latter were selected by random choice and observed in the field. Ground-truth verification was done regarding description of map zones, correctness of boundaries and especially significance of agrolandscape units. For each zone, the codified description was compared with the findings of visual observation in the field and a photograph was taken. No major error was committed during description from satellite images: no confusion in land cover types, no incorrect representation of relief, no omission of tree populations, no false estimation of shape of plots, etc. In some cases and for certain variables, small deviations were observed between descriptions from images and descriptions in the field using the same variables. They concerned presence or absence of a stream, presence of isolated hedges, proportions of slopes and plateaus, overestimation of vineyards and orchards (abandoned plots), overestimation of small plots. These errors notwithstanding, visual interpretation of satellite images produced a reliable representation of ground truth.

With respect to correctness of boundaries, it is more difficult to draw conclusions from ground observations. In fact, the field of view on the ground is often small (because of the shape of the Earth or high plant cover). This considerably hinders the observer from understanding the structure of the landscape and, in particular, organisation of basic patterns. Moreover, boundaries are often indicated by a change in this organisation. Location of boundaries on the ground is hence difficult.

Only the boundaries corresponding to strong contrasts in a few criteria are distinctly discernible in the field when the point of observation offers a synoptic view to the observer (Fig. 18.1). However, a precise global assessment of accuracy of these boundaries is impossible.

Lastly, ground observation permits verification of landscape classification over the entire region of the district. Of 62 map zones observed, only 6 (10%) were incorrectly classified: one or two characteristics of landscape do not correspond to the agrolandscape profile to which it belongs. These characteristics suffice to make it appear closer to another unit.

However, caution is needed in using the ground-truth data. The concept of landscape is strongly related to perception from space. The points of observation are so different between a satellite and a ground observer (Fig. 18.1) that the ground-truth data must be considered only as a partial tool for verification of results.

18.4 OTHER EXAMPLES

18.4.1 Auge region

Mapping landscape units of Lower Normandy region using satellite data was verified by socio-economic information. The Atlas of Normandy and a thesis in economic sciences (Laurent, 1992) were used for this purpose. This comparison validated division of Auge region into two subunits, north and south (Girard et al., 1995).

18.4.2 Rhone district

In Rhone district, the objective of remote-sensing study was to prepare a vocational plan of agricultural and forest zones. The agrolandscape units, which were georeferenced on BD CARTO of the IGN, were verified by ground truth (Boutefoy, SIRA) and compared with climatic, agronomic and other data. They aided in augmenting the databank by geographic delineation of the perimeter concerning the vocational plan and contributed through description of land-cover zones, to evaluation of the agronomic potential of various regions, identification of zones sensitive to erosion and location of zones susceptible to effects of afforestation and deforestation.

The description format was similar to that used for Yonne district. It comprised 22 variables systematically described for each of the 2547 polygons traced (directly on the computer monitor using a geographic information system) for Rhone district. They were regrouped into 72 map units and through statistical analysis into 12 simplified landscape zones (CD 18.2).

18.4.3 Aube district

The objective in the case of Aube district was to define agrolandscape zones and integrate them with soil data acquired in various parts of the district at different scales. This study enabled preparation of a database on soils and land cover for various agricultural applications and in particular identification of zones vulnerable to soil erosion (CD 18.3).

An image showing maximum coverage of bare soil was chosen from the DALI (now SIRIUS) catalogue (CD: SPOT system). We interpreted it on the basis of surface state of bare soils. The study was carried out on an image enlarged to 1:50,000 scale. Vulnerability to soil erosion was categorised in four classes after regrouping about fifteen classes. This soil vulnerability map was compared with a slope map using GIS. Vulnerability of the region to soil erosion was thus determined. This was verified by farmers and consultants of the Agricultural Chamber of Aube. Possible locations for plantation of hedges and grass strips were recommended based on this vulnerability map and boundaries of agricultural plots.

18.4.4 Champagne-Ardenne region

The Champagne-Ardenne region was also investigated by computerised interpretation of landscape units. A mosaic was prepared with several LANDSAT TM images and processed by supervised classification to obtain a single image comprising 20 agrolandscape units (CD 18.4).

18.4.5 Agrolandscapes and "small agricultural zones" or "agricultural reference units"

In various cases (Yonne, Aube, Champagne-Ardenne, Rhone), the results of agrolandscape analysis were compared with boundaries of agricultural reference units (Baize, 1993; Eudes, 1993). It was necessary to combine several agrolandscape units that were more detailed than agricultural reference units or small agricultural zones. This was done using ascending hierarchic classification and OASIS methods (see Chap. 11).

18.5 CONCLUSION

This method of studying landscapes described here has been used for various applications and various places. It assumes an adaptation of agrolandscape descriptions to the objective of the investigation, region under study and precision required.

If GIS software is used, it is advantageous to make zoning directly on the computer monitor rather than on paper prints of satellite images, since integrated maps are readily available under GIS. For working on a district with 2500 map zones, visual zoning requires a month for an interpreter. On a computer, visual interpretation can also be carried out using several colour composites and various techniques of image processing, such as filtering (see Chap. 12) which enhances boundaries of plots.

With a digital elevation model (DEM), the image and the elevation map can be overlapped (CD 2.2) and several output images prepared using the DEM and the various bands of the image.

Visual interpretation and zoning of landscape units are largely facilitated by a preliminary analysis by OASIS which automatically creates map zones on digital criteria (digital number values of pixels and neighbouring pixels). Thus mosaics of multiple images can be processed ensuring homogeneity of visual interpretation.

This method of interpretation by image-processing techniques (statistical and structural classifications) and visual methods results in division of the region under study into a number of agrolandscape units. The latter are defined by a reduced number of variables that suffice for analysing the structure of the region with high accuracy. Agrolandscapes obtained are more homogeneous and more accurate than agricultural reference units. Hence they can be more readily included in a spatial model and used in simulation studies. In fact, the problem of semantic analysis of inhomogeneous geographic units is not yet satisfactorily solved.

The landscape approach as an indicator of environmental conditions and human activity, along with a map representation, is very important for comparing various regional data sets under a geographic information system.

References

Baize D. 1993. Petites régions naturelles et paysages pédologiques de l'Yonne. Institut national de la recherche agronomique, Orléans, 191 pp.

Bertrand P. 1994. Élaboration d'une base de données localisées sur les agropaysages à partir d'images satellitaires. Application à l'étude des organisations spatiales et à la segmentation du département de l'Yonne. Mémoire de Mastère "Système d'Informations Localisées pour l'Aménagement des Territoires". Institut national agronomique, Paris-Grignon, 46 pp.

Deffontaines J-P. 1985. Étude de l'activité agricole et analyse du paysage. Espace géographique, Paris, 1: 37–48.

Deffontaines J-P. 1986. Un point de vue d'agronome sur le paysage. Lectures du paysage. Coll. INRAP Éd. Foucher, 191 pp.

Donadieu P. 1993. Du désir de patrimoine aux territoires de projets. Paysage et gestion conservatoire des milieux humides protégés: le cas des réserves naturelles du plateau de Versailles-Rambouillet et de quelques marais de l'Ouest, Thèse de Doctorat, Université Jussieu-Paris, 280 pp.

Eudes S. 1993. Étude des relations spatiales entre les sols et les options financières des exploitations. Mémoire Diplôme d'Agronomie Approfondie, Institut national agronomique, Paris-Grignon. 41 pp. + ann.

Francoual T, Gilliot J-M, Girard M-C. 1997. Réalisation d'une base de données localisée sur les agropaysages du département du Rhône. Détermination des agropaysages par interprétation visuelle de données satellitaires. Association Sol Info Rhône-Alpes, Lyon, 34 pp.

Gilliot J-M, Girard M-C. 1997. Étude de la vulnérabilité des sols à l'érosion. Chambre d'Agriculture de l'Aube, Troyes, 14 pp.

Girard C-M, Girard M-C. 1994. Aide à la cartographie d'unités paysagères par une méthode d'analyse du voisinage des pixels: application en Basse-Normandie, Photo-interprétation, 3–4: 145–154.

Girard C-M, Girard M-C, Gilliot J-M. 1995. Qualité des méthodes d'interprétation, application à la caractérisation et la cartographie d'unités de paysage. Qualité et validation des résultats. Bull. Soc. Franc. Photogr. Télédétection, 137: 62–66.

Girard M-C. 1983. Recherche d'une modélisation en vue d'une représentation spatiale de la couverture pédologique. Application à une région des plateaux jurassiques de Bourgogne. Thèse Doc ès sciences, Université Paris 7. SOLS, 12: 414 pp.

Girard M-C. 1993. L'agropaysage, concept permettant de segmenter spatialement le milieu naturel à des fins d'analyse agronomique. Rapport ministère de l'Agriculture et de la Pêche, DERF, Bureau des sols, 17 pp.

Girard M-C. 1995. Apport de l'interprétation visuelle des images satellitaires pour l'analyse spatiale des sols. Un exemple dans la région de Lodève. Étude et gestion des sols, 2 (1): 7–24.

Girard M-C, Girard C-M, Bertrand P, Orth D, Gilliot J-M. 1996. Analyse de la structure des paysages ruraux par télédétection, C.R. Acad. Agri. Fr., Paris, 82 (4): 11–25.

Hénin S. 1993. Le paysage: étude sémantique. C.R. Acad. Agri. Fr., Paris, 79 (7): 29–30.

Hénin S. 1994. Le paysage, une entité pour l'appréciation du milieu. C.R. Acad. Agri. Fr., Paris, 79 (7): 30–43.

Laurent C. 1992. L'agriculture et son territoire dans la crise, Thèse ès Sciences Économiques Université Paris 7, 454 pp. + ann.

Lizet B, De Ravignan F. 1987. Comprendre un paysage, guide pratique de recherche. INRA, 147 pp.

Mollet S. 1994. Élaboration d'une base de données des agropaysages du département de l'Yonne—application à l'étude des dynamiques financières agricoles. Mémoire Diplôme d'Agronomie Approfondie, Institut national agronomique Paris-Grignon., 55 pp.

Rougerie G, Beroutchachvili N. 1991. Géosystèmes et paysages: Bilan et Méthodes. Armand Colin, Paris, 302 pp.

Yongchalermchai C. 1993. Étude d'objets complexes, sol/plante à différents niveaux d'organisation: de la parcelle au paysage. Thèse de l'INA-PG. Sols, Grignon, 19: 232.

19

CORINE Land Cover

19.1 INTRODUCTION

On 27 June 1985, the European Union adopted the CORINE (Co-ordination of Information on Environment) program under the responsibility of the DG XI (Environment, Nuclear Safety and Civil Protection) of the European Commission. The objectives of this program, implemented during 1985 to 1990, pertained to three aspects (Cornaert and Maes, 1992):

—Collection of information on the state of the environment and development of an information system;

—Imposition of uniformity in existing nomenclatures and development of nomenclature and methodologies necessary for execution of the program;

—Co-ordination of activities undertaken in the member states or at the international level, aimed at enhancing information on environment.

This program is now managed by the European Agency on Environment. Of the various themes included in CORINE, covering geographic, biological, agricultural and other aspects concerning the environment, land cover has been the major component since its extension to Central and Eastern Europe from 1991. By the end of 1996, CORINE Land-Cover mapping (level 3) had been completed over 3,000,000 km^2, covering 17 European and 2 North African countries mapped with a single legend (Table 19.1) and 17 national databases combined as an European database. In France, the *Institut français de l'environnement* (French Institute of Environment) (IFEN, 1995; Bossard, 1996) is the chief of operations of the CORINE Land Cover program within the framework of collection and public dissemination of data necessary for environmental policy planning and choice of economic development.

19.1.1 CORINE Land-Cover mapping

Data used as the main source of information for the program are provided by the satellites SPOT and LANDSAT TM (Thematic Mapper). In fact, satellite data constitute objective information in digital and readily usable form. Image interpretation was carried out visually by photo-interpreters using paper displays of 1:100,000 colour composites, aided by topographic maps, aerial photos and ground-truth data.

The scale of the map obtained was 1:100,000 and the dimension of the smallest map unit 25 ha. This choice emerged from the minimum information content considered necessary and the possibilities of identifying changes over a time interval not exceeding 5 to 10 years on the one hand, and the time and cost constraints on the other.

The nomenclature, grouped into 44 classes in the third level (CORINE Land Cover terminology), is adopted as standard at the European level and enables comparisons between states, regions, districts and zones of study (Table 19.1).

Table 19.1: Nomenclature of 44 classes of CORINE Land Cover, level 3 (after IFEN, 1995)

Level 1		Level 2		Level 3	
1.	Artificial surfaces	1.1	Urban fabric	1.1.1	Continuous urban fabric
				1.1.2	Discontinuous urban fabric
		1.2	Industrial, commercial and transport units	1.2.1	Industrial or commercial units
				1.2.2	Road and rail networks and associated land
				1.2.3	Port areas
				1.2.4	Airports
		1.3	Mine, dump and construction sites	1.3.1	Mineral extraction sites
				1.3.2	Dump sites
				1.3.3	Construction sites
		1.4	Artificial non-agricultural vegetated areas	1.4.1	Green urban areas
				1.4.2	Sport and leisure facilities
2.	Agricultural areas	2.1	Arable land	2.1.1	Non-irrigated arable land
				2.1.2	Permanently irrigated land
				2.1.3	Rice fields
		2.2	Permanent crops	2.2.1	Vineyards
				2.2.2	Fruit trees and berry plantations
				2.2.3	Olive groves
		2.3	Pastures	2.3.1	Pastures
		2.4	Heterogeneous agricultural areas	2.4.1	Annual crops associated with permanent crops
				2.4.2	Complex cultivation patterns
				2.4.3	Land principally occupied by agriculture with significant areas of natural vegetation
				2.4.4	Agroforestry areas
3.	Forests and semi-natural areas	3.1	Forests	3.1.1	Broad-leaved forests
				3.1.2	Coniferous forests
				3.1.3	Mixed forests
		3.2	Scrub and/or herbaceous vegetation associations	3.2.1	Natural grasslands
				3.2.2	Moors and heathland
				3.2.3	Sclerophyllous vegetation
				3.2.4	Transitional wood-land scrub
		3.3	Open spaces with little or no vegetation	3.3.1	Beaches, dunes and sands
				3.3.2	Bare rocks
				3.3.3	Sparsely vegetated areas
				3.3.4	Burnt areas
				3.3.5	Glaciers and perpetual snow
4.	Wetlands	4.1	Inland wetlands	4.1.1	Inland marshes
				4.1.2	Peat bogs
		4.2	Coastal wetlands	4.2.1	Salt marshes
				4.2.2	Salines
				4.2.3	Intertidal flats
5.	Water bodies	5.1	Inland waters	5.1.1	Watercourses
				5.1.2	Water bodies
		5.2	Marine waters	5.2.1	Coastal lagoons
				5.2.2	Estuaries
				5.2.3	Sea and oceans

Organisations involved in the mapping of the French territory were not the same in all regions. Land-cover mapping of the Ile-de-France region was carried out by the *Institut d'aménagement et d'urbanisme* (Institute of Management and Urban Affairs) of the Ile-de-France region (IAURIF) in a 1:100,000 scale, conforming to the recommendations of the CORINE Land Cover program. However, the visual interpretation employed corresponded to a more detailed nomenclature (levels 4 to 5) and minimum resolution of 4 to 5 ha. Later, only the 3rd level interpretation was retained during integration of the map of Ile-de-France with the rest of the CORINE Land Cover database.

The entire Ile-de-France region was mapped using 9 SPOT scenes acquired in equivalent periods (May and June) during two successive years (1989 and 1990), during which urbanisation had just commenced. Preparation of CORINE Land-Cover map involved the following operations:

1) geometric corrections using control points,
2) re-sampling by restoration (pixel of 20 m side) in Lambert Zone I projection, to ensuring maximum accuracy, and
3) construction of a mosaic, subsequently divided into four zones, on 1:100,000 photographic contact of colour composites.

The work was completed in two phases:

1) mapping of agricultural areas of the rim districts (accuracy 25 ha) with ground control of sectors of doubtful interpretation;
2) mapping of urban zones of the Paris agglomeration and peripheral districts (accuracy 4 ha).

22 classes of level 3 were thus interpreted, in which 27 classes of levels 4 and 6 of level 5 were integrated. Maintenance of CORINE Land Cover (level 3) norms was later entrusted to a service society.

Interpretation of all the nine images of the Ile-de-France region required employment of two photo-interpreters for a period of about 6 months. Despite the experience of the photo-interpreters, the visual interpretation showed certain subjectivity, however. For a given date, risks of error varied according to the classes and for a given class, according to the date of acquisition. The accuracy of identification of classes hence depends on their nature and the date of acquisition of the satellite data. The IAURIF compared the CORINE Land-Cover map with the existing land cover modes in the region (IAURIF, 1995), for which only the polygons of more than 5 ha were taken into consideration. Correlation between the two maps was observed in 84 to 85% cases.

19.1.2 Automatic mapping

A 20 by 18 km zone covering the southern part of the Val-d'Oise district and the northern part of the Yvelines district was mapped by IAURIF using the SPOT scene (KJ 038-251) of 17 May 1989. Nineteen classes were identified in this zone.

The land-cover map of this small zone west of Ile-de-France, obtained from CORINE Land Cover program, was compared with a map of the same theme prepared by applying methods of computerised classification and aggregation in a SPOT image (see Chaps. 8, 9 and 11). The objective of this comparison was to assess the potential of computerised methods for preparing a map at level 3 and a more detailed level (level 4). The aim was not to model the results of visual interpretation, but to compare the results obtained by various methods.

Computerised mapping was carried out on a multispectral SPOT scene of 11 March 1995. This date was particularly favourable since a strong radiance contrast existed between forests (mainly broad-leaved or mixed) and agricultural plots. Moreover, within agricultural areas, annual crops corresponding to bare soils or sparse or very sparse chlorophyllian vegetation, are readily distinguished from short-duration fallow lands (dense and non-chlorophyllian vegetation) at the end of winter and perennial herbaceous vegetation—grasslands and long-duration fallow lands (dense and chlorophyllian). The results were verified using ground observations in the spring of 1995 and July 1996 and from aerial photographs of the National Geographic Institute. Quantitative validation was applicable only to invariant zones and this posed problems for classes of artificial surfaces, since in this region construction openings are common. Hence, a qualitative assessment of the results was also made. Contrarily, no risk of modifications was observed for other classes, which was confirmed by ground check-up.

Land-cover classification was conducted pixel by pixel and the zones obtained were subjected to spatial integration. The final results were entered in MapInfo© geographic information system, after geographic restitution using TeraVue program. Flow chart of the method employed is shown in Fig. 19.1.

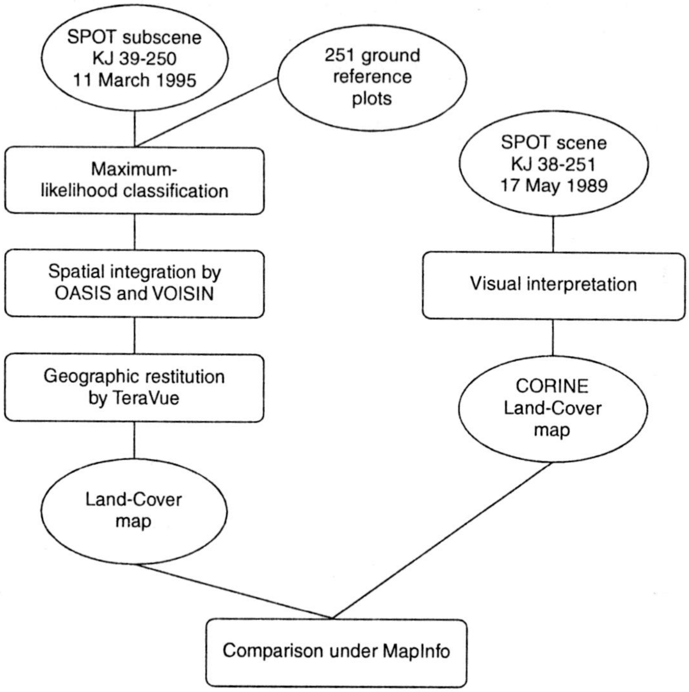

Fig. 19.1: Flow chart of the method used for computerised mapping.

19.2 DATA PROCESSING AND DISCUSSION OF RESULTS

19.2.1 Supervised classification of land cover

In the infrared colour composite, 251 reference plots identified on the ground were delineated on the computer monitor and used to determine the parameters of 17 classes. The number of reference plots for each theme varied from 4 to 30. In fact, there was no need to increase the number of reference areas for themes with no ambiguity such as water (7 references). On the other hand, it was very difficult in some cases to locate a large number of reference areas for the classes few represented in the subscene, viz., quarries and industrial units (4 references). Contrarily, more numerous reference areas were necessary for characterising heterogeneous areas such as fallow lands (24 references) or spring crops (30 references).

The entire subscene was then classified by 'maximum-likelihood method with Gaussian assumption' using MULTISCOPE (Multiscope, 1993) program which in fact pertains to the Bayesian group, with a priori choice of probability. In order to obtain a land-cover map as close to CORINE Land Cover as possible, summer crops (corresponding to various bare soils), winter crops and spring crops were combined under the theme 'arable lands'. To compare the results of a classification with a map in which no 'unclassified' category exits, it was imperative to classify maximum number of pixels of the image. A classification of 50 to 94.9% of reference plots was obtained (Table 19.2), with more than 80% correctly classified for 8 of the 13 classes. Final classification of the entire image segment resulted

Table 19.2: Results, in number of pixels and percentage (bold), of classification of reference areas after application of maximum-likelihood method.

■ Accords (number and percentage of correctly classified pixels), major diagonal of the error matrix Kappa = 0.75

	Ground-truth classes	1	2	3	4	5	6	7	8	9	10	11	12	13	rejections	Number of pixels (%)
1	Quarries	702 **89.9**	31 **4**												48 **6.1**	781 **100**
2	Arable land	401 **4.6**	6661 **76.9**						654 **7.5**	152 **1.8**		453 **5.2**	143 **1.7**	28 **0.3**	171 **2.0**	8663 **100**
3	Broad-leaved forests		5 **0.2**	1894 **87.5**		158 **7.3**		158 **7.3**	4 **0.2**			68 **3.1**	23 **1.1**		12 **0.6**	2164 **100**
4	Coniferous forests		30 **2.2**	6 **0.4**	1040 **74.9**	289 **20.8**						3 **0.2**	18 **1.3**		3 **0.2**	1389 **100**
5	Mixed forests			112 **6.2**	79 **4.4**	1587 **88.5**						8 **0.4**	8 **0.4**		0 **0**	1794 **100**
6	Old gravel pits						2934 **94.6**	16 **0.5**					2 **0.1**	28 **0.9**	123 **4.0**	3103 **100**
7	Recent gravel pits							413 **92.0**			2 **0.4**	1 **0.2**		16 **3.6**	17 **3.8**	449 **100**
8	Fallow lands		148 **6.1**	59 **2.4**	12 **0.5**	11 **0.5**			1803 **74.1**	1 **0.0**		46 **1.9**	349 **14.3**		5 **0.2**	2434 **100**
9	Grasslands		41 **2.8**						3 **0.2**	1346 **90.6**					96 **6.5**	1486 **100**
10	River							63 **4.4**			137 **94.9**			5 **0.3**	5 **0.3**	1444 **100**
11	Continuous urban fabric	6 **0.2**	65 **2.3**	70 **2.4**	4 **0.1**	4 **0.1**			122 **4.2**			1844 **63.9**	384 **13.3**	333 **11.5**	56 **1.9**	2888 **100**
12	Discontinuous urban fabric	10 **0.1**	242 **3.0**	595 **7.3**	29 **0.4**	71 **0.9**			1393 **17.2**	24 **0.3**		1497 **18.5**	4047 **50.0**	86 **1.1**	107 **1.3**	8101 **100**
13	Industrial units	86 **1.7**	5 **1.1**						2 **0.0**			365 **7.2**		4463 **87.7**	114 **2.2**	5087 **100**
	Classification	1205 **3.0**	7279 **18.3**	2736 **6.9**	1164 **2.9**	2120 **5.3**	2934 **7.4**	492 **1.2**	3981 **10.0**	1523 **3.8**	1373 **3.5**	4285 **10.8**	4975 **12.5**	4959 **12.5**	757 **1.9**	39783 **100**
	% commission errors	**41.7**	**8.5**	**30.8**	**10.7**	**25.1**	**0**	**16.1**	**54.7**	**11.7**	**0.1**	**57.0**	**18.7**	**10.0**		

in a rejection of 1.86 % or 669.6 ha of unclassified area, corresponding to pixels of very heterogeneous contents in the boundaries of discontinuous urban fabric, gravel pits under excavation or forest blocks.

Accords in Table 19.2 indicate the number (or percentage) of correctly classified reference pixels (shaded squares along the diagonal of the error matrix). Errors correspond to pixels affected to a class when they truly belong to another. Errors of commission (inclusions) computed along rows of the error matrix are distinguished from errors of omission (exclusions) computed along columns. A classification is statistically judged as very correct when agreement corresponds to 80% or more pixels of reference plots (Lillesand and Kiefer, 1994). The themes 'old gravel pits, recent gravel pits, grasslands, rivers, industrial zones' are correctly classified (87.7 to 94.9%) with a few errors (0 to 16.1%), whereas 'fallow lands' and 'continuous urban fabric' are less correctly classified (63.9 and 74.1%) and present numerous errors (54.7 and 57%). The themes 'quarries', 'broad-leaved forests' and 'mixed forests' are correctly classified (87.5 to 89.9%), but with significant number of errors (25.1 to 41.7%). Lastly, the themes 'arable lands', 'coniferous forests' and 'discontinuous urban fabric' are incorrectly classified but with fewer errors (8.5 to 18.7%).

Excluding major crop areas, the classification result is presented as a mosaic of small zones of various colours, indicating the complexity of spatial organisation of land-cover units, in particular in semi-urban regions. Such a situation is observed in the classification of Saint-Christophe sector of the new town of Cergy-Pontoise, in which forest and agricultural areas are intermixed. For catering to the needs of managers who desire integration of land-cover maps in a GIS, we should have maps that synthesise information in the form of compact map units, albeit less accurate. It is hence necessary to conduct spatial integration of the classification results obtained.

19.2.2 Spatial integration

Spatial integration of the results was carried out using OASIS and VOISIN programs (Girard, 1992; see Chap. 11). The OASIS method was applied to the results of the preceding classification. For this, one or several polygons corresponding to various classes were selected, avoiding the class 'quarries' because of its small size and confusion with 'arable land' (Table 19.2). For tracing the polygons, the classes 'continuous urban fabric and 'discontinuous urban fabric' showing small and complex units were combined into a single one. The heterogeneity index, expressed as a function of window size, varied from 5 for the most homogeneous (river) to 23 for the most heterogeneous (summer crops). Processing was successively repeated with several window sizes close to that of the most homogeneous class, viz., 3 × 3, 5 × 5 and 7 × 7. In fact, use of a larger window resulted in disappearance of small-size units corresponding to the most homogeneous classes.

These precautions notwithstanding, spatial integration resulted in erosion or even complete elimination of linear forms. Hence, we have chosen to insert, in the final map, units entitled 'watercourses' obtained from supervised classification.

Note the incompatibility between spatial integration, which results in delineation of fewer and more compact map units, and semantic accuracy (see Chap. 15). Therefore, we have retained the result corresponding to window size 5 × 5 or a precision of 100 m × 100 m on the ground.

This land-cover classification took 1 week: 3 days for supervised classification (selection and tracing of reference plots is the longest) and 2 days for spatial integration (data analysis for choice of moving windows was the longest operation). After geographic restitution, the results of classification were compared with CORINE Land Cover data using GIS.

19.2.3 Assessment of results

The quality of a map can be expressed in terms of the parameters semantic accuracy and geometric accuracy (Chap. 17). The accuracy assessment presented below corresponds mainly to semantic accuracy. The results were verified with aerial photos and ground-truth survey. 334 ground control points were chosen by systematic sampling. To ensure adequate coverage of units of small size and/or of sinuous forms, and hence undersampled by this method, 28 additional points were selected by directed sampling. In total, 362 points were used for assessment of results.

■ Semantic evaluation of automatic mapping vis-à-vis ground truth

The nomenclature chosen as reference was that of CORINE Land Cover. Some classes recognised in aerial photos did not figure in the land-cover map obtained from computerised classification. The various errors are discussed below.

□ Errors associated with automatic mapping method

Some classes such as 'Fruit trees and berry plantations, road and railway networks' were not identified since it is impossible to distinguish them by spectral criteria alone without structural and shape criteria. The boundaries of the class 'watercourses' were obtained from supervised classification. All the 15 sampling points were correct with no affectation errors. The use of a 5 × 5 window for spatial generalisation can explain overestimation of 'continuous and discontinuous urban fabric' at the expense of 'non-irrigated arable land'.

□ Errors associated with use of single-date image

Confusions between 'pastures' and 'non-irrigated arable land' and between various types of forests are eliminated to a great extent by using another image of a different date, such as June 1995, corresponding to other physiological states of vegetation.

□ Errors due to content of certain CORINE classes

The classes 'transitional wood-land scrub[1]', 'complex cultivation patterns[2]' constitute 'mixed' classes, considering a minimum area of 25 ha. Here it corresponds only to a verification point, which explains their non-identification in the classification and the affectation errors observed.

□ Errors related to the nature of some CORINE classes

These errors are associated with 'industrial or commercial units, airports, construction sites, mineral extraction sites, green urban areas, sport and leisure facilities' and 'water bodies' which in reality do not correspond to a land-cover type but a land-use category. In fact, the first class represents mineral areas, hence classified as 'continuous urban fabric', whereas 'airports' in the region are large herbaceous areas and hence classified as 'pastures' or even 'crops'. The 'green urban areas' and 'sport and leisure facilities' are likewise grouped under 'forests' or 'pastures'. In the case of 'mineral extraction sites', which include underwater gravel, some may be confused with 'water bodies' or even with 'sport and leisure facilities'.

■ Use of the concept of fuzzy subgroups

The existence of the four types of error mentioned above has led not only to a quantitative assessment but also to a qualitative evaluation of accuracy of this classification. The latter is established according to the ultimate use of the map obtained from the decision of an expert who identifies the severity of errors. For example, confusion between 'broad-leaved forests' and 'continuous urban fabric' is more serious than between 'broad-leaved forests' and 'mixed forests'. This error is expressed in the form of

---

[1]Definition: shrubby or herbaceous vegetation with scattered trees. Such formations may result from degradation of forest or recolonisation/regeneration by forest. *Cited in*: IAURIF, 1995.

[2]Definition: Juxtaposition of small plots of various annual crops, grasslands and/or permanent crops. *Cited in*: IAURIF, 1995.

a notation according to the following linguistic scale, close to that described by Gopal and Woodcock, 1994:

5: title of class exactly corresponds to the land-cover type—no error;
4: title of class pertains to a similar land-cover type—small error;
3: title of class corresponds to a neighbouring category—medium error;
2: title of class pertains to a different category—large error;
1: title of class pertains to a totally different category—very large error.

In the context of land-use type, expert judgement enables construction of a table of affectation errors between the land-cover map and the CORINE nomenclature (Table 19.3).

Table 19.3: An expert's qualitative assessment of errors of affectation between the land-cover map and CORINE nomenclature for verification points

CORINE Land Cover classes	1.1.1	1.1.2	2.1.1	2.3.1	3.1.1	3.1.2	3.1.3	5.1.1	5.1.2
Continuous urban fabric	5	4		1	1				
Discontinuous urban fabric	4	5	1	2	1		1		1
Industrial or commercial units		4	1					1	
Roads and rail networks	3			1					
Airports		1	1	1					
Mineral extraction sites	1		1						4
Construction sites			1	1					
Green urban areas	3		2	2	2		2		
Sport and leisure facilities	3		1	1	2				1
Arable land	1	1	5	3	1		1		1
Fruit trees and berry plantations	1		4	3					
Pastures	1			5					
Complex cultivation patterns	1		4	4		2			
Broad-leaved forests	1			3	5	4	4		
Coniferous forests						5			
Mixed forests					4	4	5		
Transitional wood-land scrub	1	1					4		
Watercourses	1	1						5	
Water bodies									5

It was thus possible to establish a fuzzy relationship between various units and accuracy levels (Table 19.4), based on the preceding Table and the error matrix using the 362 control points.

The units 'watercourses, old gravel pits, coniferous forests, non-irrigated arable land, industrial units' have more than 80% points classified in category 5 (no error), sometimes with a certain percentage in class 4 (small error). The unit 'mixed forests' has 98% of its points classified in categories 5 and 4. The units 'continuous and discontinuous urban fabric' and 'broad-leaved forests' have more than 70% points classified in category 5, with about 10% points in category 1 (very large error). Lastly, the unit 'pastures' shows 58% points distributed in categories 1, 2 (large error) and 3 (medium error). In the last case, these errors are due partly to the use of a single date and partly to the nature of CORINE

Table 19.4: Representation of the fuzzy relationship (expressed as percentage) between the land-cover classes and qualitative accuracy classes for comparison with control points

Classes	Qualitative accuracy notation				
	1	2	3	4	5
Continuous and discontinuous urban fabric	14	2	3	1	80
Industrial or commercial units	4	0	0	11	85
Arable land	9	0	7	2	82
Pastures	17	12	29	0	42
Broad-leaved forests	8	3	0	13	76
Coniferous forests	0	0	0	20	80
Mixed forests	0	2	0	24	74
Watercourses	0	0	0	0	100
Old gravel pits	3	0	0	0	97
For entire image	11	3	6	6	74

Land Cover classes. The entire image shows more than 74% points as very correctly classified (category 5) and 11% with very large error (category 1).

In conclusion, while the accuracy of this map vis-à-vis the control points is overall acceptable for an user (Kappa coefficient equal to 0.72), the magnitude of errors varies according to categories, which may have more or less serious consequences depending on the ultimate use of the map.

■ Comparison between land-cover map and CORINE Land-Cover map

Overall, 53% points are classified correctly (Kappa coefficient equal to 0.50) but very large disparities are observed depending on the classes.

□ Very correctly classified classes

There are no classes with high percentage of accords and low values of commission and omission errors.

□ Correctly classified classes

'Pastures' and 'water bodies' as well as 'continuous and discontinuous urban fabric' and 'industrial or commercial units' have a high percentage of accord and omission errors, while 'watercourses' and 'non-irrigated arable land' have medium percentages of accord and omission errors and low levels of commission errors. The reasons for this poor classification have already been given.

□ Incorrectly classified classes

These classes exhibit low percentage of accords, medium level of commission errors and high omission errors. They comprise 'broad-leaved forests' confused, on the one hand, with 'discontinuous urban fabric' (question of a spectral differentiation threshold between these two categories) and, on the other, with 'green urban areas' and 'sport and leisure facilities' (difficulty in identifying land-use classes and not land-cover classes). Confusions with 'arable land' are due to the effects of the 5 × 5 window.

As earlier, an accuracy assessment of errors was also done. The results are shown in Table 19.5.

For the entire image, 59% points are classified without error and 71% correspond to the accuracy classes 5 and 4. Contrarily, 18% points correspond to a very large error (category 1). Detailed analysis of the units shows that only the class 'watercourses' has 74% of its points in category 5, while the classes 'coniferous forests' and 'mixed forests' have no points in category 5 and 92% and 82% points, respectively, in category 4. The classes 'broad-leaved forests', 'old gravel pits' and 'industrial or commercial units' have, respectively, 73% and 62% points in categories 5 and 4, whereas tow-thirds of

Table 19.5: Matrix representation of fuzzy relationship (in percentage) between units of the land-cover map and qualitative accuracy categories for comparison with the CORINE Land-Cover map

Classes	Qualitative accuracy notation				
	1	2	3	4	5
Continuous and discontinuous urban fabric	44	6	4	4	42
Industrial or commercial units	38	0	0	31	31
Arable land	29	0	15	1	55
Pastures	10	17	63	8	2
Broad-leaved forests	19	5	3	45	28
Coniferous forests	0	8	0	92	0
Mixed forests	18	0	0	82	0
Watercourses	26	0	0	0	74
Old gravel pits	27	0	0	40	33
Entire image	18	3	8	12	59

points for the last two units lie in category 1. 'Arable land' and 'continuous and discontinuous urban fabric' respectively show 55% and 42% points in category 5, with 29% and 44% points in category 1. Lastly, 'pastures' have 63% points in category 3, the remaining being distributed in other classes.

Low correlation was observed between the land-cover map and the CORINE map and hence explanatory hypotheses have been simply indicated. However, for a better understanding of the sources of these errors, the CORINE Land-Cover map has also been compared with aerial photographs and ground data.

■ Verification of CORINE Land-Cover map with aerial photos and ground data

A complete analysis of various maps also necessitated verification of the 362 reference points of the CORINE Land-Cover map using aerial photography and ground investigations (Kappa coefficient equal to 0.56 and overall accuracy 0.58).

Various cases can be distinguished but the unit 'transitional wood-land scrub', which had no control points, will not be discussed here.

□ Unmapped classes

A study of the error matrix (Table 19.6) indicated certain classes (construction sites, fruit trees and berry plantations, coniferous forests, mixed forests) identified from aerial photos and on the ground but not represented in the result. In the case of 'construction sites', their nature and the difference in the year of acquisition of SPOT data and aerial photos explain why they were not mapped. For three other classes, the 1:100,000 scale of the map has certainly resulted in a simplification of boundaries and omission of small-size zones by the mapping personnel. These unmapped classes contributed to the percentage of commission errors for the units 'continuous urban fabric, non-irrigated arable land, decidous forests'. Moreover, the date of acquisition of the scene used for classification (17 May 1989) was perhaps less favourable for identification of these themes.

□ Class readily identified by its specific shape

The theme 'airports' represents this case. The results indicated 50% accords and 0% commission errors. The 50% omission errors can be explained to be due to the map scale used.

Table 19.6: Error matrix of the CORINE Land-Cover map vis-à-vis aerial photos and ground investigations

	Aerial photos	CORINE Land Cover																				
		111	112	121	122	124	131	133	141	142	211	222	231	242	311	312	313	324	511	512	Total	Accords (%)	Omis-sions (%)
1.1.1	Continuous urban fabric	**1**	16	1			1		1		4							1			25	4	96
1.1.2	Discontinuous urban fabric		**33**						1		13			3	3			2	2		57	58	42
1.2.1	Industrial or commercial units		3	**4**			1				2							1	1		12	33	67
1.2.2	Roads and rail networks		1		**1**				1	1	2										6	17	83
1.2.4	Airports			1		**2**					1										4	50	50
1.3.1	Mineral extraction zones		1				**7**												1		9	78	22
1.3.3	Construction sites	1		1							1										3	0	100
1.4.1	Green urban areas		5							1											6	0	100
1.4.2	Sport and leisure facilities		1						1	**4**					1						7	57	43
2.1.1	Arable land		2								**114**			2							118	97	3
2.2.2	Fruit trees and berry plantations		2								7				1						10	0	100
2.3.1	Pastures									1	9		**1**	3	1						15	7	93
2.4.2	Complex cultivation patterns		1																		1	0	100
3.1.1	Broad-leaved forests								1		9				**23**						33	72	28
3.1.2	Coniferous forests													1	7						8	0	100
3.1.3	Mixed forests		1								3				21						25	0	100
3.2.4	Transitional wood-land scrub																	**1**			1	100	0
5.1.1	Watercourses			1															**14**		15	93	7
5.1.2	Water bodies		1								2									**5**	8	63	37
Total		2	67	8	1	2	9		5	7	167		1	9	57			5	18	5	362		
Percentage of commission errors		50	51	50	0	0	22		100	43	32		0	100	60			80	22	0			

Accords (number of correctly classified pixels).

□ Correctly mapped classes

The criteria of shape and location, combined with texture, enable correct visual identification, as illustrated by the small number of omission errors. Such units are 'mineral extraction sites' (78% accords, 22% commission errors and 22% omission errors), 'watercourses' (97% accords, 22% commissions and 3% omissions) and 'non-irrigated arable land' (97% accords, 32% commissions and 3% omissions). The commission errors of this class are due to the scale of the map, which assigns to it zones smaller than 25 ha in area belonging to other units, especially 'discontinuous urban fabric, fruit trees and berry plantations, pastures, broad-leaved forests'.

□ Classes incorrectly classified due to inappropriate map scale

Such classes are 'continuous urban fabric' (4% correct, 50% commission errors and 96% omission errors), 'discontinuous urban fabric' (58% accords, 51% commissions and 42% omissions), 'roads and rail networks and associated land' (17% correct, 0% commissions and 83% omissions), 'pastures' (7% correct, 0% commissions and 93% omissions) and 'broad-leaved forests' (72% accords, 60% commissions and 28% omissions). The scale of map representation could be the source of omission errors, as well as commission errors from other classes, as just mentioned above.

□ Classes incorrectly classified due to their definition

The concept of land use contained in the definition of classes is also a source of errors for some units such as 'sport and leisure facilities' (57% correct, 43% commission errors and 43 % omission errors), 'industrial or commercial units' (33% accords, 50% commissions and 67% omissions), 'green urban areas' (0% correct, 100% commissions and 100% omissions) and 'water bodies (63% correct, 0% commission and 37% omission errors).

□ Classes incorrectly classified due to unknown causes

A systematic verification was done in more than 362 points over the entire region of Ile-de-France for the class 'complex cultivation patterns'. Of the 88 map zones existing in the CORINE Land-Cover, 3 are correct and the other 85 correspond to villages.

An error matrix of qualitative accuracy assessment was established and applied to the CORINE map (Table 19.7). It shows 82% of the total points classified in categories 4 and 5 and 13% in category 1.

Classes showing the smallest errors (75% points in categories 4 and 5) are 'coniferous forests, pastures, mixed forests, green urban areas and broad-leaved forests'.

Classes with the highest errors (more than 25% points in categories 1 and 2) are distributed in several classes.

1. *Errors (omission or commission) related to year differences* between the SPOT data used for CORINE Land Cover, aerial photographs and ground check-up. They are associated with the classes amenable for rapid changes: 'construction sites, discontinuous urban fabric, industrial or commercial units, transitional wood-land scrub' and 'non-irrigated arable land'.
2. *Errors (omissions) related to scale of the CORINE Land-Cover map* and small sizes of areas mapped (25 ha and linear objects less than 100 m long). These are 'roads and rail networks' and 'water bodies'.

It follows from these various analyses and comparisons that automatic land-cover mapping does not permit representation of certain classes of the CORINE Land Cover, since they correspond in fact to land-*use* types but not land-cover units. Visual interpretation of satellite data enables mapping of these classes by the shape and geographic location of objects. To improve the automatic mapping it will be necessary to employ the criteria of shape and neighbourhood, as well as a cross-check with digital elevation models for formulating decision rules (expert systems) for mapping these classes.

Table 19.7: Matrix representation of the fuzzy relationship between CORINE Land-Cover map and qualitative accuracy notations for comparison with control points.

Classes	Qualitative accuracy notation				
	1	2	3	4	5
Continuous urban fabric	0.2	0	0.11	0.65	0.04
Discontinuous urban fabric	0.33	0	0.01	0.30	0.36
Industrial or commercial units	0.30	0	0.05	0.40	0.25
Roads and rail networks and associated spaces	0.32	0.17	0.17	0.17	0.17
Airports	0.25	0	0	0.25	0.50
Mineral extraction zones	0.09	0	0.27	0	0.64
Construction sites	0.33	0	0.33	0.33	0
Green urban areas	0	0.09	0.09	0.82	0
Sport and leisure facilities	0.10	0.20	0	0.30	0.40
Non-irrigated arable land	0.32	0	0	0.01	0.67
Fruit trees and berry plantations	0.20	0	0.10	0.70	0
Pastures	0.07	0	0.07	0.79	0.07
Complex cultivation patterns	0.44	0	0	0.56	0
Broad-leaved forests	0.06	0.01	0.16	0.42	0.35
Coniferous forests	0	0	0	1	0
Mixed forests	0.04	0	0.12	0.84	0
Transitional wood-land scrub	0.80	0	0	0	0.20
Watercourses	0.26	0	0	0	0.74
Water bodies	0.37	0	0	0	0.63
For entire image	0.13	0.01	0.05	0.22	0.60

19.3 CONCLUSION

Land-cover mapping constitutes an essential information for environmental management. The CORINE Land-Cover program produces maps with the same nomenclature for all the European countries. Hence this is an important stage in enhancing information on environment. However, the scale of the map is 1:100,000 and we should not search for more precise information than that permitted by the rules applied in preparing this map. Moreover, zones of certain classes undergo very rapid temporal variations, which diminishes the value of information. Visual interpretation enables mapping of land-use classes, which is not possible from computerised classification alone.

References

Bossard M. 1986. Production de la base de données CORINE Land Cover. Séminaire IFEN 'CORINE Land Cover'. 11–13 décembre 1996, pp. 11–18.

Cornaert MH, Maes J. 1992. Land cover, an essential component of the CORINE information system on the environment. GIS implications, Proc. Central Symposium 'International Space Year' Conf., Munich, Germany, 30 March–4 April 1992. ESA SP-341, pp. 473–481.

Girard M-C, Yongchalermchai C, Girard C-M. 1992. Analyse d'un espace par la prise en compte du voisinage. Gestion de l'espace rural et système d'information géographique. Séminaire de Florac, 22–24 Oct. 1991. INRA, Paris, pp. 349–359.

Gopal S, Woodcock CE, 1994. Theory and methods for accuracy assessment of thematics maps using fuzzy sets. Photogrammetric Engineering & Remote Sensing, 60: 181–188.

IAURIF. 1995. Les 'ecozones' d'Île-de-France, dans le cadre du programme européen CORINE Land Cover affiné. D8-334, Région Île-de-France, Centre national d'études spatiales, 28 pp.

IFEN. 1995. Programme CORINE Land Cover, 8 pp.

Lillesand TM, Kiefer RW. 1994. Remote Sensing and Image Interpretation. John Wileys Sons, NY, 750 p.

Multiscope. 1993. Manuel d'utilisation, 150 pp.

20

Herbaceous Formations and Permanent Grasslands

Some applications of remote sensing in studies of herbaceous formations and permanent grasslands, mainly of temperate zones and a few examples of intertropical zones, are illustrated in this chapter. Humid grasslands and marshes will be discussed partly in this chapter and partly in Chap. 21 — Wetlands.

20.1 TERRESTRIAL HERBACEOUS FORMATIONS

Botanical and biogeographical investigations have resulted in identification of several types of herbaceous formations in the world. The first level of distinction is based on climax[1] or other origin of these formations, the second on perenniality of vegetation species and the third on major climatic types.

Various types of herbaceous formations are represented by varied floristic compositions and evolution dynamics, knowledge of which is imperative for managers (Table 20.1). In fact, while the area of secondary herbaceous formations in intertropical zones increases at the expense of tree formations, an inverse trend is observed in temperate zones (particularly in Europe), permanent grasslands being in regression to the advantage of crops as well as natural or induced reforestation. Thus, in France, 3.5 Mha (25% of the initial area) of permanent grasslands had disappeared between 1970 and 1995, of which 67% were replaced by cereal crops, 10% by fallow lands and 3% by woods (IFEN, 1996).

Various types of herbaceous areas, including temporary, artificial pastures and permanent grasslands, in the nomenclature of the *Service Central des Etudes Economiques et Statistiques (SCEES)* (Central Service of Economic and Statistical Studies), are given in Table 20.2.

Herbaceous formations are characterised by highly variable quantities of aerial or subsurface biomass depending on their climax situation, degree of perenniality and mode of usage. Decrease (or disappearance) of pasture pressure leads to large accumulation of aerial phytomass, dry fraction in particular, accompanied by an increase in litter, while quantities of root biomass are relatively stable.

Differences in the quantity of fresh phytomass (a) and percentage of dry matter in it (b) for three types of permanent grasslands corresponding to three different levels of exploitation are shown in Fig. 20.1. Decrease in exploitation results in increase in the aerial phytomass (a). Contrarily, phytomass of a non-exploited rangeland is less since it corresponds to a botanical unit other than the two grasslands. This shows the importance of differentiating grasslands and herbaceous formations according to their botanical and agronomic type. Increase in percentage of dry matter in aerial phytomass with decrease

[1]Climax: state of an ecosystem having reached a relatively stable stage of equilibrium (at least in the human scale), conditioned only by climatic and edaphic factors (Delpech et al., 1985).

Table 20.1: Major types of herbaceous formations in the world

Climax herbaceous formations	Perennial species	Subpolar frigid regions	Herbaceous tundra
		Temperate regions with subarid or dry periods	Continental steppes
		Hot regions	Primary savannas
		Montane supraforest stages	Alpine greenward
		Littoral zones, temperate regions	Salt marshes
	Annual species	Mediterranean subarid regions	Steppes
		Subtropical semi-arid regions	Pseudo-steppes
Non-climax herbaceous formations	Semi-natural secondary formations	Humid temperate regions	Permanent grasslands
		Tropical regions, non-arid zones	Savannas, some pampas
	Artificial	Mainly annual or biannual crops	Pure or mixed crops
		Mainly multi-annual crops	Temporary artificial leys

in exploitation or its absence is seen in Fig. 20.1 (b). This variation in percentage of dry matter with level of exploitation plays an important role in studying herbaceous formations and permanent grasslands by remote sensing.

20.2 PROBLEMS

Herbaceous formations and permanent grasslands perform various functions:

1. Production of grass and fodder for feeding the cattle and wild fauna;
2. Protection and conservation of environment (protection areas in sloping basins for tapping potable water, protection of soils against wind- or water-borne erosion, etc.);
3. Conservation of vegetation and animal species;
4. Contribution to aesthetic quality and diversity of landscapes, etc.

Table 20.2: Categories of herbaceous areas (after SCEES)

	Duration	Species or type	Category
HA: Herbaceous area comprised in UAA*	Cultivated grasslands of 0 to 5 years	Pure legumes or mixed legumes	Artificial leys
		Italian rye-grass or hybrid rye grass	Short-duration temporary leys
		Pure graminaceous plants or mixed graminaceous-leguminous plants or mixed graminaceous plants	Temporary leys
Outside UAA	Cultivated grasslands of 6 to 10 years	Variable floristic composition	Long-duration temporary grasslands
	Uncultivated or cultivated grasslands of more than 10 years	Arable	Arable PGA**
		Non-arable	PGAC (compulsory)
		Others (rangelands, shrubs, heaths, high-altitude pastures, marshes, etc.)	Rough grazings

*UAA (Used Agricultural Area) = crops + HA (Herbaceous Area)
**PGA: Perennial Grassy Area

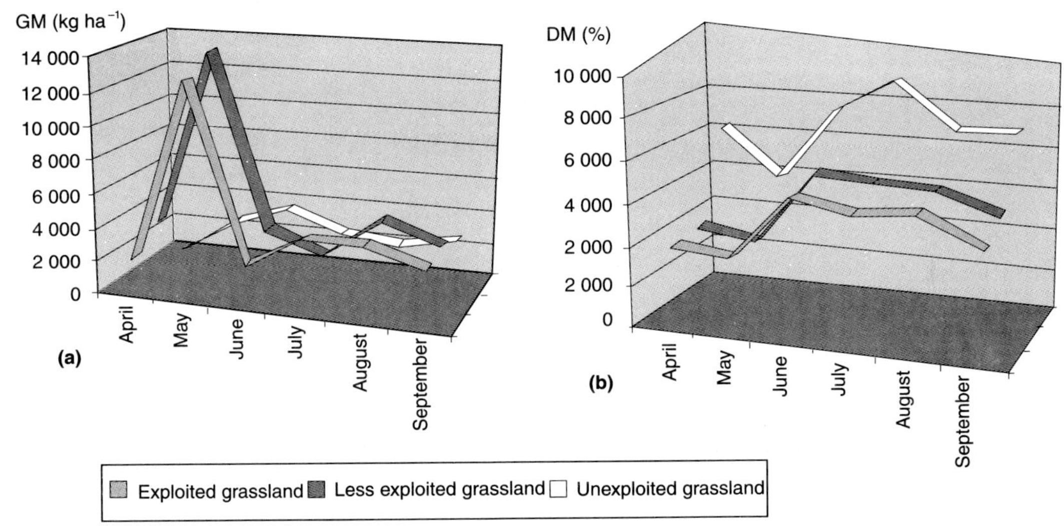

Fig. 20.1: Effect of decrease or absence of exploitation on (a) aerial phytomass (green matter, GM, in kg/ha^{-1}) and (b) percentage of dry matter, DM, contained in aerial phytomass.

Information sought by thematic specialists and managers is hence concerned with the following aspects:

1. identification of vegetation species, vegetation groups and units or types of grasslands;
2. geographic distribution and estimation of areas;
3. quantity assessment of aerial phytomass (production and productivity);

4. quality evaluation of aerial phytomass;
5. facilitating prediction of dynamics of plant communities (spontaneous or induced by usage, agricultural practices, variations in management, etc.).

The main difficulty in using remote sensing for such applications lies in the difference between the parameters investigated and those measured by remote sensing. Hence, it is necessary to establish relationships between various disciplines and the concerned typologies (Fig. 20.2). The latter are based on additional observations and measurements and on records of acquisition of data different from those of normal practice. These will be developed while discussing the applications.

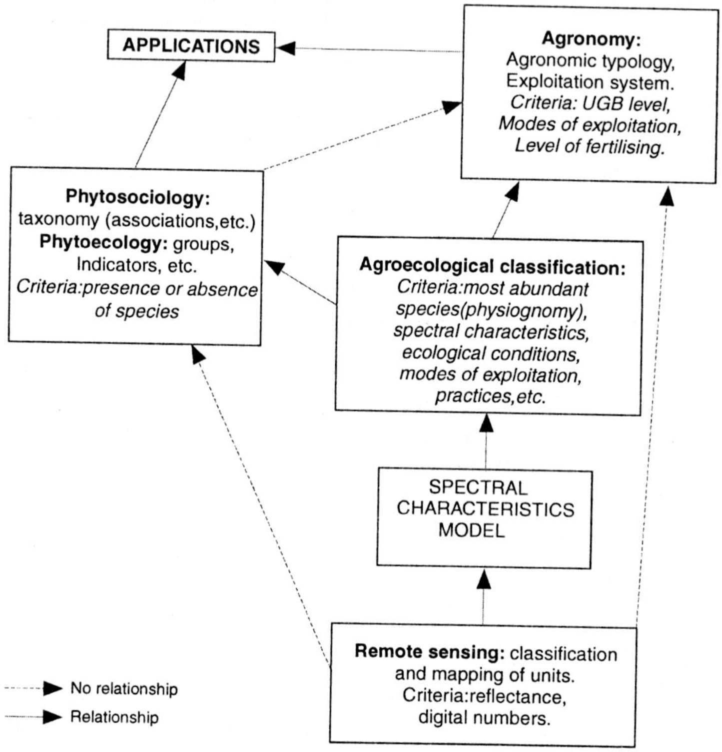

Fig. 20.2: Model of use of remote sensing data for applications in grassland and pasture studies.

20.3 APPLICATIONS

20.3.1 Distribution of herbaceous formations and permanent grasslands

Managers require information on distribution of herbaceous formations at a national level or at a more local level. This information is essential for estimating surface variations over time (see the beginning of this chapter). In fact, this variation represents the impact of human activity, viz., clearing of forests and creation of secondary grasslands in intertropical zones, abandoning permanent grasslands and cultivation or afforestation in temperate zones, etc. These changes have a direct impact on climatic changes (global change) and quality of water resources (nitrate pollution in the case of conversion of grasslands into crops), as well as on soil erosion.

The objective of application of remote sensing in grassland studies consists of separating such areas from other land-cover types. The European Union countries now are equipped with this type of information thanks to CORINE Land-Cover data (see Chap. 19). However, the precision and re-actualization of CORINE data are not necessarily suited to all applications investigated. Hence, we present below a method of studying satellite data. An example on wetlands is given in Chap. 21.

Objective: Separation of herbaceous formations and grasslands from other land-cover types.

Principle: Selection of satellite data acquired on days when the herbaceous formations and permanent grasslands exhibit very distinct spectral characteristics, i.e., specific physiognomic features. Depending on the region, a single or multiple dates of acquisition will be necessary. Thus, in Lorraine, a scene taken at the end of winter enabled identification of permanent grasslands and rangelands from other land covers (Girard and Benoit, 1990), whereas in Ile-de-France a scene of end-winter and another of spring were found necessary.

Methodology: Application of an unsupervised classification method (see Chap. 8) on the scene(s). If accurate data on spectral characteristics of various land-cover types are available, supervised classification can also be employed. The results can be refined by application of geographic or radiometric masking (see Chap. 7).

Preparation of a map necessitates georeferencement of the results obtained (see Chap. 13).

20.3.2 Identification of vegetation species and groups

A vegetation group is a collection of living species in a given environment. Some species are strictly associated with well-defined media and others may occupy entirely different environments. If we accept that vegetation groups are good indicators of environmental factors and that they react to modes of exploitation or usage by man, we can constitute groups of indicator species for these factors (Fig. 20.2). It is interesting to identify such groups by remote sensing and thus indirectly acquire spatial information on ecological factors.

A species is identifiable by remote sensing if it exhibits on a given day a specific spectral characteristic (reflectance values, digital numbers) enabling its identification. This spectral characteristic is related to the phenological stage and physiological state of the species. In practice, unambiguous identification of a particular species is rare. It is possible for some species and on limited days. For example, in rough grazings, quasi-pure populations of *Calluna vulgaris* can be identified in satellite data acquired in spring, since this species has a spectral characteristic in the near infrared band (type A, see Chap. 4, Fig. 4.6) entirely differing from that of neighbouring graminaceous species. The graminaceous species *Brachiaria dictyoneura* can be distinguished from other forage and savannah grass species using reflectance measurements made in a dry season (see *infra*: example of Llaños in Colombia).

A vegetation group consisting of several species can be identified due to specific physiognomy corresponding to the combination of phenological stages and physiological states of the most abundant species in the group. Again, this identification may be possible on a particular date and impossible on another. Such identification assumes that the botanical composition of the group and an interpretation model of spectral behaviour are available.

■ Spectral behaviour model

For multi-annual and multi specific vegetation with large coverage (usually > 70%), such as permanent grasslands in temperate zones, various situations can be recognised: chlorophyllian, senescent or florescent species. Other situations are the case (for example, after mowing) in which bare soil has a small spectral contribution (sparse savannas or crops) or the case of large shadows. These various behaviours are summarised in Fig. 20.3.

The spectral contribution of nonchlorophyllian plants in case 2 (Fig. 20.3) results in stronger reflectance in the visible and middle infrared bands and weaker reflectance in the near infrared, relative to case 1. A limited spectral contribution of soil (case 3) enhances the reflectance of vegetation in the visible band and reduces it in the near infrared. The moisture content of soil surface is assumed to be close to that of vegetation: reflectance values in middle infrared are similar. Pigments of many coloured flowers (case 4) modify the reflectance only in the visible band and the spectral characteristic is constant in the other two spectral domains. Lastly, contribution of shadows (case 5) drastically reduces reflectance in the near and middle infrared bands. This effect is counterbalanced in the visible band by increase in reflectance due to bare soil.

■ Examples

Herbaceous formations and permanent grasslands are differentiated from crops due to the fact that they are multiannual, multispecific and highly heterogeneous populations. As a matter of fact, temporal succession of vegetal species according to their coverage and phenological stage adds up to the

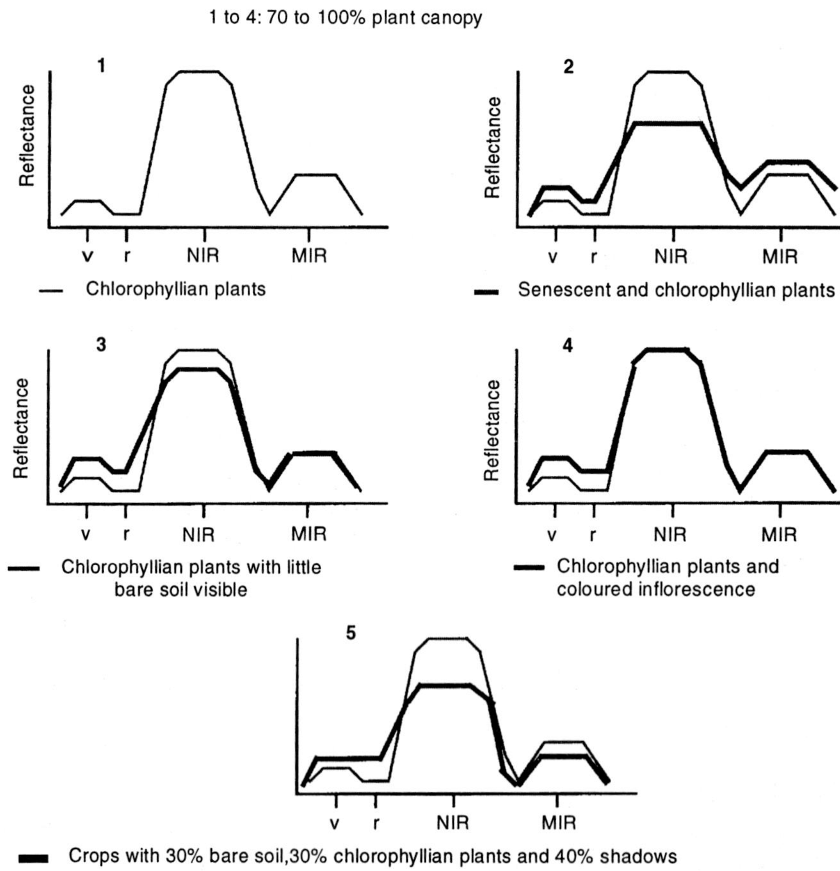

Fig. 20.3: Types of spectral characteristics.

effects of mode of exploitation. Spectral properties of vegetation covers in the visible and especially in the near infrared band (see Chap. 4) enable differentiation of grasslands with dominance of graminaceous species from grasslands with dominance of leguminous and dicotyledonous species. If the spectral characteristic of a permanent grassland is taken in April (Fig. 20.4), rye-grass and colchicum species dominate. In May, the yellow inflorescence of abundant Ranunculus increases the reflectance values in the visible band and the abundance of standing green biomass results in high values in the near infrared. Mowing in June, which decreases the quantity of chlorophyllian aerial phytomass, leads to lower reflectance values in the near infrared, whereas values in the visible band are close to those of May. In July, vegetation dominated by Yorkshire fog again becomes chlorophyllian and dense: the spectral characteristics are close to those of April.

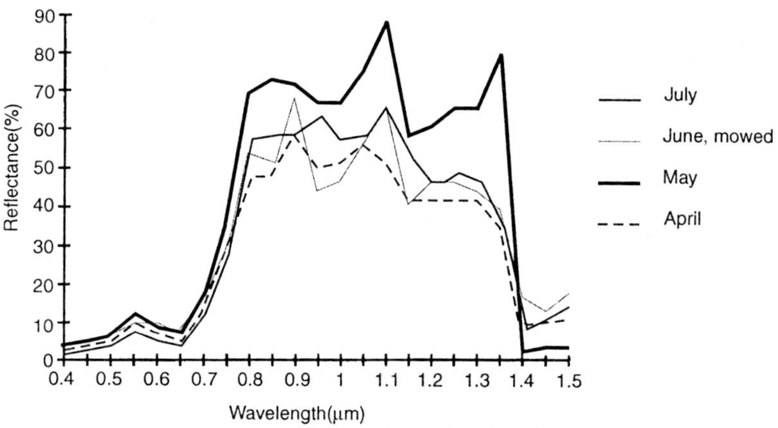

Fig. 20.4: Ground reflectance measurements for permanent grasslands from April to July.

In a given region, knowledge of phenological stages and calender of various agricultural operations is imperative (Table 20.3).

In the example given above, 4 satellite images acquired in end April, May–June, July and August–September differentiated 7 out of 10 grassland types and only the types 7, 8 and 9 could not be separated.

The large range of physiognomic variations created by combinations of states and modes of exploitation enables identification of various herbaceous formations or grassland types, provided remote sensing data acquired on precisely chosen days during a year are used. This is not always possible due to frequent cloudiness of grassy regions. Lastly, the need for diachronic data enhances the financial costs of study.

Identification of an IRC type colour composite or other combination of bands by visual interpretation on computer monitor (or in printed hard copy) can be done for units of small extent (a few pixels). Contrarily, automated classification by supervised or unsupervised methods necessitates a larger number of pixels (see Chap. 16). Since identification of vegetation species and groups requires an accurate knowledge of the terrain, supervised methods are preferable over unsupervised ones (see Chap. 8).

20.3.3 Mapping of units and estimation of areas

Geographic distribution of rangelands and permanent grasslands and estimation of corresponding areas are necessary for determining the forage resources of a region.

Table 20.3: Example of calender of agricultural operations in permanent grasslands in Lorraine (after Benoit et al., 1988)

	Months					Grassland types
	April	May–June	July	August–September	October	
	--------	---------	M	****************	********	1
	---------	---------	M	M	---------	2
	---------	M	---------	****************	********	3
	---------	M	*******	****************	********	4
	---------	M	---------	M	---------	5
	---------	E	M**	****************	*********	6
	*****	****************	****************	****************	********	7
	*****	****************	****************	****************	********	8
	*****	****************	****************	****************	********	9
------	--------	---------	---------	---------	---------	10
	D	D	D	D	Recommended dates of image acquisition	

M: mowing, E: ensilage, * * * * * * grazing --------no operation

An example of mapping the savannas and pastures in Llaños in Colombia illustrates the procedure adopted. These savannas cover an area of 3.5 Mha and constitute environments for which relationships between vegetation and ecological factors are not known. Land management in this region necessitates a precise inventory of crops, natural vegetation formations and zones of degradation. Moreover, while some natural plant communities and pastures have been accurately studied in highly localised sites, generalisation and extension of the results obtained to large areas constitutes a problem for further studies. The savannas and pastures in this region have been investigated as follows:

—Spectral reflectance measurements were conducted on the ground with a SPOT simulation Cimel radiometer (Girard and Rippstein, 1994) during January and late September to early October 1991. Simultaneously, height and density of plant cover, instantaneous production of aerial phytomass and water content were measured in reference sites. The objective of these measurements was to determine spectral characteristics of various savannas and fodder crops during dry and wet seasons, respectively, and relate them to various plant parameters.

—Multispectral SPOT data were acquired in March (dry season) and September (wet season) of 1991. Meteorological constraints did not permit simultaneous acquisition of ground truth and satellite data. It is generally difficult to obtain satellite and ground data simultaneously.

Ground truth measurement sites were used as reference zones for supervised classification of SPOT data of each date. Moreover, statistical processing (analysis of variance, PCA, AHC, etc.) of the ground truth data enabled establishing, for each period, groups of savannas and fodder crops significantly differing in reflectance. These groups served as a basis for preparing the legend of the map resulting from the two classifications (Table 20.4).

Relative estimate of areas is given by the number of pixels corresponding to each class, on the basis of a pixel of 400 m^2 in area. An absolute estimation requires geometric correction and restitution on a topographic base.

Table 20.4: Legend for the map of savannas and pastures in Llaños of Altillanura (Colombia)

Highly exploited fodder crops (degraded)	Sparse savannas (degraded)	Savannas on clay soils with *Trachypogon vestitus*
Fodder crops of <1 year	Savannas on clay soils	Savannas with *Andropogon gayanus*
Fodder crops of >1 year	Savannas on sandy soils	Savannas with termite nests
Old fodder crops	Savannas: burnt >1 year	Grassy lowlands
	Savannas: burnt <1 year	Shrubby lowlands

20.3.4 Aerial phytomass quantity estimation

Use of vegetation indices for estimating instantaneous production of aerial phytomass was mentioned in Chap. 4. Such an evaluation for herbaceous formations and permanent grasslands is beset with several difficulties:

—Difficulty in describing highly heterogeneous plant covers: erectophyll species (Gramineae, graminoids) mixed with plagiophyll species (dicotyledons and the leguminous species);

—Very large quantity of chlorophyllian aerial phytomass for some grasses, resulting in saturation of reflectance;

—Mixing of standing dry matter with green material for many types of plants, giving rise to variations in vegetation indices (Fig. 20.5);

—Sparse plant covers (some natural herbaceous formations) with a spectral contribution of soil, necessitating use of appropriate vegetation indices.

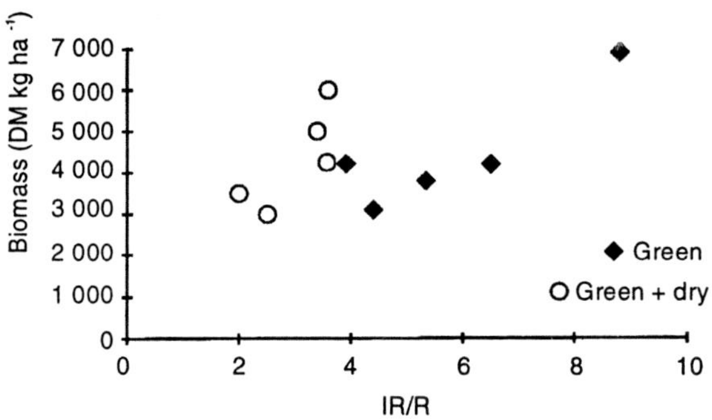

Fig. 20.5: Effect of dry plants mixed with chlorophyllian vegetation on the ratio IR/R.

For this unexploited (measured in April to October) calcicole rangeland (Fig. 20.5), a relationship between IR/R and aerial phytomass could be established only for green vegetation areas.

Asymptotic relationships between leaf area index and aerial phytomass (Fig. 20.6), vegetation index and leaf area index (Fig. 20.7) and vegetation index and aerial phytomass (Fig. 20.8) were computed for a permanent pasture.

However, regular grazing of grass does not allow observation of large quantities of aerial phytomass and reduces the validity of such a relationship (limited range of values).

Estimation of aerial phytomass production as a function of efficiency of interception of radiation, determined from remote sensing (linear relation between vegetation index and efficiency), may be envisaged. However, this assumes a temporal characterisation of 'states of efficiency' of the cover, which is difficult to achieve for very heterogeneous natural or semi-natural groups.

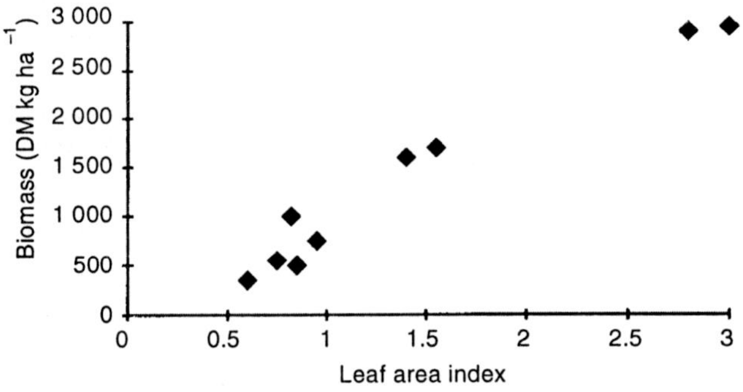

Fig. 20.6: Relationship between leaf area index and aerial phytomass for a permanent pasture in various stages of exploitation (r^2 = 0.96).

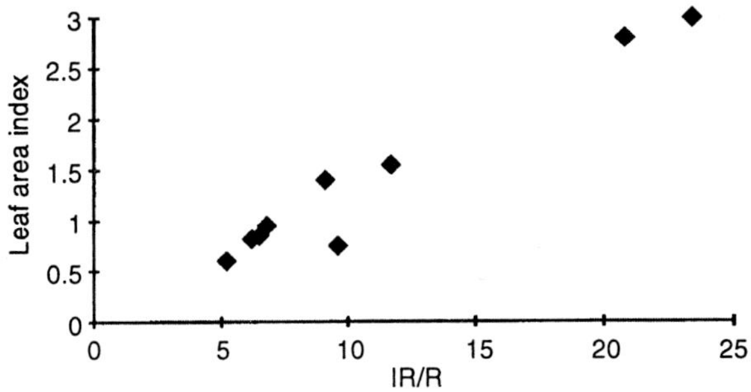

Fig. 20.7: Relationship between leaf area index and IR/R for a permanent pasture (r^2 = 0.95).

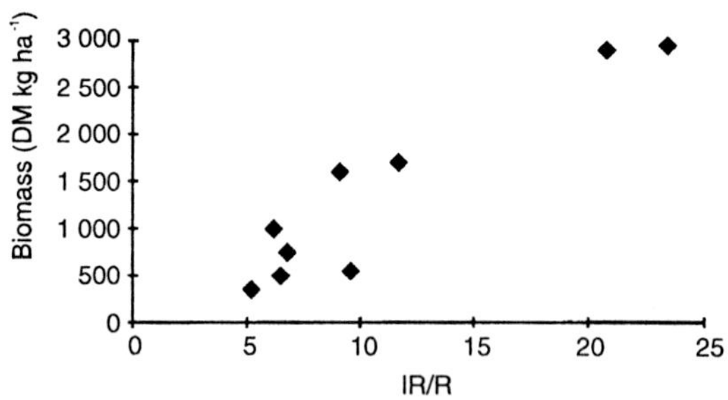

Fig. 20.8: Relationship between IR/R and aerial phytomass for a permanent pasture ((r^2 = 0.88).

Variation of fresh aerial phytomass (measured after extraction and weighing) for permanent grasslands of Lorraine with NDVI values (see Chap. 4) computed from reflectance measurements on the ground with a SPOT simulation Cimel radiometer is shown in Fig. 20.9. 200 measurements were made over the period from beginning-April to end-September for the years 1986 and 1987. A very strong dispersion of values and an insignificant exponential fit can be observed.

Contrarily, after separation of the grasslands into various types according to modes of exploitation and management (Benoit et al., 1988), the value of r^2 varies from 0.22 to 0.87 (Table 20.5) for an exponential fit.

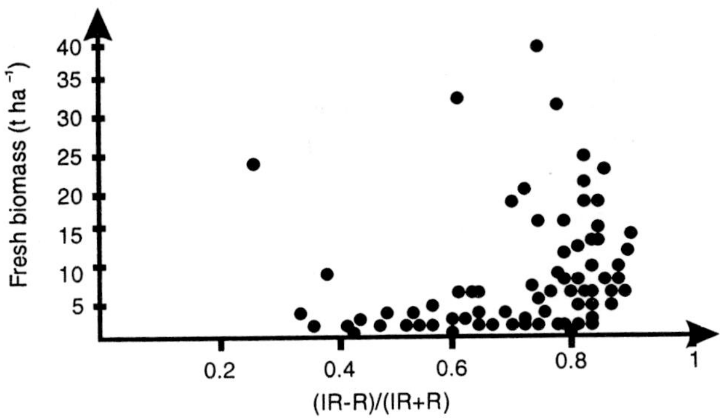

Fig. 20.9: Relationship between NDVI and fresh aerial phytomass for various permanent grasslands.

Table 20.5: Correlation (r^2 values) between vegetation index (obtained from ground Cimel radiometer measurements) and aerial phytomass for seven types of permanent grasslands of Lorraine

Grassland type	A6	A7	A8	A3	A1	A2	A9
Exponential fit for phytomass	0.87	0.55	0.55	0.41	0.69	0.22	0.56

Vegetation indices for estimating the quantity of aerial phytomass ought to be determined within preliminarily defined grassland types.

Grasslands (more particularly A6) for which the relationship between the two variables is strong are those for which there is no accumulation of standing dry nonchlorophyllian biomass due to regular removal of grass. When grass removal is large, i.e., 200 < UGB $ha^{-1}yr^{1}$< 500 (A7, A8, A9 and A1) and nitrogen fertilising more or less important, the range of measured biomass is relatively small, which explains the rather low values of correlation coefficient. When grass removal is small (A2, A3), standing dry matter accumulates and the value of r^2 becomes small.

20.3.5 Phytomass quality assessment

Evaluation of quality of phytomass, associated with identification of groups, is based on knowledge of specific indices attributed to various vegetal species according to their pastoral or forage importance. The 10- or 5-scale notation developed by Holland, New Zealand and French pastoral agronomists enables computation of the pastoral value (*PV*) of grasslands according to a formula of the type (for 10-scale notation):

$$PV = \Sigma(CS_i \times IS_i)/10$$

where CS_i is the specific contribution of species *i* (frequency of occurrence or coverage) and IS_i the specific index of species *i*.

Pastoral values thus obtained must be considered only as a means of quality assessment and comparison of grasslands. In fact, for the same vegetal species, the values of *IS* vary according to the stage of development, animal race that consumes it, etc., whereas a single *IS* value is assigned to a species once for all. These limitations notwithstanding, estimation of *PV* gives an approximate idea of the quality of biomass produced.

Types of perennial grasslands shown in Table 20.3, identifiable by remote sensing, are botanically defined by various combinations of the most abundant species. Mean *PV* values of each type, computed based on these species, differ significantly at 5% confidence limit. Hence, a spatial representation of the quality of perennial grasslands can be obtained.

Another estimate of the quality of aerial phytomass is the water content of standing grass. Water content can be related subsequently to the phenological stages. It serves as an indicator for fixing the date of first mowing in the case of permanent grasslands targeted for agro-environmental measurements. Various studies seem to indicate that water content can be estimated from the reflective middle infrared data (Fig. 20.10).

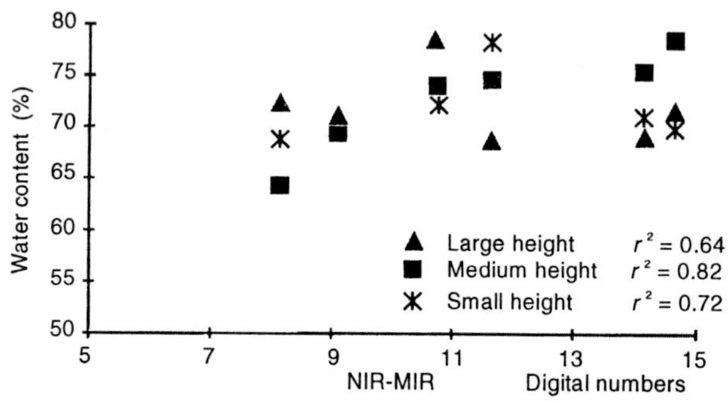

Fig. 20.10: Relationship between NIR-MIR and water content for various permanent grasslands (after Orth, 1996).

Water content of aerial phytomass of various marshes was determined using field samples and double weighing method (fresh grass and after 48 h in a drying oven). A relationship could be established between these values and the difference between ground measurements (NIR-MIR) of near- and middle-infrared bands (Cimel radiometer, SPOT-4 simulation).

The regression coefficient r^2 for all places of measurement is equal to 0.57, whereas it is 0.82, 0.64 and 0.72 when grasslands are divided according to grass height. This shows the influence of the type of canopy and the need to investigate correlation between factors within various grassland units.

An application of such results is in management of plots of permanent grasslands in marshes under agro-environmental contracts. In fact, the date of first cutting is fixed according to the dates of nesting of some bird species and applied to all plots. However, use and application of these contracts would be facilitated if the date of first cutting were regulated. A map of marshy grasslands of Cotentin and Bessin (Orth, 1996), prepared from TM 3, 4, 5 bands of a LANDSAT TM scene of 29 April 1990, showed large differences in commencement of vegetation in end-winter. These differences, due to the magnitude and duration of submersion, provide an objective basis for regulating the date of first cutting depending on plots, according to their stage of growth.

20.3.6 Monitoring and forecasting of changes in vegetation groups

Permanent grasslands almost everywhere in Western Europe and in Central Europe are subjected to underexploitation or even abandoned, due to changes in agricultural practices. This results in increase in dry matter and even in accumulation of standing dry matter, accompanied by changes in vegetation groups: conversion from calcicole pastoral grasslands to rangelands dominated by *Brachypodium pinnatum*, gradually encroached by thorny and shrubby species (wild briars, blackthorns, hawthorns, etc.). These modifications can be monitored by remote sensing (aerial photography or satellite data). Thus, in a small zone of French Vexin (west of the Paris Basin), permanent grasslands and calcicole rangelands could be identified in 1:25,000 aerial photos of 1968 and their status studied in 1:30,000 aerial photos of 1994. The following information could be inferred:

—certain level of maintenance of grassland areas in valleys (PGAC due to hydrological situation), replacement of some of them by crops notwithstanding;

—large growth of underbrush in pastures and rangelands of mid- and high slopes;

—quasi-disappearance of grasslands in plateaus.

Based on these results of photo-interpretations, a map of grasslands and rangelands could be prepared for western part of Ile-de-France using SPOT data of 1995. Comparison of this map with a map of calcicole rangelands prepared in the 1970s revealed disappearance of a number of these rangelands. This represents an adverse effect of agricultural community decisions. In fact, these rangelands were ploughed just before compulsory set-aside land establishment: they were hence counted in UAA and were left for natural colonisation (spontaneous fallow). Unfortunately, fresh growth of rangelands is a slow process: ploughing these areas increases splitting of habitats of particular fauna and flora, which are thus endangered.

Some fallow lands in the Paris region were also monitored by satellite data. For this, various types of fallow lands were identified on the ground, resulting in preparation of the physiognomic typology (bare soil, dominance of herbaceous species, mixture of grasses, shrubs and trees, etc.). Moreover, position of these types in the dynamics of vegetation groups was studied.

A diachronic study was subsequently made using LANDSAT TM data of 30 April 1984 and SPOT data of 24 April 1987 (Szujecka and Girard, 1990). A mask for eliminating land-cover types other than various types of fallow lands was designed and applied on both scenes. Each scene was then subjected to supervised classification into three classes, viz., young fallow land (sparse plant cover, bare soils present at places), herbaceous fallow land and shrubby fallow land. After geographic restitution accompanied by a resampling of pixels (imperative since the pixel size of TM is 30 × 30 m and that of SPOT 20 × 20 m) the two scenes became comparable close to pixel level. An image showing changes during 1984 to 1987 was prepared assigning a characteristic colour to each type of change. An example of the observed changes is given in Table 20.6.

Table. 20.6: Example of changes in fallow lands in Ile-de-France region between 1984 and 1987 derived from classification of satellite images

Classification in 1984	Classification in 1987	Interpretation
Non-fallow land	Herbaceous fallow land	Progressive change
Herbaceous fallow land	Herbaceous fallow land	Equilibrium within a stage
Shrubby fallow land	Shrubby fallow land	Equilibrium within a stage
Shrubby fallow land	Young fallow land	Regressive change: clearing before growth
Herbaceous fallow land	Non-fallow land	Change of land cover: construction of buildings

This example shows the role of remote sensing in land management, especially in the peripheries of urban areas where disappearance of patches of rural space makes it difficult to monitor sites characterised by multiplicity and geographic dispersion using ground data alone.

20.4 CONCLUSION

Remote sensing data provides valuable information for management of land cover and especially herbaceous formations and permanent grasslands. The most significant features are:

—capacity for spatialisation and generalisation of results acquired over limited areas;

—flexibility of processing digital data and possibility of analysing the results by geographic information systems;

—complementary information provided by diachronic data to other reference sources.

On the other hand, their usage ought to take into consideration the following constraints:

—relatively recent existence of these data;

—difficulty in acquiring pertinent data in cloudy regions (see Chap. 26);

—necessity for collecting ground data, particularly more severe in precise thematic applications;

—information concerning mainly the upper part of the plant canopy due to bias of their spectral characteristics;

—geometric resolution of data often inappropriate for the study of phenomena under investigation.

Hence, development of classification methods based on the theory of fuzzy groups and artificial intelligence seems very promising in this domain.

References

Benoit M, Girard C-M, de Vaubernier E. 1988. Comparaison du comportement spectral de prairies permanentes en Lorraine avec leur type d'utilisation. Agronomie, 8: 265–272.

Delpech R, Dume G, Galmiche P. 1985. Typologie des stations forestières. Paris, Ministère de l'Agriculture, IDF, 243 pp.

IFEN. 1996. Les données de l'environnement: milieu, 25: 4 pp.

Girard C-M, Benoit M. 1990. Méthode de cartographie des prairies permanentes: application à la Lorraine sur des données SPOT. C.R. Acad. Sci., Paris, 310 (III): 461–464.

Girard C-M, Rippstein G. 1994. Utilisation de données SPOT HRV pour la cartographie de savanes et pâturages dans les Llaños de Colombie. Bull. Soc. Française de Photogrammétrie et Télédétection, 133: 11–19.

Orth. 1996. Typologie et caractérisation des prairies permanentes des marais du Cotentin, en vue de leur cartographie, par télédétection satellitaire, pour une aide à leur gestion. Thèse INA-PG, 149 pp. plus annexes.

Szujecka W, Girard C-M. 1990. Cartographie et suivi diachronique des friches en Ile-de-France à partir de données TM et SPOT. Photointerprétation, 90: 1–3.

21

Wetlands

A programme for investigating wetlands was launched in 1996 by the Ministry of Environment, water agencies and a public interest group on hydrosystems (comprising the BRGM, CEMAGREF, CNRS, IFREMER, INRA, ORSTOM, etc.). However, among the projects undertaken under this programme, very few included remote sensing in their studies. We feel that it is useful to present some examples of application of remote sensing for the study and management of wetlands.

21.1 NATURE AND IMPORTANCE OF WETLANDS

According to the law on water (No. 92.3 dated 3 January 1992), '*wetlands mean regions, exploited or not, normally flooded or filled with fresh, saline or brine water permanently or temporarily; vegetation, if it exists, is dominated in these zones by hydrophilous plants, at least during part of the year*'. France comprises about 3 Mha of wetlands distributed in the following main types:

—wetlands of alluvial valleys (examples: Loire valley, Rhine valley and Alsatian Reid, etc.);

—wetlands of inland plains (examples: Brenne, reservoirs and ponds of wet Champagne);

—regions rich in peat and/or small wetlands (examples: Jura, Morvan);

—Atlantic-English Channel littoral wetlands (examples: Bay of Mont-Saint Michel and marshes of Sougeal, estuaries and marshes of the Seine, etc.);

—Mediterranean littoral wetlands (examples: Camargue, ponds and salt marshes of Languedoc, etc.);

—Wetlands of DOM-TOM (example: marshes of Kaw in Guyana).

Implementing a combative policy particularly aggressive during the last 40 years, these wetlands have been subjected to drainage and cultivation operations, which have greatly shrunk their area (Interministerial committee on assessment of public policies, 1994). This has serious consequences for natural heritage and economy of France. As a matter fact, wetlands play an important role in the regulation of water flow and storage in watersheds and in the protection of water quality. They also contribute to conservation of animal and plant diversity. Lastly, it has now become necessary to give more attention to wetlands in land management. Various programs, such as *Natura 2000, la directive Habitat, the law of water* accompanied by *Schémas directeurs d'aménagement et de gestion des eaux (SDAGE)* and *Schémas d'aménagement et de gestion des eaux (SAGE)*, are based on the premise that the spatial distribution and nature of wetlands are known and that powerful tools for their monitoring and assessment are available.

While many inventories and specific studies are available, they generally pertain to zones of limited areal extent, with no data about continuity between them, and provide no information on the nature of habitats and functioning of non-inventoried zones. It is hence necessary to generate data facilitating generalisation and spatial analysis of information acquired by other means and create basic maps for locating changes produced in these zones and assessing their magnitude.

A few examples of application of remote sensing for some of these problems are described below.

21.2 DELINEATION OF WETLANDS BY REMOTE SENSING

21.2.1 Marshes

Before undertaking any complex studies in wetlands, it is necessary to delineate them. This application is illustrated by an example pertaining to the marshes of Cotentin and Bessin, taken from a PhD (Orth, 1996).

The regional natural park of Marais, created in 1991, used the contours established in 1836 by the Bas Fonds syndicate based on maximum floods observed. The corresponding boundary is not very reliable and evidently obsolete, considering the management of these marshes for the last century. Its delineation using satellite images is possible only if correct information is available about the nature and functioning of marshes.

Thus, it was essential to know that these marshes resulted from filling of valleys about 10,000 years ago consecutive to a rise of sea level, followed by submersion under fresh water and major erosion of soft formations (such as marls, clays) on slopes due to human activity. The marshes are localised along many watercourses flowing into the English Channel at the foot of the Contentin isthmus. These marshes constitute a group of digitate landforms, comprising a mosaic of large sized permanent grasslands (one to several tens of hectares) flat, cut across by channels. Generally submerged during winter, they form part of a conventional bocage landscape. The bocage, situated outside the marshes in the 'highland', consists of permanent grasslands and crops, with medium size plots (2 to 8 ha) outlined by a dense hedge system. In satellite images marshes are continuous zones with homogeneous texture and structure, whereas the 'highland' is a fragmented zone with a heterogeneous texture and a cellular structure.

Few satellite data acquired during the period of submersion are available and an instantaneous measurement of the latter is too punctuate to be representative of modes of waterlogging and to facilitate tracing of marsh boundaries.

The marsh boundaries were initially determined by visual interpretation of 1:100,000 paper displays of IRC colour composites of 4 SPOT scenes (KJ 31-250 and 32-250) of May 1990, May 1992, August 1993 and November 1990. Visual interpretation was based on colours, texture and structure of the image (see Chap. 5). The effect of acquisition dates on delineation of boundaries was low (43 to 49% marshes estimated in the same area depending on the date of acquisition). In this specific case, autumn was found to be more favourable for visual delineation.

Moreover, the SPOT scene of November 1990, for which the hedge system was distinctly visible, was subjected to automated classification. In order to evaluate differences in structure between marshes and the bocage, a filtering technique (Sobel, see Chap. 12) was used on the first two PCA components (see Chap. 7) of 3 bands.

A new plane was obtained by combining the two planes resulting from this filtering. The OASIS classification (see Chap. 11), which takes the pixel neighbourhood into consideration, was applied on this filtered plane using a 17 × 17 window and to bands b3 and b2. The reference nuclei were chosen in marshes and the 'highland', respectively. The result thus obtained consisted of homogeneous map zones of these two units. Attributing a zero value to the zones corresponding to 'highland', a mask was prepared for selecting only marsh areas.

A comparison of the boundaries obtained by these two methods showed 62% boundaries classified identically. The remaining 38% comprised partly valley bottom grasslands, correctly classified as marshes by the automated classification whereas they were not identified in the visual interpretation, and partly marshes , incorrectly classified as 'highland' by the automated method (22%). These were the zones benefited by general water management and local agricultural development. These areas also have a marshy structure but host grasslands close in quality to those of 'highland'. The question of attributing these areas to marshes or not is beyond the scope of remote sensing data interpretation, but concerns the very definition of the theme.

21.2.2 Flooded and wet grasslands

The objective here was to delineate flooded and wet grasslands, among others, of the Voire valley situated in the Haute-Marne district of the wet Champagne region (Belluzzo, 1997). Two TM scenes acquired in different seasons and for different crop years (1 April 1990 and 15 May 1992) were used for this purpose. Demarcation of these grasslands would have been easier in two scenes acquired on the same dates and during the same crop year. In fact, while the botanical composition of permanent grasslands had no chance of varying between these two scenes, modes of exploitation of these areas were likely to change. With respect to crops, the land cover for the same plot would be different in the two scenes. Delineation of wet grasslands hence poses difficulties that would not have been there had the two scenes acquired on the same dates and within the same year were available.

A few 1:17,000 panchromatic aerial photos of the Voire valley region acquired by the IGN were used for verification of boundaries and classification. The work was carried out not on the entire scene but on an image segment of 1500 × 1284 pixels, for each scene. The image of May was geometrically restored using that of April in such a way that the two become comparable (see Chap. 13). The image segment comprises a portion of chalky Champagne in the north-west corner. This region is characterised by a chalky plateau, with large size open-field plots occupied by intensive annual crops, among which alfalfa crops are sometimes observed. Contrarily, the wet Champagne region mainly consists of clayey substratum, covered by smaller plots, scattered valleys and moist depressions mainly occupied by grasslands and forests with no plots clearly visible.

■ Method of study

Before applying any processing technique, it was necessary to study the characteristics of the bands of the two images for determining the means, standard deviations and coefficients of correlation with other bands (Table 21.1). Use of bands pertaining to various uncorrelated spectral domains with high

Table 21.1: Data characteristics of bands for two images (bold: high correlation coefficients)

	1 April 1990							15 May 1992					
	Correlation matrix							Correlation matrix					
	TM1	TM2	TM3	TM4	TM5	TM7		TM1	TM2	TM3	TM4	TM5	TM7
TM1	1						TM1	1					
TM2	**0.947**	1					TM2	**0.939**	1				
TM3	8.832	**0.947**	1				TM3	0.847	**0.949**	1			
TM4	0.461	0.411	0.217	1			TM4	0.390	0.300	0.092	1		
TM5	0.783	0.876	**0.905**	0.384	1		TM5	0.811	0.854	0.879	0.292	1	
TM7	0.705	0.846	**0.945**	0.124	**0.938**	1	TM7	0.748	0.842	**0.934**	0.009	**0.933**	1
Mean	63.03	27.69	28.60	53.95	54.58	24.47	Mean	66.37	31.20	30.41	85.64	59.15	24.51
Standard deviation	14.34	7.95	11.85	20.54	21.70	14.23	Standard deviation	16.30	9.98	15.04	26.79	25.14	17.52

standard deviations provides more information than the inverse case. Bands TM 3, 4 and 5, correlated least with each other, were used for classification of grasslands. Band TM 7, less correlated with TM 4 than TM 5 but having a lower dynamic range, was not used for classification but employed for detecting boundaries of agricultural fields. Band TM 2 was chosen for preparing IRC colour composites and the results of their interpretation used for verification of classification results.

A mask for isolating the grasslands was applied as shown in Fig. 21.1.

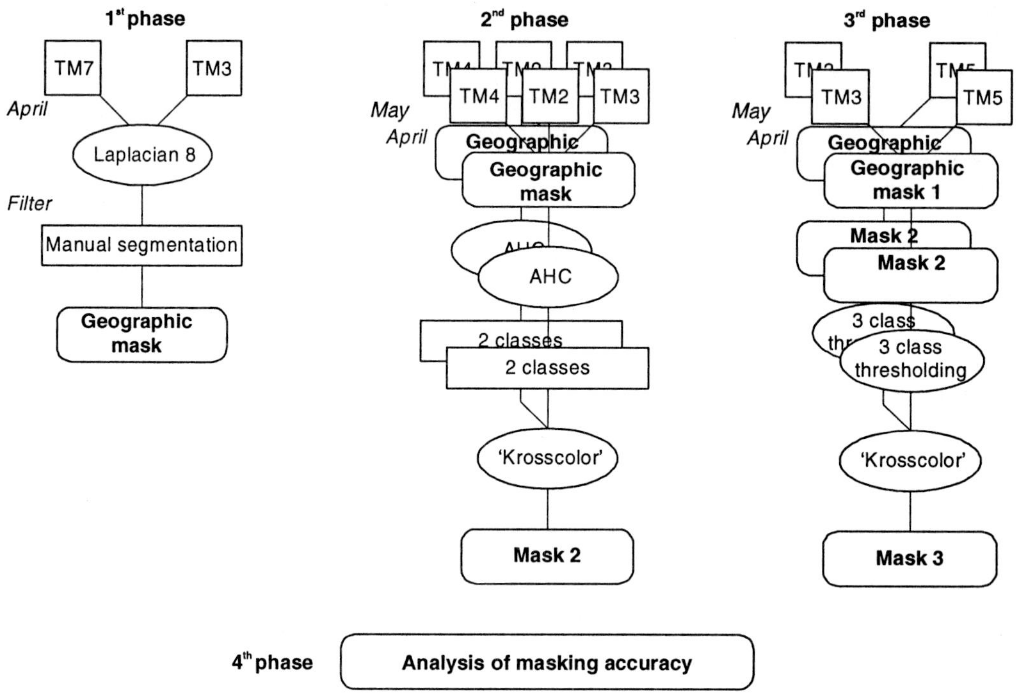

Fig. 21.1: Flow chart of mask for isolating perennial grasslands.

■ Geographic masking: first stage

Manual delineation of a geographic mask is that much better inasmuch as it is based on distinct differentiation criteria. The objective here is to identify the boundaries of agricultural plots between chalky Champagne and moist Champagne. To make the boundaries more apparent, a Laplacian type 8 filter (see Chap. 12) was applied on bands TM3 and TM7 of the images of April. For these two bands, low digital numbers corresponding to permanent grasslands (high chlorophyllian activity and water content) distinctly revealed the agricultural boundaries.

The new bands thus generated were used to prepare a three-colour composite in which zones corresponding to these distinct agricultural fields were manually demarcated. This made it possible to remove the entire chalky Champagne and also to isolate and eliminate cultivated areas in wet Champagne. Evidently, the first mask could not eliminate all cultivated plots, for example, those planted with a more or less chlorophyllian crop side by side with grassland plots, also chlorophyllian; the boundary between these two is not revealed. Moreover, all forest areas still persist.

■ Mask derived from an unsupervised classification: second stage

It was then necessary to eliminate land cover types other than grasslands, based on digital numbers. To differentiate between statistically differing units, an ascendant hierarchic classification (AHC, see Chap. 8) that uses a Euclidean distance was applied separately on the two image segments (for bands TM4, 3, 2) after application of the mask prepared earlier.

The dendrogram was divided into two classes: a class comprising chlorophyllian canopies and a class consisting of other non-chlorophyllian land cover types (verified with an infrared colour composite). In April, the first class includes grasslands and some standing crops and in May, it corresponds to the same land cover types with the addition of forests.

Diachronic study was accomplished by means of the 'krosscolour' function which combines two coded bands (Monget and Roberston, 1992). Two-dimensional histogram analysis facilitated investigation of relationships existing between the types of objects and set of vectors associated with corresponding pixels. These values were represented as a cluster of coloured points in two-dimensional space, with the separately measured values of the two bands as co-ordinate axes. Agricultural plots not eliminated by the preceding graphic mask could thus be identified. These fields were chlorophyllian on one date and non-chlorophyllian on the other, whereas the permanent grasslands were chlorophyllian on both dates. Predominantly broad-leaved forests could likewise be detected. The resultant classified band, which does not conserve the objects identified as crops and forests, is used to apply a second mask.

■ Mask derived from thresholding of bands: third stage

The preceding mask does not eliminate all the crops but still retains chlorophyllian crops in April and a few non-chlorophyllian ones in May. Criteria that ensure their differentiation vis-à-vis permanent grasslands need to be identified.

Crops are annual monospecies populations subjected to a specific type of maintenance, whereas permanent grasslands are multiannual and multispecies plant communities, with a possible accumulation of standing dry matter, subjected to varying modes of management year by year (see Chap. 20). Chlorophyllian activity analysed in TM3 and the equivalent water thickness (water content × aerial biomass), a parameter identifiable in the reflective middle infrared band (Gao and Goetz, 1995) and analysed from TM5, are the pertinent criteria for differentiation. Crops in May show a higher chlorophyllian activity (low digital numbers in TM3) and larger equivalent water thickness (low digital values in TM5) than grasslands with small variations in these bands, whereas the large diversity of modes of maintenance of grasslands gives rise to a large variance in their values.

After application of the preceding masks, bands TM3 and TM5 of both dates were separately segmented into three classes: high, medium and low values of digital numbers. Threshold values of the three classes were selected such that the corresponding zones are in the best possible agreement with the boundaries of plots identified in the aerial photos. Comparison of the results of the two dates using the 'krosscolour' program enabled identification of the last crop plots and creation of a mask for uniquely isolating the permanent grasslands (Fig. 21.2).

■ Mask accuracy assessment: fourth stage

As aerial photographic data was available only for 1990, accuracy of masking and grassland classification could be assessed only for that year. The accuracy assessment consisted of a systematic sampling of aerial photos at 72 points with a 4 × 4 cm grid (680 × 680 m on the ground). Since these photos were earlier used only for verification of plot boundaries but not for any information on their content, their use for such analysis was considered valid. The two classes retained for this analysis were the mask and the grasslands. The resultant error matrix between the aerial photo and the classified image is given in Table 21.2.

Table 21.2: Error matrix between aerial photo and classified image

		Classified image			
		Mask	Grasslands	Total	% correct
	Mask	42	6	48	87.5
	Grasslands	0	24	24	100
Photo	Total	42	30	72	
	% commission errors	0	20		

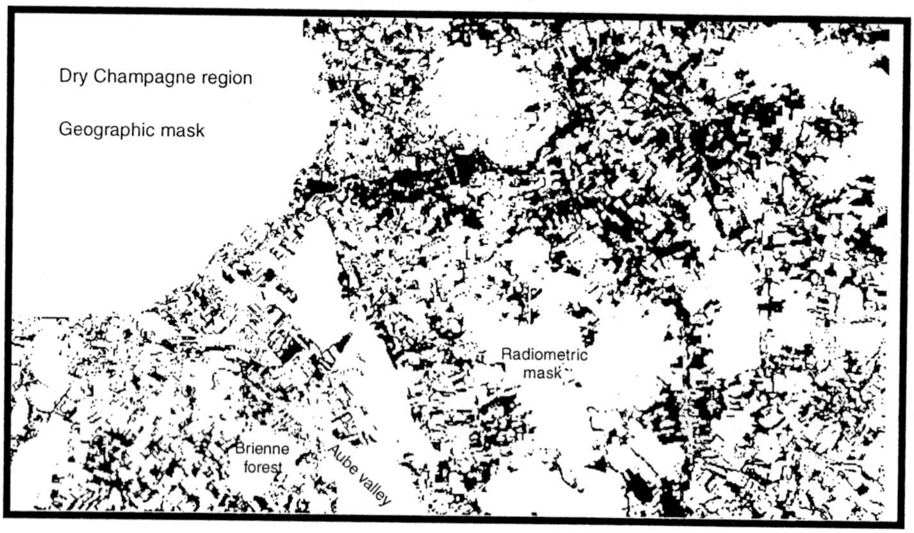

Fig. 21.2: Final mask isolating permanent grasslands.

The small number of control points warrants caution in commenting on this table. A high Kappa value, 0.86, was obtained (see Chap. 17). The mask used for isolating permanent grasslands seems to be reliable: 87.5% correct with 0% commission errors. Contrarily, while 100% of the grasslands existing in the photos were identified, a 20% commission error was observed. This would have an influence on the geographic accuracy of grassland mapping carried out subsequently.

21.3 CLASSIFICATION AND MAPPING OF WETLANDS

The typology of continental wetlands according to their organisational level can be described as follows (Table 21.3). Each class of a given level is made up of a mosaic of lower-level units. For the local level, the CORINE Biotopes map has been used as reference, retaining it at level 2 for legibility.

Some examples are given to illustrate the mapping methods employed depending on the geographic level of organisation.

21.3.1 Mapping at national or regional level

Major types of wetlands (national level) will obviously be mapped at small scale. As they are defined mainly from geographic criteria, this mapping can be envisaged only under GIS, comparing the altitude and slope data obtained from a digital elevation model with the satellite data classified according to land cover types. This is illustrated by an example of classification of a mosaic of LANDSAT TM images (for covering the entire region) of the Champagne-Ardenne region. The land-cover types used as criteria for differentiation of wetlands are based on the definition given in the Introduction. They are identified from an unsupervised classification (AHC) according to spectral characteristics (see Chap. 7). These land-cover classes are then compared with altitude and slope data (altitudes ranging from 70 to 100 m, slopes less than 1%).

Wetlands are often associated with permanent or temporary presence of water on or near the surface. Water is readily discerned from digital numbers (especially in near and middle reflective infrared bands) in satellite images provided the dimensions of water bodies are greater than those of pixels. Lastly, artificial water bodies (gravel pits, ballast pits, reservoirs) are distinguished by their rectangular forms and proximity to highly reflective mineralised surfaces, in addition to their spectral characteristics. Identification of water tables is useful for locating flooded and waterlogged zones by the criterion of 'proximity to water'. This criterion can be used under GIS by means of a 'buffer' whose dimension is chosen according to the phenomena to be detected.

Wetlands associated with watercourses and water expanses are characterised by proximity to water tables, a fairly homogeneous continuous pattern depending on valley shapes and a varied land cover consisting of peat, fen, forest and herbaceous vegetation, crops and bare soils.

Exploited peat bogs not covered by vegetation can be identified in any season due to their low values of digital numbers. Vegetation groups of peat, sedge, reeds, etc., can be differentiated from

Table 21.3: Wet zone units of various levels

National level (*)	Regional level	Local level (extract of CORINE Biotopes)
Wetlands associated with a given water expanse	Water bodies	2.1 Lagoons 2.2 Standing fresh water lakes, ponds, pools
Pond regions	Poplar plantations	2.3 Standing brackish and salt water lakes, ponds and pools 2.4 Running waters
Marshes and wet moors of plains	Riverine forests and permanent grasslands	3.1 Temperate heath and scrub 3.7 Humid grasslands and tall herb communities
Slope wetlands	Peat moors, wooded moors	4.4 Temperate riverine and swamp forests and brush 5.1 Raised bogs
Mixed wet plains associated with watercourses	Hydromorphic bare soils	5.3 Water-fringe vegetation 5.4 Fens, transition mires and springs
Wetlands of water-courses and wooded borders	Mosaic of crops and wet soils	8.1 Improved grasslands 8.9 Industrial lagoons and reservoirs, canals
	Mosaic of grasslands and crops	
	Permanent grasslands	

(*) CIEPP typology (1994)

grasslands and crops by their very large heterogeneity (combination of dry and more or less chlorophyllian vegetation) represented by large variation in digital numbers. They are also characterised in spring by delay in growth (colder microclimate), i.e., non-chlorophyllian spectral characteristic.

Alluvial wet forests (riverine forests, poplar plantations, alder and ash trees) are difficult to distinguish from other broad-leaved forests, except by a delay in spring vegetation. They can be identified only by a combination of altitude and slope criteria. The same is true of grazed or mowed grasslands that are more or less moist and intensively exploited, as well as cultivated drained zones. When bare soils are visible, they are most often of dark colour (low value of digital numbers).

The wetlands of the Champagne-Ardenne region were mapped on this basis by regrouping various land-cover classes. Two categories were distinguished according to the degree of submersion and waterlogging. An example of constituent elements of the wetlands of the Saint-Gond marsh is given in Table 21.4.

Table 21.4: Wetland units identified in satellite images for the Saint-Gond marsh

National level	Regional level		Local level
	Very wet	More or less wet	
Marshes and wet moors of plains (Saint-Gond marsh)	Wet marshes and moors Alluvial forests and thickets Riverine forest Flooded grasslands Hydrophilous grasslands Very wet grasslands Very humid bare soils	More or less wet grazed or mowed grasslands Mesohydrophilous grasslands, crops and broad-leaved trees	Ponds Peat bogs Sedge-fens Reed-fens Rushes Shortgrass Ash trees Poplar plantations Willows Alders Grasslands

Boundaries thus obtained surround the NZEFFI[1] (only wet environments), SPZ and IZBC boundaries of the BRIDGE database, as well as those of inland wet zones (inland swamps and peat) of the CORINE Land-Cover database (see Chap. 19). As a matter of fact, the definition of these units is more restrictive than that used in our method. This verification shows the importance of such a methodology for small-scale spatialisation of medium-scale limited data (1:100,000 accuracy for CORINE Land Cover).

21.3.2 Mapping at local level

■ Marshes

It is important to know to what accuracy can remote sensing data be used for identifying and mapping wetland units. The rightmost column in Table 21.4 consists of the units classified by a supervised classification method using a LANDSAT TM subscene of 15 May 1992 in the Saint-Gond marsh region. Classes and their names were chosen based on a (1:10,000) habitat map of these marshes (CPN Champagne-Ardenne, 1996). The accuracy of results varied depending on the units: the ground-truth data, albeit very precise, was not compatible with the resolution of satellite data and made delineation of training polygons difficult. Causes of confusion between classes were varied:

[1]NZEFFI: Natural Zone of Ecological, Faunal and Floristic Importance; SPZ: Special Protection Zone; IZBC: Important Zone for Bird Conservation.

—heterogeneity of plant communities: this was mainly the case for sedge-fens and peat bogs which are confused with sedge-fens and when they are wooded, with thickets;

—confusion with physiognomically similar plant communities: such was the case with reed-fens and rushes, grasslands and shortgrass and willows and poplar plantations.

These confusions can be partly reduced if a scene acquired on another date is used to take into consideration the phenological offsets between certain communities. In such a detailed study, aerial photos enabled mapping of various units; however, this necessitates ground-truth verification, on the one hand, for establishing reference data for visual interpretation of vegetation communities and, on the other, for assessment of interpretation accuracy. For this reason, costs of mapping plant communities at the local level are necessarily high, requiring purchase of aerial photos, time for interpretation, ground check up, etc.

■ Wet and flooded grasslands

The example of grasslands of the Voire region (whose delineation was discussed earlier) illustrates a classification based on diachronic data. The image segments of April and May, masked to retain only perennial grasslands, were classified separately in order to avoid radiometric corrections that introduce a bias in digital numbers of pixels (see Chap. 16). The method consisted of a supervised (parallelepiped) classification by thresholding bands TM3, 4, 5 in three or four levels using the spectral characteristics model of grasslands (see Chap. 20). Threshold values were chosen such that they determine zones conformable with the shapes of grassland plots identified in aerial photos.

Three classes were defined: (i) least chlorophyllian grasslands with low levels of biomass quantity, chlorophyllian activity and water content, (ii) highly chlorophyllian grasslands with large values of all these parameters and (iii) moderately chlorophyllian grasslands with intermediate values (Fig. 21.3). For the May data, one more class was added, viz., low-biomass grasslands with moderate values of chlorophyllian activity and water content, corresponding to grasslands that were probably subjected to an early ensilage operation.

The preceding classifications need to be superposed over one another for preparing a map and legend of permanent grasslands. The legend should also take into consideration variations in the physiological state of grassland classes (Table 21.5). It should also incorporate the possibilities and duration of floods which affect the flora—predominance of rushes and sedges (*Carex*)—and the speed of recovery of chlorophyllian activity in spring. It may only represent an instantaneous effect of exploitation

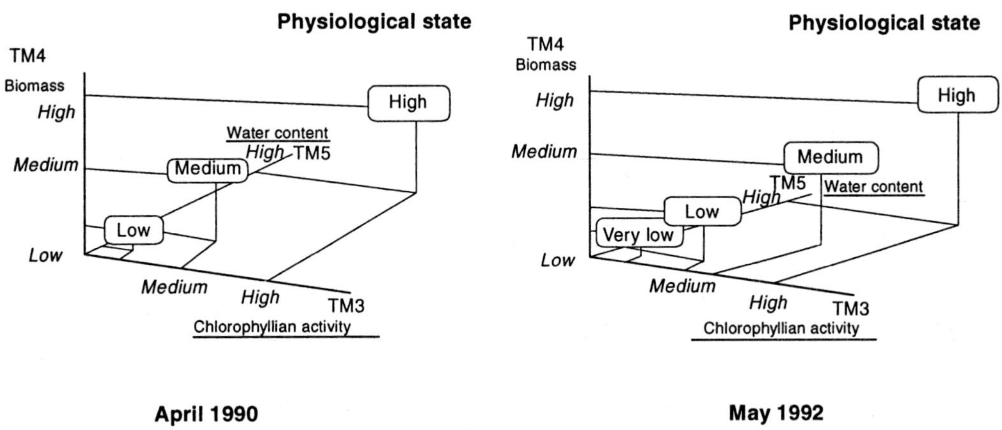

Fig. 21.3: Classification of permanent grasslands according to their physiological state, using manual thresholding of bands TM3, 4, 5 of April 1990 and May 1992.

Table 21.5: Legend for diachronic classification of permanent grasslands using variations in their physiological state

		April		
		Very low	Medium	High
	Very low	Reed-fens or flooded grasslands with predominance of *Carex*	Wet grasslands, not flooded	Well-drained grasslands ensiled since a more or less long time
	Low			
May	**Medium** **High**	Grasslands flooded for a more or less long time	Grazed grasslands with rushes	Grazed grasslands and well-drained mowed grasslands

modes: only the grasslands cut before the acquisition of the scene will appear with a low physiological state and those mown one day later have strong vegetative growth. Some specific species populations such as reeds or sedges have a spectral characteristic that facilitates their identification.

Flooded grasslands are mainly found in the Voire valley whereas reed-fens are essentially localised in the borders of lakes in the Orient forest and many ponds of wet Champagne, as well as in the Aube valley. A field survey conducted subsequent to this classification in the south-western zone, used as reference for classification, revealed the accuracy of mapping (Table 21.6). The number of reference

Table 21.6: Accuracy assessment of grassland mapping (in number of plots)

		Classification							
		Reed-fens or flooded grasslands with sedges	Flooded grass-lands	Grazed grass-lands with rushes	Non-flooded wet grass-lands	Grazed and mowed grasslands	Ensiled grass-lands	Total	% omission errors
Ground truth	Reed-fens or flooded grass-lands with sedges	8						8	0
	Flooded grasslands		5		4			9	45
	Grazed grasslands with rushes			9	2			11	18
	Non-flooded wet grasslands		3		7			10	30
	Grazed and mowed grasslands		4		3	48	7	62	23
	Ensiled grasslands					2	10	12	17
	Others	7	20					27	
Total		15	32	9	16	50	17	139	
% correct		53	16	100	44	96	59		
% commission errors		47	84	0	56	4	41		

fields varied depending on the classes (see Chap. 17). About 19% pixels classified as grasslands corresponded to other land-cover types: sometimes crops, often a combination of grasslands, road edges, trees, etc. This corresponds to 20% commission errors pertaining to the mask used for isolating grasslands (Table 21.2). They represent 'mixels' due to the 30 m × 30 m resolution of LANDSAT TM which combines the little or non-chlorophyllian objects in April and the more or less chlorophyllian ones in May. They are hence confused with classes 'reed-fens', 'flooded grasslands with sedges', on the one hand, and 'flooded grasslands', on the other.

Errors vary depending on the classes. The 'grazed grasslands with rushes' are correctly classified (0% commission errors and 18% omission errors) and 'grazed and mowed grasslands' somewhat less correctly (4% commissions, 23% omissions). 'Non-flooded wet grasslands', contrarily, seem to be incorrectly classified (56% commissions, 30% omissions) but more control points are needed to assess the accuracy. For 'ensiled grasslands' (41% commissions, 17% omissions), the mediocre quality of results is explained by the gap between the dates of acquisition (1990 and 1992) and the date of verification (1997). In fact, decision to ensile a field or not varies considerably from year to year depending on the needs of farmers.

The accuracy of local level mapping of wet and flooded grasslands can be improved by using the reflective middle infrared data of higher geometric resolution such as those of SPOT-4 (see Chap. 2) or infrared colour aerial photos.

Other examples pertaining to marsh grasslands are given in Chap. 20.

21.4 STUDY OF PLANT COMMUNITY DYNAMICS

In large river valleys (such as Loire), fluvial dynamics (modification of the major course due to erosion of banks, water leakage, cutting of meanders, etc.) influence the dynamics of plant communities. Remote sensing data can be used for several objectives in this case[2]:

—Mapping variations in fluvial geomorphology and land cover of some sites based on visual interpretation of aerial photos (1:25,000 to 1:30,000). This map will be useful in selecting profiles of phytoecological sampling and their positioning;

—Mapping natural environments and spatialisation of functional units using satellite data.

These examples clearly illustrate the importance of remote sensing for local as well as regional studies. Moreover, these data are equally useful at the beginning and at the end of an investigation, as the modes of processing and interpretation are chosen according to the objectives of study.

21.5 CONCLUSION

The examples of wetlands demonstrate the need for correct ground truth data about botanical composition and phenological stages of plant communities, on the one hand, and practices and modes of exploitation by man, on the other. Moreover, information furnished by remote sensing constitutes the basis for selection of sites and preparation of sampling plans, for generalisation and spatialisation of field data and for study of dynamics.

These examples also indicate the need to adapt, as far as possible, remote-sensing data to the objectives pursued (see Chap. 16). The data available today can be used for applications ranging from local to regional or national level studies and from small to large-scale investigations.

---

[2]Project 'Determination of free space for the Ligerian fluvial system: identification and spatialisation of functionai morphodynamic and ecological units in open and impounded valleys of Loire—social concerns and factors' as a part of the National Program of Research of Wetlands.

Since wetlands are venues for complex interactions between static and dynamic phenomena, involving biotic, abiotic, natural and anthropogenic factors, applications concerning them essentially necessitate use of geographic information systems facilitating management, comparison and synthesis of data of various sources and periods.

References

Belluzo G, Girard C-M. 1997. Identification et classification de l'occupation du sol à partir de scènes Thematic Mapper: application aux prairies d'une région de Champagne humide. Bull. SFPT, 146: 22–32.

Comité Interministériel de l'évaluation des politiques publiques. Premier Ministre, Commissariat général du Plan. 1994. Les zones humides: rapport d'évaluation. La Documentation Française, Paris, 391 pp.

Conservatoire du patrimoine naturel de Champagne-Ardenne. 1996. Cartographie des habitats naturels et des espèces de la Directive Habitats—Marais de Saint-Gond.

Gao BC, Goetz AFH. 1995. Retrieval of equivalent water thickness and information related to biochemical components of vegetation canopies from AVIRIS data. Remote Sensing of Environmnent, 52: 155–162.

Monget J-M, Robertson YC. 1992. Two-variable mapping applied to remote sensing data interpretation: a software implementation. Remote Sensing from Research to Operation. Proc. 18th Annual Conf. The Remote Sensing Society, pp. 571–580.

Orth D. 1996. Typologies et caractérisation des prairies permanentes des marais du Cotentin, en vue de leur cartographie par télédétection satellitale, pour une aide à leur gestion. Thèse INA-PG, 150 pp.

22

Crop Inventory

22.1 ECONOMIC ASPECTS

Since the launching of the first LANDSAT satellite in 1972, civil remote sensing has been extensively applied for crop inventories. The American research programs *LACIE* (Large Area Crop Inventory Experiment) and *Agristars* were devoted to this problem. In the European context, the most important program of application of spatial remote sensing is *MARS*: Monitoring Agriculture with Remote Sensing.

Crop yield assessment (Meyer-Roux and Girard, 1997) is, in fact, a very important economic operation and its procedures are difficult to control owing to the inherent variability of yields due to human but more often, meteorological factors.

In the United States, the emphasis has been on estimation of crop yields of foreign countries. Within the European Union, requirements are related to controlling production, far exceeding consumption, leading to high costs for the Community's agricultural policy. A number of mechanisms such as land management are set up. Their efficiency mainly depends on reliable forecasting of yields and stringent controls.

During the period 1986–1992, the cereal production, considered maximum, was 165 Mt for the European Community of twelve nations. This yield has been exceeded since then (Fig. 22.1). An earliest possible forecast enables estimation of expenses and eventually corrective measures for the following season. It is important to note that collecting of grain is carried out gradually after harvesting due to storage in farms and that effective statistics for market operations are thus known very late. The low yields in 1994 and 1995 associated with low global stocks led to a mandatory decrease in set-aside lands in the European Union in 1995 and 1996.

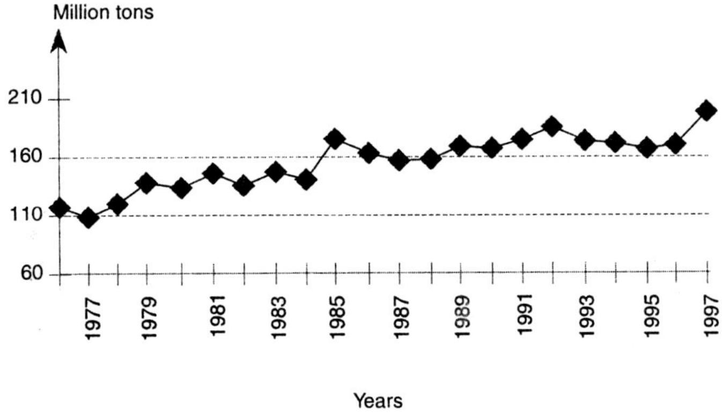

Fig. 22.1: Variation of cereal production in the twelve European nations from 1976 to 1996. Starting from 1990, production in East Germany is included (source Eurostat).

22.2 OPERATIONAL USE OF SPATIAL REMOTE SENSING

Spatial remote sensing is used operationally for assessment of foreign crop yields by USDA (United States Department of Agriculture) and by private companies who sell their services essentially for speculation on market prices. The Earth Satellite Corporation with its *Cropcast* service is widely known. The European Community also uses remote sensing operationally to obtain neutral information, independent of agricultural lobbies or member states. The Community Research Centre (ISPRA) supplies monthly information whereas the Direction General of Agriculture finances operational subcontracts such as satellite image analysis.

Remote sensing, however, remains marginal among the various techniques used for such estimations for several reasons:

—It only partly covers the required information on agriculture and a simultaneous inventory of animal and vegetation products and economic structure is often preferable;

—Remote sensing does not identify crops, and hence cannot provide information on their areal extents, except when they are well in place, whereas a cultivator can foresee his intentions and give precise information starting from the sowing stage;

—Remote sensing can contribute to forecasting yields but cannot really give their exact values; conventional methods are hence needed to obtain definitive results and often such forecasts are considered a luxury not strictly necessary;

—Costs of enquiries with cultivators are very low, especially when they are made by correspondence. The real cost corresponds to the time spent by the cultivator but it is not billed whereas remote sensing costs are;

—Many technical limitations hinder an efficient usage of remote sensing. For example, in Europe, cereal crops reveal high aerial biomass and vegetation indices for very dense chlorophyllian canopies get saturated.

Remote sensing is hence mainly used when conventional techniques cannot be applied or when they are difficult to employ harmoniously, as in Europe. Their more or less extensive usage thus depends essentially on the quality of results by remote sensing compared to the corresponding cost and quality of conventional methods.

Operationally, remote sensing is employed mainly in the United States for crop yield assessments in areas where such yields are generally low, since in this case remote sensing information is significant as vegetation indices are not saturated.

The European Commission uses remote sensing essentially for estimation of areas and indirectly for verification of agrometeorological models of yield prediction. Direct application for yields is still in the research and development stage in the European conditions of high yield and sensors and satellites presently available.

22.3 AREAL ESTIMATION IN EUROPE BY REMOTE SENSING

The method employed in project *MARS* consisted of representative sampling of 60 sites, 40 × 40 km in area, for the European Union (Fig. 22.2). A series of SPOT images, about 3 to 4 per year, were interpreted to extract the desired information.

For estimations conducted during a year no ground-truth information is needed. However, retroactively some ground-truth data enable accuracy assessment of computer aided visual interpretation.

Choice of sampling sites was not aimed at achieving national representation or obtaining a reliable absolute estimation. It constituted an early variation from the preceding year, as the data of the preceding year were obtained through Eurostat using conventional statistics. A number of constraints such as overlapping with the SPOT grid or an adequate percentage of arable land, led to the choice of a systematic type of sampling rather than random.

Fig. 22.2: Location map of 60 sampling sites in Europe.

Setting up of operational procedures enabled reduction of delay between image acquisition and its availability to the interpreter (3 days). In practice, any image taken before 20th of a month is included in estimations at the end of the month.

Visual interpretation was carried out over randomly selected segments (about 50 ha in size) of sites in the image. Interpretation results of a segment were then extended to the entire site by automated classification. These site segments were used for ground-truth verification at the end of the year.

A fairly complex model of statistical interpretation was employed to take into consideration the fact that during certain early stages many crops cannot be distinguished and that some sites may not have images. This model uses relative percentages of crops recorded in the preceding year and, in the case of lack of images, the results obtained from adjoining images of the same relatively homogeneous soil strata (Dallemand and Vossen, 1995).

The quality of results varies depending on the years and crops. Since 1992, when the method was developed, the results are better than those of conventional methods up to the end of August or September, afterwards data received from member states are better. Reference data generally corresponded to Eurostat statistics, but for two years the data issued by CCR, viz., areas and yields, were supplied to member states who used them in their own estimations and then sent them to Eurostat; hence, the methods are no longer independent.

The example presented below (Fig. 22.3) gives, for the year 1992 the results of this method for wheat compared to two datasets for 8 March and 14 October supplied by Eurostat. The results were especially interesting for the period 1992–1994 when changes in arable land were very high following

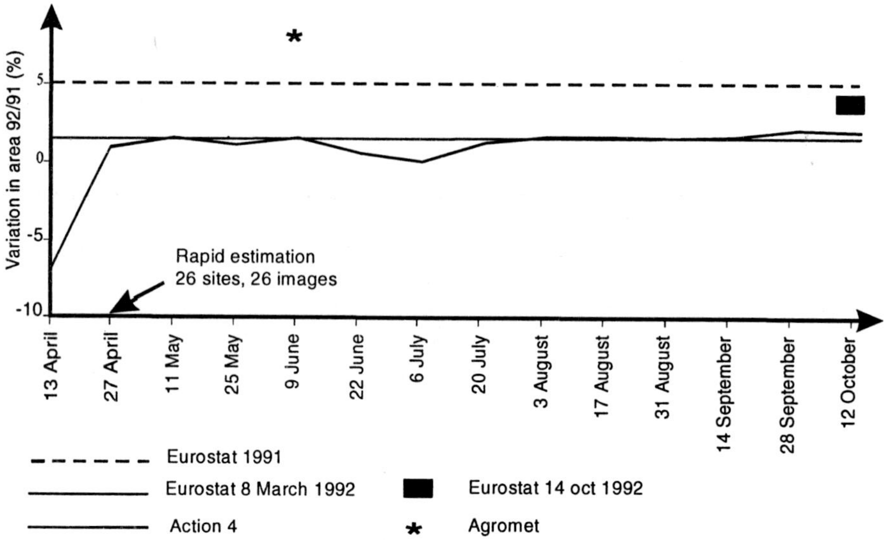

Fig. 22.3: Estimation of variation in wheat area in Europe during 1992/1991.

ground freezing. These limitations notwithstanding, the method was the only tool available for estimation of pre-existing agricultural fallow lands before setting in of set-aside lands in the Iberian Peninsula. The variations recorded subsequently are of the same order of magnitude as is the precision of the method, viz., about 2% variation from one year to another for important crops.

The site investigation technique has been extended since 1997 to the central European countries under the program PHARE and project MERA, 'MARS and Environmental Related Applications', and the SCOT society conducted a preliminary study for extending such a method to Russia.

Attempts are in progress to modify this system facilitating national representation to the most important countries. However, its cost is already high, about 1.6 million Euros per year for the European Union of 15 nations and cannot be increased significantly to achieve these new objectives.

The same method is applicable for verification of agriculturists' statements. In this case sites are not selected statistically but by the member states depending on the zones to be verified. The method provides a check of validity of agriculturists' statements vis-à-vis the results of image interpretation. Land registration and reference maps of plots necessitating subsidies are used for comparison between image and statement. Positive validation by remote sensing is judiciously acknowledged by community leaders and negative results are subsequently validated by ground-truth verification. Remote sensing thus provides a preliminary sorting on computer monitor and limits field investigations to 1% of the statements, instead of 5% required in the conventional method.

22.4 AGROMETEOROLOGICAL MODELS OF YIELD ESTIMATION

22.4.1 Yield estimation methods

There are many methods of crop-yield prediction; some are operational and some in research stage. The most common operational methods are statistical or regression methods. In practice, they consist of a statistical relationship between objective observations in the growing period of crops, i.e., before harvesting, and the final yield. This relationship is applied to the current year.

Observations may be related to crops (number of plants and heads, length of head, etc.) or to meteorological nature. In the latter case, decennial or monthly variations of meteorological parameters are used. In general, these methods are based on analysis of trends while objective field measurements incorporate technological changes. Statistical correlations based on field observations are more precise but cannot be routinely employed in Europe because too expensive.

Results of agrometeorological statistical methods are considered deceptive. A rigorous analysis of the 'AGROMET' model of 'Eurostat', used for more than ten years, showed that significant correlation between meteorological variables and yields cannot be obtained, once trends are deduced (Anonyme, 1997).

Crop-functioning models are equally operational but only in controlled conditions when many input parameters are measured. They are generally more useful for studies such as investigation of new crop varieties, resistance to cold, some diseases, increasing yield potential, etc., than for yield prediction, which is only one of the output parameters. The method used by the project MARS is based on a growth model (Vossen and Rijks, 1995), but uses it only for defining stress that is subsequently included in a statistical regression for final yield. Thus it forms a mixed model.

22.4.2 MARS method or CGMS model

The CGMS (Crop Growth Monitoring System) model is based on a large geocoded database covering all of Europe, whose parameters determine operation of a growth model for the important annual crops of Europe using meteorological data as input. This database essentially comprises crop characteristics, dates of sowing, harvesting, water needs, temperature constraints, etc., and soil data. The former are essential for the crop growth model and the latter for the hydrological model. Major work on this model consisted of creating this homogeneous database. The soil map, for example, took more than 8 years of study by the Soil Map Service of INRA of Orleans (King et al., 1994; King et al., 1995).

Usage of the method is associated with constraints of a geographic information system in which diverse types of data coexist: data related to geographic location such as meteorological information, natural boundaries as in soil maps or some agronomic data and administrative boundaries connected with crop varieties or yields. A fictitious meteorological station is created by interpolation along a 50 km × 50 km square grid (module 1) and a crop growth model is applied to homogeneous soil zones inside this square with a time interval of a day (module 2). These models, producing varying output data, are then combined according to administrative zones for which previous information on yield estimation exists, leading to prediction of yield (module 3) (Fig. 22.4).

Output data of module 2, such as delayed or early growth of vegetation, foliar coverage, can be related to remote sensing data for verification of functioning of the model (Dallemand and Vossen, 1995).

22.4.3 Remote sensing as a complement to models

Crop yield data operationally supplied to the Direction General VI at Brussels are not a direct output of the model. These raw data are controlled by trading organisations. The latter are equipped with crop-monitoring maps acquired by NOAA-AVHRR. These data are based on vegetation index map with the same grid of 50 km × 50 km. The indices are an indicator of activity and advance or delay in vegetation and hence are not directly transformable to yields. They do, however, give important complementary information (Dallemand and Vossen, 1995). The monthly MARS bulletins for the period April to

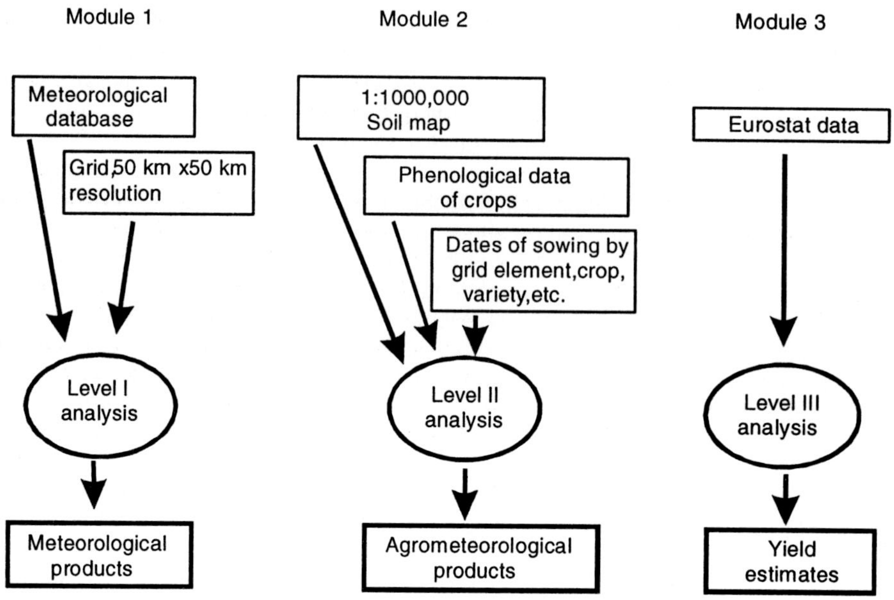

Fig. 22.4: CGMS model.

September, sent to the DG VI, Agriculture, are a synthesis of surface aspects and yields derived from a model controlled by NOAA-AVHRR indices.

22.4.4 Perspectives

Application of remote sensing has already become very important with respect to yield estimations and verification of agriculturists' statements (Meyer-Roux, 1996). Methods employed are relatively simple but demand considerable effort in organisation and auxiliary data to provide results at continental level. Their application is hindered mainly by three factors:

—technical constraints: revisit capability and resolution;

—quality and suitability of sensors for the phenomenon under study;

—cost of operations.

Resolution is not really a problem for yield predictions and in the case of area determinations if multispectral SPOT data are used. It is a problem mainly in verification of aid statements and in particular for determination of exact areas of fields. Acquisition of three or four SPOT images during a survey does not represent adequate repetition for defining the state of crops and hence for predicting crop yields. Repetition may be resolved by using radar images of sufficient resolution, about 30 m, but the corresponding methods are not yet operational and it is not certain that they will be. The ERS-1 instrument has been used for verification in the case of lack of images of visible band but such attempts represent only experiments. Moreover, present-day costs do not permit frequent coverage with high-resolution data, even limited to a representative sampling of images, irrespective of visible or microwave data. The essential tool used for these investigations corresponds to low-resolution meteorological satellites. However, the quality of sensors and production of products is adequate only for extreme vegetation conditions. The instrument VEGETATION (see Chap. 2 and CD SPOT System) operating on SPOT-4 ought to provide a significant improvement for yield estimations.

22.5 CONCLUSION

Remote-sensing techniques applied to agriculture have been extremely important and played a strategic role during the change in the European Community's agricultural policy for the years 1991, 1992 and 1993 whereas national administrations had no adequate conventional system for verifications. A lack of readily available methods could have led to serious consequences.

Methods of yield estimation in Europe are also significant but economically justifiable only when the conventional statistical data, less expensive, are not reliable or available too late. Conventional statistical methods have improved for the last ten years but remote-sensing technology has likewise progressed. It is especially imperative to conserve and improve these methods for strategic aspects such as, for example, negotiations with the central European countries at the time of membership in the European Union. Lack of reliable and neutral data can, in fact, spell trouble for the European Community and the member countries. Hence, a great effort is underway to rectify the situation.

France has played and continues to play a major role in these operations, primarily through spatial data and the SPOT program, well suited for the European agricultural requirements and constituting the main source of data. Methodologies developed by national research centres, and in particular the INRA Bioclimatology of Avignon, have been significant, in addition to those of teaching and research institutions. The Soil Map Service of INRA of Orleans has taken the responsibility of soil aspects and associated geographic information systems. Service societies such as SCOT or computer information societies have adapted themselves to meet the demands in the European context, mostly in agricultural matters. Amalgamating of investigations at various levels of organisation, ranging from detailed to continental, is no easy task but necessary if the true potentialities of remote sensing are to be realised.

References

Anonyme. 1997. Prévisions de rendement agricole. Textes présentés lors du séminaire de Villefranche-sur-Mer, France, 24–27 Oct. 1994. Office for Official Publications of the European Communities, Luxembourg.

Dallemand JF, Vossen P. 1995. Agrometeorological models: theory and applications in the MARS project. Proc. Workshop for Central and Eastern Europe, ISPRA, 21–25 November 1994. JRC-PHARE, EUR 16008 EN.

King D, Daroussin D, Tavernier R. 1994. Development of a soil geographical data base from the soil map of the European Communities. Catena 21: 37–56.

King D, Daroussin J, Le Bas C, Jones RJA, Thomasson A.K. 1995. Estimation des réserves en eau potentielle des sols à partir de la carte de sols de communautés Européennes.

Meyer-Roux J. 1996. Le programme Spot au service de l'agriculture Communautaire. *In:* D'une décennie de réalisations ... à une décennie de promesses, CNES, 15–18 avril 1996.

Meyer-Roux J, Girard C-M. 1997. Télédétection et estimation des récoltes. C.R. Academie d'Agric. Fr., 83 (3): 149–162.

Vossen P, Rijks D. 1995. Early crop yield assessment of the EU countries: the system implemented by the Joint Research Center. JRC, EUR 16318 EN.

23

Soil Mapping

23.1 REMOTE SENSING AND SOILS

The term 'soil' refers to the pedological cover: upper layer situated between the atmosphere, biosphere and fresh rock. In this flat volume (thickness of about a metre to tens of metres and area hundred to tens of thousands of km^2), several layers, known as soil horizons, are differentiated, most often occurring one above the other.

Obviously, remote sensing can give information only on the topmost part of soil. The penetrating power of an electromagnetic wave is of the order of magnitude of its wavelength and varies from a micrometre in the visible band to about a few centimetre in microwave frequencies. Thus, when clouds or plants do not conceal soil, certain soil parameters can be directly detected only from its surface state. However, a rain that moistens the surface, a wind that dries it, a slaking crust or efflorescence of a few mm in thickness suffices to completely modify the spectral characteristics of soils.

Remote sensing is useful for direct acquisition of soil information on multiple dates without altering the surface state and continuously in geographic space. The multitemporal capability permits acquisition of dynamic properties of soil, such as its temperature, water content, variation of surface roughness due to meteorological events or soil works (directly sensed parameters).

It is possible to obtain information on deeper part of soil cover from data on vegetation whose spectral characteristics are partly determined by its roots that penetrate soil (indirectly sensed parameters).

Studies on surface state of soils have been carried out by several groups; the important among them are Condit (1970), Goetz and Huete (1991) and Baumgardner (1981) in USA, Bialousz (1978) and Cierniewski (1993) in Poland, Epema and Mulders (1987) in the Netherlands, King (1993) of BRGM, Girard (INA-PG), Courault and Baret (1992) of the INRA, Escadafal, Mougenot and Pouget (1998) of the iRD and Cervelle (1977) of the Paris University in France. The following discussion is based on these papers, but it is not necessary to give all the references in this book, since most of them are found in a limited number of publications (see list at the end of the book).

23.2 DIRECTLY SENSED PARAMETERS

23.2.1 Spectral characteristics of soils in visible and near infrared bands

Reflectance curves of soils are uniform, continuous and intersecting; they do not show peaks or distinct absorption zones, but for water. These curves can hence be compared taking only one or two wavelength bands. No spectral zones requiring specific interpretation occur as in the case of vegetation (see Chap. 4, Fig. 4.13).

23.2.2 Surface states of soil cover

The concept of '*surface state of soil*' was proposed in 1978 by Aubert and Girard and developed by Escadafal in 1981 and 1989. In a wider context it incorporates information on agricultural activity, slaking crusts, salt encrustations and efflorescences, etc. A surface of soil is defined as '*the composition and organisation of a soil surface at given instant*'. It is transitional zone between the atmosphere and soil volume; it is a kind of first or zero horizon of the soil profile. Escadafal characterises it with the following parameters: colour, texture, effervescence, coarse soil components, vegetation and animal organic matter and its cover, algae, moss and other plant material; surface features such as slaking crusts, efflorescences, pores and cracks, microrelief, etc. Laboratory analytical data and microscopic descriptions can be added to this list. Measurements made for determining the surface state of soils concern only with parameters occurring on the surface and viewed from above. This definition is hence correctly well adapted to the spatial view of remote sensing, although this concept was developed independent of the latter.

A surface state of soil is a complex entity comprising various soil elements exposed to the sun or shadow and more or less covered by vegetation. The main factors of spectral characteristics of soil are those that influence its surface state, viz., colour, roughness (type of surface: slaking crusts, efflorescences, coarse elements, texture, structure and shadows), calcareous or organic matter, iron, moisture and chemical composition. The term 'soil' in the following text must be understood as 'surface state of pedological cover'.

23.2.3 Colour

Soil colour is measured in the field with reference to the Munsell code. Hues for soils vary from red (R) to yellow (Y), passing through all the intermediate spectral bands (YR). In some cases, colour may tend to blue. In most cases, the brightness value ranges from 0 to 8 and chroma from 0 to 8. Dark soils have value/chroma codes of 2/2 and 3/2 and bright soils 8/3 or 7/2. The most common soils in France are characterised by hues of 10 YR, 7.5 YR and 5 YR when red or 7.5 Y if greenish-yellow. If they are plotted in a Cartesian reference defined by their reflectance values in the 450 nm and 750 nm bands (Fig. 23.1), soil colours vary from darker to brighter along a linear cluster of points (Courault et al., 1998). Soils with the same chroma (/2, /4, /5, /6) are aligned along straight lines constituting a cluster and ranging between 5% and 10% reflectance for 750 nm band. For soils on these straight lines, reflectance increases with increase in brightness value (4/, 5/, 6/, 7/, 8/). Samples with hues of 2.5 Y, more yellow, occur together with those of hue equal to 10 YR, but with a slight shift in chroma: they appear duller. Practically no difference is observed between samples of 7.5 YR and those of 10 YR. This is conformable with results of colour analysis (see Chap. 3, Fig. 3.9).

It was seen in Chap. 3 that Munsell colour of soils can be related to their reflectance and digital numbers of satellite images (Fig. 23.2). Thus soil-colour maps can be prepared if a satellite image acquired over bare soils is available.

Colour of soil is identified from reflectance curves (Fig. 23.3). When a soil is dark (Munsell value 2 or 3), the curve is concave up to about 800 nm; if the brightness is medium (Munsell value 4 or 5), the curve is nearly linear; when brightness is high (Munsell value 6, 7 or 8), the curve is convex. If the colour is pure (Munsell chroma 6 or 8), the slope of the curve is steep between 400 nm and 600 nm; if it is dull (Munsell chroma 2 or 3), the slope is small.

23.2.4 Roughness

Surface roughness of soils depends on:

—intrinsic soil factors, such as salt efflorescence, ferruginous crusts, cracks, soil movements (Gilgai of vertisols), structure and porosity, slaking crusts, etc.;

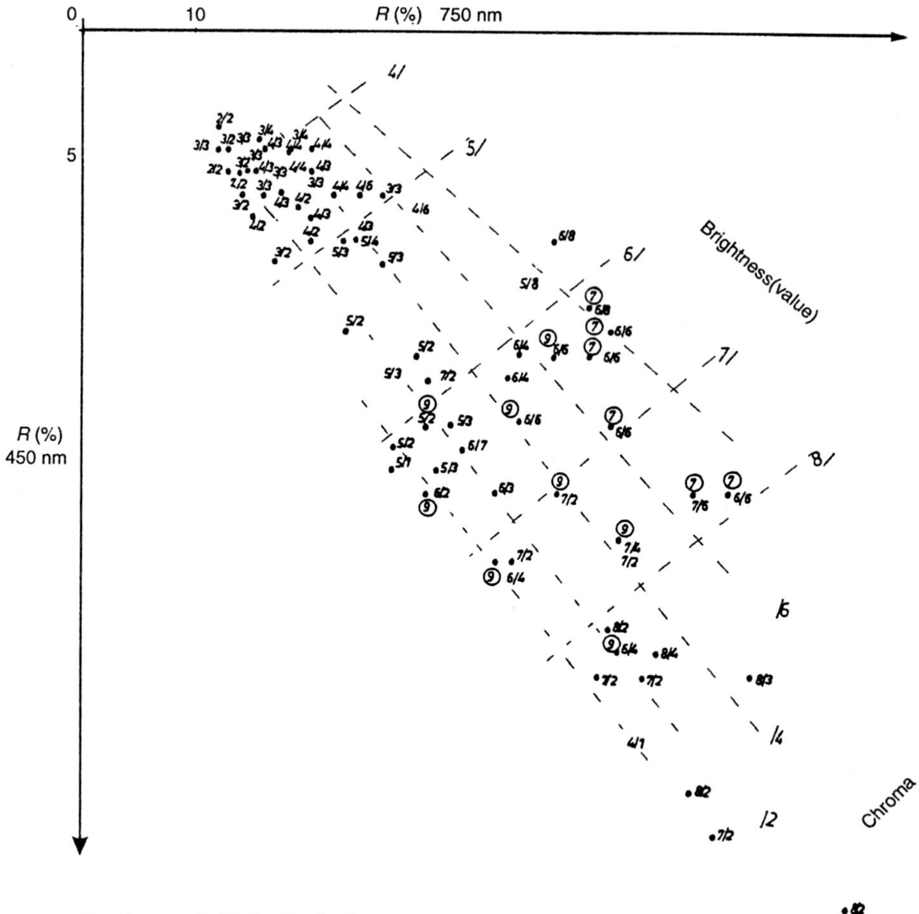

Samples are air dried and seived
(All uncircled points lie in the hue zone of 10 YR; others correspond to, 7: 7.5 YR, 9: 2.5 Y)

2 /,4/, 6/....Munsell brightness
/2, /3,/4.... Munsell chroma

Fig. 23.1: Relationship between Munsell colour and reflectance at 450 nm and 750 nm for 84 soil samples (after Courault, 1989).

—factors extrinsic to soil, such as microtopography or agricultural activity, etc.

Roughness manifestations are generally produced during specific times of the year depending on meteorological conditions (slaking crusts, efflorescences) or agricultural activities. Soil surface roughness is hence a temporary state in most cases.

Roughness may also be associated with microrelief created by sands, gravel, pebbles or blocks (see infra) permanently situated on the soil surface. Roughness in such cases is determined mainly by shadows produced by these coarse features on the soil surface. It must hence be related to the sun's height above the horizon and therefore with season, time of acquisition of scenes and latitude of the zone under study.

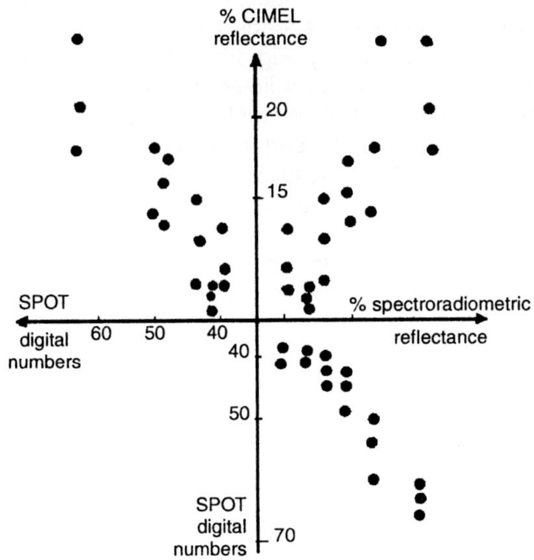

Fig. 23.2: Comparison between 18 radiometric values measured in laboratory (spectroradiometric), reflectance values (Cimel) and digital numbers in a SPOT image.

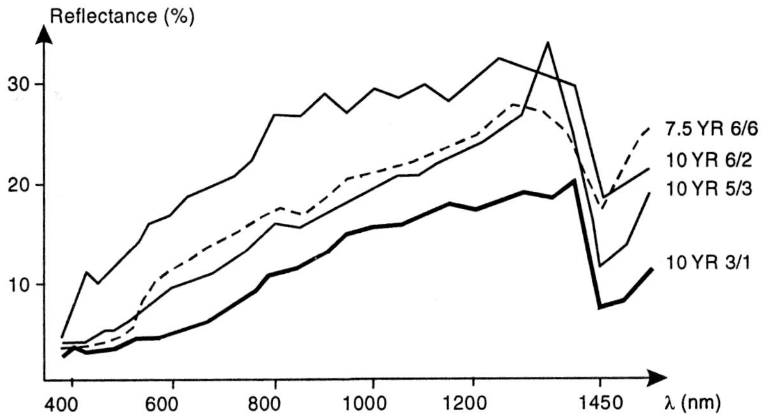

Fig. 23.3: Forms of spectral curves of soils according to colour.

For all types of roughness, the main response in terms of radiance is due to the shadow detected by the sensor. In fact, sensors are most often vertical whereas the sun is oblique. Hence, the sensor perceives a part of illuminated area and a part of shadow. Various indices have been proposed for measuring roughness, viz., standard deviation of heights relative to a mean plane, tortuosity index and shade index (Courault et al., 1993), and a number of models developed (Cierniewski and Courault, 1993).

When the proportion of shadow increases, reflectance decreases (Fig. 23.4). However, this effect is more important for a bright soil than for a brown or dark soil (Yongchalermchai, 1993).

In agronomy, Boiffin (1984) defined several phases of soil degradation, ranging from F0: non-degraded state to F2: degraded and slaked surface. In the last case, soil surface is smooth with no shadows whereas for the F0 state, it is full of mounds. It is observed that the greater the degradation, the stronger the reflectance (Courault, 1989) (Fig. 23.5). On an organic soil (with manure), with a stable structure, reflectance is lower than on a limed, brighter and less organic soil (Table 23.1). Strongest and most variable reflectance is observed for sodic soils with unstable structure.

Agricultural operations also influence the surface roughness of soils. Thus, over a given type of soil, ploughing (very rugged surface) gives rise to lower reflectance than sowing (less rugged surface) and movement of a roller makes the soil surface almost smooth, leading to strong reflectance (Fig. 23.6).

Studies of SPOT images have shown that it is possible to differentiate calcareous soils from slaking soils, since the latter indicate less distinct boundaries than limestone outcrops (Courault and

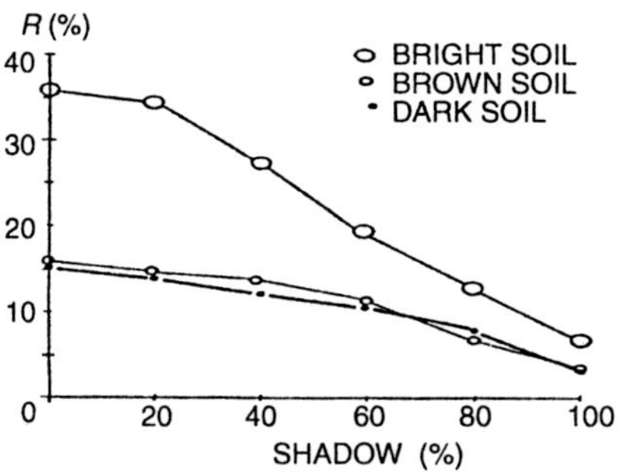

Fig. 23.4: Relationship between proportion of shadow and reflectance for three soils (590–680 nm) (after Yongchalermchai, 1993).

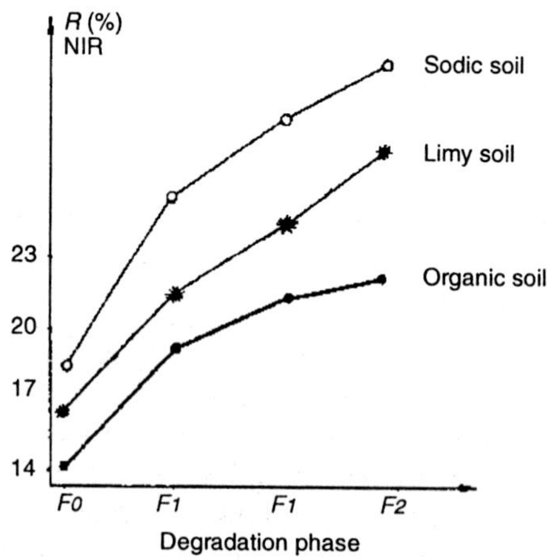

Fig 23.5: Reflectance of three different surface states of soil according to their roughness (after Courault, 1989).

Table 23.1: Characteristics of three surface states of soils

Analysis of surface state of soils	Munsell colour	Organic matter (%)	pH	Structural stability	Clay content (%)
Soil with manure	10 YR 4/4	21	6.4	Strong	20
Limed soil	10 YR 5/6	5.3	7.9	Strong	18
Sodic soil	10 YR 7/4	6.2	5	Low	20

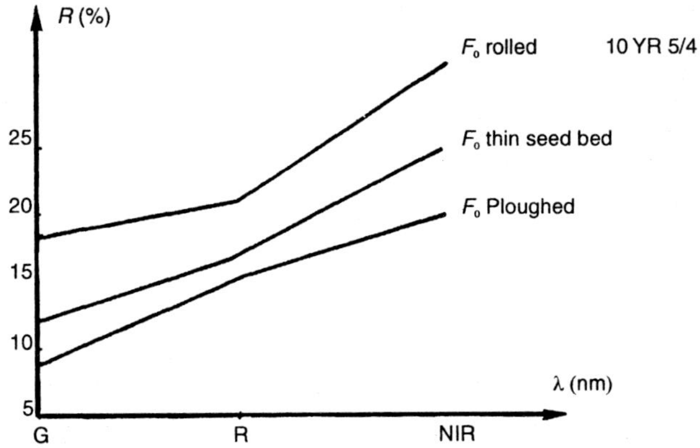

Fig. 23.6: Reflectance curves of three agricultural fields of various roughness conditions: ploughed, sown and rolled plots (after Courault, 1989).

Girard, 1990). More recently, three to four surface states of more or less slaking soils could be distinguished in loams of Caux County (Burlot, 1995).

23.2.5 Carbonate

Carbonate content influences reflectance of soils by giving very high values starting from the blue band. The reflectance curve is convex, brightness values very high but less than soils with salt efflorescence. For total carbonate contents less than 10–20%, carbonate seems to have no influence on reflectance. Contrarily, beyond this limit, the stronger the reflectance, the larger the carbonate content, other factors being constant. For values exceeding 60–70% carbonate, there is little difference in reflectance and saturation occurs in high brightness values in images and aerial photos. Special processing is often required to improve the signal.

Correlation between reflectance at 400 nm and total carbonate content for 84 soil samples illustrates this relationship (Fig. 23.7). However, the significance of this relationship, for image processing, lies in graphically deducing total carbonate content from reflectance value. It is seen that when reflectance is less than 15%, the total carbonate content may vary between 0% and 50%. This is because only a single-factor analysis is presented here. Thus, a calcareous soil may have the same reflectance as that of a non-calcareous slaking soil.

In sedimentary regions, highly calcareous zones usually correspond to outcrops of limestone, chalks or bright marls. Ploughing also often brings these materials to the surface. Consequently, one

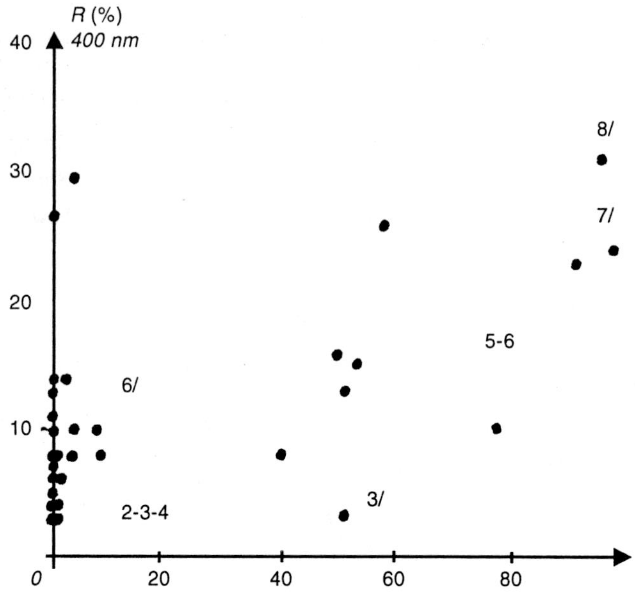

Fig. 23.7: Relationship between total carbonate content and reflectance at 400 nm. Munsell values are shown (Courault, 1989).

way of identifying them is to analyse contrast between the calcareous zone and its surroundings. Since contrast is generally high for calcareous zones, boundaries can be readily delineated.

23.2.6 Organic matter

Organic matter reduces reflectance in all bands of the visible spectrum (Fig. 23.8).

The reflectance curve of organic soils remains concave for the entire visible range. For organic matter contents less than 1.5%, there seems to be no effect on the reflectance curve. Contrarily, for

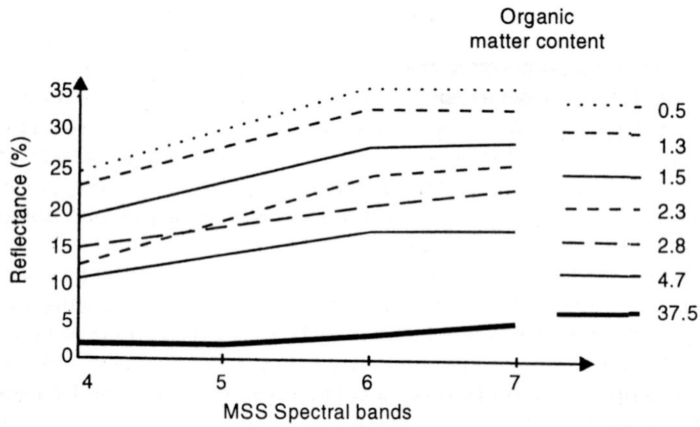

Fig. 23.8: Reflectance curves (obtained with an Exotech radiometer) of more or less organic soils.

contents greater than 8% and up to 20%, reflectance decreases. However, since the brightness values are very low in the visible bands, it is preferable to pay attention to the red and near infrared bands.

At 650 nm, the relationship between reflectance and organic matter content is exponential (Fig. 23.9). In this case also, organic matter content cannot be readily inferred from reflectance, since other soil factors interfere.

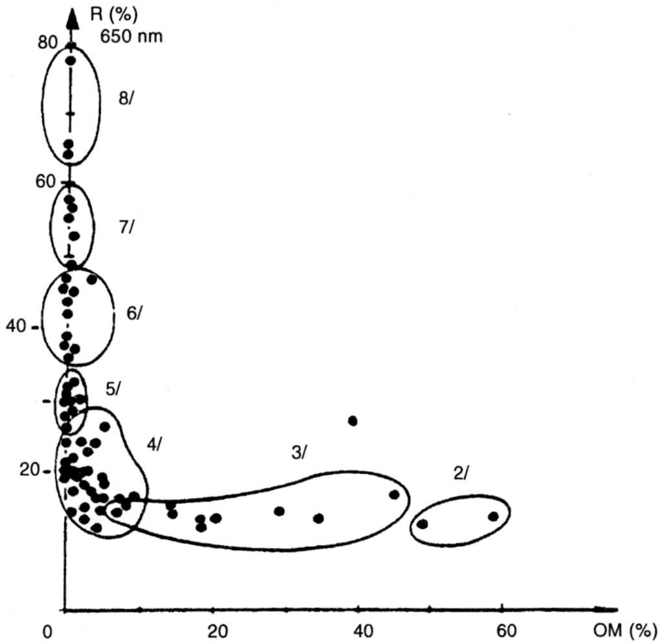

Fig. 23.9: Relationship between organic matter content and reflectance at 650 nm. Munsell values (4/) are shown (after Courault, 1989).

In temperate climatic regions, organic zones in bare soil are visible in satellite images or aerial photos only when they are cultivated. Others are covered by vegetation. Thus, the range of variation of organic matter content is much more reduced. Starting from a certain weight ratio of organic matter, the preceding relationship becomes invalid. In fact, if all soil components of a soil surface are entirely covered by organic matter only over a few microns (as in A horizon of Chernozem soils—Soil Reference Manual, Baize and Girard, 1995), its reflectance will be very low, although its organic matter content may be only 5 to 8%. An organic horizon, whose content is very high because all components are made up of organic matter, will not give higher reflectance. In fact, reflectance in the visible and near infrared bands depends only on surface conditions.

23.2.7 Iron

Since iron is characterised by low reflectance in the blue band, slope of the reflectance curve in the visible range is high. There is no stronger reflectance in the red band for iron-rich soils than for other soils. Contrarily, an absorption zone occurs between the wavelengths 860 and 910 nm. Hence, an acquisition band as closer to 400 nm as possible is needed for detecting red soils and two others around 850 nm and 900 nm for detecting the absorption zones. Unfortunately, such bands are not presently available in satellite sensors.

It has been observed from laboratory measurements that an exponential type relationship can be established between reflectance at 900 nm and iron content (Fig. 23.10). It may be noted that the signal is not affected by iron content below 2 to 3%. When iron content is more than 15%, the reflectance is small and brightness values are low. A proportional relationship exists between these two limits. Soil complexity due to mixing of soil components is again observed.

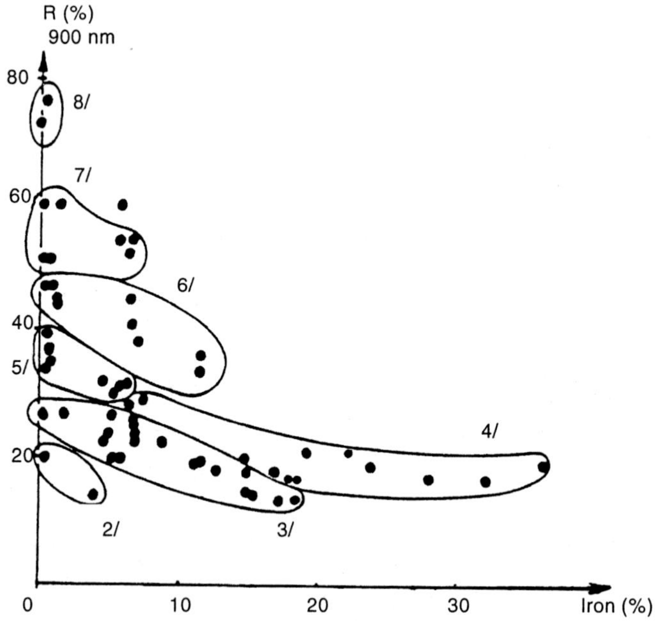

Fig. 23.10: Relationship between iron content and reflectance at 900 nm. Munsell values are indicated, for example as 2/ (after Courault, 1989).

Interpreting iron contents in soils from satellite images is not common. In intertropical regions, where iron-rich soils with variable contents frequently exist, coatings of algae or those produced from brushwood fires are often observed on iron pans. In deserts, 'desert varnish' covers all surfaces so that iron is not directly perceived (see Chap. 24). In the case of savannahs, dense vegetation canopies, more or less dry, do not permit perception of soil. Hence, detection of differences in iron content of soils assumes special importance for cultivated fields in these regions and in Mediterranean zones.

23.2.8 Moisture

Absorption bands for water situated at 950, 1150, 1450, 1950 and 2350 nm are visible in soil reflectance curves and more so as these are the principal modifications affecting such curves. However, absorption peaks at 950 and 1150 nm are fairly small, whereas those at 1450 and 1950 nm are generally distinct since water is always present in the soils, at least associated with salts, oxides or hydroxides contained in soil. However, no information is acquired by satellite sensors since radiation of these bands is strongly absorbed by atmosphere.

In general, the more the soil moisture, the lower the reflectance (Fig. 23.11) when other soil components are similar. When the moisture content increases, colour becomes darker and reflectance lower for all wavelengths.

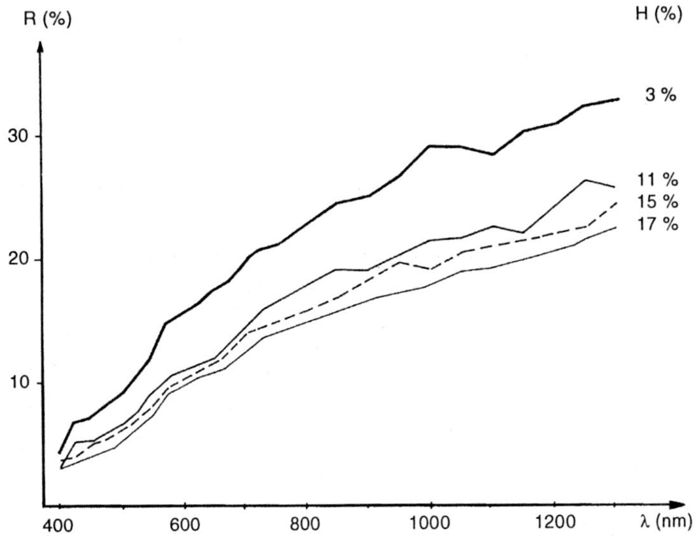

Fig. 23.11: Variation of reflectance of a soil with moisture (weight percentage).

Moisture thus detected obviously corresponds exclusively to the surface moisture. Information about moisture content can be obtained up to a depth of 10–30 cm if microwave band is used and about water movement if thermal infrared data is used (see Chaps. 1 and 26).

Field experiments under controlled conditions have shown that when soil moisture decreases (from 28.8% to 10.6% by volume), points plotted on a graph with values of reflectance in red and near infrared bands as co-ordinates fall on two linear segments (Fig. 23.12). An offset and a rotation between the two linear segments are observed. Scatter of points is smaller for the higher moisture state (colour 10 YR 4/4) than for the lower one (colour 10 YR 7/3). Regression relationships between the two wavelength bands for the two states are obtained as follows:

Moist state: $IR = 1.64\ R - 56.58 \quad r = 0.98$

Dry state: $IR = 1.33\ R - 65.38 \quad r = 0.94.$

Similar results are obtained from laboratory studies. Between moist and dry states, samples of organic soil show a very small variation in reflectance (2 to 3%), while for ferruginous, organo-ferruginous and organo-calcareous soils the variation is larger (10 to 15%); maximum variation (20 to 30%) is observed for calcareous and gypsum soils.

In reflective middle infrared band, spectral characteristics of soils are dependent on their moisture content: dry soils have higher reflectance than the same soils with moisture. This spectral region is much less sensitive to organic matter than the visible. However, as these various parameters are related with each other (peat soils have higher water contents than sandy soils), it is difficult to isolate the parameters that influence spectral characteristics in natural conditions.

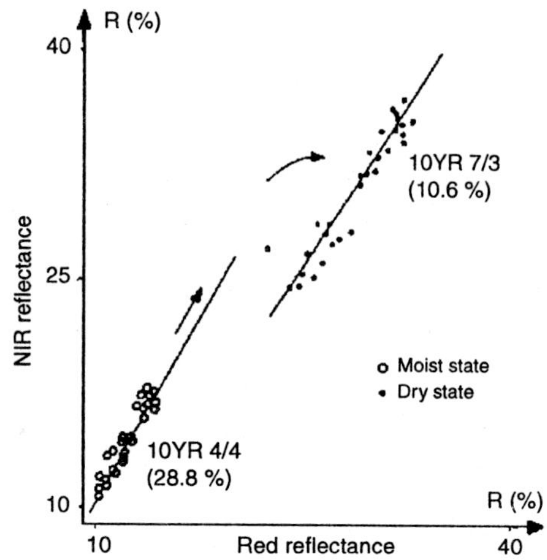

Fig. 23.12: Variation of reflectance with moisture content (%) for the same soil (with its colour given in Munsell code) (after Courault, 1989).

23.2.9 Grain size

Soils are divided into two groups according to their grain size with a limit of 2 mm. Above 2 mm, soil is referred to as coarse: gravel (2 mm to 2 cm), pebbles (2 to 7.5 cm), stones (7.5 to 25 cm) and blocks (more than 25 cm). Below 2 mm, it is known as fine soil: clay (< 2 μm), silt (2 to 50 μm) and sand (0.05 to 2 mm). Soil texture is a characteristic resulting from assemblage of particles of various sizes. In the visible range of wavelengths no interaction of radiation with particles takes place since sizes of clay particles are too large for these wavelengths, but they influence reflectance in the near infrared band. Silt and sand particles are too large to affect reflectance in the visible or near infrared bands. Some authors reported variations in reflectance according to their clay or sand contents, but then they are dependent on the water content of these components.

■ Clays

Relationships between grain size and moisture content are indirect. In fact, for a given moisture content in the ground, clays contain more water than sands. Moisture content hence is not a good criterion for evaluation. Hydraulic potential or water retention capacity should be taken for a soil sample. Clays are often associated with organic matter. Moreover, as clays are usually of low brightness, reflectance of clayey soils is relatively low.

■ Sands

Sands are drier since they retain no water in their porosity, often very coarse (see Chap. 24). On the other hand, most sands are of high brightness and hence sandy soils are relatively strong reflectors.

■ Silts

In the case of non-calcareous silts of low organic content, the surface structure is modified by rains and roughness changes to smoothness. This is the slaking process. They then appear very bright in aerial photos and satellite images. However, this high reflectance, in fact, ought to be attributed to the low roughness of silts, the cause of slaking, and not the nature of particles. The same silty soil, just ploughed or tilled, has much lower reflectance.

■ Coarse components

Surface abundance of all coarse components in soil is referred to as 'stoniness'. Most often reflectance increases with stoniness. This is mainly due to the fact that the energy reflected by these coarse components, which usually act as specular surfaces is greater than the shadows they produce. Moreover, in quarries, extraction of material gives rise to a surface coating by fine powders, most often white. Hence, quarries are generally characterised by very strong reflectance.

23.2.10 Salts

An integrated study of saline soils was carried out by Mougenot et al. (1993). Laboratory measurements (Mougenot, 1990) (Fig.23.13) revealed that absorption zones of salts, except that of NaCl, are clearly differentiated in the infrared band. Therefore, bands 5 and 7 of Thematic Mapper or the reflective middle infrared band (b4) of SPOT must be used for detection of salts by satellite images.

In laboratory measurements, halite is detected mainly from very bright hues in the blue band of the visible spectrum. Gypsum, very sensitive to water, shows three absorption peaks at 1450, 1950 and 2350 nm. Jarosite, common in mangrove zones, depicts absorption peaks at 1950 nm and, if iron is present, at 600 and 900 nm. Absorption peaks for calcite occur between 1600 and 2500 nm, with an apparent peak at 2150 nm.

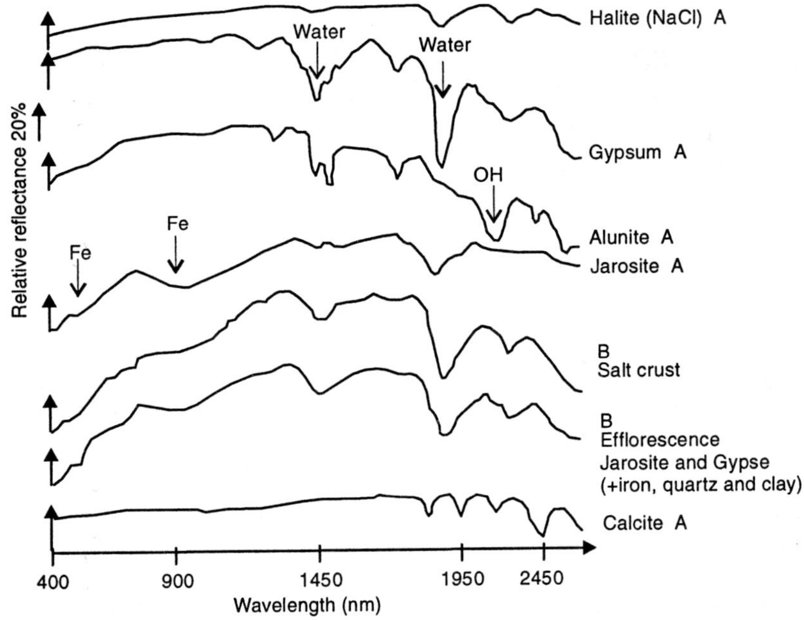

Fig. 23.13: Spectral characteristics of various salts: Halite (NaCl), Gypsum ($CaSO_4$, $2H_2O$), Alunite [$KAl_3(SO_4)_2(OH)_6$], Jarosite [$KFe_3(SO_4)_2(OH)_6$], Calcite ($CaCO_3$); (A) Laboratory and (B) Field measurements (after Mougenot, 1990).

Field measurements are more difficult to interpret since an association of several salts mixed with organic matter, various minerals and water is often observed. Measurements are mainly affected by surface states of soil such as encrustations and efflorescences. However, it is necessary that the image resolution be very good or that these crusts cover large areas as in the case of dry and saline zones.

Saline soils can often be detected indirectly if plants also exhibit salinity.

It is also necessary to use multi-date images, since characteristics of saline soils change rapidly depending on the seasons. It may be reasonably expected that detection of saline soils can be improved using new thermal and microwave sensors.

23.2.11 Multifactorial analysis

Since soils generally comprise varied quantities of organic matter, carbonates, iron, etc., of various particle-sizes (soil texture) and distributed variedly (soil structure), multifactorial analysis becomes necessary. A factorial correlation analysis was carried out using reflectance and soil parameters of 84 samples (Fig. 23.14). The analysis revealed several classes in a plane of axes 1 (45.5% information) and 2 (33.5% information).

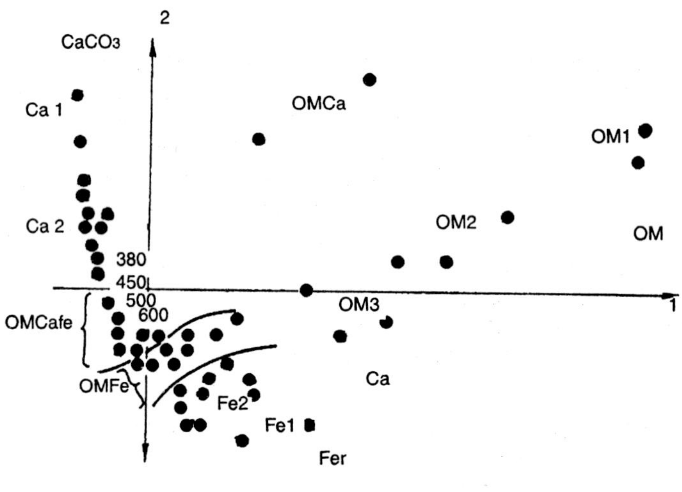

Fig. 23.14: Plane (1, 2) of multifactorial analysis for carbonate, organic matter and iron for 84 soil samples, defined by four wavelengths: 380, 450, 500 and 600 nm (after Courault, 1989).

Axis 1 indicated three groups according to organic matter content: OM1 more than 45%; OM2, 20 to 45% and OM3, 10 to 20%. Axis 2 differentiates in its positive part two groups by their total carbonate content: Ca1 more than 70% and Ca2, 50 to 70%. In the negative part of axis 2 and positive part of axis 1, samples rich in iron are distributed in two groups: Fe1 more than 30% and Fe2, 15 to 30%. In the centre of the plane, two groups containing less than 10% organic matter and less than 6% iron can be identified, one without carbonate (OMFe) and the other with less than 10% carbonate (OMCaFe). Another group containing carbonate and organic matter but no iron (OMCa) is also noticed.

This example does not represent all possible cases of the large number of soil types. Nevertheless, it shows that soil parameters of a region can be modelled by their multifactorial analysis with reflectance and digital numbers from the image under study.

Similar investigations on moisture and surface roughness resulted in the following conclusions:

—A dry organic soil has a lower reflectance than that of a less organic and moist soil. Organic matter content hence dominates over water content.

—A smooth soil has a stronger reflectance than that of a rough soil, irrespective of the moisture content. The effect of roughness is also greater than that of water content.

—However, on the ground, roughness and moisture content of soils are not absolutely independent of each other. Too moist a soil cannot be ploughed; evaporation and hence moisture content of soil surface and the sun's height (hence shadows) are likewise not independent.

In a Thematic Mapper image of southern Tunisia (Fig. 23.15), from the blue-green (TM1, 450 to 520 nm) and infrared (TM7: 2.1 to 2.35 μm) bands, Escadafal (1989) identified various surface states of soil. These are shadowed (1), gypsum (2, 3), carbonate (4 to 7), sandy, silty and calcareous (8 to 11), and quartz (12 to 14) soils. This demonstrates the significance of reflective middle infrared data.

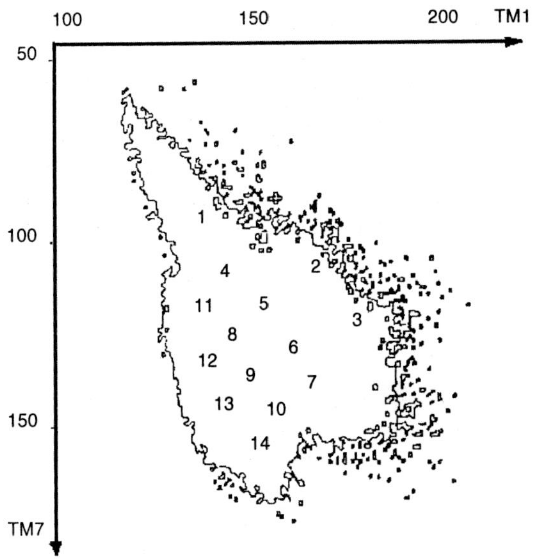

Fig. 23.15: Separation of gypsum, carbonate and quartz soils from TM 1 (blue-green) and TM7 (infrared) bands of LANDSAT (after Escadafal, 1989).

The magnitude of influence of soil parameters on reflectance can hence be given qualitatively. Colour constitutes the primary indicator since it is related to many parameters and physically inferred from radiometric measurements in the field or from aerospatial platforms. Surface roughness seems the next most important factor for signal variation. Next in importance are total carbonate and organic matter, which have opposite effects, followed by iron and moisture content in the last position.

23.2.12 Linear clusters of soils

When the proportion of shadows changes, soil reflectance values for two wavelength bands, blue (400–500 nm) and near infrared (800–900 nm) or red (600–700 nm) and infrared, plotted in a Cartesian diagram, cluster along a linear direction. This indicates that reflectance as well as digital numbers of images for the two bands are positively correlated (Table 23.2). This is referred to as a linear cluster of

Table 23.2: Parameters of five surface states of soils

Reference index of surface state	Colour of dry soil	Colour of moist soil	Slope of regression line	Ordinate at origin	Correlation coefficient
VB	10 YR 8/1	10 YR 7/2	0.75	18.7	0.98
MB	10 YR 7/1	10 YR 5/3	0.87	9.5	0.99
BS	10 YR 6/4	10 YR 5/4	0.98	6.7	0.97
MD	2.5 Y 5/2	2.5 Y 4/2	1.04	4.6	0.98
VD	2.5 Y 5/4	2.5 Y 4/4	1.24	2.2	0.95

soils (Fig. 23.16). Such a cluster consists of straight-line segments defined by the slope of regression line and intercept on the ordinate.

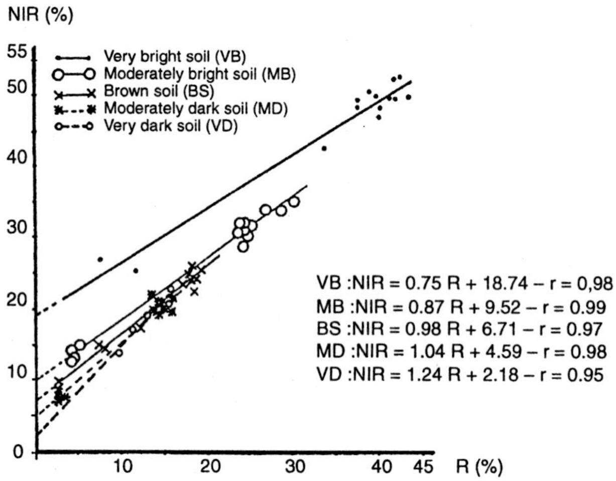

Fig. 23.16: Five linear clusters of soils (after Courault, 1989).

It is seen that surface states of very bright (VB) soils have an intercept greater than that for darker soils. The slope of the regression line for bright soils is smaller than that for dark soils.

These results show that the soil line is dependent on its colour but not on the quantity of shadow. As the slope of these straight lines varies, a soil may not have a single soil line. It is observed that soil lines are distributed as a cluster which widens towards lower values of reflectance.

23.3 MODEL OF IMAGE INTERPRETATION

Various surface states of a soil, represented by digital numbers on a visible-infrared plane, occur as a cluster globally oriented in the direction of the first bisectrix. The preceding data plotted on such a graph constitutes a *model for interpretation of soils* using images (Fig. 23.17). Most often, digital numbers are projected onto two axes representing red and near infrared bands, since these are most explicit for vegetation. For soils, a projection on infrared and green or blue is preferable.

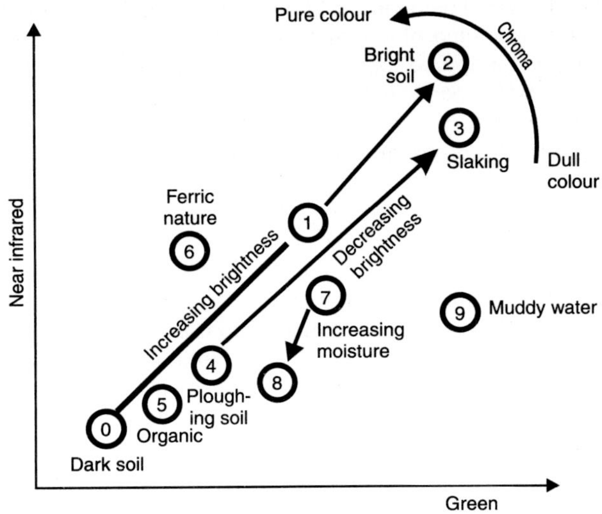

Fig. 23.17: Model for soil interpretation in images.

Very dark soils (0) are associated with the lowest values of green (G) and infrared (IR). These may be confused with pure waters or cloud shadows. Shapes of map zones most often enable differentiation of an extended form of a water body or a cloud shadow from a very dark soil surface. When a cloud is not observed in the image, identification of a soil with a cloud shadow may be difficult.

If soils are brighter (1), they are situated in the diagram at higher values for both the axes. Brightest soils occur at the extremities of the axes (2) and may be confused only with clouds or snow. For these two latter features, ambiguities can be eliminated from knowledge of geography or examination of the two images, if no map is available. It may be noted that waters charged with suspended particles (9) are generally situated outside the cluster, to the right (in the model used in Fig. 23.17).

Rough bare soils (4) are fairly close to very dark soils. They often correspond to ploughed lands for cultivated fields. When roughness decreases, points representing surface states of soil move away from the centre. They may be confused with some bright soils (3).

Highly organic bare soils (5) are close to dark soils. Calcareous soils are close to bright soils (between (1) and (2)).

Iron-rich soils (6) are not specifically distinguished in this diagram, since they are situated to the left of the cluster and are confused with vegetation.

The effect of soil moisture causes a displacement ((7), (8)) in an inclined direction relative to that of the cluster.

This cluster can be readily detected in an image by analysing a two-dimensional histogram (see Chaps. 9 and 8, Figs. 9.3 and 8.4). A principal component analysis can also be carried out, in which the first axis most often corresponds to the linear clusters of soils.

Obviously, this model needs to be modified to suit each situation. Such modifications are related to the geographic region, date of image acquisition and spectral bands available. An example is given in Figs. 7.4 and 9.4 (see Chaps. 7 and 9).

However, the linear clusters for soils given here assume that the soil is not covered by vegetation, which is common in arid zones and which happens only during half a year for cultivated soils. Hence, spectral characteristics of soils must be analysed when they are not covered by vegetation.

23.4 REMOTE SENSING AND SOIL MAPPING

In France, soil scientists have been using aerial photos for delineating boundaries between soil units since 1950s and since the beginning of soil mapping (Boulaine, 1957; Jamagne, 1967). Usage of satellite images in normal soil mapping took longer and is less common, since their interpretation necessitates computerised data processing and hence a specialised group. Also, images are not as readily available as aerial photos.

Escadafal and Pouget (1987) prepared 1:100,000 maps of surface state of soil for Tunisia using satellite images. The legend of the maps clearly explains all soil parameters used, viz.,

—predominantly gypsum soils, with shadows, stones, pebbles and encrustations;

—predominantly calcareous soils, with shadows, stones, pebbles and crusts, silt and gravel;

—predominantly quartz soils, with more or less fixed sands, sandy covers on limestone, mobile sands and sandy covers on gypsum;

—soils mostly covered by dry vegetation, dense or sparse.

At present, satellite images are found to be very useful in the programme of soil inventory, management and conservation (Ministry of Agriculture, INRA) in 1:250,000 scale for the programme regions and national territory. Concept of soil landscape is used in particular for mapping at these scales.

23.4.1 Soil landscape and remote sensing

Soil landscape is defined as follows:

'*Combination of soil horizons and landscape elements, viz., vegetation, effects of human activity, geomorphology, hydrology, substratum or bedrock, whose spatial organisation determines a (or a part of) soil cover in its entirety*'. Subsystems can be defined as soil landscape units and soil landscape elements.

A soil landscape unit corresponds to a soil system, is based on geomorphology and can be represented by a map unit. A soil landscape element corresponds to one or several spatially connected typological units, connected spatially, with a simple spatial organisation (for example, a sequence of soils).

Satellite images can be used to define most of the landscape elements such as vegetation, effects of human activity, geomorphology, hydrology, substratum or bedrock, if necessary aided by:

—vegetation maps (1:200,000),

—geological maps (1:320,000, 1:80,000 and, ultimately, 1:50,000),

—topographic maps (1:250,000, 1:100,000 and 1:50,000).

Visual interpretation (Girard et al., 1993; Bertrand, 1994; Girard and Gilliot, 1997) and computer processing (Burlot, 1995; Gilliot et al., 1998) of images provide a synthesis of various landscape elements. Each of these elements must be defined by using a systematically coded format. All such formats can then be statistically analysed and interpreted.

A similar type of interpretation can be made when only maps are available but not satellite images, but synthesis between various maps becomes more complicated (Antoni and Girard, 1996).

■ Interpretation with limited soil data

For each map zone or map unit, assumptions can be made about chorological laws relating landscape elements with existing soils, since soil differentiation factors are then available. These assumptions will be supported by ground truth data or by maps, if exist. It must be then verified by GIS whether all the assumptions are mutually coherent and spatially coherent with chorological laws (Francoual, 1997). Lastly, ground truth verifications should obviously be made. In fact, several factors of soil differentiation cannot be interpreted using satellite images or aerial photos (Yongchalermchai, 1993).

■ Interpretation with abundant soil data

If several maps or ground data exist, we can preliminarily define chorological laws relating soil cover, differentiation factors and environmental features seen in images. Using these laws, guidelines for interpretation of satellite images are formulated.

Importance of satellite images lies in obtaining an instantaneous view over a large area (3600 km^2 for SPOT) and hence comparing and, consequently, interpreting map zones situated at tens of kilometres from each other. The resolution of satellite images, 10 to 30 m, is adequate for soil landscape interpretations in 1:250,000 and 1:100,000 or even 1:50,000 scales (Gilliot and Girard, 1996). When interpretation is carried out in GIS and images restored on a topographic base, the results of interpretation can be directly used for GIS analysis.

23.4.2 Method of soil landscape elaboration

■ Interpretation of soil landscapes

Satellite images provide a possibility of interpreting various soil genetic factors in an integrated manner. A single map effectively gives the overall general structure of the physical environment of a region, including morphology, hydrology and land cover. Interpretation is facilitated since information is acquired at a single moment in a homogeneous manner over a vast field. Already published thematic maps, such as topographic, geologic and vegetation maps, can be used for enhancing the semantic content of each unit delineated in the image. These maps, which may not have been of the same scale and not prepared on the same dates, must obviously be used for extracting semantic information and transferring it to the map units based on a graphic model coherent with that of satellite images. These maps cannot be directly scanned and superposed over each other, since such superposition results in a meaningless map with no semantic coherence. Boundaries between soil units are drawn in satellite images by incorporating data from other maps.

■ Geomorphology

In temperate zones and for interpretations aimed at constructing graphic soil databases in 1:250,000 scale, morphology is one of the main factors of differentiation of soil landscapes. Morphology and hydrology are readily interpreted from satellite images. In fact, as the field of view is large, a synoptic perception of the entire geomorphology is obtained, which is rarely the case with aerial photos or ground observations; its form as well as its position vis-à-vis other entities is readily analysed. An entire watershed and distribution of talwegs in it can be easily mapped. Since zooming is possible without difficulty, any morphological feature can be analysed in detail and a more general picture obtained. Thus, fronts of talwegs, extending beyond 'temporary or permanent watercourses' indicated on topographic maps, can also be traced. It is possible to further improve this interpretation when a digital elevation model is available.

■ Winter images with bare soils

Satellite images, especially those acquired in winter when soils are bare revealing the hue of surficial horizon, give correct information about soil landscape elements.

Thus clayey soils are differentiated from marly and calcareous soils that are very stony, as well as coarser elements on soil surface.

As deciduous trees lose their leaves, various talwegs running through forests can be identified. Impact of drainage patterns on rocks and hence their resistance to weathering and soil erosion can thus be interpreted. This enables assumption of soil thickness and types of soil genesis.

Permanent grasslands as well as very humid or even peat zones can also be readily distinguished, leading to interpretation of differences in hydrological regime of soils.

Lithological interpretation of geological maps gives information on decomposition of rocks, from which, data pertaining to various soil parameters such as acidity or content of carbonaceous elements is obtained. Probable texture of material in which soil is formed can also be inferred.

■ Objectives of soil landscape maps

Objectives of preparing soil landscape maps are numerous.

1. Landscape maps prepared from a satellite image facilitate integration of various data sets already acquired at various scales and making a graphical synthesis. Images furnish scale transfer functions since they have a vast spatial field. Moreover, when interpretation is directly made under a geographic information system, zooming can be used to delineate boundaries and hence incorporate various precision levels of boundaries. Thus, information corresponding to maps of various scales can be taken into consideration.
2. Importance of a soil landscape map lies in the fact that it facilitates a clear understanding of the soil environment through various factors and their comparisons and enables definition of chorological laws by multi-criteria analysis, leading to evaluation of spatial organisation of soils.
3. Once a sketch is ready, a legend is prepared which must obviously be corroborated and explained by ground observations and soil analyses.
4. When no map exists for the entire region under study, such a sketch is useful to classify soil environment for the purpose of gathering data from specific places and hence for planning field itineraries and observation profiles. Such field studies are aimed at characterising soils defined by soil landscape analysis and thus verifying the chorological laws proposed. The latter may be confirmed or rejected. In the second case, other relationships need to be investigated. Field studies must also provide verification of questions raised during preparation of legends for soil landscape maps.

23.4.3 Example

A study was conducted in Lorraine for the regional chamber of agriculture (Girard and Gilliot, 1997) for testing the image interpretation of a 1:250,000 map.

Chorological laws used in this study were based on morphology, geology and land cover types. The soil cover seemed strongly related to the geology comprising limestone and marls, since surface formations are less extensive. The essential soil constraints are the shallow soil cover and excessive water. People had to necessarily adapt themselves to these constraints of soil use. Consequently, the land cover types, clearly sensed in images, were also used for soil studies.

As sun's elevation was low (18° 06') at the time of image acquisition (10 December 1987), the small level differences between the two sides of a talweg could be distinctly seen, since one was illuminated and the other not. Numerous talwegs in the Woevre region could thus be delineated.

■ Criteria used

Three main criteria, viz., land cover, morphology and geology, were used for this study. The following modes were used for the three criteria:

—Land cover: forest, permanent grasslands, crops, fallow lands and orchards.

—Morphology:

- *hydrology*: lakes, talwegs, secondary rivers, main valleys and low terraces;
- *relief forms*: plains, slopes (zero, small, moderate and steep), undulating morphology, low valley landscape, plateau, ledge, mounts/peaks.

—Lithology:

- *Surface formations*: alluvium, recent or ancient, colluvium and talus, sands, gravel, grits, plateau silts;
- *Calcareous sedimentary material*: clayey marl, marl, calcareous marl, hard marly limestone, gryphaeate limestone, oolitic limestone, sublithographic limestone, coral limestone and sandy limestone;
- *Noncalcareous sedimentary material*: clays, schists, sands.

Soil references (Baize and Girard, 1995) interpreted from soil landscapes in the zone under study comprise (in alphabetical order): Brunisols, Calcisols, Calcosols, Colluviosols, Fluviosols, Lithosols, Luvisols, Pelosols, Redoxisols, Reductisols, Rendisols, and Rendosols.

Most reference soils have not been defined by precise characteristics, since the latter cannot be identified by remote sensing.

■ Some chorological relationships

Some chorological relationships can be identified for calcareous, clayey and loamy substrata as well as for talwegs in the region studied.

□ Limestone

If the limestone is hard, disintegration takes place slowly and decalcified clays (Rendisols) remain. If erosion is intense, very little material remains and hence results in thin soil (less than 10 cm in thickness; Lithosols). If the limestone is softer and oolitic, Rendisols or Calcisols are obtained depending on the distribution between disintegration/alteration of limestone and erosion. If the limestone is marly or if the bedrock is marl, Calcosols or Calcisols are produced.

□ Clays (and schists)

Clays and schists give rise to clayey Brunisols, more or less thick, which may be saturated. Pelosols may also be observed in such zones. The Oxfordian clays on Woevre plateau produce redoxic Neoluvisols and clayey Redoxisols.

□ Silts

Silts are covered by forests or crops and produce Neoluvisols, redoxic or clayey. Under forests, silty zones could not be interpreted. They can be identified with difficulty by a variation in the species, which is rarely evident if forest exploitation has modified the environment, or by the vigour of plants which can be eventually identified in images acquired at the beginning of summer or in aerial photos. Winter images do not reveal this differentiation.

□ Talwegs

Talwegs are numerous in the region between Woevre and Moselle. Clayey talwegs are distinguished from hydromorphic talwegs which constitute the flow of 'lakes' in the Woevre region. These are continued by flat bottom talwegs, which are fresher and enter into the Bajocian limestone, creating several meanders. V-shaped talwegs in the Bajocian join them, most often under the forest.

■ Soil landscape map

A soil landscape map comprising 54 units (CD 23.1) was prepared using visual interpretation and aided by topographic, geologic and vegetation maps. Each unit is described by 15 variables: 2 soil references, morphology, 2 types of land cover, 2 slopes, 2 geological stages, 2 lithologies and 4 characteristics for soil types: coarse elements, texture, redoxic nature and calcification. This description is accompanied by comments indicating possible accuracy of these variables. All this information was included in a database for analysis of maps under GIS.

Obviously, it is difficult to represent all these units by allotting a separate colour to each such that they can be readily identified. A simplified soil landscape map can be prepared based only on the most common reference in the unit and defining a dozen groups.

Graphically, some units are more compact than others that are more distorted. This is characteristic of visual image interpretation. When the chorological laws are based on distinct morphological data, such as talwegs, boundaries can be readily traced very accurately. Contrarily, when they are based on data indirectly sensed from images, boundaries are more rounded or less distorted.

It is important to draw all graphic boundaries in order to integrate them with other geographic maps such as topographic map, digital elevation model or any other digital map. It is thus possible to add boundaries, to modify them and to highlight graphic and semantic information.

□ Contrast

Lastly, contrasts need to be mentioned; contrast between two sites is equal to the ratio of semantic distance to geographic distance. Semantic distance can be computed synthetically by a mathematical distance between soil references or analytically by a mathematical distance between various soil parameters such as coded observations, measurements, soil analyses, etc. Contrasts vary in the case of boundaries. Some boundaries have high contrast and probability of their detection is very high. Others may have less contrast for two reasons:

1. Difference between soil references (numerator) is small, indicating that the chorological laws are not clear to ensure delineation of boundaries from the data used. Such may be the case of boundaries between Rendosols and Calcisols, for example.
2. Geographic distance between soils (denominator) is large and transition between two units is gradual or it cannot be correctly sensed in satellite images. Such is the case of silty soils, which cannot be delineated under forest for want of establishing chorological laws. If they exist, we may not be able to locate them. In such a case, another map is used and the boundaries determined by other means are drawn on it, but without being able to justify them from satellite data.

Contrast on a map is estimated by the area of coloured zones and by differentiation of colours by the eye.

□ Validation and factorial maps

A better method of validating a map is by constructing factorial maps from the base map. As the map discussed here is a soil landscape map, it can be validated partly by thematic maps corresponding to the criteria used for chorological laws, viz., land cover, morphology and geology (CD 23.2). It is also possible to draw maps using the 15 criteria used to describe each unit. A new thematic map is thus prepared: for example, a 'soil genetic ambience map'. The latter makes use of redox and carbonate characteristics (CD 23.2).

The following soil groups were identified: cognate group of 'reductive-redoxic' soils (blue, magenta, green); cognate group of 'carbonate-saturated' soils (yellow, green); cognate group of 'noncarbonate' soils (red, pink, magenta).

■ Conclusions on soil landscape mapping

Soil landscape maps are more informative than the earlier small-scale soil maps. It is desirable to compare the time required for conventional methods and remote sensing methods of mapping virgin areas.

Such a comparison was made (Girard, 1995) between a 1:250,000 soil map, prepared by a combination of soil maps of various scales for the region west of Montpellier (Bornand et al., 1988), and a soil landscape map obtained from interpretation of two SPOT stereoscopic images of two seasons. In the case of soil units that are not under forests (the latter being difficult to interpret by remote sensing), 73.8% units were found common for both the maps. The image interpretation and data processing took not more than one month, which is smaller than the time required for integration of existing maps.

The soil landscape map constitutes a very good tool for classification of soil cover in zones for which no ground data are available. Moreover, it is a practical tool for synthesising soil data derived from maps of various scales. In fact, it provides a geographic framework for such a synthesis, either from existing maps (Antoni and Girard, 1996) or from satellite images (Yongchalermchai, 1993; Bertrand, 1994; Francoual, 1997).

Use of remote sensing and chorological laws, developed around the concept of soil landscape, enables flexible and rapid production of integrated soil maps of 1:250,000 scale for various objectives.

Diachronic images are useful for preparing thematic maps pertaining to erosion, soil movements, floods, determination and classification of wetlands and hence of hydromorphic soils.

However, for better utilisation of remote sensing, ground surveys are necessary in all cases for identifying the proper functioning of soil cover and obtaining necessary information on mapping, soil genesis and various thematic applications.

23.5 CONCLUSIONS

A fairly efficient model is now available for determination of some soil parameters from reflectance and digital number data. This model ought to become more powerful with use of microwave and thermal infrared data.

Obviously, the position of pixels in their neighbourhood must be taken into consideration, since soil cover is a continuum. Textural processing methods are used for this purpose. In the case of mapping, remote sensing and the concept of soil landscape have proven useful: it is currently employed for preparing reconnaissance soil maps in the national programme 'Soil inventory, management and conservation', aimed at covering the entire country by 1:250,000 soil maps. The main importance of remote sensing in soil surveys lies in providing systematic and repetitive geographic information.

References

Antoni V, Girard M-C. 1996. Reconnaissance régionale des sols du Tarn-et-Garonne. Établissement d'une carte des pédopaysages à l'echelle du 1/250,000, Adeprina, 37 pp.

Aubert G, Girard M-C. 1978. Vocabulaire de l'environnement pédologique tropical, ACCT, Paris, 81 pp.

Baize D, Girard M-C. 1995. Référentiel Pédologique. INRA-AFES, Paris. 332 pp.

Bertrand P. 1994. Élaboration d'une base de données localisées sur les agropaysages à partir d'images satellitaires: application à l'étude des organisations spatiales et à la segmentation du département de l'Yonne, Mémoire de Mastère SILAT, INA-PG, 45 pp.

Bialousz S, Girard M-C. 1978. Les coefficients de réflexion spectrale des sols dans les bandes de travail de Landsat, Fotointerpretacja w geograffi, Katowice, III (13): pp. 96–109.

Boifffin J. 1984. La dégradation structurale des couches superficielles du sol sous l'action des pluies, Thèse de l'INA-PG, Paris, 320 pp.

Bornand M, Barthes J-P, Bonfils P, Legros JP, Conventi S. 1988. Region Languedoc-Rousillon. Carte régionale des sols et des contraintes du milieu à l'échelle du 1/250 000. Essai méthodologique, Convention PIM CEE Régions.

Boulaine J. 1957. Étude des sols de la plaine du Chelif, Thèse Doctorat d'État, Université d'Alger, 582 pp.

Burlot F. 1995. De l'interprétation visuelle à l'interprétation automatique des images satellitaires: application aux pédo- et hydropaysages. Mémoire Mastère SILAT, Grignon, 65 pp.

Cervelle B, Malezieux J, Caye R. 1977. Expression quantitative de la couleur liée à la réflectance diffuse de quelques roches et minéraux, Bull. Soc. Fr. Mineral. Cristall., Université de Paris, 100: 185–195.

Cierniewski J, Courault D. 1993. Bidirectional reflectance of bare soil surfaces in the visible and near-infrared range. Remote Sensing Reviews, 17: 321–339.

Condit HR. 1970. The spectral reflectance of American soils. Photogrammetric Engineering & Remote Sensing, 34: 955–960.

Courault D. 1989. Étude de la dégradation des états de surface du sol par télédétection, Thèse de l'INA-PG, Grignon, SOLS, 17: 239.

Courault D, Girard M-C, Escadafal R. 1988. Modélisation de la couleur des sols par télédétection, 4e coll. Int. "Signatures spectrales d'objets en télédétection", Aussois, Esa SP-287, pp. 357–362.

Courault D, Girard M-C. 1990. Éléments d'interprétation des marques d'érosion intra-parcellaire à partir d'images SPOT. Photo-interprétation, 1 (2): 11–20.

Courault D, Girard M-C, Bertuzzi P. 1993. Monitoring surface changes of bare soils due to slaking using visible and near infrared reflectances. Soil Science Society of America Journal, 57 (6): 1595–1601.

Escadafal R. 1981. L'étude de la surface du sol dans les régions arides (sud-tunisien). Recherches méthodologiques, ORSTOM, Bondy. ES. 187, 60 pp.

Escadafal R. 1989. Caractérisation de la surface des sols arides par observation de terrain et par télédétection. Thèse de pédologie, Université Paris VI, 317 pp.

Escadafal R, Huete AR. 1991. Improvement in remote sensing of low vegetation cover in arid regions by correcting vegetation indice for soil 'noise', C.R. Acad Sc., Paris, 312 (2): 1385–1391.

Escadafal R, Pouget M. 1987. Cartographie des formations superficielles en zone aride (Tunisie méridionale) avec Landsat TM, Photo-interprétation, 4 (2): 9–12.

Francoual T. 1997. Détermination des agropaysages du département du Rhône par interprétation visuelle de données satellitaires SPOT. Réalisation d'une base de données spatialisée, Rapport INA-PG, Chambre régionale d'agriculture de Rhône-Alpes, 34 pp.

Gilliot J-M, Girard M-C. 1996. Étude de la vulnérabilité des sols à l'érosion, dans la vallée de l'Orvin, Chambre d'agriculture de l'Aube, 15 pp.

Gilliot J-M, Girard M-C. 1997. Le programme Inventaire, gestion et conservation des sols et les pédopaysages. Application à une zone de la Lorraine, Chambre d'agriculture régionale de Lorraine, 30 pp. + ann.

Gilliot J-M, Girard C-M, Girard M-C. 1998. Cartographie et analyse spatiale des zones humides de Champagne-Ardenne, Direction régionale de l'environnement. Champagne-Ardenne, 29 pp.

Girard M-C. 1995. Apport de l'interprétation visuelle des images satellitaires pour l'analyse spatiale des sols. Un exemple dans la région de Lodève, Étude et gestion des sols, 2 (1): pp. 7–24.

Girard M-C, Soyeux E, Bornand M, Yongchalermchai C. 1993. Structuration de l'espace régional et protection des ressources naturelles. C.R. Acad. Agric. Fr., 79: 37–50.

Goetz AFH. 1991. Overview: Imaging spectrometry for studying earth, air, fire and water. EARsel, Adv. Rem. Sens., 11 (1): 3–15.

Jacquemoud S, Baret F, Hanocq J-F. 1992. Modeling spectral and bidirectional soil reflectance. Remote Sensing of Environment, 41: 123–132.

Jamagne MX. 1967. Bases et techniques d'une cartographie des sols, Ann. Agro., N^{o} hors série, 18: 142.

King Ch, Delpont G. 1993. Spatial assessment of erosion: contribution of remote sensing, a review, Remote Sensing Reviews, 7: 223–232.

Mougenot B. 1990. Caractéristiques spectrales de surfaces salées à chlorures et sulfates au Sénégal, 2e journées de télédétection: Caractérisation et suivi des milieux terrestres en régions arides et tropicales. Colloques et séminaire, ORSTOM, Bondy, pp. 49–70.

Mougenot B, Pouget M, Epema GF. 1993. Remote sensing of salt affected soils. Remote Sensing Reviews, 13: 241–259.

Mulders MA. 1987. Remote sensing in soil science, Development in Soil Science, Elsevier, 15: 379.

Pouget M, Mulders MA. 1988. Description of the land surface for correlation with remote sensing data. Proceedings of the 5th symposium of Working group remote sensing, ISSS, Budapest, pp. 153–158.

Stoner E-R, Baumgardner M-F. 1981. Caracteristics variations in reflectance of soils. Soil Sci. Soc. AM. J., 45: 1161–1165.

Yongchalermchai C. 1993. Étude d'objets complexes, sol/plante, à differents niveaux d'organisation: de la parcelle au paysage, Thèse de l'INA-PG, Grignon, Sols, 19: 232.

24

Mining Geology

Although the basic objective of mineral prospecting lies in the discovery of mineral deposits, often very small in size and almost always concealed at depth, a task almost impossible for a surface technique such as remote sensing, the latter has emerged as a valuable tool for the prospector. This is due to the fact that it forms an integral constituent of a coherent approach guided by metallogenic models and is complementary to other techniques of investigation such as geophysics or geochemistry. It has hence become an effective component of synergistic data, leading to decision making of a prospector who is generally guided by ground observation data.

24.1 METALLOGENY AND TYPES OF DEPOSITS

Any problem of mineral exploration invariably reduces to two types of questions: What to search and where to find it? The first question is often guided by current economic conditions and by the price of a particular metal in the international market. Once a choice is made, metallogenic models are studied. This science of ore deposits is based on their geology and geochemistry and on the chronology of facts leading to the formation of mineral concentrations. Solution of the second question, viz., location of mineralisation, will be guided by more global concepts related to tectonic evolution of entire regions of the globe, which thus define preferential zones, metallogenic provinces, where deposits may be concentrated.

This approach is imperative for employing prospecting techniques and especially remote sensing. As the latter can only indicate indirect surface effects of mineralisation, success of its utilisation depends on identification of precise metallogenic objectives before any investigation. We can locate a thing only when we know what to search for! Experience of mining geologists is imperative for defining the interpretation keys which are essential for application of remote sensing in this domain.

24.1.1 Models of mineral deposits

Three fundamental questions are posed in metallogeny (Pelissonnier, 1976): where do the constituent elements come from (*source*), in what manner do they arrive at the point where they are presently found (*transport*) and, lastly, by what mechanisms are they deposited or concentrated (*deposition*)? Based on these three criteria, a volcano-sedimentary hydrothermal deposit, for example, refers to a mineral concentration originating from an eroded volcanic source and deposited in a sediment in which hydrothermal fluids develop later, probably due to reactivation of tectonic events.

Such theoretical models often correspond the definition of type systems based on observations on existing deposits. For example, the 'Kuroko' or 'Chypre' type of deposit is a 'gossan' model of pyrite accumulations (Fig. 24.1). This volcanic type of deposit with hydrothermal concentration is manifested on the surface by an alteration aureole rich in haematite, which is particularly favourable for detection by multispectral remote sensing in desert and semi-arid regions (Kruse et al., 1990).

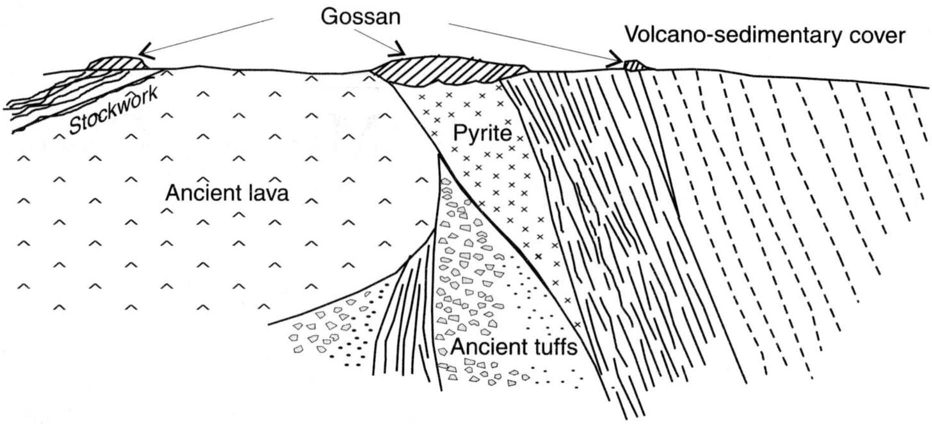

Fig. 24.1: Volcano-sedimentary 'gossan' type of Pb-Zn deposit. It is the case of sulphide accumulation followed by erosion.

24.1.2 An example: uranium deposits in France

The uranium deposits in France, for example, can be classified as three types: vein deposits in granites, mineralisations associated with sedimentary formations and those, rarer, associated with metamorphic rocks (Devilliers et al., 1980).

The intragranite vein deposits occur mainly in the Hercynian formations of the Massif Armorican (Vendée), especially in the Massif Central. They are associated with a granite rich in potassium (leucogranite), light in colour due to its acidic composition. Mineralisation occurs as vein fillings in fractures and as columnar-shaped bodies resulting from hydrothermal alteration. The accompanying mineralisation is dark-coloured pitchblende in vein deposits and bright-coloured alveolar episyenite in larger mineralised bodies. In the latter, quartz grains are completely replaced by uraninite with concentrations reaching up to 10 to 100 kg of uranium per ton (U/t) in exceptional cases.

The most classical sedimentary ore deposits are situated in the peripheries of granite massifs, which constitute the source. These deposits often occur in sandstone formations, red in colour, which are, however, buried under more recent sediments. They are hence referred to as concealed deposits. Mineralisation in them is pitchblende or coffinite in disseminated form (1 g to 3 kg U/t). Lastly, some low-grade marginal deposits occur in the Palaeozoic schistose metamorphic rocks (in Bretagne) affected by Hercynian granitic intrusions.

In other parts of the world, this classification extends to other types of deposits such as 'roll-front' of Wyoming and Texas (Raines et al., 1978). The important point to note is that the specific characteristics associated with leucogranitic vein deposits are also observed everywhere outside France. The contrasts in geochemical composition, structural state and colour may hence constitute key guides for interpretation of remote-sensing data, which are applicable in all latitudes when vegetation cover is low or negligible.

24.1.3 Regional metallogeny and metalliferous provinces

As mentioned above for uranium, the conditions favourable for occurrence of source-transport-deposition system for any mineral deposit are not the result of purely random processes. They are always logically related to a particular type of geology whose tectonic history per se is related to the major episodes of evolution of the Earth's crust.

A study of all such causes and their effects, combined with a systematic inventory of previous discoveries, leads to a model of spatio-temporal distribution of deposits and delineation of boundaries of regions in which they occur. Existence of such boundaries gives rise to the notion of metalliferous province and the basic concept of regional metallogeny.

The regional disposition of the French uranium deposits (Fig. 24.2; CD 24.1), for example, follows a west-east curved zone produced by Hercynian orogeny. The axis of this structure corresponds to the Averno-Vosgien cratonic nucleus. However, association between metallogenic province and ore deposits is not unique. Most often, similar favourable conditions may correspond to any metallic mineral assemblage.

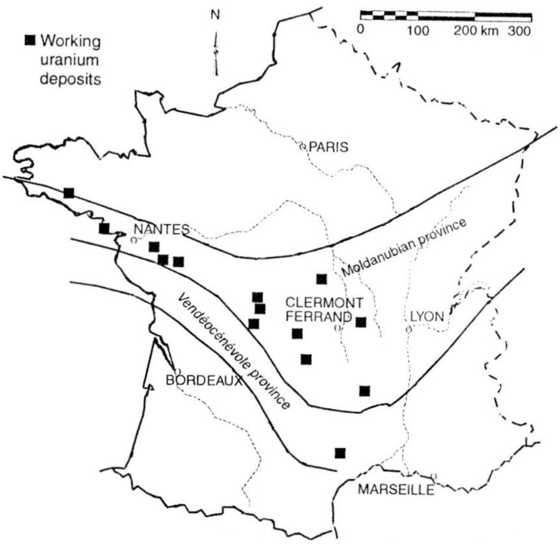

Fig. 24.2: Distribution of uranium deposits in France, compared to the geometry of the Moldanubian (Variscan orogeny) and Vendéocénévole (Mesozoic) provinces.

For example, a geographic relationship similar to that observed for uranium can be drawn for the Pb-Zn deposits associated with the Hercynian chain. The latter can be extended divergently towards south-west, combining the Armorican and Iberian peninsulas into a single metalliferous province, including the Pb-Zn deposits of the Zamora region in Spain (Marconnet, 1987). It thus follows that techniques of application of remote sensing are similar in both cases, the apparent geographic and climatic diversities notwithstanding.

24.1.4 Metallotect: an operational concept

Although the concepts of metallogeny are essential for guiding systematic mineral exploration, they only provide theoretical models that are often least related to the practice of prospecting.

In fact, prospecting methodology is more a combination of multiple techniques of investigation such as geophysics, geology, geochemistry and remote sensing, than really a theoretical approach. The prospector analyses a sequential series of diagnostic features based on a large amount of data gathered from the field. The concept of metallotect is introduced to answer to this empirical approach of mineral prospecting (Routhier, 1969).

The term metallotect is defined as any geological, geochemical, tectonic, etc. object or phenomenon which favours or contributes to the formation of a mineral deposit or concentration. This concept particularly facilitates proceeding from pure genetic models to more flexible models such as those introduced by geostatistics. In geochemistry, this leads to the concept of paragenesis, assemblage of chemical elements accompanying the metallic compound explored depending on the type of deposit.

In remote sensing, the term lineament-metallotect is used to denote the factors that may relate the density of lineaments traced on an image and localisation of deposits in the zone of observation. Thermal metallotects can be constructed if certain properties of fluid circulation in mineralised zones are related to variations in their surface temperature. Remote-sensing metallotects are generally based on the one hand, on the principle of synergy of forms (ex. lineament-lineament or lineament-contact intersections) and on the other, on the detection of 'transparent zones' that reveal the object under exploration if it is masked by plant cover, soil or other surface perturbations (Marconnet et al., 1982). This transparency can also be inferred from correlation with other factors such as geochemical or geophysical.

The work of the prospector thus lies in combining the metallotects to guide his decisions of exploitation or exploration. Systematic methods of decision making based on the principle of integration have also been developed (Leymarie et al., 1982).

24.2 MULTISCALE APPROACH IN MINERAL PROSPECTING

Mineral prospecting is essentially a multiscale operation. It is organised in successive phases that tend to focus on mineralisation. It starts at regional level based on purely metallogenic reasoning, gradually descending into detail and finally ends with the discovery of an object which most often is very limited in volume. Remote sensing, by its synoptic view from space, constitutes one of the best-suited techniques for this approach.

24.2.1 Multiscale structural model in remote sensing of vein deposits

Remote sensing is often suitable for understanding the multiscale structural phenomena, in particular in studying lineaments and their association with localisation of deposits.

Most bedrock mineral deposits are fractured as in the case of uranium deposits of Vendée and the Massif Central and the Pb-Zn deposits of the Zamora province in Spain mentioned earlier. Several successive episodes have led to mineralised concentrations, which make the deposits economic.

Mineralisation traps often occur in major shear zones such as in Margeride, the Vendée massif or in the uraniferous district of La Crouzille. Sequences of permeable breccia, lamprophyres forming a screen and open fractures serving as drains—conditions favourable for occurrence of episyenite—are observed in such zones. Remote sensing provides a wide scope for efficient exploration of structural traps due to its multiscale approach, ranging from satellite images to field investigations (Monget, 1982).

The wide range of scale of all such observations enables classification and characterisation of significant tectonic objects according to their geographic association with mineralisations:

—Orbital images ranging from 1:1,000,000 to 1:50,000, with a resolution of 30 m for LANDSAT satellites and up to 10 m for SPOT or even better for IRS, indicate faults, fractures and lineaments, 100 km to less than 1 km in size. They provide coverage of large parts of a metallogenic province and reveal major fault zones often organised around transform faults (Campredon et al., 1980).

—Aerial photos, side-looking radar and high-resolution orbital observations enable structural investigations in 1:50,000 to 1:5000 scales, with a resolution of 1 m to 50 cm. They facilitate detection of joints and fractures, varying in size from a kilometre to about 10 m, and echelon or curved patterns that mark tectonic nodes or lineament networks. Together with lineament-density maps, interpretation aids in locating favourable zones (Gilli, 1985).

—Detailed geologic mapping, in scales from 1:10,000 to 1:1000, clearly reflects faults and fracture zones of about 100 m in size. They enable precise location of metre-size objects and detection of trap structures (Rousselin, 1985).

—Lastly, opening of a mine or quarry in a 1:200 or higher scale facilitates mapping of objects of a decimetre or even a centimetre in size. Detailed sketching of joints, complex and braided joint systems, braided faults and open fissures is done at this stage. It enables working at the minute geometric level down to mineralisation traps.

Geometric comparison between lineaments and structural features indicated by these various observations depicts similarities as well as lack of correlation from qualitative, quantitative and directional points of view. In particular, major lineaments in satellite images often have no direct equivalents on the ground but their directions may indicate regional-scale fault structures. This demonstrates the utility of an extensive range of investigations.

24.2.2 Example of regional application of structural interpretation

We have seen earlier that the uranium deposits in France are distributed in narrow provinces associated with geographic localisation of Hercynian massifs. It is however possible to construct indicator metallotects at this level of global perception based on detection of linear and narrow features, i.e., lineaments, which distinctly differ from their surroundings.

This approach can be developed in particular based on observations by TIROS-AVHRR type meteorological satellites. In the example of CD 24.2, thermal and near-infrared data are used to delineate lineaments extending over distances exceeding hundred km. Changes in these contrasts depending on thermal inertia of formations and variations in the vegetation coverage during a year can be investigated based on diachronic comparison between summer and winter images. Thus maps depicting geographic relationships between lineaments and existing mineral deposits can be prepared to delineate zones favourable for occurrence of yet to be discovered mineralisations (Fig. 24.3; CD 24.1) (Durandau et al., 1982).

24.3 REMOTE SENSING: AN INDIRECT METHOD

Geological phenomena are essentially subsurface and with a few exceptions, mineralisation per se is rarely detected on the surface. Therefore, in mineral prospecting remote sensing is an indirect geophysical technique.

It is no less significant that remote-sensing data can delineate indicator metallotects either as structural forms or paragenetic systems accompanying mineralisation, whose expression may manifest on the surface or below a thin cover. Important remote-sensing parameters in this context are reflectance and temperature. Use of microwave imagery for this purpose is also described in Chapter 26.

24.3.1 Reflectance properties

■ Direct paragenetic detection: hydrothermal alteration

Hydrothermal alterations represent modifications in physicochemical properties of minerals and hence of rocks, due to the hot-water flows associated with volcanic eruptions or crystallisation of magma.

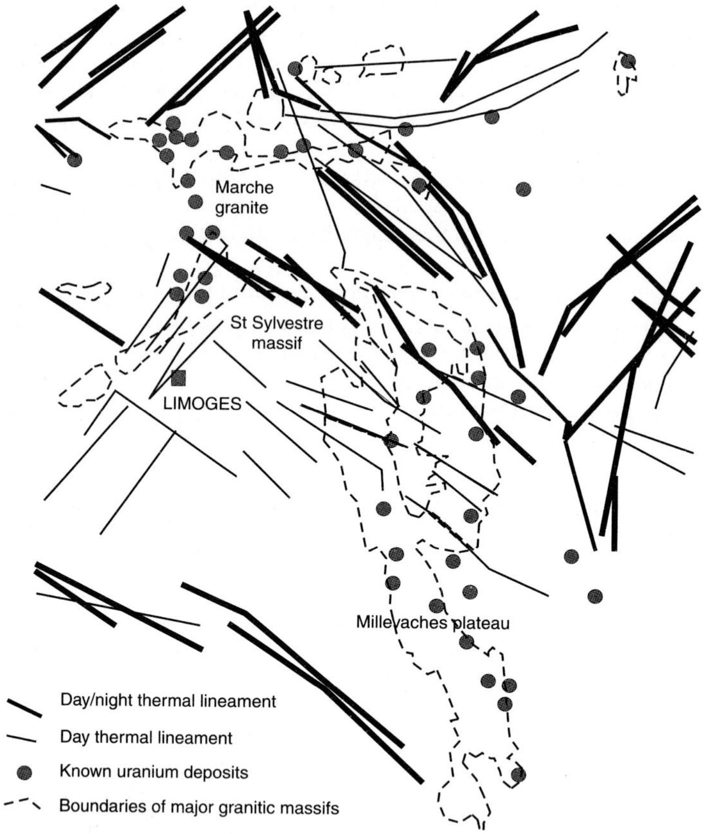

Fig. 24.3: Relationship between regional thermal lineaments and uranium deposits in the western region of the Massif Central .

Hydrothermal solutions contain various metals such as iron (Fe), titanium (Ti), copper (Cu), lead (Pb), zinc (Zn), tin (Sn) and uranium (U). Various forms of hydrothermal activity are recognised with specific composition of metal assemblages, and iron in the oxidised form often plays the role of an indicator element. When conditions are favourable, these metals may concentrate as economic deposits.

The exposed zone of such a metallic deposit is often oxidised and forms a 'gossan'. On the surface it depicts the form of an aureole and yellowish to brownish-red hues characteristic of the presence of haematite. Thus such zones can be recognised in multispectral remote-sensing data containing bands sensitive to these oxide compounds (band 7 of TM) or very narrow bands (see Chap. 4).

Alterations of crystalline or volcanic host rock at the contact of veins occur in the form of friable clays and distinct colours. Formation of white kaolinite in the alteration of leucogranites rich in potassic feldspars and impoverished in ferromagneisan minerals is a typical example. Kaolins of the Brame massif around La Crouzille in north of Limoges are well known for their use in fabrication of porcelains.

Most geological materials produce their maximum reflectance at 1.6 μm (Hunt et al., 1979). The suite of minerals often associated with hydrothermal alterations is characterised by a maximum reflectance at 1.6 μm and a marked absorption at 2.2 μm. This paragenetic association includes dioctahedral phyllosilicates (kaolinite, montmorillonite, etc.), some sulphates (gypsum, alunite) and aluminium oxides (diaspore, gibbsite).

Absorption in these minerals is caused by vibration processes and inner electrons of the molecular structure. Between 0.35 and 1.3 μm, electron transitions of minerals rich in iron, such as haematite, goethite, montmorillonite and jarosite, show characteristic features marked by minima in the wavelength bands of 0.43, 0.65, 0.85 and 0.93 μm. The minima in the interval 1.30 to 2.5 μm are more particularly related to the molecular structures of alunite, kaolinite, montmorillonite, pyrophyllite, potassic micas, diaspore, gypsum, jarosite and carbonates (see Chap. 23, Fig. 23.13).

Detailed observations in the position of absorption or reflectance peaks enable identification of indices on the content of alterations. For example, occurrence of a very narrow but low-amplitude band near 0.43 μm indicates jarosite. An intense minimum centred around 0.85 μm represents haematite whereas a similar characteristic centred on 0.93 μm indicates the presence of goethite. Thus, if the sample observed contains haematite or goethite, a band will occur with a minimum between 0.85 and 0.93 μm. On the other hand, development of an additional band appearing as a shoulder of the reflectance maximum at 0.75 μm and centred on 0.65 μm is more pronounced for samples rich in goethite rather than for those with high haematite content.

For clay minerals and alunite (Fig. 23.13; CD 24.1), characteristic absorption bands are situated more distinctly in the near infrared zone, at 1.4 and 2.2 μm respectively. However, an entire range of variants exists, some of which can be used to recognise constituents of alteration clays. Intense minima, narrow and single, at 1.4 and 2.2 μm indicate presence of pyrophyllite and potassic micas whereas dual peaks of absorption at the same position rather indicate existence of kaolinite (Fig. 24.4;

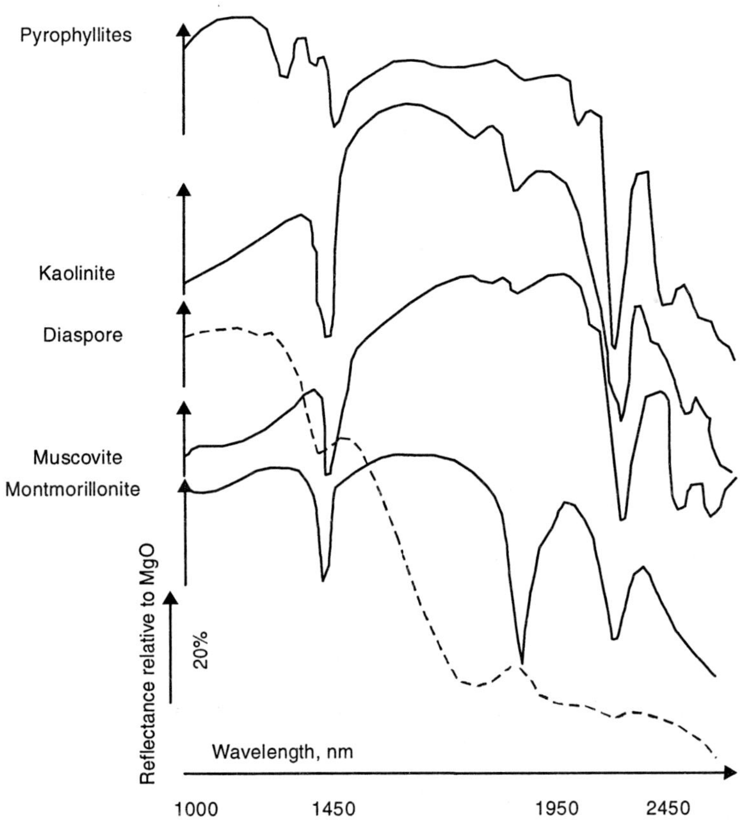

Fig. 24.4: Reflectance curves of minerals associated with hydrothermal alteration of rocks.

CD 24.1). For montmorillonite one more distinct minimum appears at 1.9 μm; it is also known as the 'water band'. Gypsum has a characteristic similar to that of alunite, but at 1.75 μm.

The spectral contrasts of pure samples are not generally observed in real conditions of existing mineralisations. For example, if hydrothermal alterations associated with gold deposits of Beaver Creek (Colorado) and Goldfield (Nevada) are considered, alunite is encountered as a major constituent of most altered samples collected. It is hence often difficult to associate a sample with a particular type of alteration (silicified, opalised, argilitic and phyllitic) based only on the form of its spectrum even if high-resolution observations, such as those of airborne hyperspectral imagery, are available. This is an obstacle in determining from distance the paragenesis of the anomaly observed and hence an uncertainty on the type of concentration. The latter can be established only from a visit to the field accompanied by precise geochemical measurements.

The observations mentioned above on the form of spectra have led to development of ratioing techniques or principal component analysis, which enable quantitative estimation of constituents (Lee et al., 1990). Absorption properties specific to such alteration products have guided the choice of band 7 (2.08 to 2.35 μm) for the LANDSAT Thematic Mapper.

■ Indirect spectral detection of geologic formations

Mineralisations, or the formations that host them, are not always associated with well-defined spectral characteristics, especially when they are covered by a sufficiently thick soil or dense vegetation.

However, favourable conditions may exist wherein a particular association of geologic bedrock-soil-vegetation has spectral contrast and thus enable indirect mapping of a formation (Siegal et al., 1977).

In the leucogranitic massif of Saraya in eastern Senegal (Blot, 1980), ferruginous soil masks the nature of underlying formations and renders their direct mapping by remote sensing difficult. Nonetheless, when the observations of the region by LANDSAT and CZCS (CD 24.3) are analysed, a distinct contrast is noticed between the soils covering the massif and those on its periphery. In the case of CZCS, highest contrast is observed for the first spectral band (blue—433 to 453 μm). This band is particularly sensitive to the concentration of quartz, which is very high in the surface soil (Fig. 24.5; CD 24.1), and reflects the influence of a granitic body situated at depth. These 'colour' contrasts are almost not visible on the ground which is flat in most cases. Thus an effect of 'transparency' is observed.

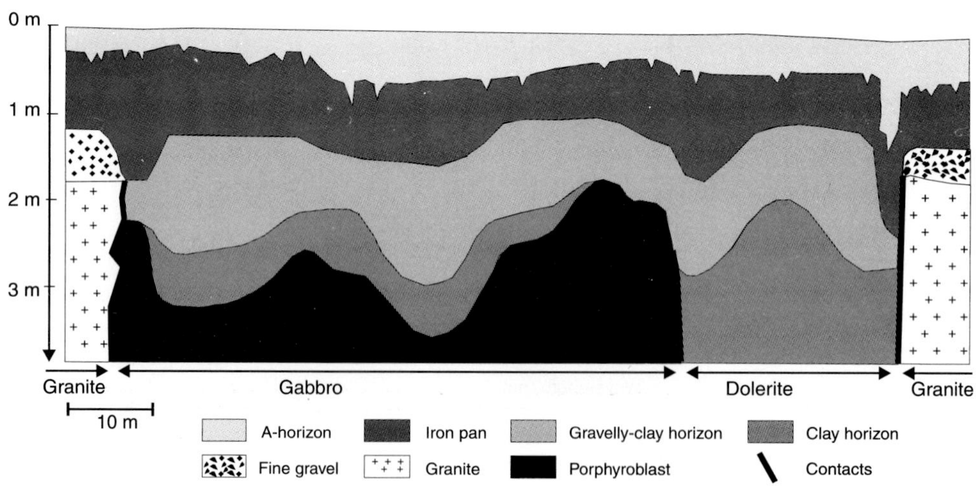

Fig. 24.5: Schematic cross-section of surface formations over Saraya granite.

At a higher level of detail in the LANDSAT MSS images (CD 24.3), structural passages can be noticed running across the massif, which have lower values of digital numbers. These zones correspond to dolerite veins, known for their association with vein-type uranium mineralisation of economic size. On top of dolerite veins, which are more clayey, soil is less silicious, thus indirectly making these formations transparent.

Although it is not possible to replace deep-level studies by surface investigations, it is seen that the interpreter of spectral contrasts should always be on the lookout for anomalous zones because they may indicate deep causes. This, of course, is possible only in zones where the medium is in natural equilibrium without anthropogenic disturbances. Examples of such unfavourable anthropogenic phenomena are land patches of annual burning in Africa.

24.3.2 Use of thermal band

Measurement of surface temperature of rocks or soils (see Chap. 26) by airborne or spatial observations are rarely used instantaneously. At least, its interpretation always takes into consideration the regime of excitation energy to which the body under observation is subjected, in view of diurnal variation of solar radiation. Moreover, knowledge of temperature variations of a rock during day and night enables determination of its thermal inertia. This volumetric property is of the same kind of intrinsic nature as other geophysical measurements such as electric conductivity, density, etc.

Computation of thermal inertia necessitates on the one hand, sequential observation of temperatures during a diurnal cycle and on the other, use of a thermodynamic model to describe thermal phenomena inside the solid medium (see later, CD 24.4).

Diurnal temperature variation of water is of low amplitude since it has very large thermal inertia. Soils, on the opposite, show large amplitudes of day/night temperature variation. For rocks and dry soils, inertia is small and amplitude is as high as the soil is porous. As the moisture content of soil or fracturing of rock increases, it gets saturated with water. Its thermal inertia increases and amplitude decreases (Fig. 24.6). In the case of chlorophyllian vegetation, self-regulation by evapotranspiration (see Chap. 1) also reduces temperature deviations.

Contrasts between a moist fractured medium, such as a granitic massif, and a dry and less consolidated medium, such as altered crushed zone where quartz sand is abundant, are very important

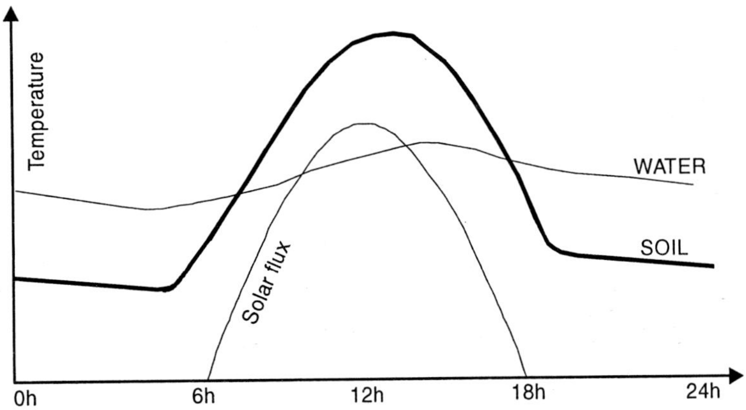

Fig. 24.6: Theoretical curve of diurnal temperature variation of soil and water body for continuous diurnal insolation.

in mineral prospecting. Such contrasts are especially more distinct during night. Zones of high thermal inertia in such cases appear hotter than their adjoining areas (Rousselin, 1985).

■ Use of night thermal contrasts at regional level

Night observations can be primarily made by small-scale satellites such as AVHRR or HCMM and used to delineate entire massifs such as the leucogranites of La Crouzille district, which are situated in North-Limousin. The granitic massif of this region consists of two main formations, the Brame granite in the west and the Saint-Sylvestre granite in the east. The Brame granite is a biotite-granito-gneiss, bounded on the west by the Nantiat fault. Its general orientation is NNE-SSW. The Saint-Sylvestre granite is a leucogranite with two micas. It occupies the entire zone south and east of La Crouzille (Fig. 24.7; CD 24.1).

Mineralisation is localised in subvertical tectonic zones, about a metre in width, filled by crushed granite. It occurs essentially in the form of autunite and pitchblende, that is, as phosphates, silicates and oxides. Three principal factors are responsible for mineral concentration. Firstly, dense tectonic activity associated with extreme alteration of contact zones results in stockwork deposits, often workable as opencast mines but of low grade. Secondly, fault intersections with lamprophyre veins give rise to localised mineralisations (veinlets) of high-grade pitchblende and, lastly, fault intersections with micaceous episyenite lenses, whose voids are filled with pitchblende, localise accumulations of high-grade ore.

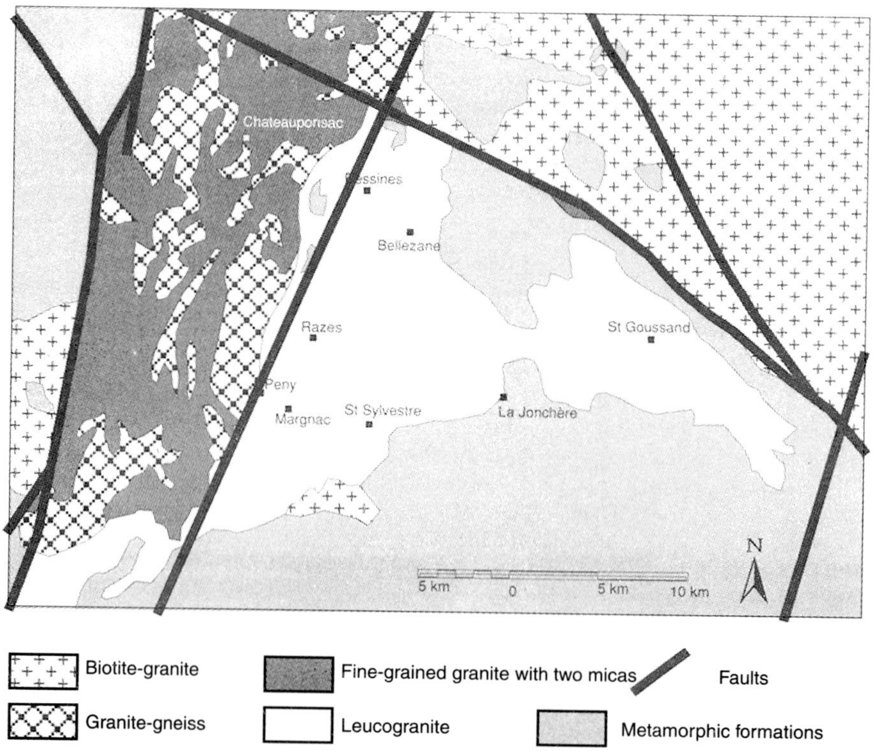

Fig. 24.7: Schematic geologic map of Saint-Sylvestre massif (after Marquaire and Moreau, 1969).

Considering all these characteristics, uranium prospecting in the La Crouzille sector is guided by two main principles: firstly, strategic exploration of favourable source rocks, viz., leucogranites, and secondly, delineation of fracture zones favourable for mineralised fillings. Remote sensing in this sector followed this approach (Durandau et al., 1982).

Satellite observations of night time thermal contrasts by HCMM (CD 24.4) showed that the Saint-Sylvestre leucogranite is marked by distinctly higher temperature than that of its surroundings. This is an indication of higher thermal inertia. This factor is probably due to a more open fracturing of this formation capable of containing larger quantities of water.

Satellite measurements were confirmed in detail by airborne thermal measurements (CD 24.4). The latter showed identical regional thermal contrasts and indicated details of vein formations of about a metre in width. They depicted differences in night temperatures between massive, cold, lamprophyres and fissure zones in which fluids circulate, which are indicated as hotter zones.

■ Night time aerial thermal measurements

Utility of night time aerial data with ARIES scanner (Tabbagh, 1973) obtained with a ground resolution of 10 m is demonstrated with a few examples (CD 24.5). All the images of our example pertain to Quercy in the southern boundary of the Permian Brive basin, zone known for its prospects of uranium deposits. Major faults such as Meyssac or Padirac control and delimit potential zones of mineralisation in the Permian sedimentary formations (Daudon, 1987). One can particularly notice the Argentat fault zone occurring in the cold and dark region, the red sandstone of high thermal inertia in Brive region appearing as bright and the Liassic formations, clearly delineated. Lastly, many sinks and avens indicating the Gramat karst plateau, which are detected as cold zones at night, appear as numerous dark spots.

Interpretation of night time thermal images, however, is also beset with some problems. Special mention may be made of meteorological effects leading to accumulation of cold air in valleys. Such an effect is observed in the thermal images of the Martel region (CD 24.5). It can mask lithological contrasts.

Night time thermal scanning is one of the most important tools for mineral prospecting (Scanvic, 1983) and detailed geologic mapping since it reveals inertial properties of rocks, which result from the nature of rocks and may also indicate effects of heat transport from deep sources.

■ Monitoring complete thermal cycle and thermal inertia mapping

Determination of thermal inertia, which requires acquisition of day as well as night data, limits practical utility of these techniques to only meteorological satellites. The low resolution of the latter, 1 km for AVHRR and 4 km at nadir for METEOSAT, narrows down their applicability to only regional metallogeny.

Let us take the Bandiagara massif in Mali (Daveau, 1959) as an example. This sandstone formation constitutes a slab sandwiched between the Niger delta in the west, towards which it dips with a mild slope, and the sandy depression of Gondo in the east over which it stands as a cliff.

The LANDSAT MSS data indicate (CD 24.5) striking spectral homogeneity of the massif. This is due to the uniform coverage of all outcrops by 'desert patina' rich in iron oxide. Observation of contrast between day and night temperatures (CD 24.5) for the same formation clearly shows, the low resolution notwithstanding, a zone situated in the northern part of the massif for which thermal inertia is much lower than that of plateau sandstones.

This thermal-inertia map depicts the true phenomenon of transparency across the layer of desert varnish that masks any lithological difference in the visible and near-infrared bands. Referring to the geological map (Fig. 24.8; CD 24.1), the zone of high inertia corresponds to an iron rich pan covering dolerite intrusions, probably mineralised. The complementary nature of spectral data in geological mapping is evident here (Bardinet et al., 1982).

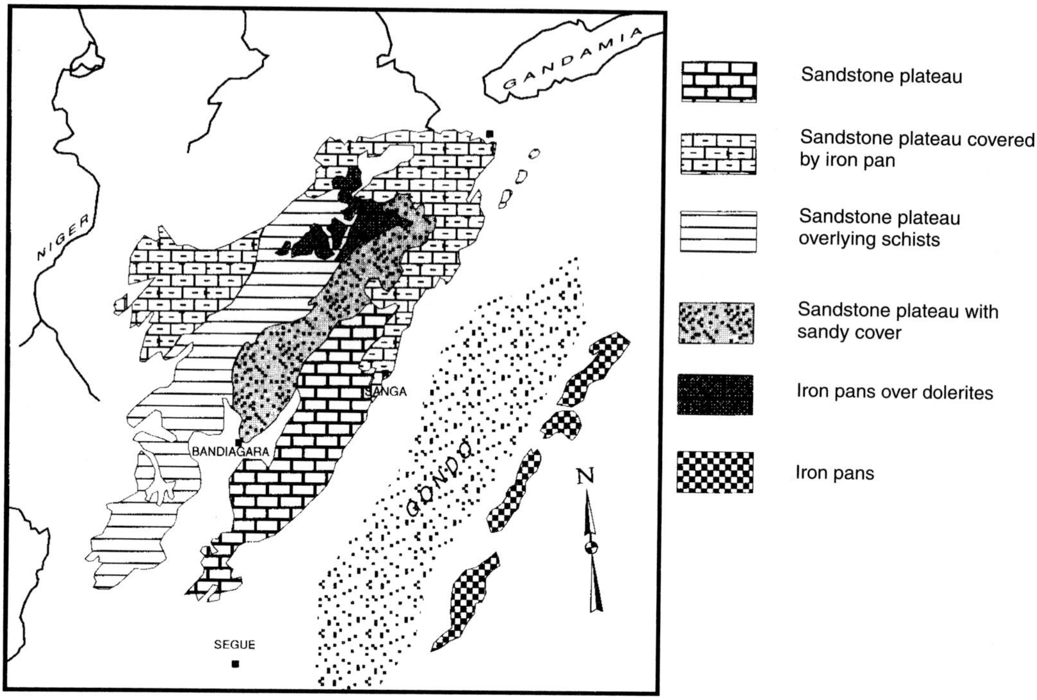

Fig. 24.8: Geological map of Bandiagara plateau (after Blanck, 1982).

24.4 CONCLUSION

Apart from a few results obtained in mapping mineralised alterations in desert zones, it can still be considered that remote sensing in general is not a tool for direct detection of minerals. Its principal advantage, however, lies in providing a synoptic base map of major lithological units, morphological forms and structures. This information becomes very useful in mineral prospecting when it is integrated with other types of ground data such as geological, geophysical or geochemical.

The quality of geological information given by remote sensing depends on the mode of instrumentation, type of terrain observed (plant cover, soil, rock outcrops, climate), method of processing, techniques of image manipulation adopted and, lastly, degree of understanding of the physical phenomena involved in the measurements. Hence it is a method in which the expertise of the interpreter plays a major role. Remote sensing can succeed only in a multidisciplinary structure in constant liaison with ground observations. It is no less significant that its capability in covering large zones of exploration makes it an essential variable for identifying regional metallotects used in economic models of prediction and development of natural resources.

In the future, enhancement in spatial and spectral resolution of remote-sensing instruments ought to enable more precise detection of mineral components of rocks if only they are exposed. The techniques of interpretation will then become similar to those of geochemistry and probably emerge as more quantitative methods. It would then be possible to achieve direct detection of mineral deposits when vegetation cover is not too dense.

References

Bardinet TC, Monget J-M, Patoureaux Y. 1992. 'Combined use of daily thermal cycle of METEOSAT imagery and multispectral LANDSAT data: application to the Bandiagara plateau, Mali'. Proceedings of an EARSEL-ESA Symposium, Igls. Austria, Apr. 1982, ESA Technical Publication SP 175: 95–101.

Blot A. 1980. L'altération climatique des massifs de granite du Sénégal, Travaux et Documents de l'ORSTOM, 114: 434.

Campredon R, Monget J-M, Simon P. 1980. Mise en évidence par télédétection d'un accident à rejeu récent dans le bassin permien de Luc-en-Provence. Comptes rendus de l'Académie des Sciences, Paris, Série D, 291: 55–57.

Daudon P. 1987. La fracturation du socle en Bas-Limousin et Quercy et ses répercussions dans la couverture sédimentaire Thèse d'État, Université d'Orléans, 215 pp.

Daveau S. 1959. Recherches morphologiques sur la région de Bandiagara, Mémoires de l'Institut français d'Afrique Noire, IFAN, Dakar, 56: 119.

Devilliers JP, Ziegler V. 1980. L'industrie minière en France. Sa situation au début de 1980, Annales des Mines, 7–8: 123–134.

Durandau A, Monget J-M. 1992. 'Interpretation of satellite and airborne thermal measurements gathered over hercynian granites (France) in relation with uranium deposits'. Proceedings of the International Symposium on Geology of Granites and their Metallogenetic Relations, Nanjing, China, Oct. 1982, pp. 611–619.

Gilli J. 1985. Analyse numérique de l'image radar du secteur Port-Gentil Azingo Lambaréné Ouest: contribution à l'étude géologique du bassin sédimentaire gabonais, Thèse de 3e cycle en Géologie, Géochimie et Techniques Avancées, Université de Nice.

Hunt GR, Ashley RP. 1979. Spectra of altered rocks in the visible and near infrared. Economic Geology, 74: 1613–1629.

Kruse FA, Kieren-Young KS, Boardman JW. 1990. Mineral mapping at Cuprite, Nevada with a 63-channel imaging spectrometer Photogrammetric Engineering and Remote Sensing, 56 (1): 83–92.

Lee JB, Woodyatt AS, Berman M. 1990. Enhancement of high spectral resolution remote sensing data by a noise adjusted principal components transform. Geoscience and Remote Sensing, 20 (3): 295–304.

Leymarie P, Baelz-Maniere S, Durandau A, Monget J-M, Sinding-Larsen R. 1982. 'Un système d'aide à la prospection minière en prospection de l'uranium'. C.R. du Symposium' AEN/AIEA de R&D: Méthodes de prospection de l'uranium, Paris, Juin 82, pp. 889–907.

Marconnet B. 1987. La télédétection spatiale, une méthode pour la prospection minière stratégique et tactique. Application aux gisements stanno-wolframifères de l'Ouest de la péninsule ibérique, Thèse en Pétrologie Structurale et Métallogénie, Université de Nancy, 244 pp.

Marconnet B, Gagny C, Leymarie P, Monget J-M. 1992. 'Phénomène de transparence d'un leucogranite à étain-tungstène sous couverture métamorphique (province de Zamora, Espagne)', Actes du Symposium International de la Commission VII de la SIPT. Toulouse, Sept., 82, Éd. GDTA, Int. Archives of ISPRS, 24-VII, 1, pp. 523–531.

Monget J-M. 1982. 'Télédétection multiscalaire des provinces uranifères françaises', C.R. du Symposium AEN/AIEA de R&D: Méthodes de prospection de l'uranium, Paris, juin 82, pp. 121–128.

Pelissonnier H. 1976. 'Classification par types en métallogénie', In Métallogénie et prospection minière, Mémoire Hors Série de la Société Géologique de France, 7: 277–283.

Raines GL, Offield T, Santos ES. 1978. Remote sensing and subsurface definition of facies and structure related to uranium deposits, Powder River basin, Wyoming, Economic Geology, 73: 1706–1723.

Rousselin T. 1985. Télédétection thermique des granites uranifères du Nord-Limousin (France), Thèse de 3e cycle en Géologie, Géochimie et Techniques Avannées, Université de Nice.

Routhier P. 1969. Essai critique sur les méthodes de la géologie, Masson, Paris, 202 pp.

Scanvic J-Y. 1983. Utilisation de la télédétection dans les sciences de la terre, BRGM, Manuels et Méthodes, 7: 158.

Siegal BS, Goetz AFH. 1977. Effect of vegetation on rock and soil type discrimination, Photogrammetric Engineering and Remote Sensing, 43 (2): 101–196.

Tabbagh A. 1973. Essai sur les conditions d'application des mesures thermiques à la prospection archéologique. Annales de Géophysique Fr., 29 (2): 179–188.

25

Remote Sensing and Coastal-zone Management

The objective of this chapter is to give several examples specific to coastal environments, based on the investigations carried out in France and in particular at the IFREMER.

25.1 INTRODUCTION: GENERAL PROBLEM

Land cover and land use of coastal regions are today undergoing rapid changes, more particularly in countries producing and exporting petroleum or mineral products. Other modes of coastal exploitation have also become preponderant. Aquaculture is fast developing as is evident from the fact that 200,000 ha of ponds have been constructed in fragile zones on the Philippine coast. Similarly, tourism has increased in the Ancient World and some island states of the Caribbean, Pacific or Indian Ocean. In the near future tourist activity will become world's premier industry in terms of volume of business.

In almost all countries with a maritime front, there is migration of populations to the coasts: 2/3 of world's population lives within 400 km from the sea. The impact of such activity is represented by:

—development of infrastructure: ports, airports, industrial, mining and energy complexes, offshore petroleum production, agro-nutritional transformation, urban complexes, hotel and tourist infrastructure, aquacultural installations, etc.

—transformation of natural zones: destruction of wet zones, reefs, natural vegetation in back lands, rectification and protection of coastline, various types of pollution of marine environment, increase in ecological hazards, etc.

Increasing awareness of such hazards and the need for more efficient management are experienced at all levels and are concretised by co-ordinated management procedures which most often are supported by legislation:

—in France, schemes for economic management of seas (SMVM), more recently territorial management directives (DTA),

—in the United States, the Coastal Management Act, Middle Atlantic Coastal Resource Council,

—at international level, programmes such as COMAR (UNESCO), Regional Seas (PNUE), etc. The European Union, concerned with this problem very early, launched specific experiments on information and integrated studies on coastal zones.

Such legislations and procedures necessitate employment of planning and management products derived from multiparameter information systems for which remote sensing constitutes a source of data.

Coastal zones present the difficulty of linear type of extension with a land-sea interface and its adjacent zones in all countries, independent of borders. This interface is under the jurisdiction of several administrations whose prerogatives are often opposed to one other. Further, like in applications described in other chapters, information furnished ought to be multispatial, multiparametric and multitemporal. Thus in metropolitan France for example, the basic geographic maps pertaining to littoral zones are of various types, various scales and adopted to various planes of projection: ex. topographic maps of the National Geographic Institute, marine maps of the Naval Oceanographic and Hydrographic Service.

In many tropical-zone countries, very few precise and up-to-date cartographic systems are available, particularly for sensitive and inaccessible regions such as mangroves and coral reefs.

25.2 SPATIAL OBSERVATION OF COASTAL ENVIRONMENTS

25.2.1 Littoral and coastal-zone objects

It is difficult to define what constitutes the littoral zone since the phenomena pertaining to this interface are both terrestrial and marine.

From an administrative point of view, the extension of the coastal zone on the upstream, coastland, side can be limited to the border of littoral communities and on the downstream (sea) side, to 12 nautical miles. Based on environmental criteria related to water exchanges and flows, it can be considered to extend from the inland boundaries of watersheds up to the zone of extension of these flows into the sea. On the geographic and topographic basis, it can range from a land zone of a defined elevation (200 m for example) to the border of continental plateau (bathymetric limit of –200 m). For the following discussions, we limit ourselves to the 'marine littoral zone', i.e., to the zone bounded on the land by high-tide mark. Two types of environment can be distinguished in this zone:

—subtidal, situated below low-tide level; venue of influences, often combined with one another, emerging: (a) from the surface (sea swell, waves, currents, winds, pollutants such as hydrocarbons, ice, etc.), (b) from the water column (turbidity, temperature, discharge plumes, water blooms, plankton, etc.), (c) from the bottom (shallow areas floors observed in clear water in optical domain, themselves consisting of various substrata, hard or soft, with or without vegetation, emerging from animal constructions such as coral reefs and with influence of the sea bottom on surface currents, detectable in the microwave band, etc.).

—intertidal, situated in the balance zone of tides, comprising bare soils (sands, mud, rocks, etc.), vegetation (plants in swamps such as mangroves, marine halophytes, algae, microflora, etc.), animals (mollusk banks, coral reefs, etc.), managed zones (oyster parks, basins, canals, breeding ponds, salt pans, etc.).

In addition to a general linear disposition, the littoral zones are characterised by small-size features constituting highly varied landscapes whose structure results from natural or artificial processes.

Lastly, the constituents of the littoral zones are highly fluctuating with varied periodicity, random (storms for example) or periodic (hydrological regimes, phenological regimes, tides, etc.).

Considering these features, it is pertinent to examine whether the Earth-observation systems possess adequate spatial resolution and observational periodicity to provide, on operational basis, data necessary for correct characterisation of surface phenomena in littoral zones, as complementary to conventional ground measurements.

25.2.2 Littoral objects and specifications of aerospatial observation systems

Preliminary analyses of adequacy between characteristics of coastal environments and technical performance of aerospatial observation systems (Klemas et al., 1980; Gierloff-Emden, 1982; Loubersac, 1983; Klemas et al., 1987) have shown that present-day remote-sensing systems are not correctly suited for coastal-zone studies. There is incompatibility between high spatial resolution and high temporal resolution required for these observations. This is a major drawback hindering the application of high-resolution remote sensing to coastal investigations that do not involve dynamic phenomena.

This incompatibility is illustrated in Fig. 25.1. An analysis of the diagram (logarithmic scales) of temporal resolution expressed in days (abscissa) and spatial resolution expressed in metres (ordinate) offered by major Earth observation systems leads to the following inferences.

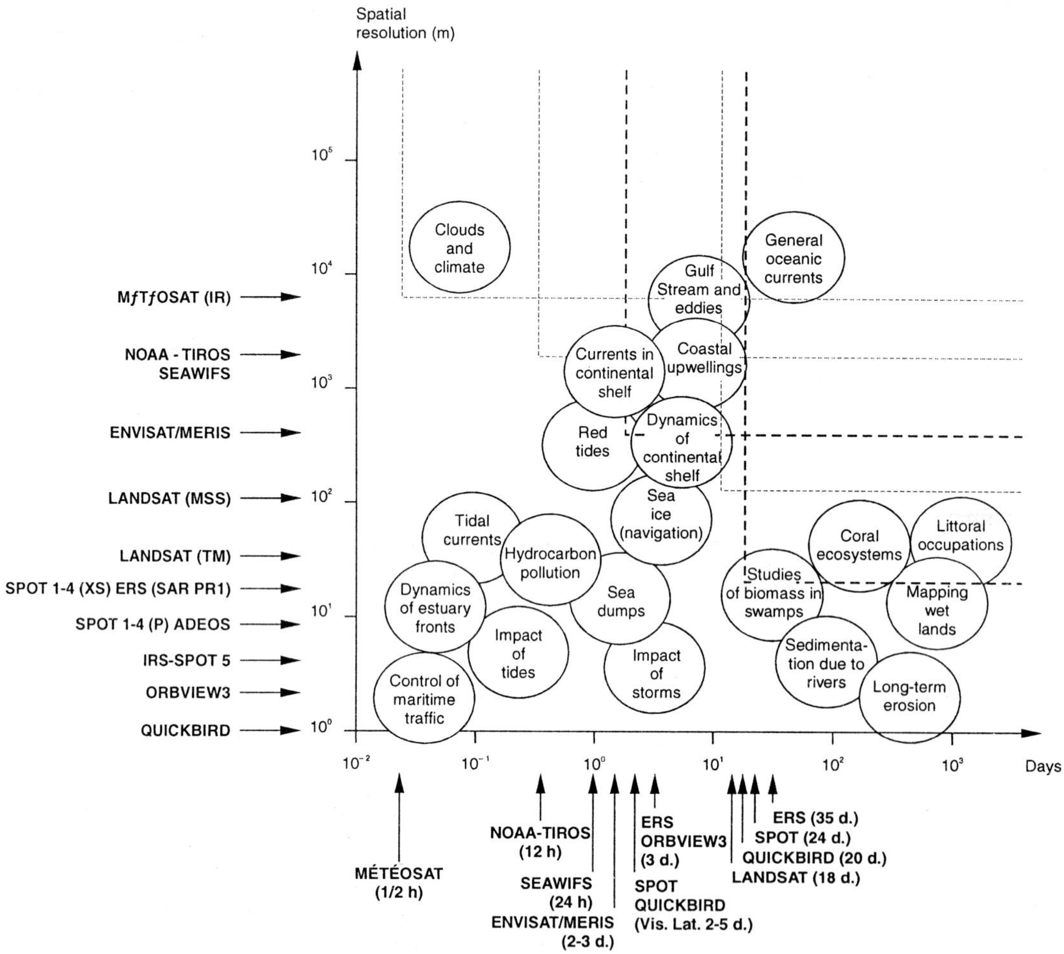

Fig. 25.1: Possibility of detection of marine and coastal phenomena by major sensors and platforms of present-day Earth observation satellites (adapted, completed and updated after Klemas et al., 1987, courtesy IFREMER).

The spatial resolution (tens of metres) of systems such as SPOT or ERS enables with difficulty operational characterisation of some of the human activities in littoral zones and certain natural phenomena such as coastal erosion (except the specific case of strong sedimentary dynamics). On the other hand, the very high resolutions of SPOT-5 (5 m) or Orbview-3 (4 m in multispectral and a metre in panchromatic) and Quickbird (3.3 and 0.8 m respectively) are closer to the resolution requirements for most applications in coastal zones.

The optimal temporal resolution, extremely variable depending on the application, is not satisfied by the presently existing satellite systems (better than 3 days with ERS for typical coastal studies). Thus operational applications of remote sensing to tidal phenomena, monitoring of dispersion due to currents and winds (dispersion of wastes, resulting pollutants, etc.) or navigation control are most often limited.

Aerial platforms are better suited for coastal studies since they provide ground resolution and revisit frequency higher than those of today's high-performance satellites and permit a greater number of onboard sensors. However, the disadvantages of aerial systems are limited synoptic view, high costs of acquisition with increasing repetition, atmospheric constraints and tedious procedures of geometric processing of data.

Operational characteristics of the major satellite or airborne sensors for various studies of coastal zones are summarised in Table 25.1 (adapted after Klemas et al., 1987). Use of certain platforms for some sensors is not technically feasible and hence not indicated. It follows from the Table that optimal observation for monitoring coastal phenomena necessitates use of almost all the spectral bands and

Table 25.1: Operational characteristics of remote-sensing systems for coastal studies

Sensor	Platform	Winds	Suspended sediment concentration	Bathymetry	Chlorophyll concentration	Circulation due to currents	Waves	Hydrocarbons	Coastal erosion	Vegetation and land use	Suspended sediments (SS)	Types of sea floor	Biomass and vegetative stress	Surface temperature	Salinity
Radar altimeter	S	3	0	3	0	3	4	0	0	0	0	0	0	0	0
Profiling lasers	A	2	0	4	0	0	4	1	3	0	1	2	0	0	0
Laser fluorescence	A	0	2	1	4	1	0	4	1	1	2	0	0	1	2
Photography	A	1	3	3	3	2	3	4	4	4	4	4	3	0	0
	S	1	2	2	2	2	2	3	4	4	4	3	2	0	0
Multispectral	A	1		3	3	2	3	4	4	4	4	4	4	0	0
scanners	S	1		2	3	2	2	3	4	4	4	3	3	0	0
Imaging	A	3	0	1	0	3	4	4	4	4	2	0	2	1	1
radar	S	2	0	1	0	2	4	3	3	3	1	0	1	1	1
IR thermal	A	1	1	0	0	3	0	4	2	1	2	0	1	4	2
scanners	S	1	1	0	0	3	0	3	1	1	2	0	1	4	0
Passive microwave	A	1	1	0	1	2	22	4	1	1	1	0	0	2	3
radiometers	S	1	0	0	0	1	1		0	0	0	0	0	2	2
Radar scatterometer	S	4	0	0	0	2	3	0	0	0	0	0	0	0	0

A—Aeroplane; S—Satellite; 4—Operational; 3—Truly potential but experimentation required; 2—Potential utility; 1—Limited utility; 0—Not usable all types available sensors, without however enabling presently reliable measurement of an important hydrological parameter such as salinity. An optimum but technically difficult solution would be to combine a multispectral scanner, a thermal scanner, a radar (especially imaging one) and a microwave radiometer, all with a spatial resolution of less than 5 m and a revisit capability of a day or less, on a single platform or on different platforms operating in phase.

The presently available multispectral data of optical and thermal bands and the SAR type radar offer great potential for a number of applications, some of which are given below.

25.3 SPECTRAL CHARACTERISTICS OF LITTORAL OBJECTS

Spectral responses of a water layer are given in Chapter 4 for optical domain and in Chapter 1 for thermal infrared and microwave bands. Examples of application of thermal and microwave data are discussed in Chapter 26. Spectral characteristics of objects of the maritime region in intertidal and subtidal zones are described below.

25.3.1 Intertidal and subtidal littoral environments

The intertidal littoral zone generally constitutes a 'mosaic' of various biological communities, substrata of varied geology and grain-size, moisture gradients associated with inundation and exudation, all disturbed by human activity (marine cultures, salt pans, various management activities, etc.). Varied pixels and mixels represent these features.

The subtidal environments correspond to littoral zones permanently inundated by brine or sea-water. The spectral response of the sea depends on the content of sediments or pigments (water colour), nature of sea floor, action of surface wind, storms, currents, pollution, etc.

25.3.2 Mineral targets of intertidal zone (optical domain)

'Pure' mineral targets such as sands, muddy sands, mud and rocks exhibit monotonous spectral characteristics between blue-violet and near-infrared (see Chaps. 4 and 23). Only level variations are detectable, which result from a number of factors such as natural colour of the substratum, water content, organic matter, grain size, etc. Contrarily, the spectral response of sediments may be considerably modified by surface deposition of biological matter such as diatoms, indicating the presence of microphytobenthos. Guillaumont et al. (1988) have shown that modification in spectral response in the presence of a microflora is directly related to the content of pigments (chlorophyll and phaeopigments). Such a result is significant since microphytobenthos, with phytoplankton and macrophyte algae, contributes to the primary production of coastal ecosystems, in particular in areas where muddy-sandy stretches exposed to low tides are vast. Their temporal variation is very high and the conventional methods of observation and measurement are difficult to employ.

25.3.3 Vegetal targets of intertidal zone (optical domain)

Spectral characteristics of intertidal vegetation are similar to those of terrestrial vegetation discussed in Chapter 4. Three classes of intertidal vegetation are distinguished: macrophyte algae, marine phanerogams and swamp halophytes.

■ Macrophyte algae

Compared to other higher forms of vegetation, algae show a large pigment diversity adapted to the variability of their luminous environment (Levavasseur, 1986). Studies of various algae groups (green, red, brown) show a low response between 400 and 500 nm (absorption by chlorophyll) and absolute values of reflectance, beyond 700 nm, dependent on the conditions of light, cover and degree of immersion. In the region 500–700 nm, reflectance indicates the diversity of pigments present (Fig. 25.2).

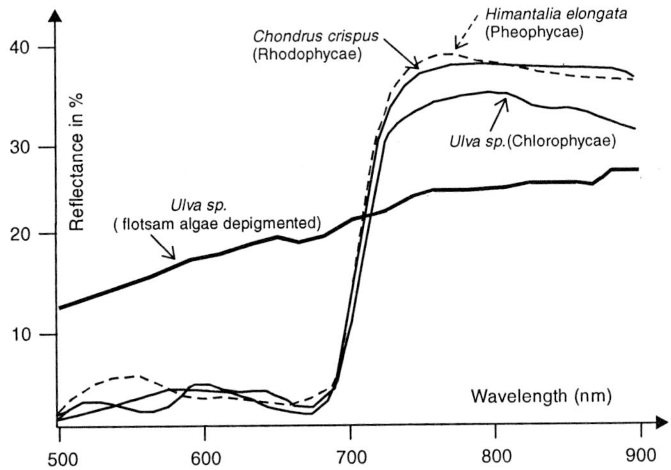

Fig. 25.2: Example of spectral reflectance curves of various algae groups (after Guillaumont, 1991).

Although the range of variation is small, Viollier et al. (1985) have shown that this diversity of pigments can be detected in wide bands of a HRV type sensor, which give the index XS1/XS2 (see Chap. 4). This index can be used to differentiate major vegetal populations, provided it is verified that the pixels analysed fulfil the condition of 'pure' target.

Using airborne remote-sensing data of high spatial (pixel < 5 m) and spectral resolutions and supported by detailed ground spectroradiometric surveys, Bajjouk et al. (1996) have shown that the 13 spectral bands of the CASI airborne imaging spectroradiometer, programmable for position and width of spectra (see Chap. 2), can be optimally selected for discriminating major macroalgae of the North Bretagne tidal flats. These positions representative of the principal species are shown in Fig. 25.3, with superposition of ground spectral curves.

On the other hand, specific growth of vegetal fronds and especially the horizontal disposition of thallus and large density of leaf canopy of algae hinder penetration of sunlight into depth. Thus, for brown algae, Ben Moussa (1987) has shown that normalised vegetation index (NDVI) reaches its maximum from a thallus density of 4 (density attained quasi-systematically in situ).

A review by Guillaumont et al. (1997) gives more information on principles and methods of remote sensing of macrophyte algae.

■ Marine phanerogams

These flower plants colonise sandy and muddy-sand zones, sometimes constituting large areas in tropical regions (turtle grass). They show in emergence a planophylic posture like macrophyte algae. In the case of seaweeds, a significant correlation is observed between biomass and normalised

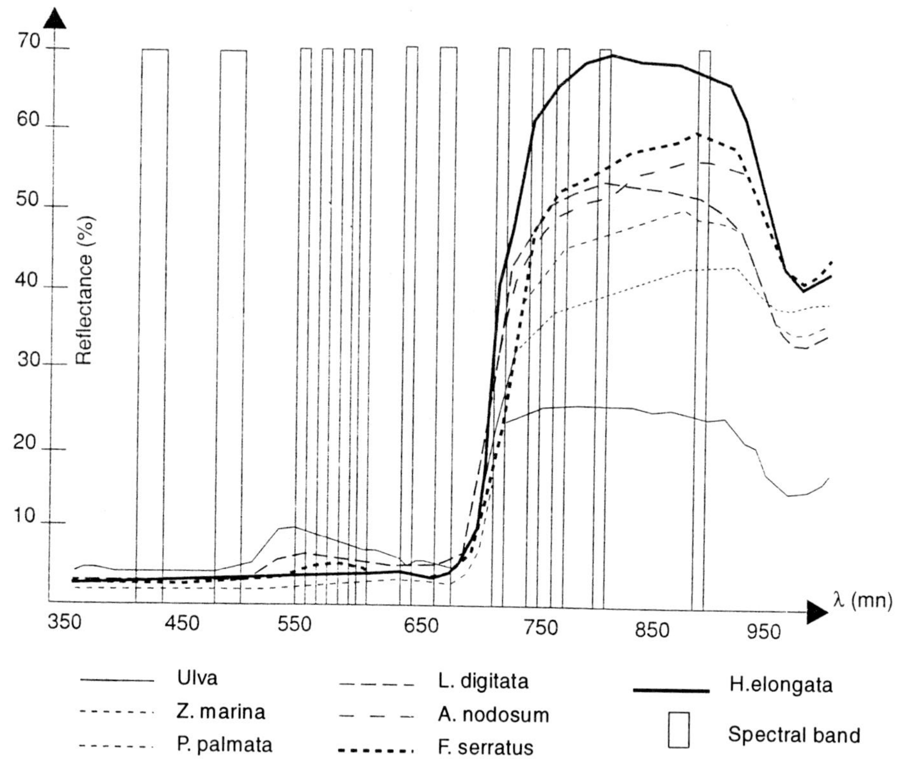

Fig. 25.3: Superposition of the CASI bands and reflectance curves obtained from field measurements (Bajjouk et al., 1996).

vegetation index. On the other hand, spectral reflectance of a marine phanerogam plant shows variations depending on the cover, which per se varies seasonally depending on the development of the plant. Further, Guillaumont (1991) has shown that correlation can be established between a normalised vegetation index and an estimate of biomass. Diachronic analysis by remote sensing is one of the methods of monitoring variations in these plants.

■ Temperate maritime swamps

The halophyte plants of maritime swamps of the temperate (and cold) zones are erectophyles, generally small in size (unlike the plants described above). Their spectral characteristics indicate phenological variations (perennial species, annual species, etc.) and the spectral response fluctuations may be rapid and large depending on natural (earing, efflorescence) or artificial (mowing, grazing) factors (Fig. 25.4).

In-situ plant populations can hence be characterised only by diachronic monitoring of spectral responses.

Significant results have been obtained in the United States (Budd et al., 1982) in favourable cases such as large stretches of practically monospecific vegetation of *Spartina alterniflora* and little modification of the landscape by man. They led to mapping and quantification of biomass of plant species of swamps, particularly through formulation of significant laws of correlation between biomass and vegetation index.

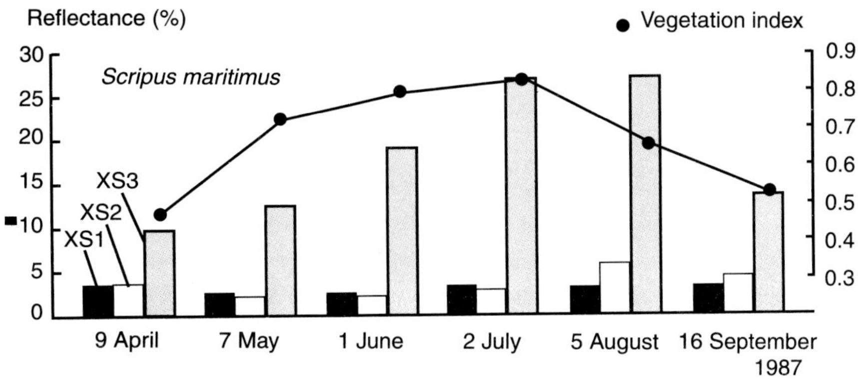

Fig. 25.4: Seasonal variation of reflectance of *Scripus maritimus*—Cimel radiometer (after Guillaumont, 1991).

For the European maritime swamps, Caillau et al. (1987) have shown that one should simultaneously take into consideration the natural seasonal fluctuations of each component of ecosystem and the calendar of activities corresponding to local characteristics (sometimes ancient ones) of land use (dams, polderland, salt works, oyster culture, fish farms). Variation of these swamps depends directly on the management of water resulting from these activities in full development.

■ Tropical maritime swamps

Tropical maritime swamps generally correspond to mangrove formations, fragile forest ecosystems since their genetic diversity is very much reduced (about sixty ligneous species constitute their flora) and their ecological tolerance is very low (salinity gradient and daily duration of immersion in particular). These are one of the most productive populations; dry aerial biomass by weight is of the order of 300 t ha^{-1} in wet regions and primary productivity in the dense mangroves of Malaysia reaches up to 12 to 15 t ha^{-1} y^{-1}.

Several important studies have been conducted, especially in the optical domain, to prepare inventories and cartographic zonings and to determine the characteristics of plant communities (Aschbacher et al., 1995). Some of the investigations in France are those of Blasco et al. (1983), Populus et al. (1986), Mougenot (1990) and Cuq et al. (1993).

In areas where a dry season exists, the mangrove zones are associated with dry brine marshes situated behind them and are called 'salitrals'. The succession of landscape units of a tropical maritime swamp in New Caledonia is shown in Fig. 25.5. Zones 1 and 2 correspond to mangrove forest proper; zone 3 is a transition region between forest and bare soil colonised by one (or more) species whose spectral characteristics are similar to those of temperate swamp plants (see the preceding section). Zones 4, 5 and 6 correspond to bare muddy-sand sediment their radiometric response is similar to that of mineralised zones but can be affected by algal cover of vegetal origin.

Combined use of the normalised vegetation index XS3 – XS2/XS3 + XS2 and the brightness index $(XS1 + XS2)^{1/2}$ (see Chap. 4), based on spectral measurements, enables classification of the principal units of this type of landscape: shrubby mangrove, transition zone and bare soils proper (Loubersac, 1987). An application in pre-selection of aquaculture sites is given below.

25.3.4 Subtidal zone: hydrocarbon pollution

Spectral response of a water layer in the optical, thermal infrared and microwave bands is discussed in Chaps. 1, 4 and 26. In this section, we will consider only surface pollution and, in particular,

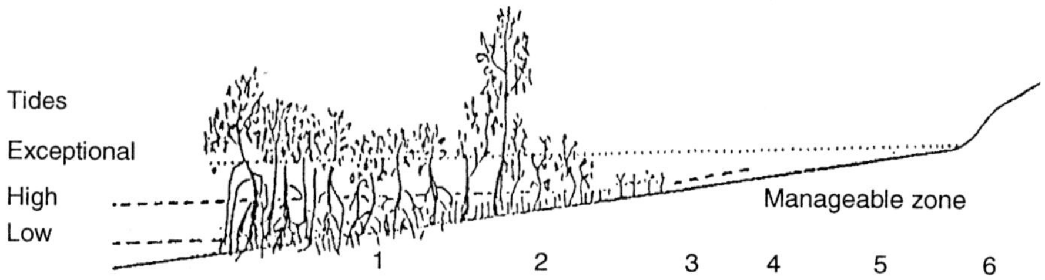

Fig. 25.5: Swamps of Mara, New Caledonia: landscape units in the intertidal zone.

hydrocarbon spilling and modification of remote-sensed signal at the sea surface due to the presence of such products.

Hydrocarbons, generally floating as a thin layer on the ocean surface can be detected by passive as well as active remote-sensing techniques covering almost the entire electromagnetic spectrum. From the beginning of research activities in remote sensing, numerous studies have been devoted to this subject: Stewart et al., 1970 (visible and near-infrared); Munday et al., 1971 (multispectral including thermal); Hollinger, 1974 (passive microwave); Guinard, 1971 (radar); Fantasia et al., 1974 (Laser). Integrated analysis of these works has resulted in dividing the electromagnetic spectrum into several parts in which hydrocarbons show sufficiently stable spectral characteristics relative to the sea surface, enabling their discrimination.

In passive remote sensing, the division of wavelength (λ) bands is as follows:

—λ between 300 and 400 nm: Reflectance of hydrocarbons is greater than that of the sea, but can be used only in day with clear sky.

—λ between 420 and 550 nm: Solar reflection on hydrocarbons is masked by scatter due to water.

—λ between 650 and 900 nm: Hydrocarbon reflection is greater than that of sea.

—λ between 3 and 5.5 μm and 8 to 14 μm: In this last portion of the spectrum (see Chap. 1), the emitted radiation is proportional to the temperature and surface emissivity of the body under observation. As the temperatures are assumed to be equal and emissivity of oil is lower than that of sea water, hydrocarbons appear 'colder': detection possible in day as well as night.

—λ between 3 mm and 3 cm: Hydrocarbons show higher radiant temperature than that of sea and variable with oil thickness: detection of petroleum pollution possible in day as well as night and estimation of the volume of oil through estimation of oil-layer thickness.

In active remote sensing, division of bands is as follows:

—λ between 250 and 600 nm: Hydrocarbons under light excitation (laser) can re-emit a signal at a wavelength, offset relative to that of excitation (fluorescence). As the backscattered wave and excitation wavelength depend on the type, density and age of hydrocarbons, it is theoretically possible to identify the type of pollution detected.

—λ between 3 and 30 cm: Since hydrocarbons have the characteristic of attenuating capillary waves of quite large amplitudes produced by wind (> 4 knots), a (lateral) radar signal will be more backscattered by unpolluted sea water than by oil layer (Wismann, 1993; Bjerde et al., 1993).

25.4 EXAMPLES OF APPLICATION TO LITTORAL MANAGEMENT

Some examples illustrate the role of remote sensing as an aid in administration and management of coastal zones. Six cases are presented below: base mapping of coral environments, pre-selection of aquacultural sites in tropical zones, thematic mapping of seaweeds, detection of hydrocarbon pollution, characterisation of surface states of sea and monitoring temperature variations of the sea surface.

25.4.1 Mapping of coral environments

Acquisition of precise and up-to-date geographic information on reef zones, lagoons or, more generally, shallow-water zones characterising intertropical regions is most often a difficult, time consuming and expensive task, especially in the case of low islands or atolls. In fact, two situations are normally encountered:

—For the terrestrial part, low altitudes, soft formations (friability of soil and intensive erosion), difficulty in acquisition of aerial data over far-off islands and small number of reference points for photogrammetry complicate setting up of a complete geodetic network.

—For the maritime part, shallow depths necessitate a dense grid of sounding profiles. Sea-floor variations and isolated reef knolls are often randomly distributed. Possibilities of precise location are limited and optical or radio-electric ranges small.

In view of the above, an intensive base mapping needs to be carried out in tropical coastal regions, both for environmental management and navigation. To facilitate such investigations, the Naval Hydrographic and Oceanographic Service of France mainly employs satellite remote-sensing data.

In French Polynesia for example, management of maritime regions, consisting of lagoons, atolls and high islands, poses problems due to development of pearl-culture activities, conflicting among themselves and with other activities (such as fishery, tourism). Hence, spatial remote-sensing data have been used for preparing 'spatial maps' of these islands in regions for which conventional data were not available or were fragmentary.

Based on the criteria of availability, identical standards (ground resolution, radiometry, format, etc.), diachronic acquisition and compatibility between graphic and semantic precision for the proposed applications, SPOT satellite data have been used. The method developed comprises:

—rectification of the image acquired under a basic preprocessing level (1A or 1B) to level 2A (UTM);

—association of a precise geographic grid (reference points obtained in situ);

—segmentation of the image into 3 major zones: marine region, bare soils, vegetation;

—estimation of bathymetric levels (method explained in Chap. 4) and calibration of the model with the aid of reference points of known bathymetric levels;

—completion of zones drawn from the preceding stages of processing by addition of external features (topographic data, roads, etc.) to obtain a spatial map (see CD 25.1: prototype spatial map of Manihi island, original 1:50,000, Loubersac et al., 1994).

Besides direct cartographic importance, the digital form of remote-sensing data and their geocoding allows restricting the number of cartographic reference points that enter in the geographic information system for management.

25.4.2 Aquaculture management (raising tropical shrimps)

Shrimp culture is highly concentrated in littoral regions of developing countries, in zones that are not always readily accessible and have incomplete or obsolete maps. This activity is mostly characterised by construction of large ponds on land (a few thousand square metres to more than 20 ha of single area). Shrimp raising is today gradually emerging as a destroyer of natural environment to the extent of self-endangering by destruction of biotopes that are necessary for itself.

In fact, the most favourable zones for construction of shrimp-culture basins are 'salitrals', large areas of flat terrain without vegetation, in the proximity of mangrove environments (Fig. 25.5). Initially, farms are set up in 'salitrals' within mangroves where juveniles are raised to nourish them and where the water is enriched and subsequently sent into the basins. These 'salitrals' are readily manageable and impact of management on environment is very often minimal. Such sites are favourable for intensive shrimp culture.

When the bottom pressure becomes more, these farms spill over the mangrove, gradually destroying it, with risk of destroying precious biological equilibria, including those that completely or partly furnish their needs of juveniles. Such an overflowing of farms is due to high bottom pressure, as well as due to incorrect location and characterisation of favourable zones, i.e., 'salitrals', in the available maps. The methods described earlier for tropical maritime swamps have been successfully applied to high-resolution spatial images for various objectives: to locate favourable zones, quantify their areas, identify their forms, furnish qualitative information on soil types and drainage patterns, determine land cover in river basins on upstream side of sites and accessibility of zones through land or sea routes and pre-select sites for pumping sea water necessary for shrimp culture.

These features, combined with external data, directly not available from remote sensing, such as physico-chemical quality of water, productivity of zones, logistic constraints, economic constraints, etc. enable preparation of pre-selection maps of sites in medium scale (mainly 1:50,000). Such maps have been used for the first time in the inventory and management plans of aquacultural sites in New Caledonia (Loubersac, 1987; Populus et al., 1990; CD 25.2).

25.4.3 Thematic mapping of seaweeds

Increasing industrial demand for algal material, and especially the fucal, has led to estimation of exploitable stocks of seaweed, particularly in intertidal zones. Maps of species classification, estimates of coverage and biomass quantity in situ and empirical models relating the vegetal cover measured and biomass available have been developed from the results of investigations (see above; CD 25.3; after Bajjouk, 1996).

Unlike brown algae whose extent and spectral characteristics are generally stable over time, green algae exhibit very large variations in development and distribution, often associated with pollution by excess nitrates. Monitoring and control of their variation are difficult by conventional methods. Use of species-discrimination methods in remote-sensing data (Populus et al., 1994) offers perspectives of application in monitoring eutrophication of coastal environment, especially as support to modelling these phenomena.

25.4.4 Detection of hydrocarbon pollution in sea

Techniques and methods for remote detection of hydrocarbons dumped fraudulently or accidentally in sea are mainly developed since the 1970s in the United States, Canada, Sweden and France. These methods make use of various spectral bands (see above).

Thermal infrared bands (Fig. 25.6) have been and are mainly used in France by 'Douanes' and 'Marine Marchande' to detect fraudulent dumps of hydrocarbons (degassed) and to dissuade polluters.

A satellite equipped with radar enables detection of hydrocarbon slick any time. ERS-1 provided shocking images of massive accidental pollution such as the one resulting from the accident of the 'Aegean Sea' in the Spanish Corogne Bay in 1992. An example of radar detection of hydrocarbon pollution offshore of Portugal is shown in Fig. 25.7. It must be noted that the present-day frequency of acquisition of such information by satellites is not yet compatible with the operational constraints encountered in the struggle against pollution.

25.4.5 Monitoring surface state of sea by radar imaging

A good knowledge of regime of swells due to wind (amplitude and direction) is necessary for their forecasting (in aid of navigation and tracking of ships) and for studies of sediment transport from the

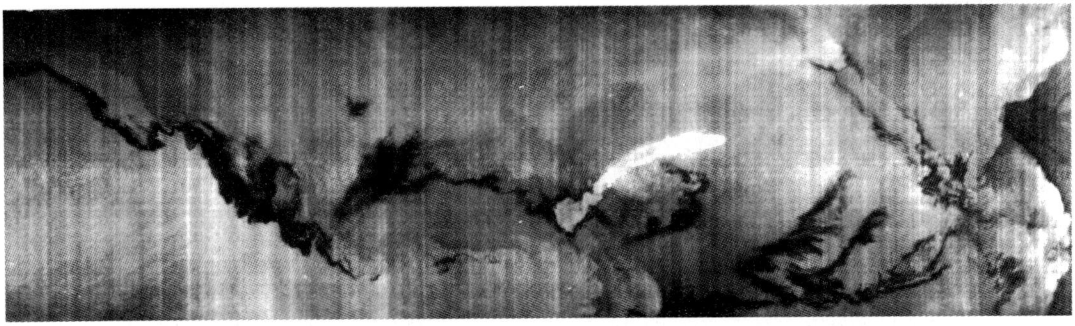

Fig. 25.6: Example airborne detection (thermal IR) of a hydrocarbon dump (dark shaded) in sea. The ship throwing the dump into the water and its wake are clear ('hot' zones) (permission IFREMER).

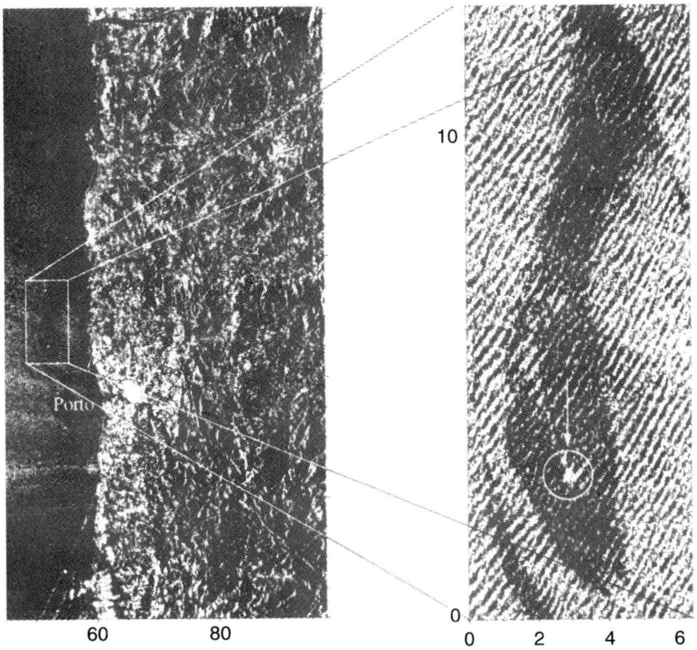

Fig. 25.7: ERS-1 radar image of a hydrocarbon pollution site offshore of Portugal (black layer due to low backscatter to sensor). The white point in the southern part of the layer is probably a ship (after Kerbaol and Chapron, pers. comm., permission IFREMER).

coast. Analysis of spectra of swell derived from processing of SAR images permits characterisation of modifications in wavelength and direction of swell when it approaches the coast.

An ERS-1 SAR image of surface sea state under the effect of a storm is shown in Fig. 25.8. Superposition of the SAR image and the coastline is given on the right. In the bottom left, a segment of SAR image depicts swell with nearly west-east orientation. In the top left, a SAR image segment indicates the impact of swell on the shore of the Audierne bay and especially the cells of surf at the coast, in white hue. Dynamic monitoring of such phenomenon is important since it directly determines the coastal sediment transport and rate of coastal erosion.

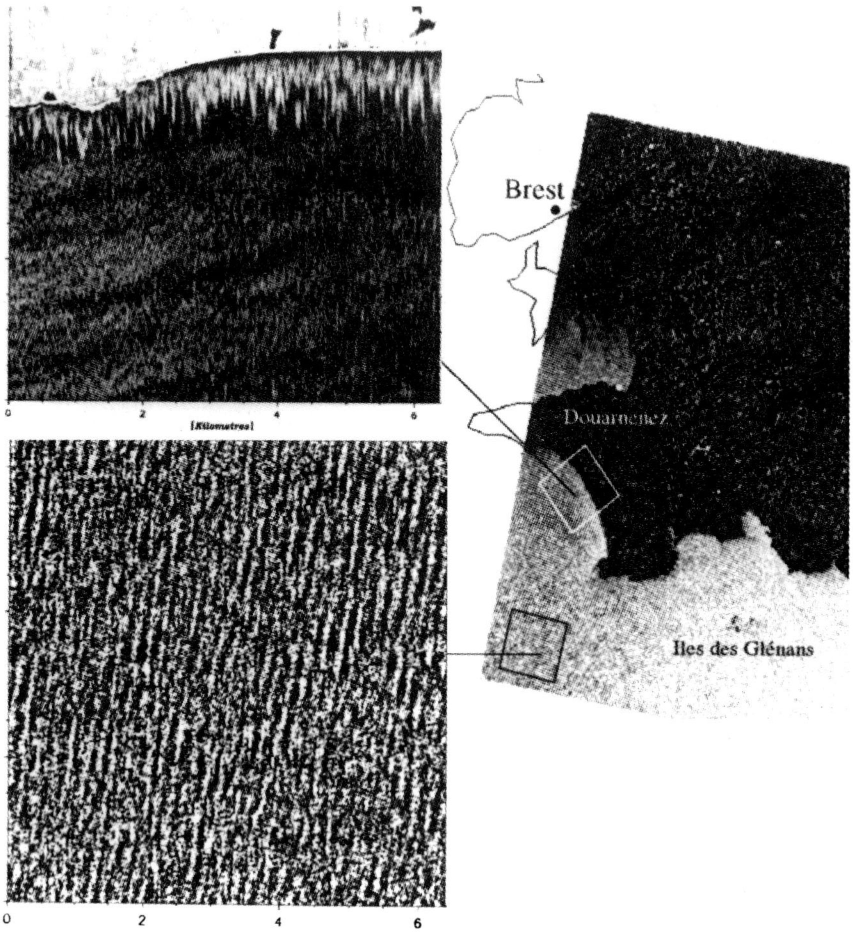

Fig. 25.8: Detection of surf cells of coastal waves under the effect of a storm in South Bretagne. ERS-1 SAR image of 13/9/1993 (after Kerbaol and Chapron; also see Forget et al., 1996).

25.4.6 Detection and monitoring of sea-surface temperature variations in littoral environment

Oceanographic applications of detection of sea-surface temperature by remote sensing are numerous: pelagic fishery (Klimley and Bulter, 1988; Petit et al., 1994), monitoring dynamics resulting from mixing and geostatistical processing of data (Gohin and Langlois, 1993), detection of frontal structures (Le Vourch et al., 1992).

An example of synthesised analysis of sea-surface temperature variations associated with ecological disturbances, eutrophication and seaweed proliferation in the water mass of the Bretagne coast is presented here. In this region, abnormal growth of green algae is observed generally from May to July and the proliferation is directly related to the excess of nutrients of agricultural origin, combined with an increase in solar radiation. Studies have shown the topography and morphology of the littoral zones (growth on shallow floors), hydrodynamic activity (growth in protected zones), nature of sediments (proliferation on sandy zones) and temperature (proliferation in hot zones in spring) as factors that explain development of seaweeds (Piriou et al., 1993).

The AVHRR (NOAA) data enables spatiotemporal characterisation of warming of surface water by an interannual synthesis of its stages. For each year and for each image available, deviations of temperatures relative to winter reference temperatures have been computed for a 12 × 12 km grid and the interannual mean of deviations obtained. (CD 25.4 shows results of synthesis for three periods: early May, end May and early July, heating of coastal zones of Bretagne in three periods and the mean of observed thermal deviations relative to a winter reference; after Piriou et al., 1993; IFREMER/CEE/ Bretagne region).

This type of information provides a better understanding of dynamics of heating of bays and protected zones in areas where seaweed proliferation has started and helps in explaining variations in commencement of the above phenomenon depending on the site (heating later in Lannion and Morlaix bays in the west than in Saint-Brieuc bay in the east). Thus it forms an aid to predictive modelling through better understanding of the phenomenon.

25.5 CONCLUSIONS

Since the beginning of the 1970s, especially with the launching of the first Earth observation satellite (ERTS 1 precursor to LANDSAT MSS missions), several space missions, viz., SEASAT in 1978, NIMBUS 7 of the same period, LANDSAT TM series since 1983, SPOT series since 1986, ERS series since 1992 and SEAWiFS in 1997, have furnished large amount of information on oceans and coastal seas.

Scientific and technical advances emerging from such data have been briefly indicated and illustrated above. During the same period important developments have also taken place in airborne remote sensing, as evidenced by organisation of international seminars on the subject since 1994.

It should be noted that specific characteristics of littoral environments impose constraints on spatial and temporal observations and measurements which remote-sensing techniques cannot necessarily satisfy.

Future perspectives of development in coastal applications of aerospatial remote sensing revolve around three axes:

—very high spatial resolution (about 1 m) with new-generation satellites,

—very high spectral resolution (programmable) as surface topography (CIDAR) which new airborne missions may provide,

—development of sensors for water colour measurements (see Chap. 4),

—use of minisatellites ensuring a high repetition of acquisition through offset cycles of orbits.

On the other hand, independent of the technological and methodological developments specific to remote sensing, a definite future exists in mixing and combining the georeferenced data with conventional geographic data and digital modelling. That is why, at the levels of conceptualisation and development of techniques, methodologies and applications, investigations are oriented towards close linkage and synergy between data, methods and tools of remote sensing, hydrodynamic digital modelling, hydrosedimentary or biological studies of ecosystems and geographic information systems. Such synergy is essential for developing operational regional information systems for coastal zones, to be used in decision making, in optimised choice and communication of environmental information.

References

Aschbacher J, Tiangco P, Giri CP, Ofren RS, Paudyal DR, Ang YK. 1995. Comparison of different sensors and analysis techniques for tropical mangrove forest mapping, IGARSS'95 Congress, 3: 2109–2111.

Bajjouk T, Guillaumont B, Populus J. 1996. Application of airborne imaging spectrometry system data to intertidal seaweed classification and mapping, Hydrobiologia, 326/327, 463/471.

Ben Moussa H. 1987. Contribution de la télédétection satellitale à la cartographie des végétaux marins: Archipel de Molène, Thèse de doctorat d'Université Aix Marseille II, 122 p.

Bjerde KW, Solberg S, Solberg R. 1993. Oil spill detection in SAR imagery, IGARRS'93 Congress, p. 943–945.

Blasco F, Lavenue F, Chaudury MU, Kerr Y. 1983. Simulations SPOT au Bengladesh. Étude des mangroves des Sunderbans, Rapp CNES/GDTA, 31 pp.

Budd JTC, Milton EJ, 1982. Remote sensing of salt marsh vegetation in the first four proposed Thematic Mapper bands, Int. Jnl. Rem. Sens, 3 (2): 147–161.

Caillaud L, Guillaumont B, Manaud F. 1987. Essai de discrimination des modes d'utilisation des marais maritimes par analyse multitemporelles d'images SPOT, Rapport ATP Télédétection Spatiale. IG/SRETIE/MERE/7161.

Cuq F, Courmelon F, Madec V. 1993. Planification côtière de Guinée Bissau, 3 vol. Cartes HT, UICN/DGFC-MDRA, Édité en français, anglais et portugais.

Fantasia JF, Ingrao HC. 1974. Development of an experimental airborne laser remote sensing system for detection and classification of oil slicks. Proc 9th Int. Symp. Rem. Sens. of Env., Ann Arbor MI, pp. 1711–1745.

Forget P, Rousseau S, Cauneau F, Chapron B, Kerbaol V, Cuq F, Bonnafoux G, Blerard C, Garello R, Grassin S, Bonicel D, Hajji H. 1996. Expérimentation radar GLOBESAR en baie d'Audierne, Rapport technique LSEET, Univ. Toulon. Ref. 94/CNES/0380.

Gierloff-Emden HG. 1982. 'Remote sensing for coastal areas', Symp IGARRS'82, Munchen WA-8, pp. 11–18.

Gohin F, Langlois G, 1993. Using geostatistics to merge in situ measurements and remotely-sensed observations of sea surface temperature, Int. Jour. of Rem. Sens., 14 (1): 9–19.

Guillaumont B. 1991. 'Utilisation de l'imagerie satellitaire pour les comparaisons spatiales et temporelles en zone intertidale', Estuaries and Coasts: spatial and temporal intercomparisons, ECSA 19 Symposium, Elliot and Ducrotoy Eds, Olsen and Olsen, pp. 63–68.

Guillaumont B, Bajjouk T, Talec P. 1997. 'Seaweeds and remote sensing: a critical review of sensors and data processing', *In:* Progress in Phycological Research, Vol. 12, Chapman and Round Eds. Biopress, pp. 213–282.

Guillaumont B, Gentien P, Viollier M. 1988. 'Mesures radiométriques haute résolution du microphytobenthos intertidal', Proc. 4th Int Coll. On spectral signatures of objects in rem. Sens., Aussois ESA SP 287, pp. 333–336.

Guillaumont B, Lavavasseur J. 1988. Variations saisonnièrence de la réflectance en fonction de la phénologie des plantes des marais, 3e Conférence Internationale sur les Zones Humides, Rennes, sept. 1988.

Guinard NW. 1971. 'The remote sensing of oil slicks', Proc. 7th Int Symp. Rem. Sens. of Env. Ann Arbor MI, pp. 1005–1026.

Hollinger JP, 1974. The determination of oil slicks thickness by means of multifrequency passive microwave technique, Nav. Research Lab. CG-D-31-75. Washington DC.

Klemas V, Gross MF, Hardisky MA. 1987. Evaluation of SPOT data for Remote sensing of physical and biological properties of estuaries and coastal zones, Symp. Int. Spot 1: utilisation des images, bilans, résultats, Paris Éd, Cepadues, pp. 1035–1040.

Klemas V, Philpott WD, 1980. The use of satellites in environmental monitoring of coastal waters, Final rpt. Univ Delaware NASA NSP-1433.

Klimley AP, Butler SB, 1988. Immigration and emigration of a pelagic fish assemblage to seamounts in the Gulf of California related to water mass movements using satellite imagery, Mar Ecol. Prog. Ser, 49: 11–20.

Le Gouic M. 1988. 'Utilisation de SPOT en hydrographie, Symp. Int. Spot 1; utilisation des images, bilans, résultats, Paris Éd. Cepadues, pp. 1063–1068.

Le Vourch J, Millot C, Castagne N, Le Borgne P, Olry J-P. 1992. Atlas des fronts thermiques en Mer Méditerranée d'après l'image satellitaire, Mémoire de l'Institut océanographique de Monaco, n° 16, VI, 152 pp.

Levavasseur G. 1986. Plasticité de l'appareil pigmentaire des algues marines macrophytes. Régulation en fonction de l'environnement, Thèse de Doctrat d'État, Paris VI, 210 pp.

Loubersac L. 1983. Coastal Zone Inventory by high resolution satellites, Alpbach Summer School, 27 July–5 August 1993, ESA SP 205, pp. 87–94.

Loubersac L. 1987. 'SPOT, un outil d'aide à la présélection de sites favorables à l'aquaculture Bilan et perspectives du projet PEPS 'ALIAS Calédonie', Symp. Int. Spot 1; utilisation des images, bilans, résultats, Paris, Éd. Cepadues, pp. 1041–1049.

Loubersac L, Andrefouet S, Chenon F, Morel Y, Varet H. 1994. Information géographique dérivée des données de la télédétection spatiale de haute résolution sur les lagons des îles hautes et des atolls. Application aux environnements des îles de la Polynésie française: état et perpectives, Mémoires de l'Institut océanographique de Monaco, 18: 75–84.

Mougenot B. 1990. Caractéristiques spectrales de surfaces salées à chlorures et sulfates au Sénégal, 2e journées de télédétection: Caractérisation et suivi des milieux terrestres en régions arides et tropicales, Colloques et séminaires, ORSTOM, Bondy, pp. 49–70.
Munday JC, MacIntyre WG, Penney ME. 1971. 'Oil slicks studies using photographic and multispectral scanner data', Proc. 7th Int Symp. Rem. Sens. of Env., Ann Arbor MI, pp. 1027–1043.
Petit M, Dagorn L, Lena P, Slepoukha M, Ramos AG, Stretta JM. 1994. 'Oceanic landscape concept and operational fisheries oceanography', Mémoires de l'Institut Océanographique de Monaco, 18: Les nouvelles frontières de la télédétection océanique, pp. 85–97.
Piriou J-Y. 1993. Cartographie, des zones sensibles à l'eutrophisation; cas des côtes bretonnes. Rapport IFREMER/ CEE/Région Bretagne, Contrat 6510-90, 2 vol.
Populus J, Guillaumont B, Ruiz O, Tallec P. 1994. 'Biomass assessment of green algae proliferations with high resolution airborne images', Proc. of the first International ETRIM Airborne Remote Sensing Conference and Exhibition, Strasbourg, France, 11–15 Sept. 1994, 3: 153–164.
Populus J, Hertz R. 1986. Cartographie des magroves de la côte sud-est du Brésil avec Landsat TM, Photo Interprétation, 85-2 fasc. 4 pp. 31–38.
Steward S, Spellicy R, Polcyn F. 1970. Analysis of multispectral data of the Santa Barbara oil slick, Publ. 3340-4-F, Willow Run Lab. Univ. Michigan, 57 pp.
Viollier M, Belsher T, Loubersac L. 1985. 'Signatures spectrales des objets du littoral, Proc 3rd Int. Coll. on spectral Signatures of Objects in Rem. Sens., Les Arcs, France, ESA SP 247, pp. 253–257.
Wismann V. 1993. Radar signatures of mineral oil spills measured by an airborne multi-frequency radar and the ERS1 SAR, IGARRS'93 Congress, 3: 940, 942.
Zacharias M, Niemann O, Borstad G. 1992. An assessment and classification of a multispectral bandset for the remote sensing of intertidal seaweeds, Canadian J. of Rem. Sens. 18 (4): 263–274.

26

Applications of Thermal-infrared and Microwave Data

26.1 THERMAL-INFRARED DATA

It was shown in Chapter 1—Physical basis—that instantaneous values of surface temperature can be related to diurnal evapotranspiration. This relationship has been used for numerous applications, described below, using satellite data: estimation of water consumption by maize, detection of dry and irrigated zones and mapping evapotranspiration in southwestern France.

26.1.1 Detection of drought in France

Detection of drought necessitates continuous monitoring of surface temperatures at a regular time interval. Due to problems of cloud coverage impeding acquisition of thermal-infrared images, combinations of data of five days of NOAA-AVHRR images have been used for this study. This analysis was conducted by CNES at Toulouse by selecting the maximum of vegetation index of AVHRR images acquired during a period of 5 days from March to October (the most important period for crops). These data comprise 48 images per annum. This method eliminates cloudy or defective images. These images were geometrically rectified (see Chap. 13), superposed and corrected for atmospheric effects by the empirical split-window method (see Chap. 1) and for temporal drift between the satellites. Twenty meteorological stations distributed throughout the territory were chosen for representing major regional climatic variations. Each of these sites was characterised by a 9 km × 9 km sector (or 9 × 9 NOAA pixels) for which surface temperatures were extracted from NOAA images of the various years under study and for which temporal climatic variations measured at the station were also available.

Analysis of cumulative deviations ($Ts_{NOAA} - Ta_{meteo}$) from March to October between various stations revealed (Fig. 26.1):

—latitude effect: marked difference between north and south sites (Mirecourt, Rennes, Perpignan);

—land-cover effect: areas where bare soils are predominant show higher surface temperatures than those of zones which are covered by dense vegetation (as in Mirecourt) or are marshy (Saint-Laurent-de-la-Prée);

—effect of proximity to coasts (case of Quimper): generally higher wind, humidity or rain tend to reduce deviations ($Ts - Ta$).

For a given station, variations in the observed temperature deviations between various years are indicative of hydrological deficiencies and enable detection of stress periods.

Marked differences are thus observed between the years (1989–1990) of highest drought and the year (1988) of maximum rain in the Toulouse region. However, these differences are less significant for the Avignon site where a large number of parcels are regularly irrigated and less affected by droughts (Fig. 26.2).

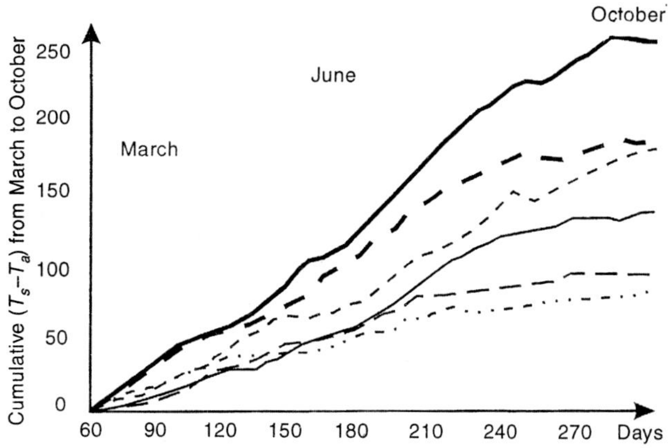

Fig. 26.1: Variation of cumulative deviations (*Ts – Ta*) from March to October for six stations (after Courault et al., 1994, in *Agronomie*).

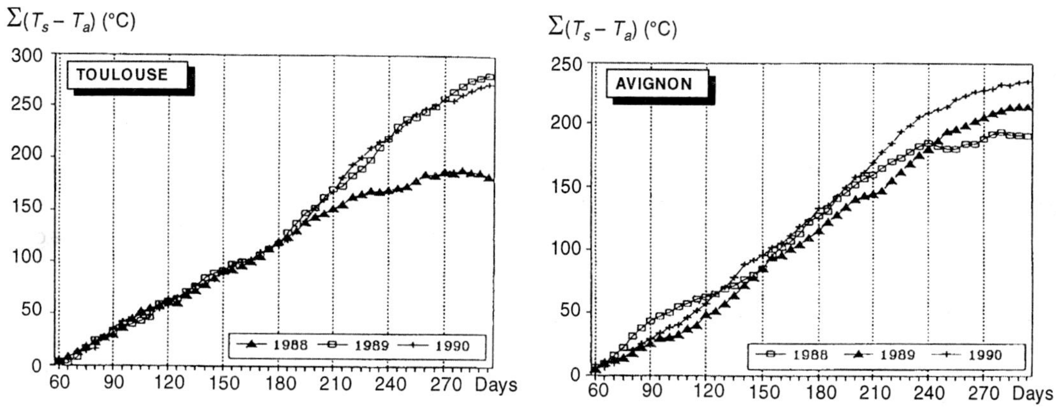

Fig. 26.2: Comparison of cumulative deviations (*Ts – Ta*) from March to October for Toulouse and Avignon, 1988 to 1990 (after Courault et al., 1994).

For estimating the evapotranspiration (ETR) of each site using the agrometeorological model 'Magret', one can calibrate the simplified model (Chap. 1, eqn 16) to the scale of France for each year of study and thus map evapotranspiration by means of NOAA images (Courault et al., 1994). Meteorological data necessary for the model *Tair and Rn* are interpolated by krigging.

Maps resulting from the sum of surface temperatures are used to prepare ETR maps. Maps of *Ts* deviations between years are prepared and used for preliminary prognosis of zones most affected by drought (CD 26.1). It can thus be seen that Southwest, Charente and Normandy are the regions that seem to have been most affected by lack of water in 1989 and 1990. The computed values of ETR for Charente confirm these observations and can be used to quantify the deficit:

ETR 88 = 503 mm, ETR 89 = 356 mm, ETR 90 = 317 mm.

This first example clearly illustrates the significance of thermal infrared data of NOAA-AVHRR in obtaining an overall estimation of water content of crops over large areas. This approach is useful to detect and confirm location of zones most affected by drought during the preceding years and is complementary to the method of monitoring vegetation indices. The advantage of surface temperatures compared to vegetation indices is that the former represent instantaneous surface hydrological conditions whereas the latter react to stress undergone by crops only after several days, which often makes their interpretation more difficult.

One of the major problems lies in validation of these estimates. At this spatial level, only local validation is possible in sites for which ground measurements are available (possible measurements over parcels). At the regional level, validation is most difficult; measurements by airborne sensors may be envisaged but they are cumbersome to employ and often limited in time. It is to be noted that such experiments of validation were attempted in the projects HAPEX-MOBILHY (Andre et al., 1988) and HAPEX-Sahel (Goutorbe et al., 1993).

26.1.2 Estimation of exchanges between soil, vegetation and atmosphere

Various approaches exist for using surface temperature in models of evapotranspiration estimations. The principle is as follows. In most cases these 'deterministic' models simulate, by means of more or less complex equations, surface temperatures observed by the satellite. Such simulations give certain parameters that are difficult to measure or estimate but important for computation of fluxes such as stomatic resistance of vegetation covers or initial reserves of water in soil, etc. These methods consist of comparing *Ts* in these transfer models by inversion techniques. A preliminary simulation is conducted with a manipulation of parameters fixed a priori. The surface temperature estimated is then compared with that observed by satellite. Attempt is made to minimise deviations between these two temperatures by modifying input parameters of the model for each new simulation.

Some models operate at diurnal scale and necessitate only pinpoint data for calibration of these parameters. Such is the case of a model of Penman-Monteith type surface, combined with a simplified description of atmospheric boundary layer. This model has been used, for example, to map evapotranspiration in south-western France with the aid of NOAA-AVHRR images, the unknown parameter being resistance of the cover. The resultant map indicates spatial variations of ETR. Evapotranspiration values observed for the forest of Landes are of the order of 3 mm on average and for vineyards and orchards 1.5 mm. Values obtained for crops such as wheat and maize vary between 2 and 2.6 mm per day for the date under consideration (Lagouarde and Brunet, 1991).

Other models, on the other hand, require several dates for continuously estimating various flows during the cycle of a crop. In the latter case, deviation between observed and simulated temperatures of various surfaces is minimised for the entire period by considering all dates for which remote-sensing data are available.

26.1.3 Mapping rainfall distribution in Sahel

Two methods applicable to the Sahel region in Africa are described here. The first is based on analysis of cloud temperatures and the other of surface temperatures.

In the Sahel region rain is often associated with thick cloud formations with a cold top. Analysis of METEOSAT images (6 images per day) gives a map of frequency of occurrence of such clouds with a threshold fixed at -40°C. Association between the number of these occurrences and rain observed in some stations is fairly strong, with a high correlation coefficient (Fig. 26.3a). Applying this relationship to the entire image, a cumulative-rain map is proposed for the agricultural season of Western Africa (Fig. 26.3b).

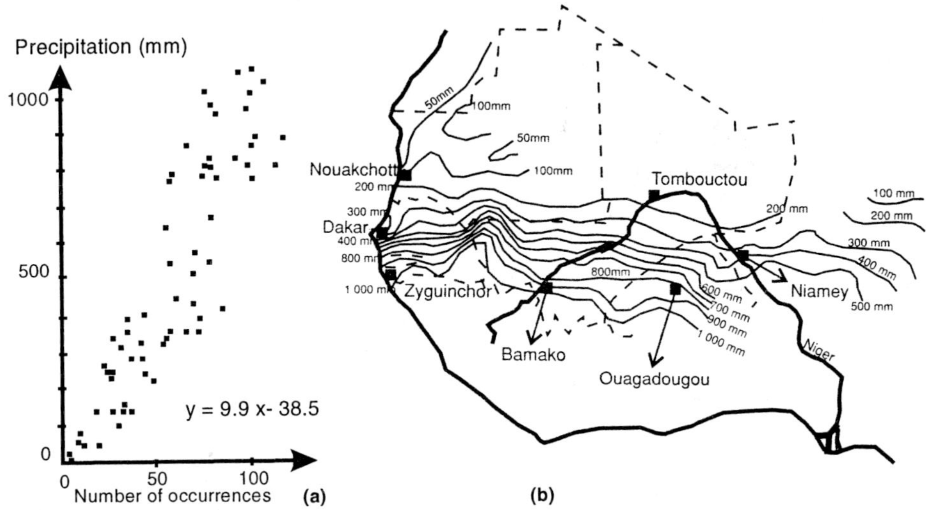

Fig. 26.3: Mapping the rain (measured at a few stations): a) occurrence of clouds with a cold top detected in Meteosat IRT images; b) application in Western Africa (after Lahuec et al., 1986).

Another method employed in Senegal gave a direct relationship between surface temperatures measured with the IRT band of METEOSAT and rainfall measured at some ground meteorological stations over several years (Assad et al., 1986). Regions affected by rain show lower surface temperatures due to evapotranspiration whereas zones with lowest rainfall retain high *Ts*. A linear relationship was established between these two variables (Fig. 26.4a), which enabled mapping cumulative rainfall during the agricultural period (CD 26.2). A comparison of isohyets drawn manually from ground rain-gauge network is satisfactory for dry and wet seasons (Fig. 26.4b).

26.1.4 Characterisation of frost zones

Thermal images acquired at night are important for mapping frost zones. An example is given for the Rhone valley wherein NOAA-7 data acquired at 3 h 00 TU on 15 and 23 March 1982 were used to detect zones sensitive to frosting (CD 26.3). Spring frost is particularly feared since it leads to large damage to crops. The first image acquired in low-wind condition on 15 March shows scattered zones of variable degrees of cooling. The second image of 23 March corresponding to high-wind (mistral) conditions shows much more homogeneous zones, indicating the influence of wind on night cooling (hazard of frost is less for strong winds). The average vertical thermal gradient is in fact as much steep as the wind is low and the sky is clear, a condition corresponding to the so-called radiation frost (or white frost). When wind blows, air in the lower layers of atmosphere is stirred, inversion of thermal profile is less pronounced and the net radiation is less negative. A low wind hence suffices to limit the cooling to the vicinity of soil. This principle is sometimes used to fight against frost by mechanical stirring of the lower layers of air.

26.1.5 Analysis of topoclimates from surface temperatures

In montane regions climatic observations are rare often due to sparse network of meteorological stations. Thermal-infrared remote sensing can furnish information on topoclimate of these regions

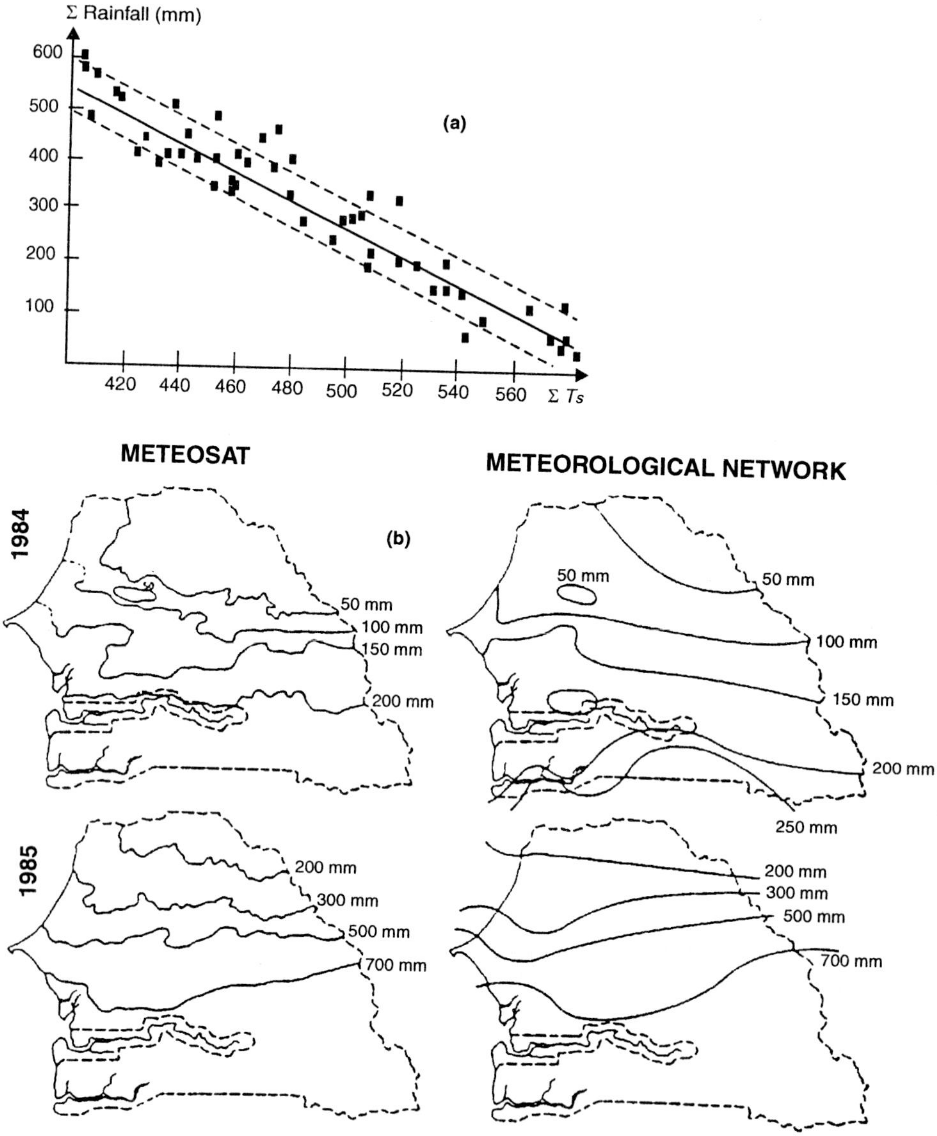

Fig. 26.4: a) Relationship between cumulative surface temperature (METEOSAT) and rainfall of some stations in Senegal for the years 1984 and 1985; b) comparison of isohyets with ground rain-gauge data (after Assad et al., 1986).

with high spatial resolution. In the north-west Massif Central, forest cover is dense on slopes and the high evapotranspiration gives rise to a surface temperature close to that of air. Relationships between surface temperature and altitude show gradients of about 0.7°C/100 m, of the same order of magnitude as those of air (Fig. 26.5). When density of vegetation cover is less, effects due to slope and orientation are more marked and the surface temperature then strongly depends on the radiation received. Thermal inversion phenomena may be observed: the cold air accumulated by gravity at the bottom of valleys

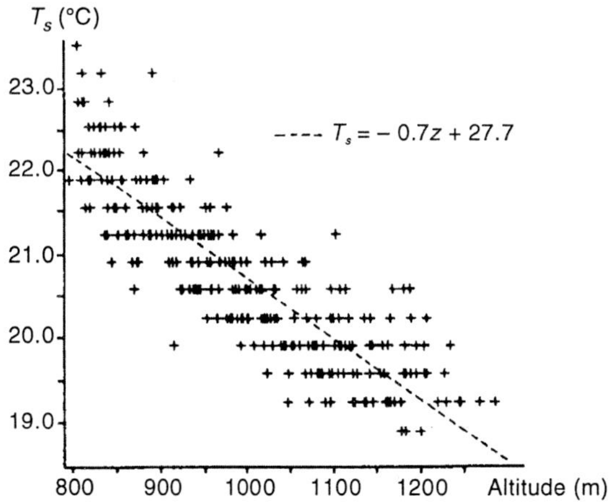

Fig. 26.5: Relationship between surface temperature (HCMM thermal images) of 17 July 1978 over Mount Forez and altitude derived from digital elevation models (after Lagouarde et al., 1983).

during night has no time to be warmed up in the morning under low wind conditions and temperatures increasing with altitude are observed. The thermal images acquired over Mount Ventoux illustrate these effects of topography on surface temperature (CD 26.4). Temperature increases with altitude up to about 100 m and then decreases for higher altitudes.

26.1.6 Use of surface temperature and vegetation index

Numerous authors have combined data acquired in various wavelength domains to obtain more information on surface characteristics. Such studies include relationships between normalised vegetation index (NDVI = NIR – R / R + NIR; Chap. 4) and surface temperature since they facilitate evaluation of variations in the structure of plant cover (by NDVI) and monitoring hydrological state of surfaces (with *Ts*). Most often a negative correlation is observed between these two variables. When a surface depicts a high vegetation index, it generally indicates a surface in which vegetation is dense and active and evapotranspiration is high, tending to reduce surface temperature. This relationship is used for various applications such as detection of zones of forest fires or estimation of surface moisture, etc.

A theoretical representation of the relationship between degree of coverage (which can be expressed as a function of NDVI) and temperature difference (*Ts* – *Ta*) is shown in Fig. 26.6. In a given region, points are generally distributed within a trapezium whose four corners correspond to the following extreme situations: 1) vegetation totally covering the ground and well watered; 2) vegetation totally covering the ground and dry; 3) bare soil saturated with water; 4) dry bare soil. A stress index, known as water deficit index (*WDI*) has been developed from this scheme: *WDI* = *AC* / *AB* which aids in detecting, for the same quantity of in-situ vegetation, differences in hydrological state along the position of the point inside the trapezium (see Fig. 26.6).

26.1.7 Conclusion

The few examples described above show the diverse range of applications of thermal-infrared data: evaporation mapping, detection of frost or stress zones, estimation of rainfall in Sahel, detection of

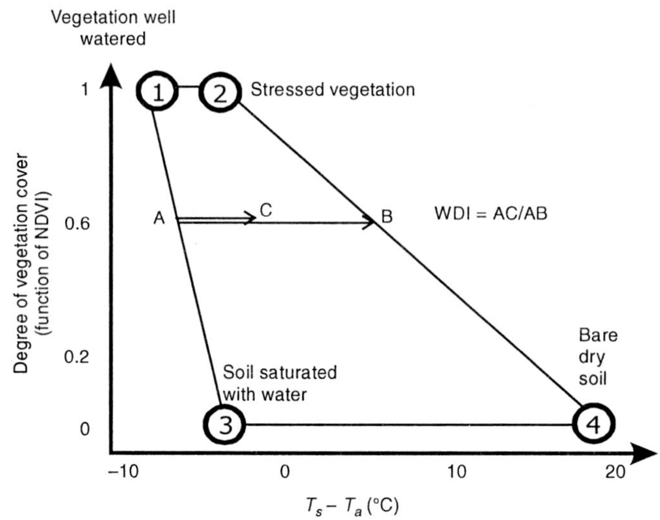

Fig. 26.6: Theoretical diagram defined by the relationship between (*Ts* – *Ta*) and degree of ground coverage (after Moran et al., 1994).

forest fire hazards, etc. Role of remote sensing lies in furnishing objective criteria for comparing zones with one another or for temporal monitoring to obtain information on spatial variability, which is often missed in conventional measurements by meteorological networks (as in the case of montane or desert zones). While meteorological satellites such as NOAA or METEOSAT provide such temporal monitoring due to their high revisiting capability, their spatial resolution is low and methods of transfer from regional to local scale are yet to be developed, awaiting launching of future sensors with higher resolution.

26.2 MICROWAVE DATA: USE OF RADAR IMAGES

A radar unit derived from a military system was used in France for civil purposes only in April 1973. The data were furnished in analog form (films). This was the period when the first scatterometers were developed (group of F.T. Ulaby at the University of Kansas, USA).

Applications of microwave remote sensing have been particularly developed in the recent years since the launching of the European Remote Sensing Satellite (ERS-1) on 16 July 1991, followed in 1992 by JERS (Japanese Remote Sensing Satellite) and in 1995 by ERS-2 and Canadian RADARSAT of high geometric resolution. Microwave remote sensing has been mainly used in moist and cloudy zones in intertropical or temperate regions to compensate for the impossibility of acquiring data in the visible and near infrared bands.

It should be noted that physical phenomena associated with this spectral region are not yet well understood and much research needs to be carried out based on ground, airborne and satellite experiments (see Chap. 1). For example, the data acquired by SEASAT in 1978 proved to be a valuable source of information, its short life span (4 months) notwithstanding. Processing and interpretation of data acquired by radar, differing from those of the visible and infrared bands, necessitate knowledge of characteristics pertaining to this particular technology.

26.2.1 Characteristics of radars

■ Spectral domain of imaging radars

The spectral domain of imaging radars, as shown in Table 1.4 (Chap. 1), lies between 30 and 2 cm or 1 and 15 GHz. This frequency range is determined by technological constraints:

—For frequencies lower than 1 GHz, technology for fabrication of radar systems with spatial resolution of less than 20 m, i.e., compatible with current applications in remote sensing is not developed;

—Beyond 15 GHz, costs of fabrication are too high or required materials do no exist.

For example, atmospheric attenuation is very high in the millimetre band (7.5 to 1 mm or 40 to 300 GHz) due to absorption by oxygen and water vapour. Only objects of dimension close to that of the wavelength can be detected and this feature hinders applications to earth resources. Moreover, the limit of 40 GHz corresponds to a change in technology that enhances costs of equipment.

Within these constraints, experiments conducted by various groups have resulted in a consensus on a large range of applications of these spectral bands. Band L (1 to 2 GHz) is important in oceanography and useful in topographic and geological studies. Band C (4 to 8 GHz) is optimal for studying soil moisture and band K (10 to 15 GHz) for vegetation.

That is why satellites ERS-1 and ERS-2 are equipped with a radar system with synthetic antenna (Synthetic Aperture Radar, SAR) for band C (5.3 GHz) in the 'wave' mode and in the image mode. The RADARSAT system also uses band C but with the possibility of a wide choice of beam widths and incidence angles, ensuring swaths of 35 to 500 km and resolutions from 10 to 100 m (Table 26.1). The European airborne survey EMAC 94/95, preparatory to the European program ENVISAT, used a polarising SAR with band L (1.25 GHz) and C (5.3 GHz) for investigating snow and ice in Scandinavia and surface moisture of soils in Belgium. JERS-1 is equipped with a band L (1.3 GHz) radar.

■ Sensitivity limits of radar systems

Systematic analysis of backscattered signal from various types of surfaces has been developed by ground measurements. The backscattered signals, reduced to coefficient σ_0 (in dB), are calibrated absolutely and hence comparable with one another. Distribution of σ_0 values of various ground objects is schematically shown in Fig. 26.7.

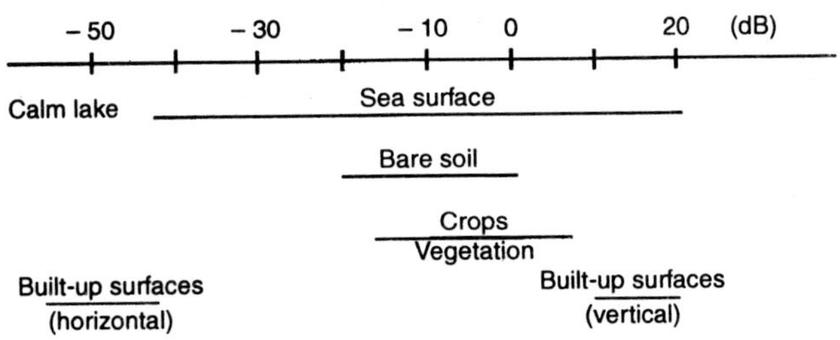

Fig. 26.7: Values of σ_0 for various surfaces.

It is evident that sensitivity limits of radar need to be precisely defined depending on the application envisaged.

■ Side looking radar (Real aperture lateral radar)

This system corresponds to older radar technology (Fig. 26.8) and continues to be used for airborne missions. It comprises a transmitter-receiver antenna placed outside the platform and oriented laterally.

Table 26.1: Examples of satellites with microwave sensors

	ERS	JERS	RADARSAT
Date of launching	–1: 25/07/1991 –2: 20/04/1995	11/02/1992	04/11/1995
Altitude	785 km	570 km	800 km
Synthetic Aperture Radar	Image mode		
Spectral band	C	L	C
Frequency	5.3 GHz	1275 MHz	5.3 GHz
Polarisation	V-V	H-H	H-H
Specific characteristics	Possibility to study waves and wind speeds	Revisiting capability 44 days	Various angles of view, 24-day cycle, revisiting at 1.5 day at 45° N possible
Angle of incidence on the ground swath	23°	35°	10° to 60°
Spatial resolution	25–30 m	18–18 m	10 to 100 m depending on angle of view
Swath	100 km	75 km	35 to 500 km
Other onboard sensors	ERS-1 - radar altimeter - ATSR scanner (4 bands centred on 1.6, 3.7, 10.8 and 12 μm) ERS-2 Same as above plus GOME (Global Ozone Monitoring Equipment)	OPS (Optical Sensor): 3 bands in visible and NIR, 4 bands in infrared	

The real aperture of the antenna beam (angle θ) determines the segment of ground, perpendicular to the direction of platform's motion, viewed almost instantaneously (a few milliseconds). Forward motion of the aircraft determines the length of the scene recorded.

Resolution of the system is better in the transverse direction (due to the frequencies used) than in the longitudinal (limited by the width of antenna).

In fact, two ground objects separated by a distance Δd such that $\Delta d \leq c\tau/2$ (where c is the velocity of light, τ the duration of impulses emitted by radar) will be confused as one, since the backscattered waves overlap each other. This distance is the limit of optical resolution (slant range resolution) while transverse ground resolution (r_t) is equal to $c\tau/2\cos\theta$, where θ is the depression angle of the radar (angle of radar beam with the horizontal). Longitudinal resolution (r_l) of radar is equal to the width of its aperture beam to its contact with ground. Since the aperture angle is approximately equal to the ratio of the wavelength used to the size of antenna, $r_l \approx (\lambda / L) \times d$ (where λ is the wavelength, L antenna length and d the inclined distance between the sensor and the object) and longitudinal resolution can be improved by increasing the antenna size. However, limitations are imposed by the bearing capacity of the platform and costs. Acquisition of images is done in two modes:

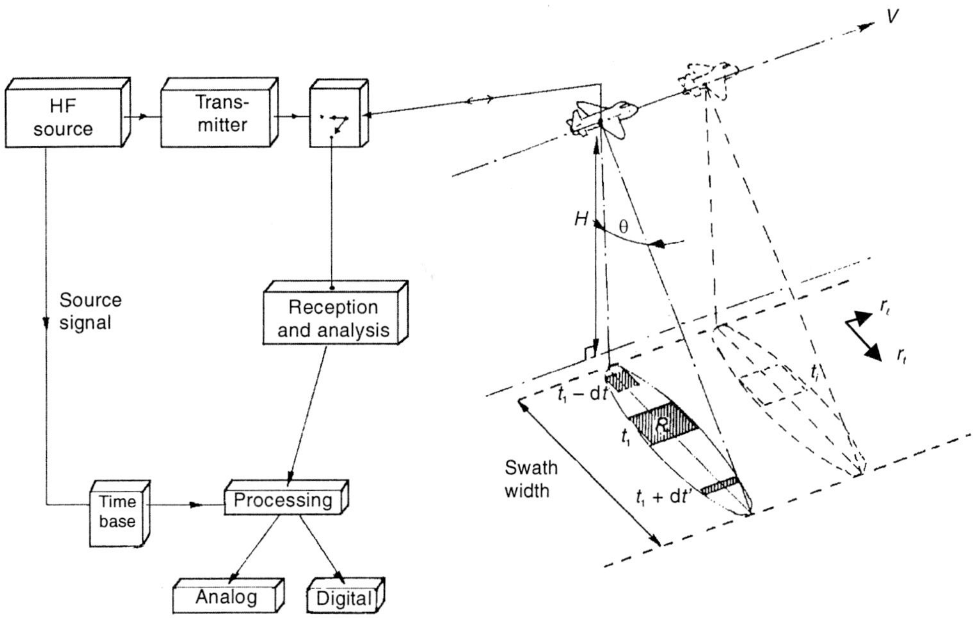

Fig. 26.8: Principle of operation of side looking radar.
H: Platform altitude; v: velocity of platform's motion; θ: aperture angle of antenna beam; t_i: instantaneous response of elementary surface; R: resolution of elementary surface.

—The analog mode, older, is based on conversion of the backscattered signal into a more or less intense light spot on a cathode-ray oscilloscope. A sensitive black-and-white film placed in front of the screen on a support rotates with a velocity corresponding to that of the platform motion and registers the light spots.

—The digital mode is used since 1978. Digital recording of backscattered signal enables calibration of data. It is thus possible to compare information acquired on different dates.

Data acquired by real aperture lateral radar have advantage since they necessitate only simple processing tools and show good spectral quality (resolution of 1 dB). On the other hand, they have the disadvantage of poor spatial resolution: the transverse resolution depends on the nadir distance (which determines the incident angle and its complement the depression angle), whereas the longitudinal resolution varies depending on the wavelength and antenna size.

■ Synthetic Aperture Radar

For improving the spatial resolution and overcoming the constraints of antenna size, Doppler effect is used. The successive positions of a real antenna, related to the forward motion of the platform, are considered elementary sources, giving rise to a 'synthetic' antenna. When the platform arrives at A_1 (co-ordinates $-L/2$, H), the ground point S (co-ordinates 0, 0) commences to receive the first pulses (Fig. 26.9).

When the platform is at A_n, after traversing a distance $A_1 A_n = L$, point S ceases to be illuminated. The principal lobe of the antenna of longitudinal resolution r_l illuminates the ground at an angle θ such that $\theta = k\ (\lambda / r_l)$. An integration of the signals is obtained by applying a coherent processing to the bipolar in-phase and quadrature views recorded for the successive positions A_1 to A_n of the platform.

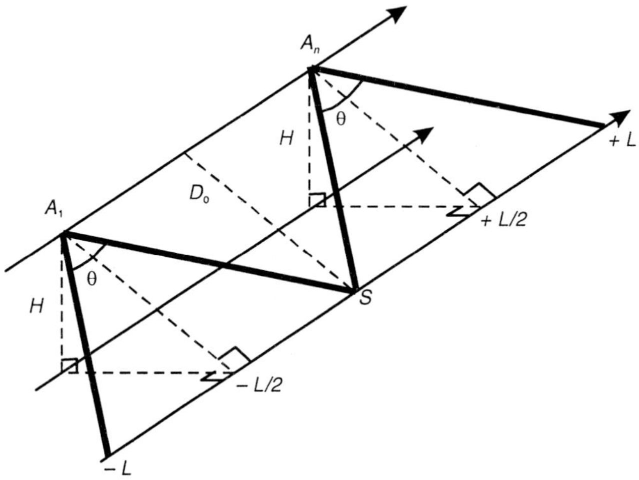

Fig. 26.9: Principle of SAR (after Paquet, 1997).

Each impulse of the signal consists of a train of coherent waves. The backscattered waves are received by an antenna over several seconds of flight of the platform, recorded and integrated. An object occurring in the path of the antenna beam reflects part of the waves towards the latter. Depending on the movement of the platform, the object-antenna distance will be equal or not to an integer number of wavelengths. The object will be 'seen' only when this distance is equal to an integer number of wavelengths. The object-antenna distance, initially large, passes through a minimum when the object is nearest to the antenna (following the motion of the platform) and increases as the platform moves away. The antenna receives the reflected waves, combines them with a reference train of coherent waves and makes the two wave trains interfere with each other. When the reflected signal coincides with the reference signal, the interference leads to high signal amplitude and the point is recorded as bright. In the inverse situation, the signal amplitude is small and the point recorded as dull. A ground point hence corresponds to a sequence of bright and dark segments, of various wavelengths, constituting a one-dimensional interference figure, known as radar hologram. Each ground object is represented by several points in a hologram. When the hologram is illuminated by a coherent light source, each bright segment constitutes an independent source of coherent light. A single point exists behind the hologram at which the light waves corresponding to the initial object are combined to form an image of the object.

Focused radars, which analyse a large number of reflections of the same object, are differentiated from non-focused ones that record a limited number of reflections.

The limit of longitudinal resolution r_l of a focused is given by the relation $r_l = L/2$, where L is the wavelength of the real antenna. The resolution is no longer dependent on the wavelength or altitude. For non-focused radar, $r_l = ½ \sqrt{\lambda d}$ is independent of antenna size but dependent on the wavelength and altitude.

26.2.2 Image quality

Images obtained by real-aperture or synthetic-aperture radars differ entirely from photos or images of other spectral bands. Topographic features have major effect on the distribution and intensity of the backscattered signal. Moreover, very large variations in grey levels are observed between surfaces of high and low backscatter coefficients. Speckle effect is produced due to surface irregularities of dimensions similar to the incident wavelength, even if the surface is homogeneous. This forms a noise which increases with the amplitude of the backscattered signal and which can be attenuated by filtering. Five specific features of radar images can be identified:

—Distortions or inversions of the position of objects may be produced depending on their relative proximity, their size and system resolution. As can be seen from Fig. 26.10, object *D* situated in front of object *ABCE* on the ground is observed within *ABCE* in the image. In montane regions, relief is sharp and peaks appear bent towards the sensor (Fig. 26.11).

—Radiometric values are affected by variations in incidence caused by changes in topography. The intensity of response of segment *BA*' is greater than that of segment *AB*, independent of composition of these segments (Fig. 26.11).

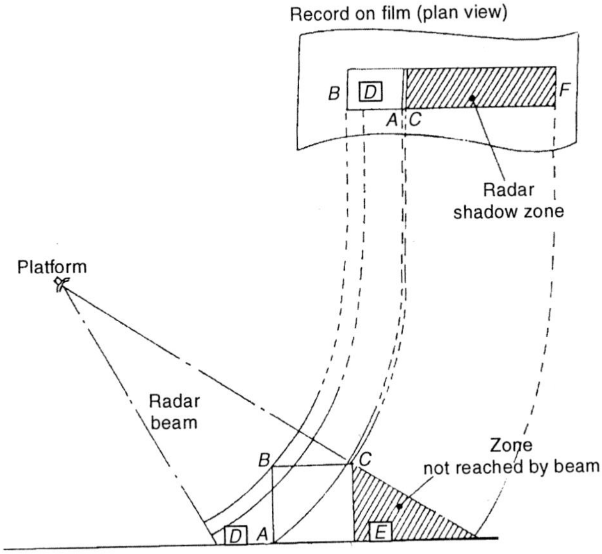

Fig. 26.10: Distortions and inversions in radar images.

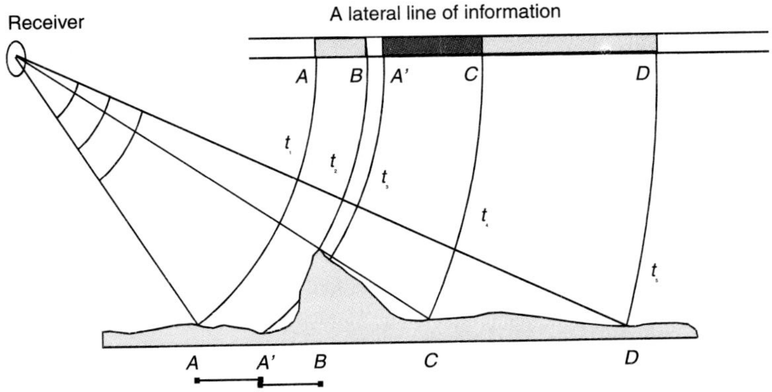

Fig. 26.11: Variations due to topography.

—Lengths are not preserved: segment *BA*' appears much shorter than segment *AA*' although their ground lengths are identical.

—'Shadow' effects (segment *BC* on ground, segment *A'C* in image) represent absence of backscatter during certain time. They appear as dark zones in the record (Fig. 26.11) since even radiation scattered by atmosphere is not there as in the case of visible band images. In virgin areas, the presence and disposition of such dark zones have been used to infer criteria of relative topography.

—'Corner' effect corresponds to a very large backscatter signal for some objects. Such is the case of some buildings whose face irradiated by radar beam produces a specular double reflection. The backscatter signal then is greater than that of adjacent objects, which may be either diffuse—bare soils or plant cover—or specular—calm water, paved surfaces (Fig. 26.12).

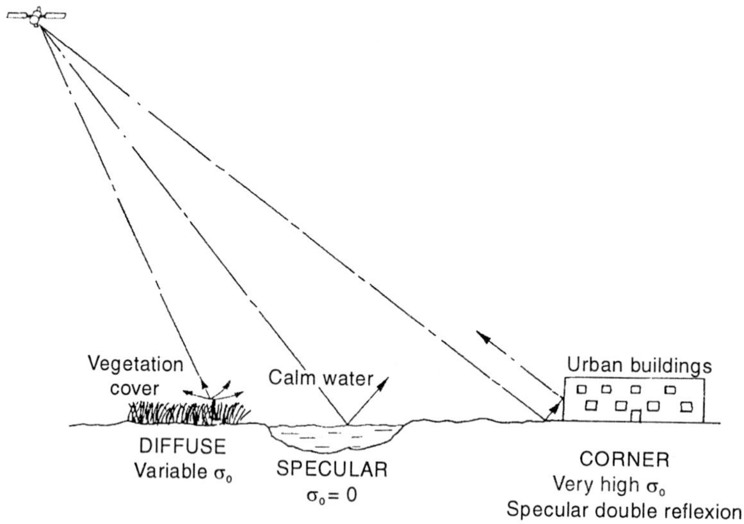

Fig. 26.12: Corner effect in radar images.

26.2.3 Applications of radar data

■ Mapping and topography

Some regions of the Earth cannot be photographed and hence cannot be mapped by conventional techniques due to intense cloud coverage. This is one of the reasons for small proportion (42%) of regions mapped in scales equal to or less than 1: 50,000. Systematic airborne radar surveys carried out during the years 1970–1980 in Brazil and Gabon have led to application of the technique in new areas such as the entire Amazon basin but its cost is prohibitive and preparation of mosaics compatible with geometric requirements of mapping is difficult. The large amounts of satellite radar data now available with systems ERS-1, ERS-2, JERS and RADARSAT (Table 26.1) respond to the cost and technical constraints, while ensuring rapidity and simultaneous acquisition of large areas, although topographic applications (production of digital elevation models, DEM) have been developed only recently. Three main techniques are used for topographic mapping by radar data (Polidori, 1995 and 1997).

☐ Radargrammetry

Radargrammetry is an adaptation of photogrammetry to radar geometry. It comprises the same two stages: construction of a stereoscopic model with the use of reference points, if possible, and stereoscopic analysis consisting of identification of pairs of homologous pixels in the two images. Radargrammetry differs from photogrammetry in geometric equations and mode of stereoscopic fusion that are specific to it. One of the first operational programs for analytical restitution was proposed in 1984.

☐ Interferometry

The objective of interferometry is analysis of phase differences between radar signals received from a given zone or from two positions of the same antenna (case of ERS-1 data, before the launching of ERS-2, acquired with a repetition cycle of 3 or 35 days) or by two different antennas. Such an analysis was conducted in the tandem ERS-1/ERS-2 mission with a very short time interval (less than 1 h) since the same instrument was present on both satellites. In favourable conditions of acquisition, this phase difference gives rise to an interferogram whose interference rings are very sensitive to ground elevation variations. Since the phase of the signal is known only in the 2π mode, it must be 'unfolded', i.e., converted to absolute phase, often a difficult task. Applications of interferometry will be seen in the sections on geology and hazard monitoring.

☐ Radarclinometry

Radarclinometry is based on the fact that any change in slope in the direction perpendicular to platform's trajectory induces (except in specific symmetries) a local variation in backscatter. The reciprocal is not true, however, since backscatter varies according to properties related to the nature of irradiated surface. Nonetheless, this relationship can be inversed in the case of very homogeneous media such as deserts, ice sheets or dense moist tropical forests. This inversion is done by radarclinometry by converting the intensity (corrected as much as possible for parasitic phenomena that modify it from the backscatter coefficient) into incidence angle (i.e., angle between incident wave and the local normal to the irradiated surface). Assuming the local form of terrain, since the relation is not unique, the incidence angle can be converted into slope and slope into altitude by integration. Considering the numerous assumptions required, the results obtained by this method show that it is more applicable to morphological description of terrain than precise computation of altitudes.

Digital elevation models (DEM) obtained from radar images are still few and hence performance of radar mapping is difficult to evaluate since the accuracy of DEM is considerably affected by influences of sensor, landscape or technique used. Simulation of radar images is used to verify various techniques of three-dimensional restitution.

■ Geology

Possibilities of radar applications in geology, in terms of accuracy of images and diversity of landscapes, were demonstrated by SIR-A and SIR-B systems onboard the space shuttles in 1981 and 1984, respectively. Various satellite data now available permit envisaging many operational applications.

The advantages of radar applications in geology are mainly related to the effects of surface roughness or relief on backscatter. Through them, geologists identify lithological units that are exposed or occur under a thin sand cover. Structural features corresponding to folds, fractures or veins can also be marked. Distinction between flexures and faults could be made in the ERS-1 data acquired over Levant fault in the Middle-East (Chorowicz et al., 1995). In desert climate, this marking can be done directly while in temperate or inter-tropical climate, the effects of morphology on vegetation are used. These data are incorporated in base maps (geological inventories) or used for detailed investigations such as mineral exploration, improving water resources, development of infrastructure (siting railways or roadways). Interpreted lithological characteristics are often complementary to those of satellite images of the visible and near infrared bands. Thus some rocks show a smooth surface in the visible band due to 'desert patina' but surfaces of various degrees of roughness in microwave data; this feature enables their identification. That is why combined analysis of satellite data of various sensors is important and operationally feasible.

However, many parameters that influence the backscatter signal hinder operational use of radar data in geology. For example, suppression of relief, mentioned in the preceding section, may result in erroneous estimation of dips, false reversal of layers or irregular drawing of faults. Moreover, although the dielectric constant of many rocks is known in mineralogy, use of radar images based on petrologic models is not yet developed (Deroin and Scanvic, 1995). In fact, due to the influence of geometric and dielectric parameters on backscatter mentioned earlier, laboratory models cannot be extended to field conditions.

In conclusion, for geological applications, radar cannot replace optical remote sensing but is complementary, since information it restores is less detailed but close to the original. An excellent field of application of microwave data is structural geology in the scales of 1:100,000 to 1:250,000.

■ Oceanography

Operational uses of radar images are most developed in oceanographic studies (sea ice, surface state of sea, petroleum slicks, etc.) and, in particular, in coastal oceanography for the contact zone between ocean and emerging lands. This zone has great ecological significance and constitutes venue for varied and conflicting stakes (see Chap. 25). In general, coastal zone boundaries coincide with those of the continental shelf (Forget and Cauneau, 1995). Depths rarely exceed 250 m before steep gradients are indicated at the level of continental slope and reach values between 2000 and 4000 m. Hence 7.5% of the surface of oceans and seas corresponds to coastal zone. Entire seas form its part: the English Channel, North Sea, Baltic Sea, China Sea, Yellow Sea, the Sonde platform, and the vast gulfs and bays.

It must be recalled that the littoral belt is characterised by a strong interaction between the atmosphere (wind), hydrosphere (water and its movements: waves, currents, etc.) and solid constituents of the continental lithosphere (Forget and Cauneau, 1995). The littoral zone depicts numerous and varied forms: beaches, dune systems, cliffs, rocky coasts, salt marshes, lagoons, deltas, etc. These environments undergo large variations that justify continuous monitoring and assessment, for example, for surveillance of beach erosion or identification of shoreline. Proximity to coast and shallow depths of open seas, in combination with the intensity of tides and associated currents, distinguish them from high seas.

Remote-sensing data in the visible band are very useful for littoral inventory and surveillance; however, radar data also show great significance due to their all-time and day-and-night capability, particularly for intertropical or Arctic/Antarctic zones. In middle latitudes, their advantage lies mainly in determination of parameters not possible with optical systems or in monitoring exceptional events such as floods, storms, black tides, etc. In fact, delineating coastline from the visible-band data is easier, less noisy and of finer resolution than from microwave data. Contrarily, observation of sea roller (crest length, amplitude variations following crests, 'group' effects) and its movements and wave surges, which enable delineation of certain bathymetric boundaries, as well as monitoring black tides constitute important applications of microwave remote sensing.

Microwave data are best suited for monitoring hydrocarbon pollution, although other types of images (visible, thermal) also assume certain importance. Systems of automated detection of slicks are under development, but wind may severely hamper detection. Identification of physical (thickness, age) or chemical characteristics of slicks seem possible from multiband (visible to microwave) data. However, this application is not yet in operational stage since multifrequency and multipolarisation radar systems are required. Time of access to data (less than 24 h) also constitutes a problem to be solved to make the technique operational. Theoretical researches are undertaken to determine mechanisms of attenuation of roughness by hydrocarbon layers in the presence of wind and waves.

On the other hand, principles of detection of bathymetry by radar data are known but preliminary attempts of modelling using controlled airborne experiments have so far resulted in an underestimation of the phenomena and quantitative application in near future is not foreseen. The SAR seems to constitute a valuable tool for description of internal waves and their dynamics, although ground data are necessary for inferring certain characteristic parameters such as wave amplitudes. Microwave data are capable of detecting surface expression of internal oceanic dynamics (internal fronts in continental slopes and shoals). Similarly, in favourable conditions, coastal currents can be studied using effects of roughness, form of slicks and refraction of sea roller.

Lastly, detection of ships and halieutic applications are under study. Resolution is the main limiting factor. In fact, most fishing boats are not accessible to ERS imagery whose transverse resolution is 26 m and longitudinal resolution 30 m. Detection of wakes, whose remanence is identified but not explained, assumes calm sea conditions, not always satisfied. Detection of fish banks has been studied by airborne radar but operational usage of satellite data is not feasible for reasons of spatial resolution on the one hand and due to the sporadic nature of surface manifestation of fish banks, on the other. Due to their specific properties, microwave data can be useful in fishing by their capacity to locate thermal fronts and contrasts and, hence, oceanic regions where this resource is likely to be present.

Study and surveillance of sea ice constitute a domain of operational applications, since any-time and day-and-night capabilities of microwave remote sensing are valued by conditions of cloud coverage and duration of polar nights. In these regions, density and movement of ice are important markers in identification of general circulation in oceans. These data constitute useful information for sedimentologists in studying transport of solid particles, for biologists in exploitation of biological resources and for climatologists in investigation of depression zones. Further, monitoring sea ice variations gives information on long-term climatic events (global warming, greenhouse effect). Lastly, monitoring of ice has also economic importance. The need to know maritime current conditions, especially for access to Baltic ports, imposes a strict surveillance of ice masses. Similarly, proper functioning of offshore drilling platforms requires permanent checking of the state of waves and proximity of ice masses. Satellite radar images, operational and repetitive, represent a significant information support for these applications.

■ Hydrology and snow gauging

Knowledge of Alpine type of snow or ice surfaces is very important for hydrological applications. In fact, snow surfaces change in area from 47 million km^2 to 4 million km^2 from winter to summer (Dedieu, 1995). It is hence necessary to have seasonal maps of snow and estimate the corresponding water equivalent. These applications are not yet in operational stage. In fact, mapping of snow is possible when it is wet on the surface but distinction between dry snow and non-snowy zones is difficult, unlike in the case of passive microwave systems. The equilibrium line of glaciers is visible in SAR images, facilitating reconstruction of annual mass balances; on the other hand, direct estimation of equivalent water of snow cover is not proven, although polarimetry seems to be a promising tool.

Current researches are oriented towards use of multispectral and multipolarisation systems. Specialists presently consider that a weekly repetition of observations is required with a resolution of 30 m during the period of snow melting. X-band (8 to 12 GHz) seems preferable over C-band due to its higher sensitivity to surface roughness and a high incidence angle (45°) is essential in montane regions.

Ice movements (30 cm d^{-1} to 2 m d^{-1}) have been monitored by interferometry using ERS data. However, detection of melting ice vis-à-vis snow and temporal offsets of dry snow due to wind effects are sources of errors.

The most advanced countries in research on such themes, for which scientific problems and economic activities are closely related, are Finland, Switzerland, Austria and Canada.

■ Hazard monitoring

Revisit capability of satellites together with any-time and day-and-night nature of radar are the basic advantages for hazard monitoring. This is particularly established for monitoring floods in regions (or periods) of intense cloud cover, for which visible and infrared data cannot be acquired. For example, ERS data were used for monitoring floods of Oder river in Poland and Germany in 1997, and RADARSAT data for assessing the magnitude of floods in Manitoba in May 1997.

Like the visible and infrared data, microwave images also enable surveillance of volcanic activity. They are irreplaceable when volcanic activity occurs in high latitudes, as in Alaska and Iceland, under a glacier in particularly unfavourable conditions of observation.

Interferometry, specific to microwaves, finds novel applications in monitoring earthquakes, fault movements and landslides.

Cost of data and their processing and limitations in operational use of remote-sensing data for some applications are less important in hazard studies in view of human lives and socio-economic consequences involved. It is also true that this is a technology of rich countries and many hazards are produced in poor countries that do not have the means of paying for such information.

■ Agriculture and forests

Application of microwave remote sensing for agriculture and forestry is least exploited and has progressed little in the recent years, although radar imagery has great potential in this domain. The main reasons are absence of multi-frequency and multi-polarisation facilities and unfavourable incidence angles for the presently available data. Three major types of applications can be identified.

□ Monitoring and quantification of hydrological processes in soils

Various levels of investigation of aqueous state of soils are 'entry' (precipitation), transfer (on surface by runoff, at depth by infiltration) and 'exit' (evapotranspiration, recharge of aquifers, river discharge, useful reserves, etc.). At each level, various parameters need to be considered: climate, morphology, soils, geology, vegetation cover, etc. Relationships of these parameters with water movement are analysed and progressively modelled.

Radar contributes to evaluation of surface characteristics and analysis of relationships between backscatter signal and the phenomena under study. In-situ experiments have been conducted to select optimal instrumental conditions for determining surface moisture (frequency around 5 GHz, incidence angle between 7° and 17°) and to identify the percentage capacity of the field from the first 5 cm of soil as being the expression of water content influencing the dielectric constant of soil and hence the backscatter signal (Fig. 26.13).

Satellite radar images provide information over vast areas and hence reveal regional variations of surface moisture. However, causes of variation of backscatter signal are such that data interpretation most often remains qualitative. Use of diachronic data for creating colour composites, for example, indicates contrasts between dry and moist soils, bare or covered by various types of vegetation. In the Sahel region, band C and incidence angle of 23° of SAR and ERS were found favourable for mapping temporal variations of soil moisture provided that preliminary knowledge about the state of soil and vegetation is available. In a totally different scale and other climatic conditions, scatterometer data of ERS have been used for monitoring melting of permafrost in the Boreal zone.

□ Monitoring of crops and yields

Considering the operational nature of crop and yield monitoring by data acquired in the visible and infrared bands, radar data are significant only for their any-time nature and high frequency of acquisition. In this case, they provide complementary information, filling gaps in crop inventories for very cloudy climatic periods. However, while differentiation of certain crops by microwave data has been

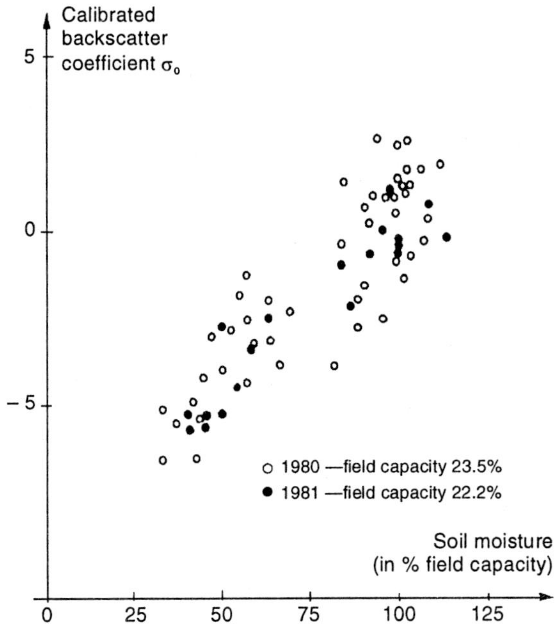

Fig. 26.13: Relationship between backscatter signal and soil moisture (after King, 1979).

demonstrated (Fig. 26.14), confusions exist between species (in particular in the case of cereals) whereas geometric resolution of radar data is much lower than that of the visible and infrared data.

Lastly, the cost of data and their processing constitutes a serious limitation in their operational usage for this type of applications.

□ Inventory and estimation of forest populations

It is mainly the structures and ages of population, rather than forest species, that are differentiated from radar data. Identification of forest species and groups is much more difficult than with satellite data of the visible and near infrared bands, particularly when they represent heterogeneous populations. That is why microwave data are essentially used for surveillance of deforestation, especially in tropical humid forests where cloud conditions hinder acquisition of data in other spectral bands. Even in this case, interactions of waves with objects are not fully understood and data is interpreted qualitatively in images earlier filtered for contrast enhancement and speckle reduction.

26.2.4 Conclusion

Many applications of microwave data still remain unexploited since studies are necessary to understand the influence of various characteristics of objects on backscatter signal and the signal needs to be processed before its interpretation. Moreover, the number of satellite microwave systems remain limited, their recent enhancements notwithstanding, and they offer restricted choice of spectral bands, polarisations, incidence angles, etc. Lastly, for microwave remote sensing to become operational, service societies have to develop. In fact, organisations directly concerned with applications just mentioned above are interested in thematic information and not in remote sensing data.

All this assumes efforts in research, teaching and formation as well as increased participation of thematic specialists in specification of radar systems and products.

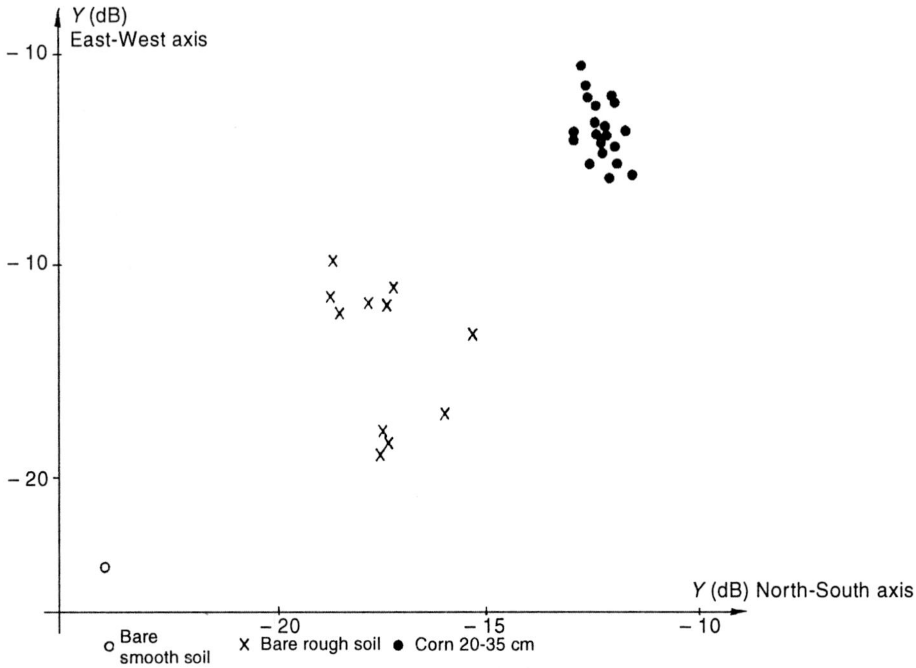

(a) April 1980

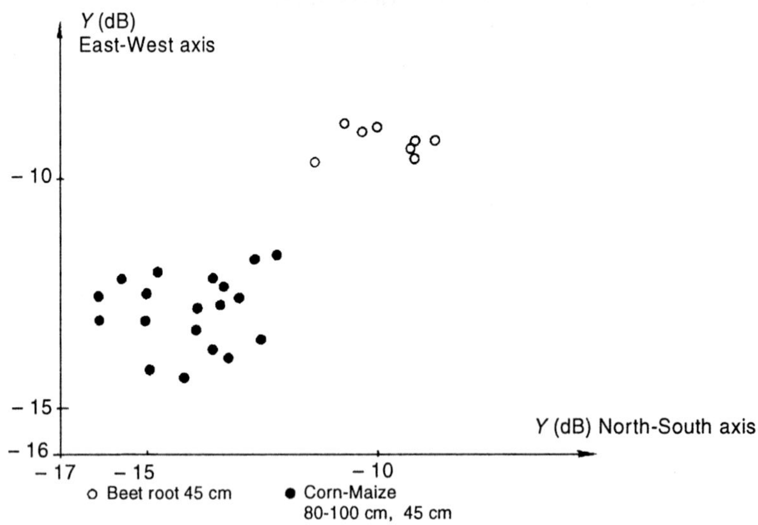

(b) June 1980

Fig. 26.14: Backscatter signal of various crops on different dates (Flight VIGIE band X, 1980; King, 1979).

References

André J-C, Goutorbe J-P, Perrier A, Becker A, Bessemoulin P, Bougeault P, Brunet Y, Brutsaert W, Carlson T, Cuenca R, Gash J, Gelpe J, Hildebrand P, Lagouarde J-P, Llyod C, Mahrt L, Mascart P, Mazaudier P, Noilhan J, Ottle C, Payen M, Phulpin T, Stull J, Schmugget T, Taconet O, Tarrieu C, Thepenier R, Valencogne C, Vidal-Madjar A, Weilla A. 1988. HAPEX-MOBILHY: First results from the Special Observing Period, Ann. Geophys., 6: 477–492.

Assad E, Freteaud J-P, Kerr Y, Lagouarde J-P, Seguin B. 1986. Utilisation de la thermographie infrarouge dans l'estimation de l'évaporation à l'échelle régionale. Application au Sénégal, Agron. Tropicale, 40 (4): 279–285.

Chorowicz J, Kofi B, Chalah C, Chotin P, Collet B, Poli JT, Rudant J-P, Sykioti S, Vargas G. 1995. Possibilités et limites de l'interprétation géologique des images (SAR) ERS1, Bulletin SFPT, 138: 82–95.

Courault D, Clastre P, Guinot J-P, Seguin B. 1994. Analyse des sécheresses de 1988 à 1990 en France à partir de l'analyse combinée de données satellitaires NOAA-AVHRR et d"un modèle agrométéorologique, Agronomie, 14: 41–56.

Dedieu J-P. 1995. Application du radar spatial à l'étude des neiges et des glaces, Bulletin SFPT, 138: 80.

Deroin J-P, Scanvic J-Y. 1995. Apport de l'imagerie radar à la cartographie géologique: exemples et réflexions, Bulletin SFPT, 138: 96–109.

Forget P, Cauneau F. 1995. L'imagerie radar en milieu marin côtier et littoral: état de l'art et perspectives, Bulletin SFPT, 138: 73–79.

Goutorbe J-P, Label T, Tinga A, Brouwer J, Dolman AJ, Engman ET, Gash JGC, Hoepffner M, Kabat P, et al. 1993. Hapex-Sahel: a large scale study of land-atmosphere interactions in the semi-arid tropics, Annales Geophysicae, 12: 53–64.

King Ch. 1979. Contribution à l'utilisation des micro-ondes dans l'étude des sols, Thèse INA-PG, 122 pp.

Lagouarde J-P, Valery P, Belluomo P, Soulier M-A. 1983. Cartographie des topoclimats forestiers. Mise au point d'une méthodologie d'analyse de l'effet du relief sur les thermographies: application aux données HCMM sur le nord-est du Massif Central, Agronomie, 3 (10): 1011–1018.

Lagouarde J-P, Brunet Y. 1991. 'Suivi de l'évapotranspiration réelle journalière à partir des données NOAA-AVHRR lors de la campagne HAPEX-MOBILHY', 5e coll int. Mesures physiques et signatures en télédétection, Courchevel, ESP SP 319: 569–572.

Moran MS, Jackson RM 1991. Assessing the spatial distribution of evapotranspirtion using remotely sensed inputs, J. of Environmental Quality, 20 (4): 725–737.

Polidori L. 1995. Apport de la simulation d'images à la validation des techniques de cartographie radar, Bulletin SFPT, 138: 16–25.

Polidori L. 1997. Cartographie radar, Gordon and Breach Science Publishers, Canada, 287 pp.

Glossary

b: Spectral bands of SPOT are designated as: b1 (or XS1) for green, b2 (or XS2) for red, b3 (XS3) for infrared and b4 for reflective infrared. b0 is used for the first band (blue) of Vegetation instrument.

Band: Part of an image comprising a group of pixels pertaining to a single spectral band or derived from a computation based on a single series of spectral bands.

Base: Distance covered between two successive aerial or satellite photos or images.

Bias: It measures the deviation between expectation of a series of measurements and the nominal value, in other words, between the centre of gravity of the group and the expected value, i.e., the reference value or nominal value. In Fig. 17.1, it is represented by the distance between the centre (*x*) of the target (nominal value) and the centre (*y*) of the group of measured points. It can be estimated by the intergroup distance vis-à-vis the reference value.

BIL (*Band Interleaved by line*): Storage format of spectral bands on satellite information base, line by line.

BIP (*Band Interleaved by Pixel*): Storage format of spectral bands on satellite information base, pixel by pixel.

BSQ (*Band Sequential*): Storage format of spectral bands on satellite information base, scene by scene.

CAP: Centre of Archival of Preprocessing of SPOT image, where radiometric and geometric corrections are applied to images.

CCD sensors: Charge-transferring devices (*Charge-Coupled Devices*) or receivers. Several thousand units are assembled in a single integrated circuit. The energy arriving at each unit creates a proportional electric charge. The analogue value of charge is transmitted to a sampler which converts it into a digital number.

CCT (Computer Compatible Tape): Magnetic tape comprising 9 tracks of 1600 bpi and 240 feet long. It can store 331 Mbits. A LANDSAT image consists of about 300 Mbits and a SPOT image, 220 Mbits. In the beginning of the tape, input device (ID) is recorded, which includes the reference number of the band, sensor, date of image acquisition, mission, trajectory and range, preprocessing executed and adopted format.

Chorology: Chorology (from Greek *khoros*, meaning country or district, and *logos*, meaning logic or science) is the study of relationships existing between characteristics of semantic (thematic) units (internal factors) and their distribution in three-dimensional landscape (external factors).

Chroma: It is one of the three variables of the Munsell code used as reference for colour coding. The higher the chroma, the smaller the grey content of a colour.

Compact: Said of a map unit whose perimeter is small compared to its area. The opposite is referred to as *digitate*.

Complex units: Several complex units exist:

Juxtaposition: Unit for which no chorological laws are defined.

Association: Unit for which defined chorological laws cannot be expressed graphically.

Sequence: Unit for which defined chorological laws are dominated by a preponderant factor such as slope, topography, lithology or duration.

Combination: Unit for which a certain number of laws that characterise chorological laws are defined: organisation of soils, soil types, size, reciprocal positions, etc.

Contrast: Ratio of the typological (mathematical) distance of two sites to the geographic distance between the same sites. This value characterises the boundary state between the two sites.

DEM (Digital Elevation Model): Digital model of elevations in a region.

Digital characteristic: The set of digital numbers of each pixel for each band existing in an image. Normally, for better perception, these values are connected by straight-line segments. It does not represent a spectral characteristic since the various bands are not calibrated relative to each other.

Digital number: Radiance (luminescence) value of a pixel expressed as an integral number, often over a scale of 256 values.

Digital-number spectrum: Set of digital numbers for each band, or for several bands, which correspond to a geographically defined object in an image.

Exactitude: Narrowness of accord between a measurement or estimate of a parameter and its nominal value (IGN, 1997). It includes the two concepts of bias (b) and precision (p) by the equation:

$$p^2 + b^2 = e^2.$$

It is often indicated by mean square error. The nominal value, by definition, serves as the reference. To be exact, a measured value must be precise as well as without bias.

Field of investigation: Area targeted in an investigation or study.

Field of view: Largest area that can be sensed from a single point of observation.

Geographic Information System (GIS): It is the information system that facilitates collection and organisation, management, analysis and integration, preparation and presentation of geographic data derived from various sources, ultimately contributing to spatial management.

GPS (Global Positioning System): A system of 24 satellites orbiting around the Earth, which transmit very accurate data about their position and distance relative to a receiver on the Earth. The receiver, usually held in hand, sees the satellites that are above the horizon if the sky is clear and no obstacles exist around (thick forest cover, walls, metal objects, people). Depending on the number of satellites received (3 to 12), the GPS receiver indicates its geographic position in two- or three-dimensions. A precision of about 1 cm can be achieved.

Graphic analysis: The part of spatial analysis devoted to shapes of map units, their position, neighbourhood, dispersion, etc., as well as their representation: colour contrasts, drawing shapes, etc.

Group: A set of combined pixels derived from a classification.

Hue: It is one of the three variables of the Munsell code used as reference for colour coding. It indicates the respective quantities of red, blue, yellow, etc., of which the colour is composed.

IFOV (Instantaneous Field of View): Area from which a sensor receives a single signal whose value is formed by summation of component values of the area.

Instantaneous field of view on ground: IFOV

Laser (Light Amplification by Stimulated Emitted Radiation): Amplification of light by stimulation of emitted radiation, through resonance of stationary electromagnetic waves between two parallel mirrors.

Level of analysis: Dataset that defines, for a map or a map unit, the type and number of variables used as well as precision of measurements and observations.

Level of perception: It corresponds to the tools and types of measurements or estimates employed for detecting a given level of organisation.

Lidar (LIght Detection And Ranging): Detection with the help of light. This system operates similar to radar and records reflection, from the Earth's surface, of a wave emitted by a source.

LUT (Look Up Table): Colour table.

Map unit: Group of geographically distributed map zones having the same semantic content.

Map zone: Graphical unit having a closed boundary and a localised semantic content assumed homogeneous at a definite probability level.

Map: Orthogonal projection, on a plane, of a group of map zones representing organised groups of objects defined by their boundaries (*container*), shapes, contrasts and semantic contents.

Group of containers defined as an extension and representing sets of organised groups of objects that are characterised by a comprehensively defined group of contents.

It is the result of a classification of objects; it shows spatial distribution and organisation of the objects and their conceptualisation.

Mapping: It comprises the scientific studies and operations, art work and techniques employed on results of direct observations or investigation of data for the purpose of preparing and establishing maps, plan views and other modes expression, as well as their usage (International Cartographic Association united with UNESCO, Paris, 1966).

It includes techniques that enable representation of objects after preparation of maps. In soil mapping, it mainly consists of two-dimensional representation of soil cover which is a three-dimensional system, incorporating the factor of time duration.

It simultaneously encompasses operations of spatial analysis (map production) and graphic representation (cartography *sensu stricto*). Map production is the process of understanding the organisation of objects in a landscape, based on development of models according to chorological laws.

Mixel: A heterogeneous pixel which comprises several objects and whose spectral characteristic is determined by the spectral composition of all the objects geographically contained in the resolution element.

It is known or assumed that the composition is derived from various, spectrally different, objects such as vegetation, bare soils, water, clouds, shadows, etc. A very high digital number is observed on images when passing from one semantic unit to another or over boundaries of units differing in spectral characteristics.

Nadir: Point vertically below the optical centre of a camera.

Nucleus: Group of pixels used a priori to characterise an object for which a classification is to be designed. This term is similar to 'training zone'.

Pachy: Term in the Soil Reference Manual (*Referentiel Pedologique*) used to indicate an abnormally thick soil.

Path: Ground track vertically above which a satellite passes. It is also the projection of the satellite orbit on the Earth surface. For SPOT, each path is represented by two K-marks indicated on either side of the satellite trajectory.

Pattern: A group of pixels or objects, different but spatially organised, which repeat in a particular manner. A pattern is hence a 'homogeneous unit formed by units that are heterogeneous in nature'. This constitutes a basic structural element.

Pixel: Term derived from contraction of *picture element*.

Precision: Narrowness of accord between a measurement or estimate and the mathematical expectation of this measurement or estimate (David and Fasquel, 1997). It measures fluctuations of a series of measurements around its mathematical expectation; it is given by the standard deviation from the mean of the series of measurements. In Fig. 17.1, it is shown as a double arrow which gives an estimate of the 'diameter' of the group of measured points. It can be determined from the intragroup distance.

Pushbroom sensor: Group of sensors that acquire information on a row in a single sweep. The sensors are pushed by the platform and hence the name.

Quarter rule: Any unit represented on a map must have at least ¼ cm^2 or, for elongated zones, more than 2 mm in width and 1.5 cm in length. This old rule corresponds to a legibility limit.

For topographic maps with hatches, this rule indicates that the hatches situated on the steepest gradient are spaced at an interval of one-fourth of their length.

Radar (RAdio Detection And Ranging): Microwave sensor which generates an image of the ground by means of temporal scanning.

Radiance or **luminance (energy)**: Intensity emitted by unit apparent area along a direction q for a point source of area d*A*, through a solid angle w.

Raster: When a data plane is considered, a grid placed above it delineates cells or pixels. Each cell is made to correspond to a digital dataset, which is said to be in raster format. Unlike vector storage, no real boundaries exist for each grid. Raster form is the most common for images.

Reflectance: Ratio of the energy in a wavelength band reflected by an object to the energy received from the Sun by this object and for the same wavelength band.

Remote sensing: Data and techniques used for determining the physical and biological characteristics of objects using measurements made from a distance, without physical contact of the latter (*JO* of 11 December 1980).

Row: Along the trajectory of a satellite, images are divided over a certain north-south distance; the marker of their centres constitutes a row (or line), indicated by letter J for SPOT satellites.

Scale: Ratio of a distance on the map (represented) to the same distance on the ground (representative).

Semantic analysis: Any part of spatial analysis which gives attributes for characterising pixels or zones: digital number values, tree heights, soil colours, species present in a plant canopy, etc.

Soil landscape: Combination of soil horizons and landscape elements, viz., vegetation, effects of human activity, geomorphology, hydrology, substratum or parent rock, whose spatial organisation in its entirety determines a soil-cover or a part of it. A soil landscape often comprises several soil-landscape units.

Its subgroups are: soil-landscape units and soil-landscape elements.

Soil-landscape unit: Each unit corresponds to a soil system, often based on geomorphology, and can be represented by a map unit (it can be divided into soil-landscape elements).

Soil-landscape element: Each corresponds to one or several typological units, spatially connected, having a simple spatial organisation (a sequence, for example).

Spatial (geometric) resolution: The smallest area for which a single value is obtained for the variable studied. The spatial resolution of a map can be determined by the ratio of the spatial field of the map to the number of sites studied.

Spatial analysis: Spatial analysis consists of studying spatial relationships between various objects distributed in a plane (or volume). Objects can be studied in terms of their abundance, estimated by area or percentage of surface. It is also important to evaluate the nature of neighbourhood between various types of objects.

Spatial field: Largest area that can be analysed including all the sites studied.

Spectral band: Wavelength interval defined by two threshold values, the beginning and the end (for example, 510-590). A spectral band measures only one energy (reflectance, radiance or digital number) value, which is equal to the double spatial and radiometric integral between the two limits, contained in a pixel.

Spectral characteristic: The set of radiance or reflectance values of an object for a set of spectral bands.

Stripping: Parallel strips visible in LANDSAT images due to differences in sensitivity of photodetectors,

Structure: An aggregate formed by objects and their relationships. For thematic analysis in image processing, it represents a system of organisation that depicts relationships between objects studied (pixels, zones, etc.) but not their characteristics derived simply from their attributes or digital numbers. OASIS, VOISIN, convolution filters, etc. are considered, from this point of view, structural methods.

In soil mapping, it represents description, analysis and evaluation of textures produced by agglomeration of constituents of varied grain-sizes and nature.

Surface state of soil: Composition and organisation of soil surface at a given moment. It takes into consideration slaking crusts, salt efflorescence, stony nature, cultivation works, cracks and other surface features, as well as coverage of soil by algae, moss and other vegetation.

TDRS (Tracking and Data Relay System): System that enables the earth observation satellites LANDSAT-4 or 5 to send images received from other satellites to ground stations.

Texture: Pattern formed by objects independent of their inter-relationships.

For mathematical analysis in image processing, the term is almost equivalent to the term *structural* used in thematic analysis.

In soil mapping, it represents arrangement of constituent 'grains', defined by their grain-size and nature, considered independent of each other.

Thematic group: Population of pixels which are a posteriori, i.e., after a classification, combined into a single group and carry the same name as the nucleus used to represent them during classification.

Tomography: Spatial representation of the material situated at a given depth from the surface of soil cover.

Training zone: see **Nucleus**.

Value: Parameter of a colour, especially used in Munsell code. A 'value' is made up of the same quantity of energy in blue, green and red spectral bands.

Wavelength: It is expressed in micrometers or nanometers.

Whiskbroom scanner: A sensor which acquires data through a rotating or oscillating mirror inclined at 45 to the vertical and situated perpendicular to the direction of motion of the platform.

General References

Specialised Scientific Journals on Remote Sensing

Major research papers are published in various journals exclusively devoted to remote sensing. Some of these are listed below. Some of these publications can be accessed from Internet currently or in near future. With the possibilities of information transfer through electronic networks, utility of publishing on paper becomes questionable in somuch as it is often essential to give images on several illustrations of remote sensing applications.

Advances in Space Research

Bulletin de la Société Française de Photogrammétrie et Télédétection, 2, avenue, Pasteur, 94160 Saint-Mandé, France.

Bulletin d'information de l'Institut géographique national, 6–8, avenue Blaise Pascal, Cité Descartes, Champs-sur-Marne, 77455 Marne-la-Vallée Cedex 2.

Canadian Journal of Remote Sensing/Journal candien de télédétection, CASI, 222, rue Somerset Ouest, suite 601, Ottawa, Canada K2P 0J1.

EARsel, Advances Remote Sensing, Robin Vaughan (Editor in Chief), APEME, University of Dundee, Dundee, DD1 4HN, Scotland, Grande-Bretagne.

Fotointerpretacja w geograffi, Katowice, Pologne.

IEEE Transactions on Geoscience and Remote Sensing, IEEE Centre, P.O. Box 4122, Hong Kong.

International Journal of Remote Sensing, University of Dundee, Dundee, DD1 4HN, Scotland, Grande-Bretagne.

Microwave Remote Sensing Active and Passive, Addison-Wesley Cy.

Photogrammetric Engineering & Remote Sensing, ASPRS, 5410 Grosvenor Lane, suite 210, Bethesda, MD 20814-2160, États-Unis.

Photo-interprétation, Éditions ESKA, 27, rue Dunois, 75013 Paris, France.

Remote Sensing of Environment, Elsevier Science Publisher, P.O. Box 211, 1000 AE Amsterdam, Pays-Bas.

Remote Sensing Reviews, Harwood Academic Publishers, 1 Bedford Street, London WC2E 9PP, Grande-Bretagne.

Scientific Journals Comprising Papers on Remote Sensing

The following periodicals on various subjects also include scientific articles on remote sensing.

Agronomie

Agronomy Journal

Agronomie tropicale

Annales Geophysicae

Annales de géophysique de France

Annales des Mines
Applied Optics
Bulletin de la Société française de minéralogie et cristallographie
Bulletin de la Société de préhistoire Nord
Comptes-rendus de l'Académie d' agriculture de France
Comptes-rendus de l'Académie des sciences
Economic Geology
Étude et gestion des sols
Geoscience and Remote Sensing
Journal Optic Society of America
Journal of Atmospheric and Oceanic Technology
Journal of Environmental Quality
Limnology and Oceanography
Mémoires de l'Institut océanographique de Monaco
Progress in Phycological Research
Science du sol
Soil Science Society of America Journal
Sols
Water Resources Research

Ph.D.Theses

Ph.D. theses produced every year from the French and foreign universities constitute primary sources of information on remote sensing. In acknowledgement of the cooperation of our collaborating scientists and former students, the recent theses on remote sensing, recently awarded at the INA-PG and other institutes, are listed below.

Courault D., *Étude de la dégradation des états de surface du sol par télédétection. Analyses spectrales, spatiales et diachroniques,* Thèse INA-PG, 1989, 239 p.

Escadafal R., *Caractérisation de la surface des sols arides par observation de terrain et par télédétection.* Aplication: example de la région de Tataouine (Tunisie), Thèse de pédologie, Université Paris VI, ORSTOM, Paris, 1989, 317 p.

Gilliot J.-M., *Traitement et interprétation d'images satellitaires Spot: Application à l'analyse des voies de communication,* Thèse de doctorat, Université ParisV, 1994, 197p.

King Ch., *Contribution à l'utilisation des micro-ondes dans l'étude des sols,* Thèse INA-PG, 1979, 122 p.

Orth D., *Typologies et caractérisation des prairies permanentes des marais du Cotentin, en vue de leur cartographie, par télédétection satellitaire, pour une aide à leur gestion,* Thèse INA-PG, 1996, 149 p. plus annexes.

Robbez-Masson J.-M., *Reconnaissance et délimitation de motifs d'organisation spatiale. Application à la cartographie des pédopaysages.* Thése de doctorat, ENSA Montpellier, 1994, 161 p.

Rogala J.-P., *Approche numérique de l'espace agricole,* Thèse docteur-ingénieur INA-PG, 1982.

Yongchalermchai C., *Étude d'objets complexes, sol/plante, à différents niveaux d'organisation: de la parcelle au paysage.* Thèse de l'INA-PG, 1993, *Sols*, n°19, 232 p.

Proceedings

Several colloquia, seminars, summer schools etc. are organised, at the rate of almost one even per week, throughout the world. This is a testimony to the active research in remote sensing. Proceedings of some of these events, organised in France by the CNES, CNRS, French Association of Photogrammetry and Remote Sensing or International Society of Photogrammetry and Remote Sensing, are cited below:

Actes de télédétection IRT, La-Londe-les-Maures

BRGM, *Manuels et Méthodes*

Colloque International. Signatures spectrales d'objets en télédétection.

CNES, École d'été

CNIG, Groupe de travail

CNRS, École d'été de physique spatiale: Principes physiques et mathématiques de la télédétection.

IFEN, Groupes de travail

IGARSS Digest, Washington.

INRA, Les colloques

International Symposium of Remote Sensing of Environment

iRD-ORSTOM, Journées de télédétection, Études et Thèses

Other Publications

CNES Magazine (trimestriel)

Collection de l'École normale supérieure de jeunes filles

Earth Observation Quaterly, ESA

Éditions Cépadues

Spot magazine (semestriel)

Télédétection, journal du réseau de télédétection de l'AUPERLF-UREF

Télédétection satellitaire, Éditions Paradigme.

Books

General books or on specific topics of remote sensing are given below.

On General aspects

Bonn F. and Rochon G., *Précis de télédétection. Vol. 1: Principes et méthodes. Vol. 2: Applications thématiques,* Presses de l'Université du Québec/AUPELF, 1992, 485 p.

Caloz R. and Collet C., *Précis de télédétection. Vol. 3: Système d'information géographique et traitement numérique,* PUQ, à paraître.

Chevallier R., *La photographie aérienne,* Armand Colin, 1971, 227 p.

Girard C.-M. and Girard M.-C. *Applications de la télédétection à l'étude de la biosphère,* Masson, 1975, 186 p.

Girard M.-C. and Girard C.-M., *Cours de photo-interprétation,* INA-PG, 1970, 208 p.

Girard M.-C., and Girad C.-M., *Télédétection appliquée. Zones tempérées et intertropicales,* Masson, 1989, 260 p.

Lillesand T.M. and Kiefer R.W., *Remote Sensing and Image Interpretation,* 3rd edition, John Wiley & Sons, 1994, 750 p.

Monget J.-M., *Cours de télédétection,* CTAMN, Octobre 1994, Sophia-Antipolis.

Smith J.T.Jr, *Manual of Color Aerial Photography,* Am. Soc. of Photogrammetry, 1968.

Wilmet J., *Télédétection aérospatiale. Méthodologie et applications,* SIDES, 1996, 300 p.

On Specific Topics

Cocquerez J.-P. and Philipp S., *Analyse d'images: filtrage et segmentation,* Masson, 1995, 457 p.

Coster M. and Chermant J.-L., *Précis d'analyse d'images,* Presses du CNRS, 1989, 560 p.

Gonzalez R. and Woods R., *Digital Image Processing,* Addison Wesley, 1992, 716 p.

Guyot G., *Climatologie de l'environnement. De la plante aux écosystèmes,* 1997.

Jensen J.R., *Introductory digital image processing. A remote sensing perspective,* Prentice-Hall, 1986, 379 p.

Kunt M., *Traitement numérique des signaux,* Dunod, 1981, 402 p.

Lliboutry L., *Sciences géométriques et télédétection,* Masson, 1992, 289 p.

Mather P.M., *Computer processing of remotely-sensed images,* An introduction: 2nd Edition, Wiley, 1999, 292 p.

Mulders M.A., *Remote sensing in soil science,* Elsevier, 1987, 379 p.

Paquet G., *Détection électromagnétique: fondements théoriques et applications radar.* Masson, 1997, 320 p.

Photographie aérienne et urbanisme, Centre de recherche d'urbanisme, Paris, 1969.

Polidori L., *Cartographie radar,* Gordon and Breach Science Publishers, Canada, 1997, 287 p.

Pratt W., *Digital image processing,* 2nd edition, Wiley, 1991, 698 p.

Rosenfeld, Kak, *Digital picture processing,* Academic press, 1982.

Rousselet M., *Graphisme 3D,* ETSF, 1985, 223 p.

Serra J., *Image analysis and mathematical morphology,* Academic press, 1982.

Worboys M.F., *GIS, a computing perspective,* Taylor & Francis, 1995, 376 p.

Wyszecki G., and Stiles W.S., *Color Science: concept and methods, quantitative date and formulae,* Wiley, 1982, 950 p.

Useful Internet Sites

Nowadays enormous amount of information is becoming available or INTERNET. This is particularly very useful in the case of remote sensing data which is based on numerous images. A list of important sites on remote sensing is given here. These sites are regularly updated and hence provide recent information. This is particularly so in the case of development of satellite sensor and platforms.

For any details regarding the book, CD-Rom or the latest version of the OASIS program, contact:
DMOS@lacan.grignon. inra. fr
Girard@lacan.grignon. inra.fr

Information Available in French

http://www.spotimage.fr
SPOT Image: for any information site in french and english
5, rue des Satellites, BP 4359, 31030 Toulouse Cedex 4
Tél.: 05 62 19 40 40
http://www.spotimage.fr/accueil/siriuswelcome.htm

http://Lacan.grignon.inra.fr/resources/resources.htm.
For *Vade-mecum books and courses on Remote Sensing télédétection* et *Cours de télédétection.*

http://sol.ensam.inra.fr/silat
For enquiries about the specialised "Mastere Systems" on local information for area management: French data on remote sensing and GIS (Bac + 6 level)

http://www.ign.fr/sfpt
Société française de photogrammétrie et télédétection.

http://www.cnig.fr
Conseil national de l'information géographique.
136, bis, rue de Grenelle, 75700 Paris 07SP

http://www.ign.fr/GP/photaer/exemples.html
Institut géographique national: commande de photographies aériennes.

http://
État des cartes de sols pour la France.

http://viviane.roazhon.inra.fr/snas/index.html
Analyse de terre pour la France.

http://www.gdta.fr/
Groupement pour le développement de la télédétection spatiale.

http://www.ccrs.nrcan.gc.ca
Centre canadien de télédétection. Quelques sites en télédétection y sont donnés.

http://www.aupelf-uref.org
Le réseau de télédétection de l'AUPELF-UREF.

Information Available in English

http://www.auslig.gov.au/acres/prod_ser/
SPOT-images receiving Centre at Hobart and Alice Springs.

http://www.ivv.nasa.gov
NASA Information Centre which regularly issues LANDSAT-TM images.

http://hdsn.eoc.nasda.go.jp/guide/homepage.html
Japanese National Remote Sensing Centre.

http://www.noaa.gov
NOAA/National Geophysical Data Center, 325 Broadway, E/GC4, Dept 993
Boulder, CO 80303, USA.

http://www.spaceimaging.com/
SpaceImaging society, istributor of IKONOS images.

http://www.ceo.org/
Centre for Earth Observation-Ispra: European Remote Sensing Program (Chap. 17).

http://www.esa.int/
All information concerning the European Space Agency.

http://www.digitalglobe.com/
Earthwatch Society, distributor of QuickBird satellite (4m). Available archives available on http://archivedigitalglobe.com/

http://www.earsel.org
European Association of Remote Sensing Laboratries.

http://seawifs.gsfc.nasa.gov/SEAWIFS/LICENSE/checklist.html.
Orblange Society, distributor of Seawifs data.

http://www.vgt.vito.be
SPOT 4-Vegetation program of CNES

http://www.terra_story_barc.asp/terraserver.microsoft.com
Microsoft Global Service of High resolution imagery.

http://www.inpe.br/english/index.htm
INPE National Space Centre of Brazil, Distributor of data.

http://eospso.gsfc.nasa.gov/
NASA EOS program.

http://edcwww.cr.usgs.gov/earthshots/slow/tableofcontents
USGS collection of images and case histories.

http://edcwww.cr.usgs.gov/
EROS Data Center, World distributor of data.

http://makalu.jpl.nasa.gov/
AVIRIS Microwave data server at JPL/NASA.

http://usgs.gov/products/satellite/tm.html
USGS.information about LANDSATTM data

http://southport.jpl.nasa.gov/
Archives of synthetic aperture radar JPL/NASA.

http://radarsat. space.gc.ca/
RADARSAT society in Canada.

http://rsi.ca
RADARSAT International data centre for information on radars and applications.

http://ewse.ceo.org
Shows existing organisations and material (CD-ROM etc.)

Data Processing Software

http://lacan.grignon.inra.fr/resources/resources/htm
TeraVue (Remote sensing).

http://www.erdas.com
Erdas (Remote sensing).

http://www.ermapper.com
ER/Mapper (Remote sensing).

http://www.esri.com
ESRI (GIS).

http://www.clarklabs.org
Idrisi (GIS).

http://www.mapinfo.com
MapInfo (GIS).

http://www.khoral.com
Khoros (Image processing).

http://www.noesisvision.com
Noésis (Image processing).

Index

A

B

C

D

E

F

G

H

I

J

K

L

M

N

O

P

Q

R

S

T

U

V

W

X

Y

Z

CD-Rom Image Index